Metrology in Urban Drainage and Stormwater Management: Plug and Pray

Metrology in Urban Drainage and Stormwater Management: Plug and Pray

Edited by

Jean-Luc Bertrand-Krajewski, Francois Clemens-Meyer and Mathieu Lepot

IWA PUBLISHING

Published by **IWA Publishing**
Unit 104–105, Export Building
1 Clove Crescent
London E14 2BA, UK
Telephone: +44 (0)20 7654 5500
Fax: +44 (0)20 7654 5555
Email: publications@iwap.co.uk
Web: www.iwapublishing.com

First published 2021

British Library Cataloguing in Publication Data
A CIP catalogue record for this book is available from the British Library

ISBN: 9781789060102 (paperback)
ISBN: 9781789060119 (eBook)

This eBook was made Open Access in August 2021.

Contents

Chapter 3
Water level and discharge measurements ... ***35***
Frédérique Larrarte, Mathieu Lepot, Francois H. L. R. Clemens-Meyer, Jean-Luc Bertrand-Krajewski, Damjan Ivetić, Dusan Prodanović and Bram Stegeman

Chapter 4
Measuring the water balance in stormwater control measures ... ***105***
Tim D. Fletcher, Jean-Luc Bertrand-Krajewski, Jérémie Bonneau, Matthew J. Burns, Peter J. Poelsma and Jasmine K. Thom

Preface

Optimism is a precious state of mind when starting an enterprise like writing a book with 50 international contributors without budget nor hard deadlines. Between the very first discussion related to this initiative, during the conference dinner at the UDM – Urban Drainage Modelling conference 2015 in Montreal, Canada, and the presentation of this book more than five years have passed. Of course, in those five years many things happened in the personal lives of people contributing to this book: job changes, hospitalization, bike accidents and last but not least: the Corona crisis. Nevertheless, the enthusiasm, energy and dedication of our colleagues that found time in their busy schedules to share their expertise kept things going. It is no exaggeration to state that all three phases of ambitious projects were encountered and sustained: uninformed optimism, informed pessimism and, finally, informed optimism.

This book, titled *Metrology in Urban Drainage and Stormwater Management: Plug & Pray*, aims at bringing together experience obtained by many people and organizations on the subject over the past decades, while summarizing and popularizing the latest research results to practitioners and engineers. Monitoring in urban drainage and stormwater management systems is challenging for many reasons. Apart from the obviously unfavourable physical/chemical/biological conditions encountered, keeping multidisciplinary teams going in organizational and sometimes politically complex environments that do not always offer optimal conditions, is no trivial task at all. We aim to supply scientists and practitioners with information, methods, examples, case studies and a comprehensive list of references that can be used to avoid (costly) mistakes/pitfalls from the initiative phase of a monitoring project down to the operation of monitoring networks and the validation of data collected for various predefined applications.

Obviously, a book on a subject like this cannot cover all related topics in great detail, therefore we do not claim to present all possible techniques and/or experiences; even worse, we don't even pretend that there are no mistakes or errors in the book. Therefore, the reader is encouraged to supply the editors with remarks, corrections and/or supplementary material that may be incorporated in future updates.

IWA Publishing is sincerely acknowledged for allowing this book to be Open Access, which is in line with the motivation of the editors and the contributing authors to freely share their experience and knowledge. We hope the

doi: 10.2166/9781789060119_xiii

book will contribute to increasing the quality of monitoring projects in the future and will be helpful in increasing appreciation of the added value monitoring has for managing urban drainage and stormwater management systems on the one hand, and supply data as a basis for a better understanding of the processes that drive the behaviour of these systems, on the other.

Jean-Luc Bertrand-Krajewski, Francois Clemens-Meyer and Mathieu Lepot

Acknowledgements

The editors heartily express their most sincere thanks to:

Irene Meyer-Clemens, for her huge and highly appreciated work to (re)-draw and harmonize the style of tens of graphs throughout the chapters.

Frédéric Cherqui (INSA Lyon, France), for his efficient last-minute reviewing of the final drafts of several chapters.

Manuel Froidevaux (High School of Engineering and Architecture, Fribourg, Switzerland), Qingchuan Zhu (INSA Lyon, France) and Roberto Pintos (INSA Lyon, France) for their careful readings of the successive versions of Chapter 8.

INSA Lyon for granting Francois Clemens-Meyer a visiting professorship in 2017 and for granting a contribution for the Open Access status of this book.

The TU Delft Library, which made available the on-line infrastructure for working on a book with multiple contributors spread all over the world and for granting a contribution for the Open Access status of this book.

A long list of municipalities, waterboards, consultancy firms, research institutes and universities across the world which made material available for this book.

And all our contributing colleagues, of course, for their shared enthusiasm for the subject and for making available their time and expertise.

JLBK wishes to address his very special gratitude to Marie-Anne, for her great patience and support, especially during the intense days of last reviews and final editing of the book just before Christmas 2020. Depending on the working mood, piano works of, among others, Johann Sebastian Bach, Philip Glass, Ludovico Einaudi and Arvo Pärt have been indispensable and stimulating.

FCM expresses his gratitude to his colleagues at Deltares and many of his PhD and MSc students for supplying materials, examples and illustrations.

ML would like to thank Jean-Luc and Francois for their invitation to take part in such an editing adventure.

Jean-Luc Bertrand-Krajewski, Francois Clemens-Meyer and Mathieu Lepot
December 2020

List of Contributors

Jakob Benisch
TU Dresden,
Institut für Siedlungs- und Industriewasserwirtschaft,
Dresden, Germany.

Simon Bloem
Eawag-aquatic research,
Dübendorf, Switzerland.

Jean-Luc Bertrand-Krajewski
University of Lyon, INSA Lyon, Laboratory DEEP,
Villeurbanne, France.

Franck Blumensaat
ETH Zurich, Zurich,
Switzerland, and
Eawag-aquatic research,
Dübendorf, Switzerland.

Jérémie Bonneau
INRAE, UR RiverLy,
Villeurbanne, France.

Elodie Brelot
GRAIE, Villeurbanne,
France.

Daniel Bourne
RPS Group, Derby,
United Kingdom.
(presently Stantec,
Redditch, United Kingdom).

Matthew Burns
The University of
Melbourne,
Waterway Ecosystem
Research Group,
School of Ecosystem
and Forest Sciences,
Melbourne, Australia.

Flora Branger
INRAE, UR RiverLy,
Villeurbanne, France.

Véronique Chaffard
IRD - Institut de Recherche
pour le Développement,
IGE - Institut des
Géosciences de
l'Environnement,
Grenoble, France.

Nemanja Branisavljević
University of Belgrade,
Faculty of Civil Engineering,
Department of
Hydraulic and
Environmental Engineering,
Belgrade, Serbia.

Frédéric Cherqui
University of Lyon,
INSA Lyon,
Laboratory DEEP,
Villeurbanne, France.

Francois Clemens-Meyer
TU Delft/Deltares, Delft,
The Netherlands,
and NTNU,
Dept. Civil & Environmental
Engineering,
Trondheim, Norway.

Justin Costelloe †
University of Melbourne,
Department of Infrastructure
Engineering,
Parkville, Australia.

Christian Ebi
Eawag-aquatic research,
Dübendorf, Switzerland.

Uwe Eichelmann
TU Dresden,
Institute of Hydrology and
Meteorology,
Tharandt, Germany.

Thomas Einfalt
Hydro & meteo GmbH,
Lübeck, Germany.

Tim Fletcher
The University of Melbourne,
Waterway Ecosystem
Research Group,
School of Ecosystem and
Forest Sciences,
Melbourne, Australia.

Günter Gruber
TU Graz, Institute of Urban
Water Management and
Landscape Water Engineering,
Graz, Austria.

James Hale
RPS Group, Derby,
United Kingdom.

Zoe Hancock
RPS Group, Derby,
United Kingdom.

Zoran Kapelan
TU Delft,
Faculty of Civil Engineering and Geosciences,
Department of Water Management,
Delft, The Netherlands.

Björn Helm
TU Dresden, Institut für Siedlungs- und Industriewasserwirtschaft,
Dresden, Germany.

Stefan Kroll
Aquafin NV,
Aartselaar, Belgium.

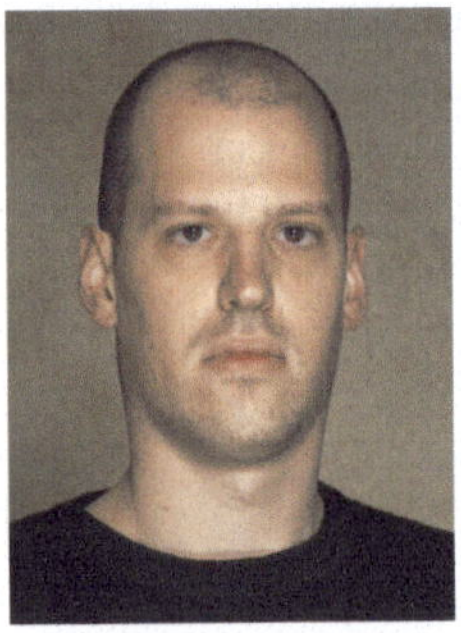

Damjan Ivetić
University of Belgrade,
Faculty of Civil Engineering,
Department of Hydraulic and Environmental Engineering,
Belgrade, Serbia.

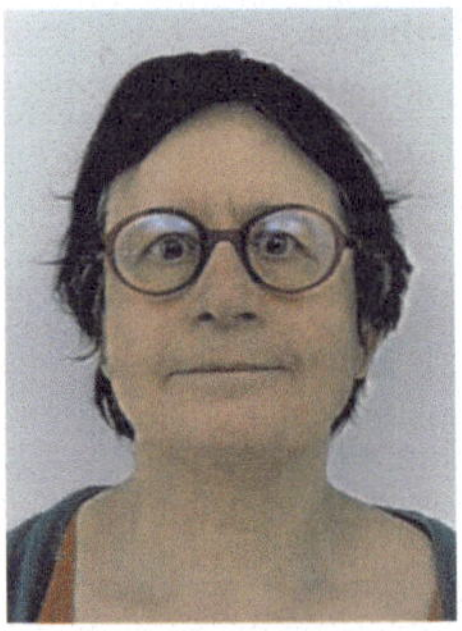

Frédérique Larrarte
Université Gustave Eiffel,
Département Géotechnique, Environnement,
Risques Naturels et Sciences de la Terre - Laboratoire SRO,
Marne-la-Vallée, France.

Robert James
The University of Melbourne,
Waterway Ecosystem Research Group,
School of Ecosystem and Forest Sciences,
Melbourne, Australia.

Vivien Lecomte
GRAIE,
Villeurbanne, France.

Mathieu Lepot
TU Delft, Faculty of Civil Engineering and Geosciences,
Water Management Dept.,
Delft, The Netherlands,
and Un Poids Une Mesure,
Lyon, France.

Dominik Leutnant
Muenster University of Applied Sciences,
Institute for Infrastructure Water Resources Environment (IWARU),
Münster, Germany.

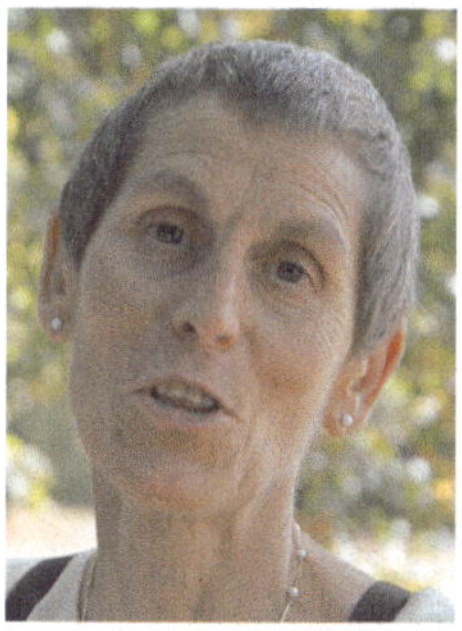

Isabelle Mallet
Métropole de Lyon,
DINSI - SI métiers,
Lyon, France.

David McCarthy
Monash University,
EPHM Lab,
Melbourne, Australia.

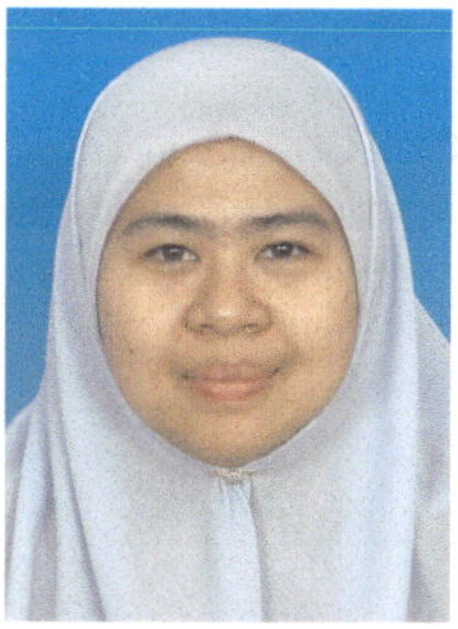

Nur Shazwani Muhammad
Universiti Kebangsaan Malaysia,
Department of Civil Engineering,
Bangi, Malaysia.

Nadia Nikolova-Peters
Anglian Water Service,
Huntingdon,
United Kingdom.

Susana Ochoa-Rodriguez
RPS Group, Derby,
United Kingdom

Peter Poelsma
The University of Melbourne,
Waterway Ecosystem Research Group,
School of Ecosystem and Forest Sciences,
Melbourne, Australia.

Dusan Prodanović
University of Belgrade,
Faculty of Civil Engineering,
Department of Hydraulic and
Environmental Engineering,
Belgrade, Serbia.

Pete Skipworth
Environmental Monitoring
Solutions, Sheffield,
United Kingdom.

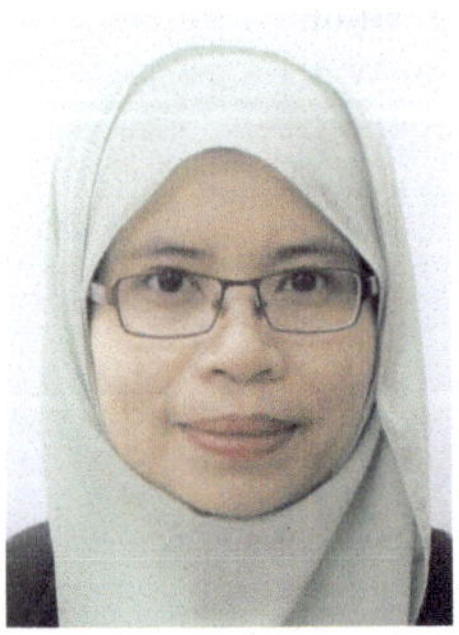

Salwa Ramly
Department of Irrigation and
Drainage, SMART Control
Centre, Kuala Lumpur,
Malaysia.

Bram Stegeman
Delft University of
Technology,
Faculty of Civil Engineering
and Geosciences,
Water Management
Department, Delft,
The Netherlands.

Alma Schellart
University of Sheffield,
Civil and Structural
Engineering,
Sheffield,
United Kingdom.

Simon Tait
University of Sheffield,
Civil and Structural
Engineering,
Sheffield,
United Kingdom.

Baiqian Shi
Monash University,
EPHM Lab,
Melbourne, Australia.

Jasmine Thom
The University of
Melbourne,
Waterway Ecosystem
Research Group,
School of Ecosystem and
Forest Sciences,
Melbourne, Australia.

Mathias Uhl
Muenster University of Applied Sciences, Institute for Infrastructure Water Resources Environment (IWARU), Münster, Germany.

Gilles Wallwork
RPS Group, Derby, United Kingdom

Johan Van Assel
Aquafin NV, Aartselaar, Belgium.

Wan Hanna Melini Wan Mohtar
Universiti Kebangsaan Malaysia, Department of Civil Engineering, Bangi, Malaysia.

Job Van Der Werf
Delft University of Technology, Faculty of Civil Engineering and Geosciences, Water Management Department, Delft, The Netherlands.

Patrick Willems
KU Leuven, Department of Civil Engineering, Leuven, Belgium.

List of Acronyms

Acronym	Meaning
2D	Two dimensions
3D	Three dimensions
ACF	Auto Correlation Function
ADF	Acoustic Doppler Flowmeter
AET	Actual EvapoTranspiration
AI	Artificial Intelligence
AIC	Akaike Information Criterion
AMP	Asset Management Plan
API	Application Programming Interface
APUSS	Assessing Infiltration and Exfiltration on the Performance of Urban Sewer Systems
AR	Auto Regressive
ARI	Average Recurrence Interval
ARIMA	Auto Regressive Integrated Moving Average
ARMA	Auto Regressive Moving Average
ASCII	American Standard Code for Information Interchange
ATEX	ATmosphere EXplosible
AWS	Anglian Water Services
BDOH	Base de Données pour les Observatoires en Hydrologie (Database for Hydrologic Observatories)
BIPM	Bureau International des Poids et Mesures
BMPs	Best Management Practices
BOD	Biological Oxygen Demand
CAPEX	Capital Expenditures
CCTV	Closed Circuit Tele Vision

CD	Compact Disc
CDR	CorelDRAW file format
CEN	Comité Européen de Normalisation - European Committee for Standardization
CEO	Chief Executive Officer
CEPT	European Conference of Postal and Telecommunications Administration
CFD	Computational Fluid Dynamics
CFM	Correction Function Model
CIMO	Commission for Instruments and Methods of Observation
CNN	Convolutional Neural Network
COD	Chemical Oxygen Demand
CRC	Cyclic Redundancy Checks
CSO	Combined Sewer Overflow
CST	Control STation
CSV	Comma Separated Values
CSW	Catalogue Service for the Web
CUAHSI	Consortium of Universities for the Advancement of Hydrologic Science
DACF	Differentiated Auto Correlation Function
DBF	dBase Database File format
DBMS	Data Base Management System
DCM	Doppler Current Meter
DEM	Digital Elevation Model
DEST	Detection of Exfiltration from Sewers using Tracers
DGUV	German Social Accident Insurance
DID	Department of Irrigation and Drainage
DIY	Do It Yourself
DSD	Drop Size Distribution
DST	Daylight Saving Time
DTS	Distributed Temperature Sensing
DV	Data Validation
DVD	Digital Versatile Disc
DWA	German Association for Water, Wastewater and Waste
DWF	Dry Weather Flow
EC	Electrical Conductivity
EM	Electro Magnetic
EMF	Electro Magnetic Flowmeter
EMP	Electro Magnetic Pulse
EMV	Electro Magnetic Velocity meter
ENSO	El Nino Southern Oscillation
ERT	Electrical Resistivity Tomography
ET	EvapoTranspiration
ET_0	Reference EvapoTranspiration
ETSI	European Telecommunications Standards Institute

EU	European Union
FAIR	Findable, Accessible, Interoperable and Reusable
FAT	Factory Acceptance Test
FCC	Federal Communications Commission
FDI	Fault Detection and Isolation
FDS	Flood Detection System
FDV	Fulldome Digital Video
FELL	Focused Electro Leak Location
FEWS	Flood Early Warning System
FMECA	Failure Mode Effects and Criticality Analysis
FOV	Field Of View
FTE	Full Time Equivalent
GB	Giga Byte
GDPR	General Data Protection Regulation
GIS	Geographical Information System
GML	Geography Markup Language
GMT	Greenwich Mean Time
GPR	Ground Penetrating Radar
GPRS	General Packet Radio Service
GPS	Global Positioning System
GRAIE	Groupe de Recherche, Animation technique et Information sur l'Eau
GSM	Global System for Mobile Communication
GUM	Guide for Uncertainty in Measurements
H&S	Health and Safety
IAHR	International Association for Hydro-Environment Engineering and Research
IBM	International Business Machines
IEC	International Electrotechnical Commission
IoT	Internet of Things
IP	Ingress Protection
ISM	Industrial, Scientific and Medical
ISO	International Organization for Standardization
IT	Information Technology
IUD	Integrated Urban Drainage
IWA	International Water Association
JCGM	Joint Committee for Guides in Metrology
JSON	JavaScript Object Notation
KML	Keyhole Markup Language
KMZ	Zipped KML file format
KNMI	Royal Dutch Meteorological Institute
LED	Light Emitting Diode
LID	Low Impact Development
LiDAR	Light Detecting And Ranging

LOCF	Last Observation Carried Forward
LoD	Limit of Detection
LoQ	Limit of Quantification
LORAN	Long Range Area Network
LoRaWAN	Long Range Wide-area network
LPU	Law of Propagation of Uncertainties
LPWAN	Low Power Wide Area Network
LS-PIV	Large Scale Particle Image Velocimetry
LS-PTV	Large Scale Particle Tracking Velocimetry
M2M	Machine to Machine
MA	Mobile Average
MC	Monte Carlo
MCM	Monte Carlo Method
MCP	Monitoring Cycle Plan
MDM	Model Delivery Milestone
MFB	Mean Field Bias
MIF	MapInfo Interchange Format
ML	Machine Learning
MSFM	Multi Sensor Flow Meter
MST	Monitoring STation
NB-IoT	Narrow Band Internet of Things
NC	National Classification
NH_4	Ammonia
NIST	National Institute of Standards and Technology
NO_3	Nitrate
NOCB	Next Observation Carried Backward
NoSQL	Not only SQL
NTP	Network Time Protocol
NTU	Nephelometric Turbidity Unit
O&M	Observation & Measurement/Operation & Maintenance
OFWAT	UK Water Services Regulation Authority
OGC	Open Geospatial Consortium
OIML	International Organization of Legal Metrology
OLED	Organic Light-Emitting Diode
OLS	Ordinary Least Squares
OPEX	Operating Expenditures
OTHU	Observatoire de Terrain en Hydrologie Urbaine (Field Observatory on Urban Hydrology)
PC	Personal Computer
PD	Processed Data
PDF	Probability Density Function/Portable Document Format
PDP	Programmed Data Processor

PE	People Equivalent
PER	Packet Error Rate
PET	Potential EvapoTranspiration
pH	potential of Hydrogen
PHP	Hypertext Preprocessor
PI	Performance Indicator
PIV	Particle Image Velocimetry
PLC	Programmable Logic Controller
PLS	Partial Least Squares
PPE	Personal Protective Equipment
PPM	Parts Per Million
PS	Pumping Station
PTV	Particle Tracking Velocimetry
PVC	Poly Vinyl Chloride
QPE	Quantitative Precipitation Estimate
QUEST-C	Quantification of Exfiltration from Sewers with the help of Tracers Continuous Dosing
RAID	Redundant Array of Inexpensive Disks
RAMS	Risk Assessments and Method Statements
RAR	Roshal ARchive file format
RD	Raw Data
RDBMS	Relational DataBase Management System
RF	Radio Frequency
RMSE	Root Mean Squared Error
RPM	Rotation Per Minute
RPT	RePeaTer
RS-232	Recommended Standard 232
RS-485	Recommended Standard 485
RTC	Real Time Control
RTU	Remote Telemetry Unit
RZ	Riool Zuid (Southern sewer system)
SAM	Sewer Assessment with Multi-sensor Systems
SAT	Site Acceptance Test
SCADA	Supervisory Control and Data Acquisition
SCM	Stormwater Control Measure
SD	Secure Digital
SELV	Safety Extra-Low Voltage
SHP	Shape file format
SIM	Subscriber Identification Module
SMS	Short Message Service
SNR	Signal to Noise Ratio
SOP	Standard Operational Procedure/Protocol
SOS	Sensor Observation Service

SQL	Structured Query Language
SRD	Short Range Devices
SSH, HTTP	Secure Shell, Hypertext Transfer Protocol
SSL	Secure Sockets Layer
SUDS	Sustainable Urban Drainage Systems
SVG	Scalable Vector Graphics
TAB	MapInfo file format
TBR	Tipping Bucket Rain gauge
TCP/IP	Transmission Control Protocol/Internet Protocol
TDR	Time Domain Reflectometry
TIFF	Tagged Image File Format
TRA	Task Risk Assessment
TSS	Total Suspended Solids
TXT	Text file format
UA	Uncertainty Assessment
UDSM	Urban Drainage and Stormwater Management
US	Ultra Sound
USB	Universal Serial Bus
USGS	United States Geological Survey
UTC	Coordinated Universal Time
UV	Ultra Violet
UV/Vis	Ultra Violet/Visible
UWO	Urban Water Observatory
VES	Vertical Electrical Sounding
VHF/UHF	Very High Frequency/Ultra High Frequency
VIM	International Vocabulary of Metrology
VSI	Vital Sewerage Infrastructure
WaPUG	UK Wastewater Planning Users Group
WDD	Dutch Waterboard 'Waterschap de Dommel'
WERG	Waterways Ecosystems Research Group
WFD	European Water Framework Directive
WFS	Web Feature Service
WG	Working Group
WIS	Water Information System
WLS	Williamson Least Squares
WMO	World Meteorological Organization
WMS	Web Map Service
WR	Weighing Rain gauge
WSUD	Water Sensitive Urban Design
WWTP	WasteWater Treatment Plant
XML	Extensible Markup Language
ZIP	PKZIP file format

Chapter 1

General introduction and book layout

Jean-Luc Bertrand-Krajewski[1], *Francois H. L. R. Clemens-Meyer*[2,3,4] *and Mathieu Lepot*[4,5]

[1]*University of Lyon, INSA Lyon, Laboratory DEEP, Villeurbanne, France*
[2]*Deltares, Unit of Hydraulic Engineering, Delft, The Netherlands*
[3]*Norwegian University of Science & Technology, Faculty of Engineering, Department of Civil & Environmental Engineering, Trondheim, Norway*
[4]*TU Delft, CITG – Civil Engineering and Geosciences, Delft, The Netherlands*
[5]*Un Poids Une Mesure, Lyon, France*

ABSTRACT

This introductory chapter indicates why well-defined, high-standard, and reliable monitoring is a key aspect in the necessary evolution of urban drainage and stormwater management and why it should become routine practice. It provides a framework, guidelines, and recommendations to define monitoring objectives and means. It also presents the structure and the chapters of the rest of the book.

Keywords: Metrology, monitoring, urban drainage and stormwater management.

1.1 INTRODUCTION

In this book, urban drainage and stormwater management (UDSM) infrastructures (Figure 1.1) designate both traditional grey infrastructures (sewer systems and their related facilities) and growing green infrastructures (green roofs, infiltration trenches, raingardens, etc.) also referred to as sustainable urban drainage systems (SUDS), low impact development (LID), etc. (Fletcher *et al.*, 2015). UDSM has contributed significantly to sustain urban areas as safe, healthy, and comfortable places to live, as such the need to maintain and operate them is beyond discussion. However, due to increasing pressures in

doi: 10.2166/9781789060119_0001

Figure 1.1 Urban drainage and stormwater management systems. *Source*: GRAIE.

terms of climate change, growth of urban population worldwide, need for reducing use of material and energy, environmental and societal demands and restrictions of these infrastructures, the need to better understand the processes that make them work increases over time. This applies to existing systems, in terms of gaining knowledge how to optimize their effectiveness and efficiency, on the one hand, while on the other hand there is a need to transfer experience obtained in designing, operating, and maintaining these systems into guidelines for the design of future systems. The obvious manner to achieve this is by quantifying the relevant processes through observation. However obvious this may seem, such an approach has been applied on a relatively modest scale over the last few decades and is so far limited mostly to the scientific community and to a few forefront practitioners only. The reason for this can only be guessed. Obviously over the years a(n) (informal) code of practice evolved in which inefficiency and acting on very limited knowledge were accepted, while at the same time design rules are largely based on historically developed empirical rules, in most cases supported by some form of extensive hydrological/hydraulic modelling. Indeed, this approach results in systems that do function, be it at an unknown level of effectiveness and efficiency on the one hand, while being responsible for unnecessary, but apparently accepted, hindrance and inefficiency on the other. Further, there exists the impression that the added value of monitoring is largely overlooked in practice as monitoring is often framed as 'expensive, while the results may even contain an unwelcome truth'. This book aims at bringing together and making available knowledge and experience (for the good and the bad) of monitoring UDSM systems obtained over recent decades to enable stakeholders to make decisions on monitoring on a more rational and well-informed basis.

As in any field of (applied) science, models or abstract perceptions of the processes under study are used, along with observations on these processes. The latter can be either qualitative or quantitative, the former can be used for initial validation of process descriptions while the latter are merely used to quantify or provide a reference for models/process descriptions.

In the field of UDSM, the models applied are based on a wide range of engineering sciences like hydraulics, hydrology, (bio)chemistry, geohydrology, and meteorology.

Over recent decades, significant progress has been made in the further development of models and model concepts as well as in obtaining quantifying observations. Schilperoort (2011) reports an interesting evolution: within a time span of 20 years the time needed for technicians, engineers and scientists to obtain one 'correct' data point from a monitoring project in an urban drainage system has decreased by a factor of 1000 (from roughly 15 minutes in the early 1990s to 1 second in 2010). Monitoring, that is, collecting data and information, has made swift progress due to developments in adjacent fields of science and technology, most notably sensor technology, data communication and data processing capabilities. Although these fields are still progressing, the area of water quantity related observational methods seems to have reached a certain maturity in the sense that a balance can be made on the progress of the last two decades. The book that lies before you will therefore focus on metrology in UDSM, restricted to making observations on the quantitative load on UDSM systems (i.e., rainfall and for wastewater systems the wastewater load) and the response of these systems to these loads (i.e., discharges, water levels, in- and exfiltration, evaporation, etc.). Therefore, it will not go into observations on water quality parameters but will be restricted to the hydraulic and (geo)hydrological processes in UDSM systems. Future extensions into water quality or the rapidly developing subject on techniques for inspection of the status of UDSM related assets, are likely to evolve though.

The next two sections discuss metrology and monitoring applied to UDSM. The last section presents the structure and the chapters of the book and its companion website.

1.2 METROLOGY IN A BROADER SENSE REFLECTING ON UDSM

Metrology is the science of 'measuring', a formal definition is given by the International Bureau of Weights and Measurements as '*the science of measurements, embracing both experimental and theoretical determinations at any level of uncertainty in any field of science and technology*' (see: https://web.archive.org/web/20110927012931/http://www.bipm.org/en/convention/wmd/2004/ (visited 23/12/2020)). Another, but largely similar, definition is formulated in the VIM (International Vocabulary of Metrology): '*Metrology includes all theoretical and practical aspects of measurement, whatever the measurement uncertainty and field of application*' (Joint Committee for Guides in Metrology [JCGM], 2012).

Metrology as a science can be traced back to the French revolution that supplied the political motivation for standardization of units to be used in France. One of the most known and acknowledged results from this is the International System of Units (SI system) as we know it today. The main activities within metrology are: (i) define standard units, (ii) materialize these standards and (iii) arrange a system of traceability for units applied in practice to the formal standard units. With respect to UDSM, the latter activity can be identified to be the most relevant, and can be summarized in the question: 'How can we be sure that monitoring data obtained are in line with the standards they reportedly claim to have?' Basically, all activities, decisions and choices made in planning, designing, operating, and reporting on monitoring UDSM systems should be in line with contributing to answer this main metrological question to ensure the sought for data- and information-yield.

Metrology can further be sub-categorized into (i) scientific, fundamental metrology, (ii) applied (technical, industrial) metrology and (iii) commercial/legal metrology. In the field of UDSM one deals, in most cases, predominantly with the second category because UDSM is not a field of fundamental science but should rather be regarded as a hodgepodge of a range of scientific and technological fields. However, in practical cases the third category is applicable as well (e.g., when the operation of systems is delegated to commercial third parties or in cases where a dispute arises on responsibility and accountability for damage due to e.g., flooding). Overall, metrology provides a sound basis on which uniformity with respect to quality assessment and control, and communication with respect to monitoring can be defined.

Like any technical domain, metrology has its own terminology. In this book, the definitions proposed in the VIM (JCGM, 2012) are used. The most important definitions are also given in the appendix of this book, illustrated with examples taken from UDSM.

1.3 MAIN ELEMENTS IN UDSM METROLOGY

As illustrated in Figure 1.2, monitoring should obviously serve some objectives. However obvious this may seem, in many practical cases such objectives are not, or only vaguely, defined. For example: in practice objectives like '*obtaining insight into the functioning*' are frequently formulated. But insight is highly subjective and depends not just on the information it provides. If there is no metric to test whether a monitoring activity has achieved its goal, then there can be no sound basis to evaluate the effectiveness or efficiency of the applied monitoring. Therefore, it is emphasized that it is of utmost importance to formulate an agreed and well-defined objective, prior to starting a monitoring project.

For example: when the objective is to obtain data to calibrate a model, such an objective can be translated into demands being put on the parameters to be monitored (depending on the model chosen and the level of detail sought), the number and exact locations of the monitoring points, and demands on sampling frequency and allowed uncertainties of the measuring data obtained. These subjects directly refer to Chapters 6, 7 and 8 of this book. Another objective may be to evaluate the volume of diluted sewage spilt by CSOs (combined sewer overflows) into receiving water bodies, in such a case, of course, the location and geometry of the CSO construction has to be known along with the allowed uncertainty of the volume. These two conditions narrow down the options at hand, depending on the geometry either a device for discharge measurement can be installed in a conduit (see Chapter 3), or a discharge-flow depth relationship over some device (a weir, an orifice, a Venturi flume) has to be established, depending on the uncertainty allowed, the proper equipment and sampling frequency have to be defined. Examples of more complex objectives (quality based real time control) can be found in literature (Schilperoort, 2011; van Daal-Rombouts, 2017), as well as effects of maintenance on performance of urban drainage systems (van Bijnen, 2018), and long term observation for scientific purposes (see e.g. OTHU – Observatoire de Terrain en Hydrologie Urbaine, http://www.graie.org/othu/ (visited 23/12/2020)).

Taking the objective(s) as a starting point, the information need can be formulated, which in turn is the basis to derive several fundamentally important parameters, such as:

- The quantities to be monitored.
- The minimum timespan or number of events of interest that has to be covered.
- The number and the locations of sensors, considering practical issues like:
    - Accessibility.
    - Availability of power supply and data communication.
    - Sampling frequency.
    - Required or accepted level of uncertainty in the results needed for the objective.

Based on this information a first estimate of the means needed can be made (purchasing or renting equipment, construction activities, maintenance, personnel needed in terms of quantity and qualifications), allowing for a first check of efficiency and effectiveness. If during this stage it becomes obvious that the means needed are more than can be justified, a decision has to be made either to accept

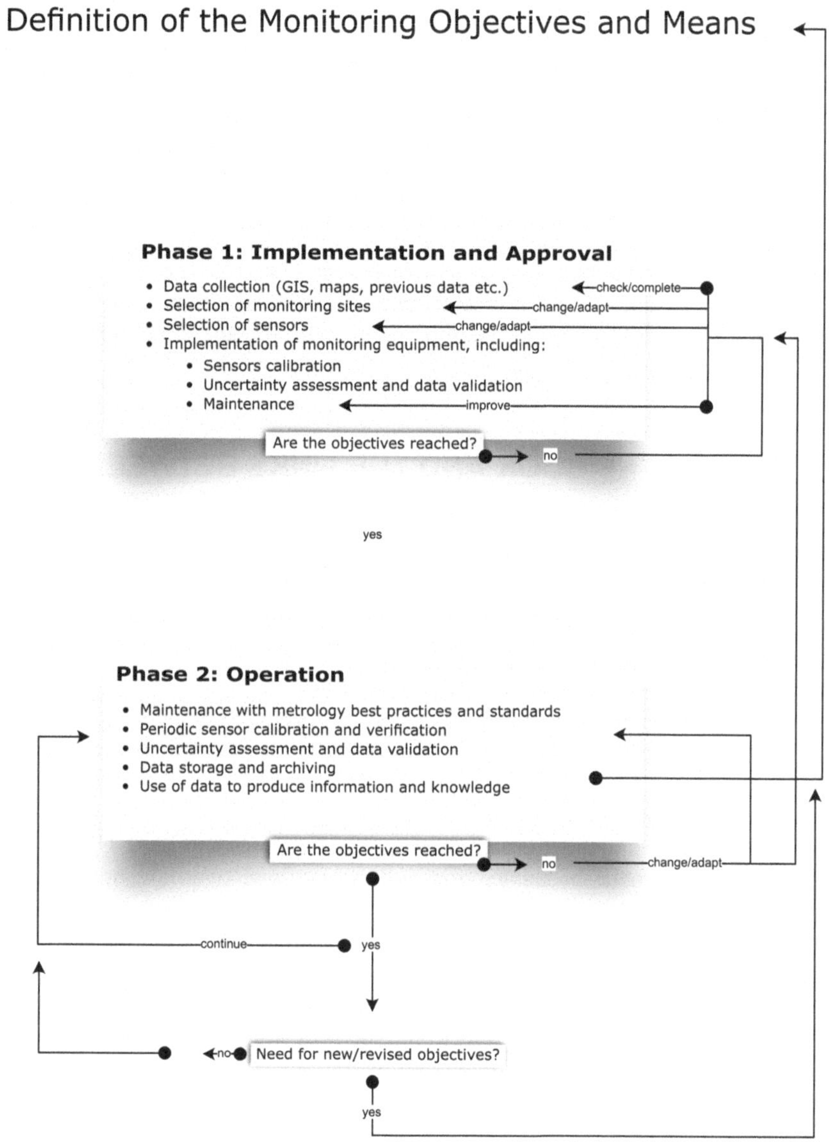

Figure 1.2 Main activities in monitoring. *Source*: Francois Clemens-Meyer (Deltares/TU Delft/NTNU).

lower standards or abandon the initiative completely. If, for example, the objective is to obtain information to justify an investment of 1 million euros and the estimated means to obtain this information amount to costs in the same order of magnitude, questions should be raised. Of course, it has to be emphasized that not all objectives can easily be expressed in monetary terms, therefore in many cases a managerial/political decision has to be made in balancing monitoring costs against achieving environmental or societal objectives. Basically, the task of a scientist/engineer or practitioner is limited to estimating as accurately as possible the expected costs, quality and quantity of the information obtained from a monitoring initiative.

In the case where the initiative passes the first phase, in the sense that the planned activities are expected to achieve the information within the envelope of available means to answer to the objectives set, the operational phase starts.

During operation, the main activity is to make sure that the installed system keeps performing at the desired level, which implies frequent evaluation of all components involved, regular checks of data quality, and calibration of measuring devices along with periodic data analysis to ensure the information obtained meets the standards as defined in the objectives. Ultimately this may lead to adaptations in the set-up, or the equipment applied. Especially in long-term monitoring campaigns, the objectives may be subject to change, requiring a redesign or (major) adaptations (e.g., Walcker *et al.*, 2018). An iterative approach is thus essential, requiring regular evaluation of the obtained data/information. At the present state of the art, we have not yet reached the level of 'plug and play', just installing a system and leaving it alone remains at present, to some extent, a matter of 'plug and pray'.

1.4 STRUCTURE OF THE BOOK AND THE LINKS BETWEEN CHAPTERS

This book contains 12 chapters along with several appendices and a comprehensive body of references for further reading.

The chapters which follow this introduction chapter are listed below.

- Chapter 2 'Sensors for rain measurements' deals with measuring systems and components (including sensors) and globally addresses two main fields: namely rainfall inputs on UDSM systems on the one hand and the systems response to these inputs on the other.
- Chapter 3 'Water level and discharge measurements' is devoted to monitoring inside urban drainage systems, mainly focused on water level, velocity and discharge measurements at different locations in centralized networks.
- Chapter 4 'Measuring the water balance in stormwater control measures' deals with measuring methods, systems and components (including sensors) for decentralized UDSM (SUDS) where several processes have to be accounted for: inflow, outflow, overflow, infiltration, exfiltration, intrusion, evaporation and evapotranspiration.
- Chapter 5 'Data communication and storage' is devoted to a subject that has proven to be of omnipotent importance in the rapid advances monitoring has made in many fields, and certainly in UDSM, namely data communication and storage. The fact that the objects under study normally offer a suboptimal environment for electronic devices makes this part of monitoring extra challenging. Chapter 5 provides a state-of-the-art overview of solutions tried and applied in practice along with an overview of available tools and standards.
- Chapter 6 'Design of monitoring: from macro to micro design'. When designing monitoring networks, a wide range of topics have to be taken into account, including very practical issues to ensure obtaining the information sought, given the objectives of the monitoring project and the available budget in terms of investment and manpower (maintenance, data analysis, sensors calibration). The interlinkage between number and location of measuring devices, sampling frequency and uncertainties of the instruments and methods applied is dealt with in depth.
- Chapter 7 'Operation and maintenance'. Once a monitoring network is in operation, appropriate efforts are required to ensure data/information quality and yield during the operational lifetime. Chapter 7 deals with all aspects involved in this, making a distinction between different systems meant for special purposes (e.g., those answering some specific scientific question vs. systems meant for real time control of wastewater systems that are expected to serve for extended periods of time). Also,

the issue of 'safety and health' is touched upon. A full section is devoted to methods and tools for sensors calibration which is of crucial importance to deliver reliable monitoring results.

- Chapter 8 'Uncertainty assessment' holds a comprehensive treatise on the necessary estimation of uncertainties in measurements. It includes a step-by-step approach illustrated with detailed examples to allow for practical applications, in order to be sure about any uncertainties and communicate them with confidence.
- Chapter 9 'Data validation and data quality assessment'. As measuring devices are prone to being attacked by Murphy (Dickson, 1981), a perfect data-yield is not to be expected under all circumstances. Performing regular data validation is a prerequisite for judging their usability for the purpose for which they are meant to be applied. Further data validation will result in information upon which maintenance activities of the monitoring system are founded. Chapter 9 holds a comprehensive treatise on methods and algorithms developed in the field of UDSM in order to

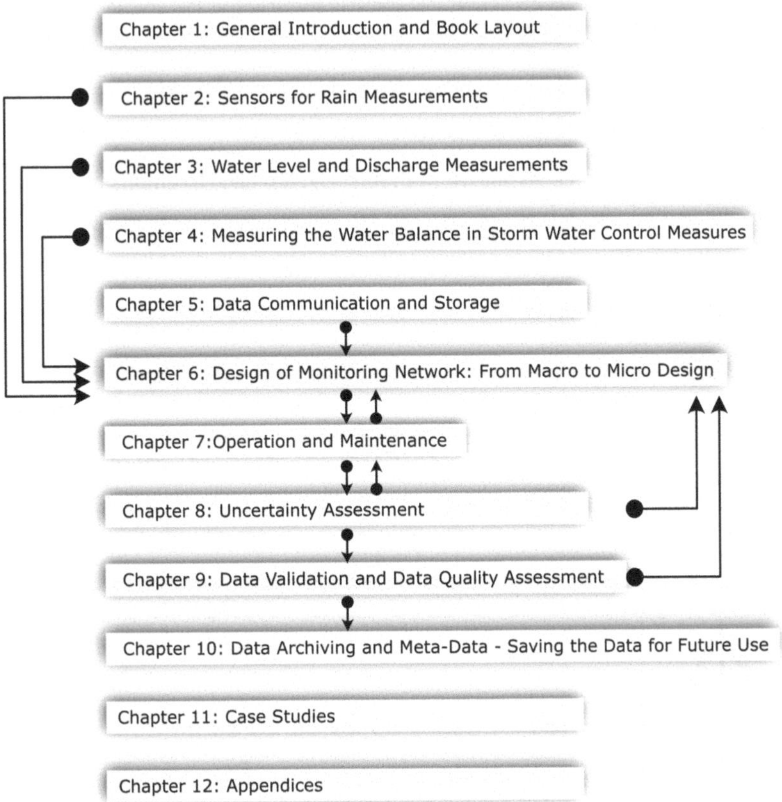

Figure 1.3 Main links between chapters – Chapters 2, 3, 4 and 5 give input for designing a monitoring set-up (Chapter 6). During operation (Chapter 7), data are obtained, that after assessment of their uncertainties (Chapter 8) and their validation (Chapter 9) may give rise to changes in the design (Chapter 6). The data obtained are archived and enhanced with meta data for future use (Chapter 10). Chapters 1, 11 and 12 are related to all other chapters. *Source*: Mathieu Lepot (TU Delft).

effectively and efficiently perform data validation, along with methods on how to report and apply the results.

- Chapter 10 'Data archiving and metadata – Saving the data for future use'. The importance of storing data along with measures to guarantee future accessibility are often overlooked in monitoring projects. As the obtained data often represent considerable value in terms of invested means and scientific effort, this subject is dealt with in some depth in this chapter. Apart from the obligation for researchers and scientific institutions to archive all experimental data, an overview of the different systems applied to date is discussed, as well as means for data sharing between researchers and practitioners.
- Chapter 11 'Data collection in urban drainage and stormwater management systems – Case studies'. To ensure that the content of this report finds its way to both scientists and practitioners, a comprehensive treatise on a range of case studies is discussed in this chapter, illustrating the application of the more theoretical Chapters 1 through 9.
- Chapter 12 'Appendices' provides detailed appendix information for all previous chapters (e.g. VIM definitions, tables, useful links, etc.).

Basically, each chapter can be read/studied separately from the other chapters. However, there are obviously strong inter-relations between several chapters, the most important of which are as indicated in Figure 1.3.

1.5 MESSAGE BOXES

Throughout the chapters, special boxes, as shown in Figure 1.4, are used to emphasize various aspects: key messages, ideas, warnings, and check lists.

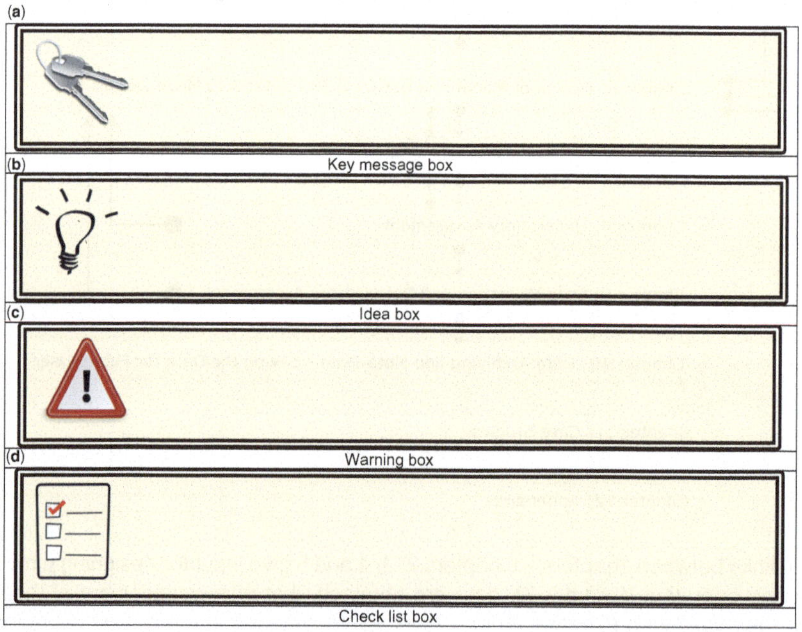

Figure 1.4 (a) Key message, (b) Idea, (c) Warning and (d) Check list boxes. *Source*: Mathieu Lepot (TU Delft).

REFERENCES

Dickson P. (1981). *'Murphy's Law' – The Official Rules.* Arrow Books, Auckland (New Zealand), pp. 128–137. ISBN 0-09-926490-0.

Fletcher T. D., Shuster W., Hunt W. F., Ashley R., Butler D., Arthur S., Trowsdale S., Barraud S., Semadeni-Davies A., Bertrand-Krajewski J.-L., Mikkelsen P. S., Rivard G., Uhl M., Dagenais D. & Viklander M. (2015). SUDS, LID, BMPs, WSUD and more – The evolution and application of terminology surrounding urban drainage. *Urban Water Journal*, **12**(7), 525–542, doi: 10.1080/1573062X.2014.916314.

JCGM (2012). *JCGM 200:2012 – International Vocabulary of Metrology — Basic and General Concepts and Associated Terms (VIM) (Vocabulaire International de Métrologie — Concepts fondamentaux et généraux et termes associés (VIM).* Joint Committee for Guides in Metrology, Geneva (Switzerland). Available at: https://www.bipm.org/documents/20126/2071204/JCGM_200_2012.pdf (accessed 26 April 2021).

Schilperoort R. P. S. (2011). *Monitoring as a Tool for the Assessment of Wastewater Quality Dynamics.* PhD thesis, Delft University of Technology, Delft, The Netherlands. ISBN 978-90-8957-021-5.

van Bijnen J. A. C. (2018). *The Impact of Sewer Condition on the Performance of Sewer Systems.* PhD thesis, Delft University of Technology, Delft, The Netherlands. ISBN 978-94-6233-987-3.

van Daal-Rombouts P. M. M. (2017). *Performance Evaluation of Real Time Control in Urban Wastewater Systems.* PhD thesis, Delft University of Technology, Delft, The Netherlands. ISBN 978-94-6233-707-7.

Walcker N., Bertrand-Krajewski J.-L., Vacherie S., Lepot M., Castebrunet H., Barraud S. & Lipeme Kouyi G. (2018). Une nouvelle station de mesure pour l'acquisition de séries chronologiques en hydrologie urbaine [A new monitoring station for the acquisition of timeseries in urban hydrology]. *Techniques Sciences Méthodes*, **3**, 55–64, doi: 10.1051/tsm/201803055 (in French).

Chapter 2

Sensors for rain measurements

Patrick Willems[1] and Thomas Einfalt[2]

[1]*KU Leuven, Hydraulics and Geotechnics, Urban and River Hydrology and Hydraulics, Leuven, Belgium*
[2]*Hydro & meteo GmbH, Lübeck, Germany*

ABSTRACT

Rain measurements based on rain gauges, disdrometers, weather radars and microwave links provide essential input data for urban drainage and stormwater modelling, management, and planning. Their quality strongly depends on the sensor type and calibration, but also on the data post-processing that includes quality control and data adjustment after comparison with reference observations. This chapter provides an overview of traditional techniques and recent developments, and practical advice on the selection of the type of instrument, the installation and calibration aspects to be considered, and the measurement data processing and adjustment needs.

Keywords: Disdrometer, metrology, rainfall, rain gauge, weather radar.

SYMBOLS

a numerical coefficient in the Marshall-Palmer formula
b numerical coefficient in the Marshall-Palmer formula
f_{HH} phase of the horizontally polarized radar reflectivity
f_{VV} phase of the vertically polarized radar reflectivity
K_{dp} specific differential phase
L_{dr} linear depolarization ratio
R rainfall intensity (m/s)

doi: 10.2166/9781789060119_0011

Z radar reflectivity (m^3)
Z_{dr} differential reflectivity
Z_{HH} horizontally polarized radar reflectivity
Z_{VV} vertically polarized radar reflectivity
ρ_{hv} co-polar correlation coefficient

Key messages on sensors for rain measurements

- KM 2.1: *Local vs. global* – rain gauges and disdrometers measure the rain locally while radars can measure over the full catchment.
- KM 2.2: *Calibration* – all devices require calibration. Radar data must be adjusted with local measurements.
- KM 2.3: *Location* – pay attention to the location where the device will be set up to avoid or reduce bias in measurements.

2.1 INTRODUCTION

Rainfall is a key driving force in urban drainage and stormwater management (UDSM) as it generates runoff, in- and ex-filtration, discharges and floods. Its measurement is thus of paramount importance, at timescales ranging from a few minutes (runoff processes and discharges at catchment scale and in UDSM facilities) to decades (statistics on rainfall regime and local climate), for a diversity of actions (system operation, planning, design, modelling, etc.).

Precipitation is mainly measured with rain gauges or/and weather radars, but other measurement techniques exist as well. Where these are unavailable, satellite data may be helpful. This chapter mainly focuses on rain gauges and radars. However, disdrometers and microwave links will be briefly introduced.

2.2 RAIN GAUGES

Rain gauges provide rainfall registrations at ground level or nearby, depending on the height of the rain gauge installation, e.g. in a ground pit, at the ground surface, on a roof top, etc. Rain gauges typically provide measurements of liquid precipitation mass (e.g. in grams) collected on the receiving area of the gauge, during a certain duration. Given the surface of the collecting area, the duration and the amount of water, that information on mass can be converted into rainfall depth (in mm or L/m^2) or rainfall intensity (in mm/h or $L/h/m^2$). In order to measure solid precipitations (such as snow or ice fall), the rain gauges can be heated to melt those precipitations and, therefore, measure them as equivalent liquid precipitation. There are several types of rain gauges:

- Weighing rain gauges (Figure 2.1(a), (c) and (d), WR).
- Tipping bucket rain gauges (Figure 2.1(b), TBR).
- Rainfall height recording gauges (Figure 2.2, only old installations).
- Graduated cylinders.
- Simple buried pit collectors (Figure 2.3).

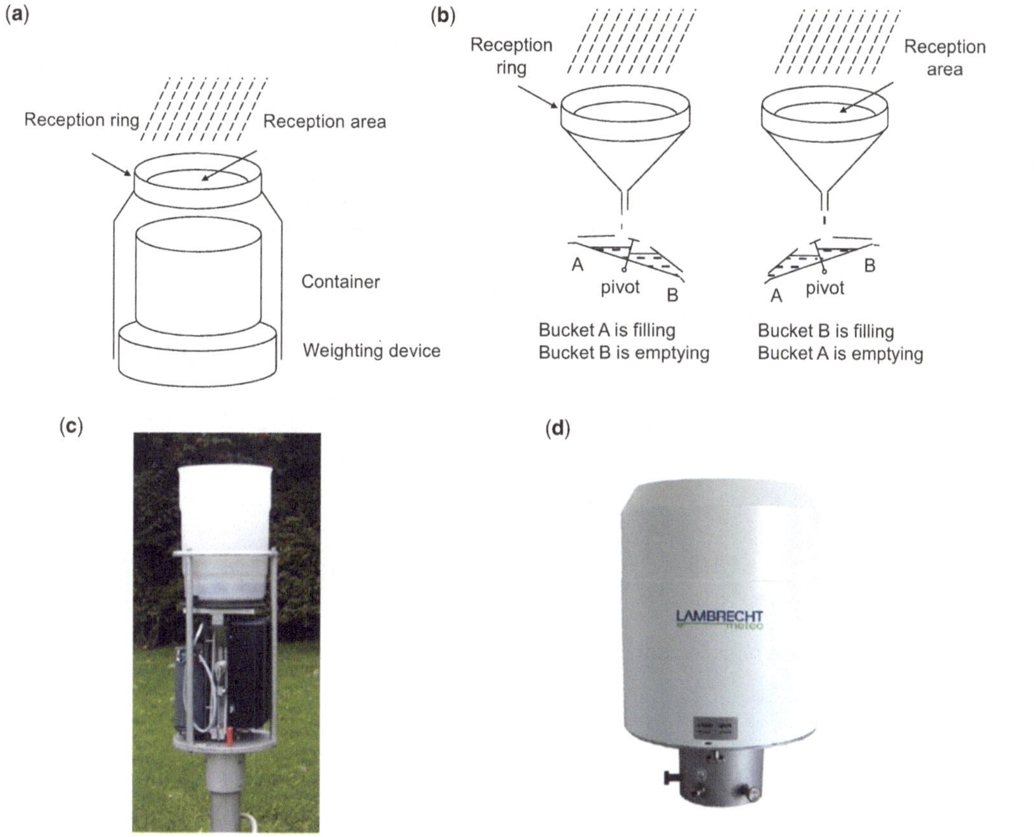

Figure 2.1 Rain gauges: Principles (a) WR, (b) TBR), open (c) and closed (d) WR. *Source:* (a) and (b) adapted from Bertrand-Krajewski *et al.* (2000); (c) Thomas Einfalt, (hydro & meteo GmbH); (d) http://www.lambrecht.net.

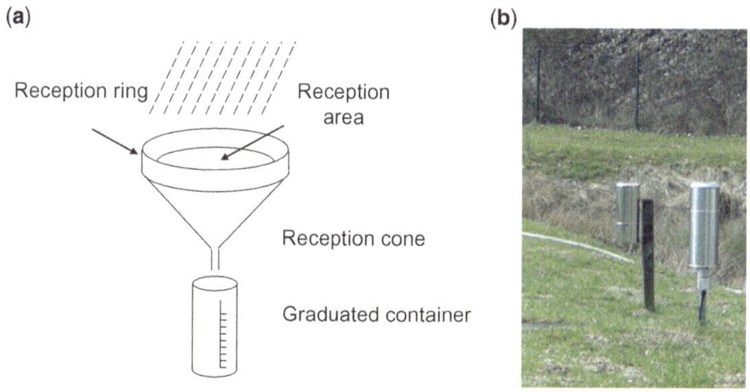

Figure 2.2 (a) Manual rain gauge principle and (b) a closed one as installed in the field. *Source*: (a) adapted from Bertrand-Krajewski *et al.* (2000); (b) Thomas Einfalt (hydro & meteo GmbH).

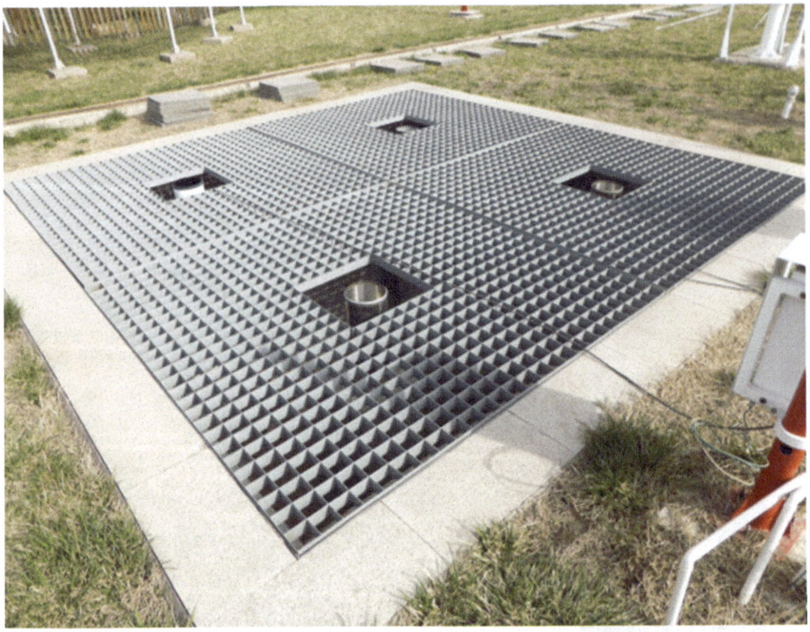

Figure 2.3 Pit rain gauges or collectors. *Source*: Thomas Einfalt (hydro & meteo GmbH).

WR and TBR (Figure 2.1) offer automatic measurements, which lead (depending on the device and its settings) to a measurement up to every minute. WR have a storage bin for the liquid precipitation, which is weighed to record the mass of water by using a vibrating wire connected to a data logger. Thus, the measurement starts at amounts less than 0.01 mm. TBR consist of a reception area that collects the precipitation and brings it through a funnel into a small bucket. When the mass of the rainwater collected in the bucket exceeds a given value, the bucket tips, actuating a switch (such as a reed switch) which is then electronically recorded (e.g. in a data logger). Each tip typically corresponds to a rainwater mass which is then converted to an equivalent depth of 0.1 to 0.5 mm, which is also the rainfall depth resolution of the gauge. By counting the number of tips in a given time interval, e.g. 5 or 10 minutes, the rainfall intensity can be calculated. Other, including more advanced, types of rain gauges exist as well, but these are not commonly used yet. This chapter does not aim at giving a complete description of all types of rain gauges; it rather focuses on the practical and relevant aspects for the further use of the meteorological data. More details about different types of rain gauges, their properties and accuracies can be found in (among others): World Meteorological Organization (WMO) (2018a), Sevruk (1996), Strangeways (2007) or Wauben (2006).

The WR have the advantage over the TBR that they do not underestimate high rainfall intensities and require much less maintenance because no regular funnel cleaning is required (no clogging possible) and in winter, snow effects (time delayed recording only after melting) are also not encountered. In addition, WR calibration (see Section 7.6) is also much simpler. These gauges are, however, more expensive than the TBR. TBR are not as accurate as the weighing or other types of rain gauges because: (i) the rainfall may start with a non-empty bucket, and (ii) the rainfall may stop before the bucket has tipped once more. The start and the end of rain events may be inaccurately measured with TBR.

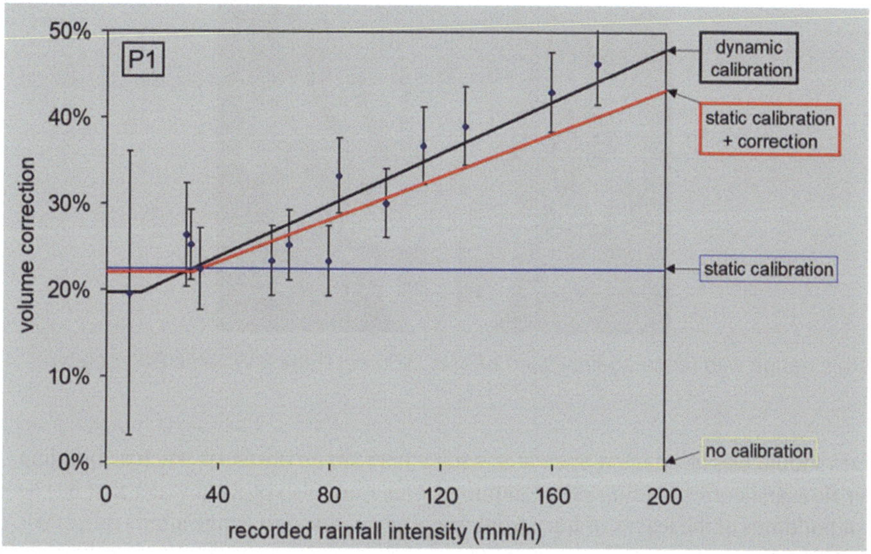

Figure 2.4 TBR gauge calibration, comparing no calibration, static calibration, and dynamic calibration. *Source*: Patrick Willems (KU Leuven).

Moreover, tipping bucket rain gauges typically underestimate the rainfall depths for the higher intensity storms: water can overflow outside the bucket and therefore not be measured. The higher the rainfall intensity, the higher the amount of water lost. Such rain gauges require dynamic calibration to correct for those possible underestimations, depending on the rainfall intensity (Adami & Da Deppo, 1985; La Barbera *et al.*, 2002; Luyckx & Berlamont, 2001; Marsalek, 1981; Niemczynowicz, 1986). Such calibration is often already carried out by the manufacturer. However, a periodic dynamic re-calibration of the rain gauges by the user is necessary (e.g. in Humphrey *et al.*, 1997; Kvicera & Grabner, 2006; Luyckx & Berlamont, 2001), which is preferable compared to the simple static (or volumetric) calibration performed by most manufacturers and users (illustration of the difference in Figure 2.4). Static calibration means that the calibration is carried out with, and hence correction factor(s) are based on, a single test for a given rainfall intensity. Dynamic calibration means that the correction factors are derived by tests at different rainfall intensities, hence as a function of the rainfall intensity (see an example of TBR calibration in Section 7.6.4.6). Several studies (e.g. Habib *et al.*, 2008; Willems, 2001) have shown that these effects can strongly influence the results of runoff simulations.

Rain gauges also have other limitations. First, they only indicate rainfall in a localized area, i.e. the receiving area of the rain gauge which is frequently between 200 and 400 cm^2. Because drops will stick to the sides of the gauge or funnel of the collecting device, rainwater amounts are slightly underestimated for TBR. Moreover, rain gauges are known to encounter difficulties measuring rain in windy conditions (Figure 2.5) as they are mostly not equipped with suitable and necessary windscreens (this causes underestimation, of up to 20%) and can have serious underestimations for high intensity rainfall events (Braak, 1945; Neff, 1977; Sevruk, 1996; WMO, 2018a). The presence of a wind shield or fence on the gauge can reduce this influence (Alter, 1937; Duchon & Essenberg, 2001; Larkin, 1947; Yang *et al.*, 1999); see an example of such a fence in Figure 2.5. Another solution is to level the rain gauge orifice with the ground so that wind effects are minimized, as shown in Figure 2.3.

Figure 2.5 Rain gauge with fence against wind effects. *Source*: Patrick Willems (KU Leuven).

Rain gauges should be placed in an open area where there are no obstacles, such as buildings or trees, to disturb the air flow and corresponding rain conditions. One must also prevent water that has been collected on the roofs of buildings or the leaves of trees from dripping into the rain gauge after a rain event, resulting in inaccurate readings. For rain gauges that measure at ground level, the vulnerability to turbulence is reduced. In this case, the surrounding surface may cause splashing of the raindrops into the gauges and again special care must be given to the selection of the surrounding surface (see Figure 2.3).

Rain gauges with a funnel (especially TBR, but also older recording WR gauges) are sensitive to blockage by e.g. leaves from trees, bird droppings or bird nests. Operational malfunctions could also include failure of the logger or the transmission hard- and/or software. Regular maintenance, check-ups and data validation are highly recommended to check the status of the rain gauge(s). Quality control of the measurement and logging mechanisms of the gauges is strongly advised as well as thorough verification of the rainfall data they provide. Willems (2001) provided an uncertainty assessment for typical TBR measurements in Belgium. Wagner (2009) provided a literature review on the different possible error sources while correcting the rain measurements affecting the TBR measurements and the possible correction methods.

Environmental conditions for rain gauges

Wind, snow and densely constructed area

A densely built-up urban area can affect the rain measurement by increasing or decreasing the amount of rain at the location of the rain gauge. Wind, as a global variable or local effect (in the surroundings of large buildings) can significantly affect rain measurements. If the rain gauge location is prone to freezing conditions or snow events, choose a rain gauge with a melting option.

Local measurements

Rain gauges measure the rain intensity and dynamics at the measuring location. This location should be carefully selected and checked to ensure the recorded data are representative of the catchment of interest.

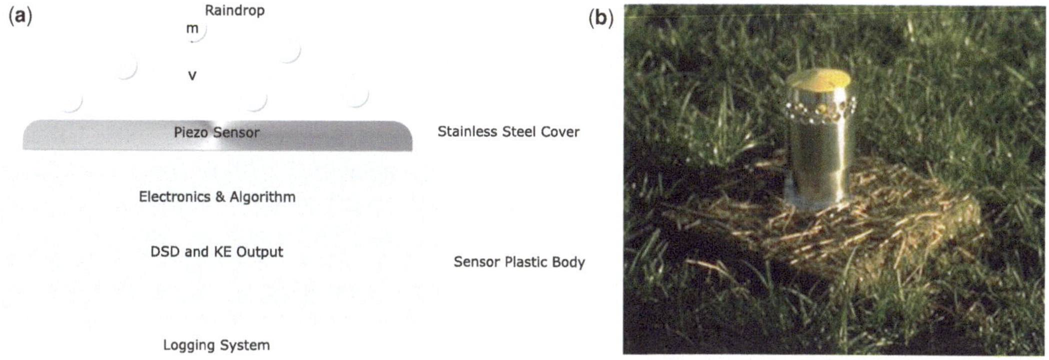

Figure 2.6 Acoustic disdrometers: (a) measuring principle (DSD: drop-size distribution, KE: kinetic energy) and (b) Joss-Waldvogel acoustic disdrometer installed in the field. *Sources*: (a) adapted from Abd Elbasit *et al.* (2011); (b) Thomas Einfalt (hydro & meteo GmbH).

2.3 DISDROMETERS

An alternative way to measure rainfall at the local scale consists of measuring the rain drops with disdrometers. Historically, the first widely used disdrometers were acoustic impact disdrometers (Figure 2.6), measuring the noise generated by the impact of a falling drop and relating this to the rainfall intensity (Joss & Waldvogel, 1967), similar to listening to the rain while driving a car or sitting under a roof window. Nowadays, most disdrometers are optical (Figure 2.7). They are made of one (or several) transmitter(s) and receiver(s) with a sampling volume(s) in between them. The transmitter generates one or several laser sheet(s) and the receiver measures either the occluded light (Battaglia *et al.*, 2010; Frasson *et al.*, 2011; Löffler-Mang & Joss, 2000) or the scattered light (Ellis *et al.*, 2006) from a drop falling through a sampling area of roughly a few tens of cm^2. The received signal is then processed to estimate the size (equivolumetric diameter) and velocity of the hydrometeor which can be a raindrop, a snowflake or a hailstone.

Two-dimensional (2D) video disdrometers have also been developed (Kruger & Krajewski, 2002) but they are not yet used operationally. Some experimental set-ups have also been deployed to reconstruct the 3D raindrop field of frames of a reference volume (1 m^3) (HYDROP Experiment, Desaulnier-Soucy *et al.*, 2001; WMO, 2018a).

Like rain gauges, disdrometers can be biased by wind. The other main limitations are: (i) the estimation of size and velocity of a drop relies on theoretical drop shapes that are often different in reality (Battaglia *et al.*,

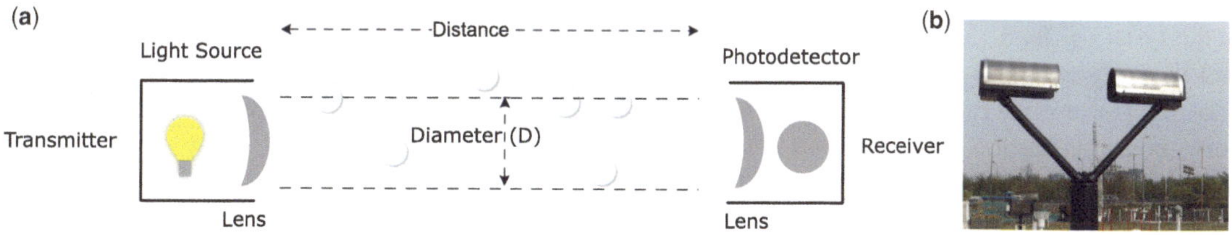

Figure 2.7 Optical disdrometers: (a) measuring principle and (b) example of set-up in the field. *Sources*: (a) adapted from Belenguer *et al.* (2020); (b) Thomas Einfalt (hydro & meteo GmbH).

2010), (ii) a significant sampling error for small time steps occurs because of the small sampling area, up to 15% error in the rain intensity for 1 min time steps and decreasing for larger ones (Jaffrain & Berne, 2012a, 2012b), and (iii) there is a non-homogenous laser beam pattern for disdrometers computing the occluded light (Frasson *et al.*, 2011). Several studies have compared various disdrometers but also compared those disdrometers with more conventional devices such as rain gauges (Brawn & Upton, 2008; Jaffrain & Berne, 2011; Krajewski *et al.*, 2006; Miriovsky *et al.*, 2004; Thurai *et al.*, 2011; Tokay *et al.*, 2001). They concluded that disdrometers are as reliable as the standard devices for point rainfall measurements.

Dense networks of disdrometers have recently been deployed which can show the importance of taking the small-scale drop size distribution variability in the Z-R or R-K_{dp} relation (see Section 2.4.3) into account and more generally of improving knowledge in this field (Jaffrain & Berne, 2012a, 2012b; Tapiador *et al.*, 2010).

2.4 WEATHER RADAR

2.4.1 Introduction

In order to avoid the main disadvantage of rain gauges and disdrometers (local measurement), weather radars (RAdio Detection And Ranging) are nowadays commonly used. Derived from military technology (from World War 2), rain detection replaced aircraft detection: military operators noticed that the images contained echoes from rainfall and other obstacles. After the war, radar technology was further developed, also in a scientific environment, with specific interest for the meteorological use of radar technology.

Radar technology is hereafter discussed in two sections:

- Temporal and spatial resolution of radar data (Section 2.4.2).
- Radar data quality, rainfall estimation, and radar data adjustment (Section 2.4.3).

More information about the measurement principle is presented and detailed in ISO (2019) and WMO (2018b). The contents of the following subsections largely follow the information compiled by Thorndahl *et al.* (2017), with additions from ISO (2019).

2.4.2 Temporal and spatial resolution of radar data

A weather radar (example installations in Figure 2.8) emits microwaves as pulses, and the encountered objects in the atmosphere reflect the emitted microwaves. The radar antenna then measures the amount of reflection and the distance to the radar, based on the travel time of the pulse between emission and reception. Simultaneously, the radar rotates around its axis in order to cover the complete area around a radar site up to the maximum range. To scan the atmosphere in three dimensions, the radar measures at several elevations, i.e. angles pointing into the atmosphere (Figure 2.9).

The microwaves of S, C and X bands are used in most cases and the scale and observation characteristics of the system differ depending on the characteristics of each band (see Table 2.1). S-band systems are large, and their observation range is wide, while X-band systems are compact and their observation range is narrow. The useful, qualitative range of S-band and C-band radars are typically limited by Earth's curvature, whereas at X-band the limit is normally attenuation dependent.

The temporal resolution of radar data is governed by the scanning strategy of the radar. A radar scans the atmosphere at different elevations (Figure 2.9) to generate a full azimuthal volume scan (Figure 2.10). This requires up to several minutes depending on rotational velocity and the number of scanning elevations. A radar collects instantaneous samples of rain intensity estimated from the measured reflectivity, unlike rain

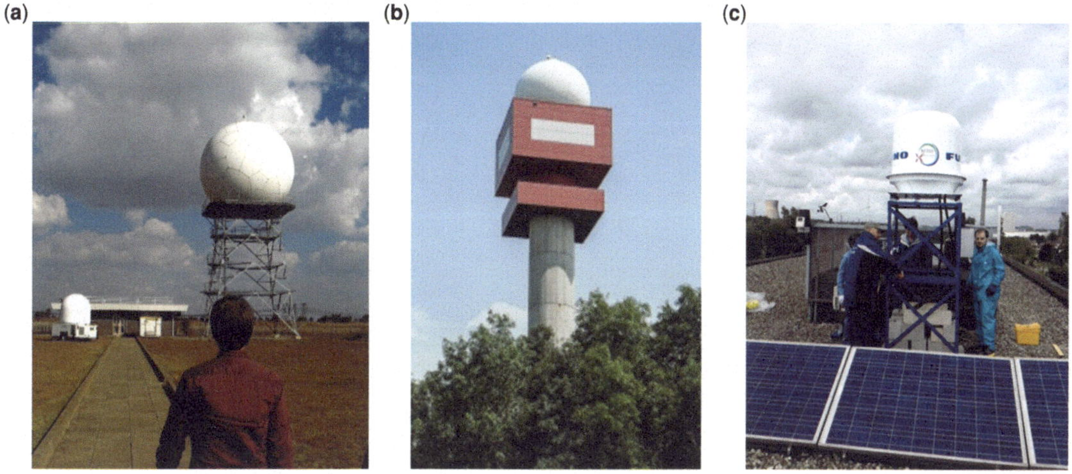

Figure 2.8 Examples of radars. (a) Irene S-band radar of South African Weather Service; (b) C-band radar of Royal Meteorological Institute of Belgium at Jabbeke in Belgium; (c) X-band radar by KU Leuven and Furuno in the city of Gent in Belgium. *Sources*: Thomas Einfalt (hydro & meteo GmbH) and Patrick Willems (KU Leuven).

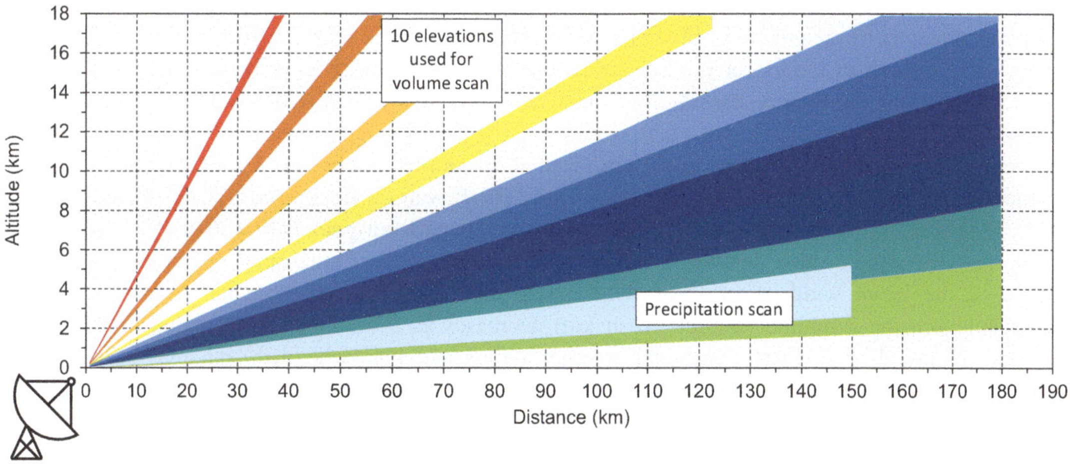

Figure 2.9 Atmospheric scanning strategy of German Weather Service. *Source*: adapted from DWD (2018).

Table 2.1 Typical operating resolutions and maximum ranges for different types of weather radars used in hydrological applications (from Thorndahl *et al.*, 2017).

	X-band	C-band	S-band
Spatial resolution	100–1000 m	250–2000 m	1000–4000 m
Temporal resolution	1–5 min	5–10 min	10–15 min
Maximum quantitative range (see Section 2.4.3)	30–60 km	100–130 km	100–200 km

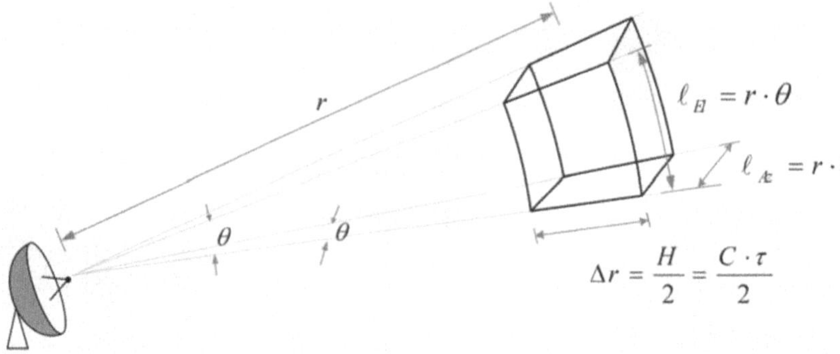

Figure 2.10 Radar scans with radial resolution Δr and azimuthal resolution l_{Az}. *Source*: adapted from Ochoa-Rodriguez *et al.* (2019).

gauges which accumulate rainfall over a given time interval. Some radars operate with intermediate dedicated Doppler scans for each volume scan, hence doubling the time between two consecutive reflectivity scans. Operational meteorological S-, C-, and X-band radars usually provide reflectivity scans with a temporal resolution of 5–15 minutes (Table 2.1), whereas research radars dedicated to high resolution rainfall monitoring in specific areas and specific elevations are reported to provide data resolutions down to 15 seconds (e.g. Mishra *et al.*, 2016; van de Beek *et al.*, 2010).

The main strength of radars for rainfall estimation is their capability to provide spatially distributed rainfall information. The spatial resolution of radar rainfall data is basically determined by the hardware and physics. The radial resolution (or range resolution, Figure 2.10) is a function of the pulse length and for operational radars goes down to 250 m. Radial resolutions between 3 and 100 m have been documented for research radars by e.g. Leijnse *et al.* (2010), van de Beek *et al.* (2010), Lengfeld *et al.* (2014), Mishra *et al.* (2016), Thorndahl *et al.* (2017) and Ochoa-Rodriguez *et al.* (2019).

The spatial resolution also depends on the azimuthal (or angular) horizontal resolution, which is a function of the beam width determined by the size and design of the antenna. In contrast to the radial resolution, the azimuthal resolution (in km) decreases as a function of the radial distance from the radar (Figure 2.10). Most operational weather radars use parabolic dish antennas with a beam width of approx. 1 degree, thus functioning with an azimuthal horizontal resolution close to 1 degree (see http://www.eumetnet.eu/opera, visited the 09/04/2021). As an example, at a distance of 100 km (resp. 55 km) from the radar, the 1° beam is ~1750 m (resp. 1000 m) wide. Small local X-band radars with (non-parabolic) horizontal fan beam antennas typically have larger opening angles between 2 and 3 degrees, but also a smaller maximum range compared to meteorological radars, due to integration of rainfall over a large vertical distance (e.g. Pedersen *et al.*, 2010; Thorndahl & Rasmussen, 2012).

Examples of radar reflectivity maps with four different spatial resolutions covering an approximately 12 km × 12 km area over the city of Aalborg, Denmark, are shown in Figure 2.11. This example illustrates the importance of high spatial resolution data to capture the spatial variability of rainfall which is of critical importance over an urban area (high spatial resolution is important to understand the variability within the urban catchment of rainfall intensities and depth, floods and peak discharges, to better calibrate and test simulation models in particular sewer overflows, etc.) (e.g. Ochoa-Rodriguez *et al.*, 2015, 2019; Rico-Ramirez *et al.*, 2015).

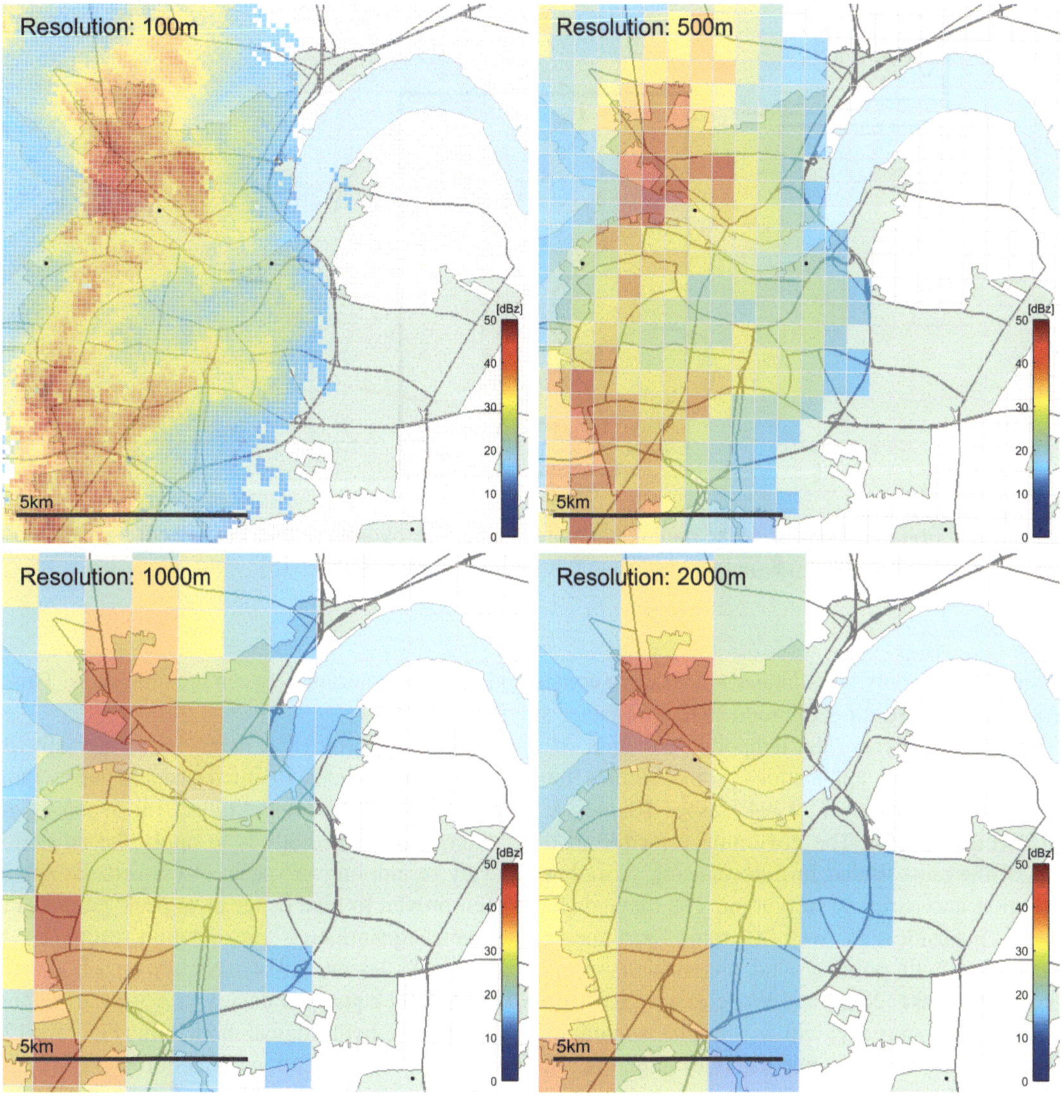

Figure 2.11 Example of radar reflectivity maps at four different Cartesian spatial resolutions over Aalborg, Denmark (Lat: 57.05, Lon: 9.92). The radar data were acquired with a Furuno WR-2100 dual-polarimetric X-band radar (Nielsen *et al.*, 2014) at 1 minute temporal resolution at 16:20:00 UTC on July 25, 2016. Black circles are rain gauges of the Danish Water Pollution Committee network. *Source*: Thorndahl *et al.* (2017).

2.4.3 Radar data quality, rainfall estimation, and radar data adjustment

The temporal resolution of radar data is determined by the recurrence time of the measurement by the radar at the same location. This ranges typically between 5 and 15 minutes for operational weather radars. Because radar measurement is an instantaneous measurement, information between the measurement times is

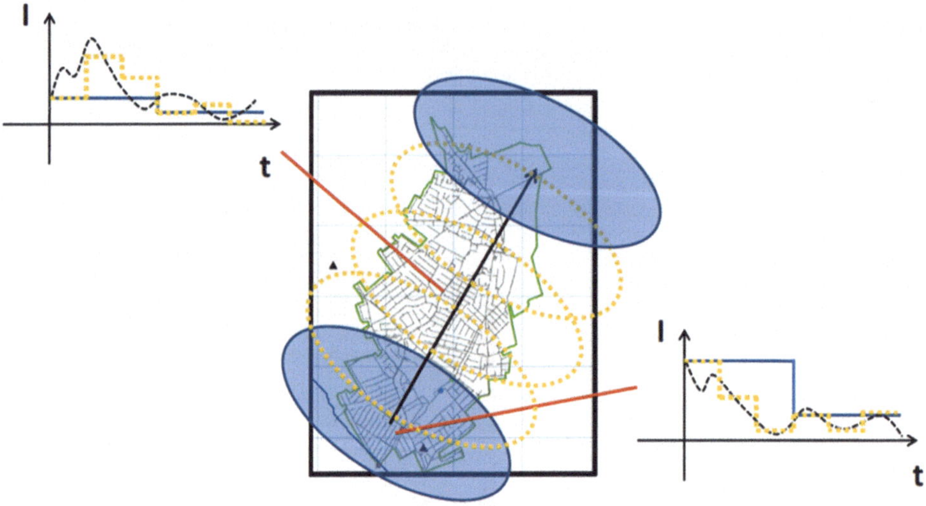

Figure 2.12 Enhancement of radar rainfall estimates for urban hydrology through optical flow temporal interpolation and Bayesian gauge-based adjustment. *Source*: Wang *et al.* (2015).

missing and can only be estimated – or is often considered as being constant between the time steps. This is frequently the case for data products from weather services.

In order to increase the temporal resolution of operational meteorological radar data, especially for urban hydrological applications, some authors have developed methods to interpolate between radar images (e.g. Fabry *et al.*, 1994; Jasper-Tönnies & Jessen, 2014; Thorndahl *et al.*, 2014; Wang *et al.*, 2015) (Figure 2.12). The governing principle in these interpolation methods is to apply the advection field of the rain, similar to a nowcasting procedure, and by resampling in space, to convert the spatial resolution into temporal resolution. The methods have been proven to give better local peak estimates of rainfall intensities as well as more accurate accumulated quantitative precipitation estimates in comparison with point ground observations. Jasper-Tönnies & Jessen (2014), Nielsen *et al.* (2014), Seo & Krajewski (2015) and Wang *et al.* (2015) have successfully converted data with a 5- or 10-minute resolution into a product with a 1-minute resolution for use in urban hydrological modelling (Figure 2.12).

Considering the advective nature of rain, it is also obvious that this advection correction yields a better estimate of the real precipitation. An accumulation of instantaneous radar data with e.g. a 5-minute sampling time interval may result in a 'fishbone' pattern (Figure 2.13).

The use of radar data requires that the data are of good quality. There are numerous items such as radar hardware calibration, clutter removal, overshooting/vertical profile correction, etc. (Li, 2020; Michelson *et al.*, 2005; Villarini & Krajewski, 2010), which have to be considered and may have to be corrected before radar reflectivity data can be converted into reliable rainfall intensities. A thorough quality check and potential correction are therefore required. Disturbances for a good radar measurement may be undesired reflections of mountains, high towers, air planes, ships, or wind turbines, attenuation by heavy rain or hail, snow or melting snow instead of rainfall, anomalous propagation conditions and others. Methods to test for these problems exist, and they are partly reduced by dual-polarization information from the new generation radars. The preprocessing of radar data by meteorological services usually only covers some of the above points.

(a)

(b)

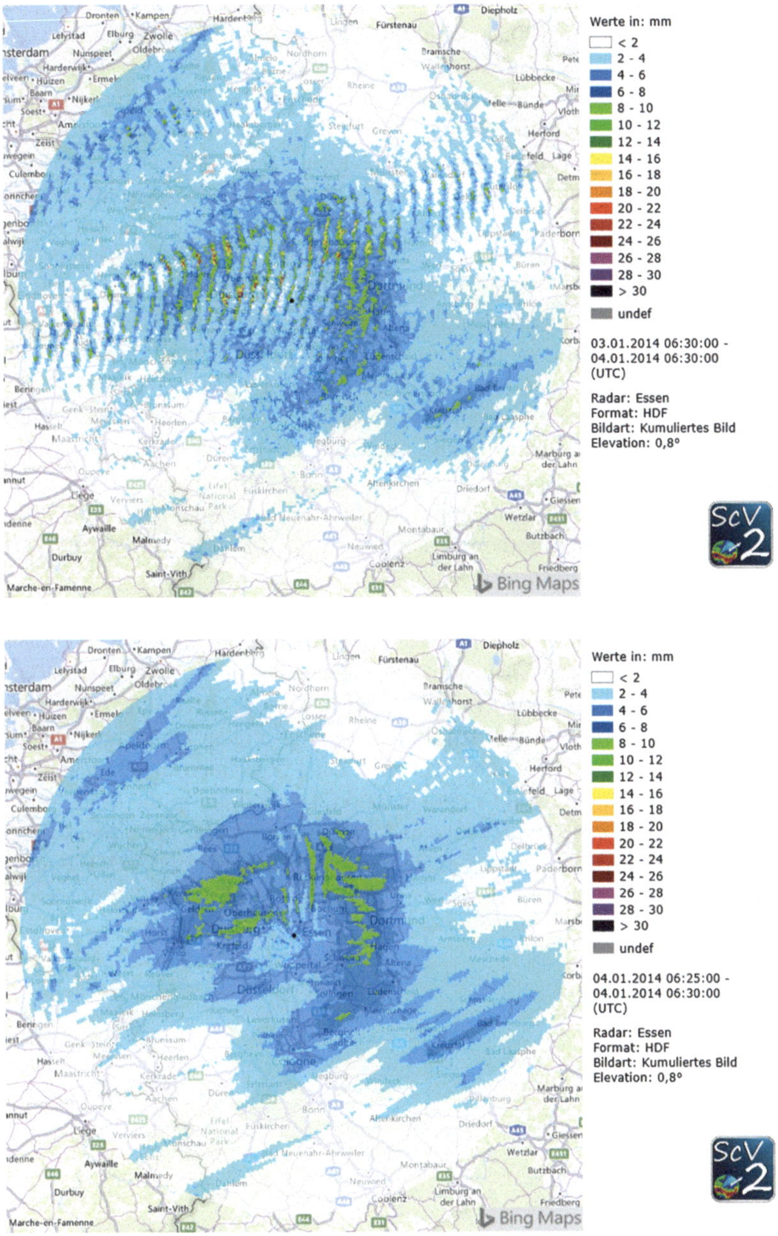

Figure 2.13 (a) Daily sum of rainfall depth (in mm) estimated from radar data measured from January 3, 2014 06:30 UTC to January 4, 2014 06:30 UTC in Essen, Germany with a 5-minute time step. Clearly visible is the 'fishbone' structure, due to the movement of the rainfall between the scanned images. (b) Image as above, but with a 1-minute temporal interpolation between each measured time step. *Source:* Thomas Einfalt (hydro & meteo GmbH).

Observed radar reflectivity can be converted into rain intensities, but comparison and adjustment with ground observations is required to produce valid quantitative precipitation estimates (QPE). This is most often referred to as radar rainfall adjustment or radar-rain gauge merging.

Rain gauges used for adjustment also need to be of high quality. Frequently observed shortcomings of rain gauge data are missing data, time shifts or differently set clocks, clogging of the gauge, data transmission drop outs, gauge calibration errors, local wind effects around gauges leading to measurement errors, or gauge sampling errors (e.g. Ciach, 2003; Gires *et al.*, 2014; Villarini *et al.*, 2008). To avoid random or systematic errors, such effects need to be eliminated before rain gauge data are used to adjust radar rainfall. Automatic procedures for these tasks exist (Einfalt & Frerk, 2011), but data controlled and corrected by experienced observers still give better results.

The relation between measured radar reflectivity, Z (in mm^6/m^3 or dBZ) and rain intensity, R (mm/h) as the target unit for hydrology depends on the drop size distribution (DSD) of the observed precipitation. As documented by numerous authors (e.g. Marshall & Palmer, 1948; Uijlenhoet, 2001), the most commonly used conversion for single polarization radars is to apply a two-parameter power-law relationship to describe the relation between rain intensity and radar reflectivity, named the *Z-R* relationship: $Z = aR^b$ (Figure 2.14). Since the power-law parameters vary with the DSD shape, i.e. the type of rain, they are not constant in time. One solution is to adjust the *Z-R* relationship continuously by use of ground observations. It is however more common to apply a fixed *Z-R* relationship and perform a posteriori bias adjustment.

Whereas traditional *Z-R* conversion is well documented in numerous applications of radar, there are recent advances in the application of dual-polarized radars which enable more accurate QPE assessment using polarimetric parameters (e.g. Anagnostou & Anagnostou, 2008; Anagnostou *et al.*, 2004; Bringi & Chandrasekar, 2001; Bringi *et al.*, 2011; Li, 2020; Mishra *et al.*, 2016; Ochoa-Rodriguez *et al.*, 2019; Scarchilli *et al.*, 1993; Simpson & Fox, 2018). Polarization of a radar signal characterizes the orientation of the electric field (both transmitted and received). Dual-polarimetric radars transmit a radar signal alternately in horizontal (H) and vertical (V) polarization. Depending on the shape of the rain drops, two different signals are received: reflectivities Z_{HH} and Z_{VV}. Additionally, the phase of the horizontally and vertically polarized return signals, f_{HH} and f_{VV}, are measured (Illingworth, 2004). Four parameters can be defined based on the polarimetric measurements: differential reflectivity Z_{dr}, linear depolarization ratio

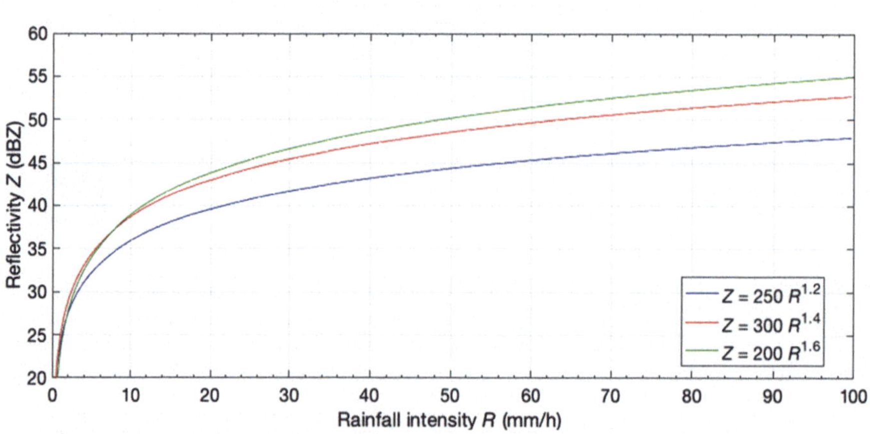

Figure 2.14 *Z-R* relationships of Marshall & Palmer (1948). *Source*: adapted from Marshall & Palmer (1948).

L_{dr}, co-polar correlation coefficient ρ_{hv} and the specific differential phase K_{dp} (Illingworth, 2004). K_{dp} is proportional to the product of rainwater content and the mass-weighted mean diameter (Bringi & Chandrasekar, 2001) and thus can be used to estimate rainfall intensities. The advantage of using K_{dp} for rainfall intensity estimation is that it is more sensitive to the raindrop shape, thus rainfall intensity can be estimated from K_{dp} in the case of rain/hail mixture. As soon as the hydrometeors are spherical or quasi-spherical, K_{dp} is about 0°/km (hail, light rain). Another advantage of using K_{dp} is that it is independent of radar calibration and not sensitive to attenuation, an issue of particular importance at X-band frequency. K_{dp} can only be estimated for medium to high rainfall rates (Otto & Russchenberg, 2011).

Many different methods for adjusting rain intensities estimated from radar reflectivities have been developed, and several profound review papers on different adjustment/merging techniques related to hydrological applications exist (e.g. Goudenhoofdt & Delobbe, 2009; McKee & Binns, 2016; Ochoa-Rodriguez *et al.*, 2019; Wang *et al.*, 2013). For specific details we refer to these. The most widely applied methods are presented hereafter.

A basic method of adjusting radar rainfall data to gauge data implies a temporally and spatially uniform relationship between both measurements. It has been proposed by Smith & Krajewski (1991) as the concept of mean field bias (MFB) adjustment. The concept is to estimate the ratio between accumulated rainfall depth in several ground observation points (rain gauges) and accumulated radar rainfall in the corresponding points (grid cells). Under the above-mentioned simplifying assumption, the whole radar field is multiplied by one MFB factor.

The processing of QPE for urban applications requires that the DSD is considered to be spatially heterogeneous and also not constant in time. For this, more sophisticated methods have been developed which either preprocess the station data by an interpolation process or perform an interpolation of the obtained relations between gauges and collocated radar pixels.

The optimal temporal integration period or spatial aggregation level is, to a large extent, dependent on the representativeness of the gauges (gauge network density) to capture the temporal and spatial variability of the rain (e.g. Gires *et al.*, 2014). It is difficult to recommend specific gauge network densities for radar rainfall adjustment since the optimal value will depend on storm type, homogeneity of the rain gauge network, orographic features of the rain, adjustment methods, etc. McKee & Binns (2016) suggest conducting a sensitivity analysis to identify the effect of gauge density on rainfall estimation.

Spatial variability adjustment approaches and geostatistical merging of radar and rain gauge data have been developed to account for range dependence issues as well as heterogeneous DSDs. They are widely applied for QPE. The concept here is to merge the spatial variability of the radar rainfall fields into the interpolated rain gauge precipitation fields to increase the spatial resolution of this product. The interpolation can be performed by many different spatial interpolation methods e.g. variations of kriging (Goudenhoofdt & Delobbe, 2009; Krajewski, 1987; Ochoa-Rodriguez *et al.*, 2019; Sinclair & Pegram, 2005), by inverse distance weighting or Thiessen polygon weighting (Haberlandt, 2007; Johnson *et al.*, 1999) or Bayesian methods (Ochoa-Rodriguez *et al.*, 2015, 2019; Todini, 2001; Wang *et al.*, 2015). The kriging-based methods rely on variograms for describing the spatial dependence in rainfall fields and are in general more computationally demanding than inverse distance weighting methods. The latter are therefore often used in real-time operation. A recent new development is the use of convective rain cells as singularity elements to be integrated in the rain gauge – radar interpolation method for a more accurate estimation of fine-scale extreme rainfall intensities, which are of importance for many urban water applications. Example results of the application of some rain gauge – radar adjustment, merging and interpolation approaches are shown in Figure 2.15.

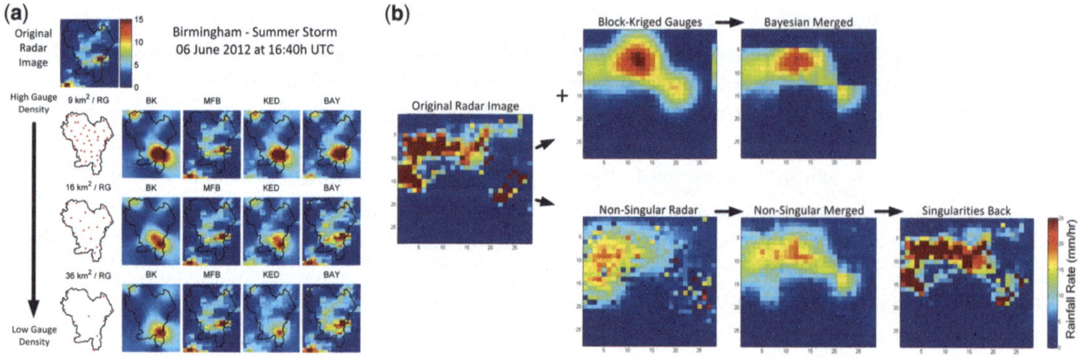

Figure 2.15 Example results of radar data adjustment approaches. (a) Original radar image is shown in the top row for an event measured in Birmingham on 06 June 2012, with rain gauge (RG) location and density on the left side, and the following four columns with (i) block-kriged (BK) interpolated rain gauge records, (ii) mean field bias (MFB), (iii) kriging with external drift (KED) and (iv) Bayesian combination (BAY). Radar records were available at 1-km^2/5-min resolution, while rain gauge records were available at 2-min temporal resolution. (b) Singularity-sensitive gauge-based radar rainfall adjustment methods by Wang *et al.* (2015). *Sources*: (a) adapted from Ochoa-Rodriguez (2017) and (b) adapted from Wang *et al.* (2015).

2.4.4 Summary

Weather radar nowadays is an established tool to provide rainfall information with a high temporal and spatial resolution. The shortcomings of radar, in particular the uncertain estimate of the absolute amount of precipitation, has to be taken care of by combining the radar information with rain gauge measurements in an appropriate manner. Radar rainfall estimates are – as an indirect measurement – prone to uncertainties associated with variability in drop size distribution, partial beam filling, overshooting, and signal attenuation. These effects need to be considered and corrected for where possible. A new way to reduce these uncertainties is by using polarimetric signals, another way is by reducing the distance to the radar, by increasing the density of the radar network. The quality of radar data adjustment in turn depends on the density and quality of the rain gauge network. The optimal temporal integration period or spatial aggregation level for radar adjustment is directly related to the ability of the rain gauge network to capture the temporal and spatial rainfall variability. For practice applications in urban hydrology, it is most often important to apply a powerful software package which is able to cover the above-mentioned work tasks for a high quality of radar measurement data.

2.5 MICROWAVE LINKS

Another emerging method for rainfall monitoring is the use of microwave links. A microwave link consists of two antennas, one sending and one receiving unit, typically a few hundred metres up to 15 km apart. Microwave links are used for mobile telephone communication, they operate around a frequency of 7–40 GHz and the link length is mostly limited to a maximum of 5–10 km. There is quite a dense network available in some countries. However, in others, this might be limited. As an estimate of the density of the microwave link network, a density of at least 0.3 links/km^2 can be assumed for European countries, according to Chwala *et al.* (2012). For example, in the Netherlands (35,500 km^2) the total number of link paths is at least 8000 and for many of those link paths, the microwave links measure in both directions (Overeem *et al.*, 2011).

The information sent over these links also has to travel through rain, which causes attenuation of the signal. The magnitude of the received power is mostly stored by the network operators and can thus be used to calculate the total integrated attenuation over the link path, from which the path averaged rainfall intensity can be estimated.

Research in this field started with research set-ups (e.g. Grum *et al.*, 2005; Holt *et al.*, 2003; Krämer *et al.*, 2005; Leijnse *et al.*, 2007a; Rahimi *et al.*, 2003, 2004, 2006; Ruf *et al.*, 1996; Upton *et al.*, 2005), but in recent years, data from commercial cellular networks have been used to estimate rainfall intensities (e.g. Chwala *et al.*, 2012; Leijnse *et al.*, 2007b; Messer *et al.*, 2006; Overeem *et al.*, 2011; Zinevich *et al.*, 2008). A limitation is the availability of the data, which could range from (near) real-time up to only on a daily or even weekly basis. Moreover, it can be hard to gain access to commercial microwave link data.

The microwave links can be used as a standalone estimator of the rainfall (e.g. Leijnse *et al.*, 2007b; Zinevich *et al.*, 2008), or can be combined with rain gauges and even radar measurements (e.g. Cummings *et al.*, 2009) to give better rainfall estimations at ground level. The links can also be used as an attenuation indicator for attenuation correction of the radar measurements. The network of these links is mostly more dense over urban areas than elsewhere, which could thus be an advantage for the correction of the radar estimates for urban hydrological applications since there are mostly only a limited number of rain gauges available in the city centre. Other advantages of microwave links are that they are mostly clutter free and very close to the ground compared to radar scans. Finally, the power law equation used to compute rainfall intensity from attenuation is almost linear, whereas the *ZR* relation employed in radar meteorology is nonlinear.

2.6 SUMMARY AND TRANSITION

This chapter introduced the most common rain measuring devices: a few for local measurements (rain gauges and disdrometers) and the weather radar for larger measurements over urban catchments. Both present advantages and disadvantages and can always be combined together to reduce their weaknesses (Figure 2.16).

A municipality or utility can easily purchase, install, calibrate, and operate a network of weighing or tipping bucket rain gauges with specifically trained employees. Access to radar data, either in real time

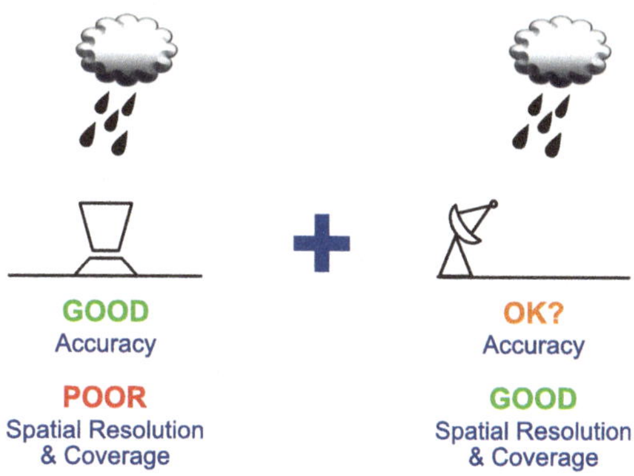

Figure 2.16 Advantages and disadvantages of local rainfall measurements (rain gauges and disdrometers) and weather radar data, where rainfall adjustment, interpolation or merging methods aim to combine the advantages. *Source*: Patrick Willems (KU Leuven).

or off-line, can be obtained by contracting with: (i) either national or regional meteorological services, or (ii) a service provider company. It is less common practice for a municipality or utility to own and operate their own radar. However, possible future dissemination of small X-band radars may change the practice in the future, likely in association with a service provider to calibrate and validate radar data.

Once rain has been measured, runoff processes will start. The following chapter is therefore devoted to discharge measurements within sewer pipes to accurately quantify flows inside the network itself.

REFERENCES

Abd Elbasit M. A. M., Yasuda H. & Salmi A. (2011). Application of piezoelectric transducers in simulated rainfall erosivity assessment. *Hydrological Sciences Journal*, **56**(1), 187–194. doi: 10.1080/02626667.2010.546359.

Adami A. & Da Deppo L. (1985). On the systematic errors of tipping bucket recording rain gauges. *Proceedings of the International Workshop on the Correction of Precipitation Measurements*, 1–3 April 1985, Zürich, Switzerland.

Alter C. J. (1937). Shielded storage precipitation gauges. *Monthly Weather Review*, **65**(7), 262–265. doi: 10.1175/1520-0493(1937)65<262:SSPG>2.0.CO;2.

Anagnostou M. N. & Anagnostou E. N. (2008). Chapter 12: Performance of algorithms for rainfall retrieval from dual-polarization X-band radar measurements. In *Precipitation: Advances in Measurement, Estimation and Prediction*, S. C. Michaelis (ed.), Springer Verlag, Berlin Heidelberg (Germany), pp. 313–342. ISBN 978-3-540-77654-3.

Anagnostou E. N., Anagnostou M. N., Krajewski W. F., Kruger A. & Miriovsky B. J. (2004). High-resolution rainfall estimation from X-band polarimetric radar measurements. *Journal of Hydrometeorology*, **5**(1), 110–128. doi: 10.1175/1525-7541(2004)005<0110:HREFXP>2.0.CO;2.

Battaglia A., Rustemeier E., Tokay A., Blahak U. & Simmer C. (2010). PARSIVEL snow observations: a critical assessment. *Journal of Atmospheric and Oceanic Technology*, **27**(2), 333–344. doi: org/10.1175/2009JTECHA1332.1.

Belenguer F. M., Martínez-Millana A., Salcedo A. M., Sánchez V. M. & Anaya M. J. P. (2020). Disdrometer performance optimization for use in urban settings based on the parameters that affect the measurements. *Symmetry*, **12**(2), 303, 19 p. doi: 10.3390/sym12020303.

Bertrand-Krajewski J.-L., Laplace D., Joannis C. & Chebbo G. (2000). *Mesures en hydrologie urbaine et assainissement [Measurements in urban hydrology and sewer systems]*. Editions Technique et Documentation, Paris (France), 794 p. ISBN 978-2-7430-0380-4. (in French).

Braak C. (1945). *Invloed van den wind op regenwaarnemingen [Wind influence on rain gauge measurements]*. 's-Gravenhage (The Netherlands): Royal Dutch Meteorological Institute of the Netherlands, Mededelingen en Verhandelingen, Rijksuitgeverij, volume 48, report 102. (in Dutch).

Brawn D. & Upton G. (2008). Estimation of an atmospheric gamma drop size distribution using disdrometer data. *Atmospheric Research*, **87**(1), 66–79. doi: 10.1016/j.atmosres.2007.07.006.

Bringi V. N. & Chandrasekar V. (2001). *Polarimetric Doppler Weather Radar*. Cambridge University Press, Cambridge (UK), 664 p. ISBN 978-0521019552.

Bringi V. N., Rico-Ramirez M. A. & Thurai M. (2011). Rainfall estimation with an operational polarimetric C-band radar in the United Kingdom: comparison with a gauge network and error analysis. *Journal of Hydrometeorology*, **12**(5), 935–954. doi: 10.1175/JHM-D-10-05013.1.

Chwala C., Gmeiner A., Qiu W., Hipp S., Nienaber D., Siart U., Eibert T., Pohl M., Seltmann J., Fritz J. & Kunstmann H. (2012). Precipitation observation using microwave backhaul links in the alpine and pre-alpine region of southern Germany. *Hydrology and Earth System Sciences Discussions*, **9**(1), 741–776. doi: 10.5194/hessd-9-741-2012.

Ciach G. J. (2003). Local random errors in tipping-bucket rain gauge measurements. *Journal of Atmospheric and Oceanic Technology*, **20**(5), 752–759. doi: 10.1175/1520-0426(2003)20<752:LREITB>2.0.CO;2.

Cummings R. J., Upton G. J. G., Holt A. R. & Kitchen M. (2009). Using microwave links to adjust the radar rainfall field. *Advances in Water Resources*, **32**(7), 1003–1010. doi: 10.1016/j.advwatres.2008.08.010.

Desaulnier-Soucy N., Lovejoy S. & Schertzer D. (2001). The HYDROP experiment: an empirical method for the determination of the continuum limit in rain. *Atmospheric Research*, **59–60**, 163–197. doi: 10.1016/S0169-8095(01)00115-6.

Duchon C. E. & Essenberg G. R. (2001). Comparative rainfall observations from pit and aboveground rain gauges with and without wind shields. *Water Resources Research*, **37**(12), 3253–3263. doi: 10.1029/2001WR000541.

DWD (2018). *Radar Products*. Deutscher Wetterdienst, Offenbach (Germany). Available at https://www.dwd.de/EN/ourservices/radar_products/radar_products.html (accessed 06 October 2020).

Einfalt T. & Frerk I. (2011). On the influence of high quality rain gauge data for radar-based rainfall estimation. *Proceedings of the 12th ICUD – International Conference on Urban Drainage*, 11–15 September, Porto Alegre, Brazil.

Ellis R. A., Sandford A. P., Jones G. E., Richards J., Petzing J. & Coupland J. M. (2006). New laser technology to determine present weather parameters. *Measurement Science and Technology*, **17**, 1715–1722. doi: 10.1088/0957-0233/17/7/009.

Fabry F., Bellon A., Duncan M. R. & Austin G. L. (1994). High resolution rainfall measurements by radar for very small basins: the sampling problem re-examined. *Journal of Hydrology*, **161**(1–4), 415–428. doi: 10.1016/0022-1694(94)90138-4.

Frasson R. P. D. M., da Cunha L. K. & Krajewski W. F. (2011). Assessment of the Thies optical disdrometer performance. *Atmospheric Research*, **101**(1–2), 237–255. doi: 10.1016/j.atmosres.2011.02.014.

Gires A., Tchiguirinskaia I., Schertzer D., Schellart A., Berne A. & Lovejoy S. (2014). Influence of small scale rainfall variability on standard comparison tools between radar and rain gauge data. *Atmospheric Research*, **138**, 125–138. doi: 10.1016/j.atmosres.2013.11.008.

Goudenhoofdt E. & Delobbe L. (2009). Evaluation of radar-gauge merging methods for quantitative precipitation estimates. *Hydrology and Earth System Sciences Discussions*, **13**(2), 195–203. doi: 10.5194/hess-13-195-2009.

Grum M., Krämer S., Verworn H.-R. & Redder A. (2005). Combined use of point rain gauges, radar, microwave link and level measurements in urban hydrological modelling. *Atmospheric Research*, **77**(1–4), 313–321. doi: 10.1016/j.atmosres.2004.10.013.

Haberlandt U. (2007). Geostatistical interpolation of hourly precipitation from rain gauges and radar for a large-scale extreme rainfall event. *Journal of Hydrology*, **332**(1–2), 144–157. doi: 10.1016/j.jhydrol.2006.06.028.

Habib E., Meselhe E. A. & Aduvala A. V. (2008). Effect of local errors of tipping-bucket rain gauges on rainfall-runoff simulations. *Journal of Hydrologic Engineering*, **13**(6), 488–496. doi: 10.1061/(ASCE)1084-0699(2008)13:6(488).

Holt A. R., Kuznetsov G. G. & Rahimi A. R. (2003). Comparison of the use of dual-frequency and single frequency attenuation for the measurement of rainfall along a microwave link. *IEE Proceedings of the Microwaves Antennas and Propagation*, **150**(5), 315–320. doi: 10.1049/ip-map:20030616.

Humphrey M. D., Istok J. D., Lee J. Y., Hevesi J. A. & Flint A. L. (1997). A new method for automated dynamic calibration of tipping-bucket rain gauges. *Journal of Atmospheric and Oceanic Technology*, **14**(6), 1513–1519. doi: 10.1175/1520-0426(1997)014<1513:ANMFAD>2.0.CO;2.

Illingworth A. (2004). Improved precipitation rates and data quality by using polarimetric measurements. In *Weather Radar – Principles and Advanced Applications*, P. Meischner (ed.), Springer, Berlin, Heidelberg (Germany), pp. 130–166. ISBN 978-3540003281.

ISO (2019). *ISO 19926-1:2019 Meteorology — Weather Radar — Part 1: System Performance and Operation*. World Meteorological Organisation, Geneva (Switzerland), 93 p.

Jaffrain J. & Berne A. (2011). Experimental quantification of the sampling uncertainty associated with measurements from PARSIVEL disdrometers. *Journal of Hydrometeorology*, **12**(3), 352–370. doi: 10.1175/2010JHM1244.1.

Jaffrain J. & Berne A. (2012a). Influence of the subgrid variability of the raindrop size distribution on radar rainfall estimators. *Journal of Applied Meteorology and Climatology*, **51**(4), 780–785. doi: 10.1175/JAMC-D-11-0185.1.

Jaffrain J. & Berne A. (2012b). Quantification of the small-scale spatial structure of the raindrop size distribution from a network of disdrometers. *Journal of Applied Meteorology and Climatology*, **51**(5), 941–953. doi: 10.1175/JAMC-D-11-0136.1.

Jasper-Tönnies A. & Jessen M. (2014). Improved radar QPE with temporal interpolation using an advection scheme. *Proceedings of ERAD 2014 – The 8th European Conference on Radar in Meteorology and Hydrology*, 1–5 September, Garmisch-Partenkirchen, Germany, 6 p.

Johnson D., Smith M., Koren V. & Finnerty B. (1999). Comparing mean areal precipitation estimates from NEXRAD and rain gauge networks. *Journal of Hydrologic Engineering*, **4**(2), 117–124. doi: 10.1061/(ASCE)1084-0699(1999)4:2(117).

Joss J. & Waldvogel A. (1967). Ein Spektrograph für Niederschlagstropfen mit automatischer Auswertung [A spectrograph for raindrops with automatic interpretation]. *Pure and Applied Geophysics*, **68**, 240–246. doi: 10.1007/BF00874898. (in German).

Krajewski W. F. (1987). Cokriging radar-rainfall and rain gage data. *Journal of Geophysical Research Atmospheres*, **92**(8), 9571–9580. doi: 10.1029/JD092iD08p09571.

Krajewski W. F., Kruger A., Caracciolo C., Golé P., Barthes L., Creutin J.-D., Delahaye J.-Y., Nikolopoulos E. I., Ogden F. & Vinson J.-P. (2006). DEVEX-disdrometer evaluation experiment: basic results and implications for hydrologic studies. *Advances in Water Resources*, **29**(2), 311–325. doi: 10.1016/j.advwatres.2005.03.018.

Krämer S., Verworn H.-R. & Redder A. (2005). Improvement of X-band radar rainfall estimates using a microwave link. *Atmospheric Research*, **77**(1–4), 278–299. doi: 10.1016/j.atmosres.2004.10.028.

Kruger A. & Krajewski W. F. (2002). Two-dimensional video disdrometer: a description. *Journal of Atmospheric and Oceanic Technology*, **19**(5), 602–617. doi: 10.1175/1520-0426(2002)019<0602:TDVDAD>2.0.CO;2.

Kvicera V. & Grabner M. (2006). Dynamic calibration of tipping-bucket rain gauges and rainfall intensity data processing. *Proceedings of EuCAP 2006 – 1st European Conference on Antennas and Propagation*, 6–10 November, Nice, France, pp. 1–5. doi: 10.1109/EUCAP.2006.4584771.

La Barbera P., Lanza L. G. & Stagi L. (2002). Tipping bucket mechanical errors and their influence on rainfall statistics and extremes. *Water Science and Technology*, **45**(2), 1–9. doi: 10.2166/wst.2002.0020.

Larkin H. (1947). A comparison of the Alter and Nipher shields for precipitation gauge. *Bulletin of the American Meteorological Society*, **28**(4), 200–201.

Leijnse H., Uijlenhoet R. & Stricker J. N. M. (2007a). Hydrometeorological application of a microwave link: 2. Precipitation. *Water Resources Research*, **43**(4), W04417, 9 p. doi: 10.1029/2006WR004989.

Leijnse H., Uijlenhoet R. & Stricker J. N. M. (2007b). Rainfall measurement using radio links from cellular communication networks. *Water Resources Research*, **43**(3), W03201, 6 p. doi: 10.1029/2006WR005631.

Leijnse H., Uijlenhoet R., van de Beek C. Z., Overeem A., Otto T., Unal C. M. H., Dufournet Y., Russchenberg H. W. J., Figueras i Ventura J., Klein Baltink H. & Holleman I. (2010). Precipitation measurement at CESAR, The Netherlands. *Journal of Hydrometeorology*, **11**(6), 1322–1329. doi: 10.1175/2010JHM1245.1.

Lengfeld K., Clemens M., Münster H. & Ament F. (2014). Performance of high-resolution X-band weather radar networks – the PATTERN example. *Atmospheric Measurement Techniques*, **7**(12), 4151–4166, doi: 10.5194/amt-7-4151-2014.

Li X. (2020). *Radar-Based Fine-Scale Rainfall Estimation and Probabilistic Nowcasting of Urban Flooding*. PhD dissertation, KU Leuven, Leuven, Belgium.

Löffler-Mang M. & Joss J. (2000). An optical disdrometer for measuring size and velocity of hydrometeors. *Journal of Atmospheric and Oceanic Technology*, **17**(2), 130–139. doi: 10.1175/1520-0426(2000)017<0130:AODFMS>2.0.CO;2.

Luyckx G. & Berlamont J. (2001). Simplified method to correct rainfall measurements from tipping bucket rain gauges. *Proceedings of the Specialty Symposium on Urban Drainage Modeling at the World Water and Environmental Resources Congress*, 20–24 May, Orlando, FL, USA, pp. 767–776. doi: 10.1061/40583(275)72.

Marsalek J. (1981). Calibration of the tipping bucket raingauge. *Journal of Hydrology*, **53**(3–4), 343–354. doi: 10.1016/0022-1694(81)90010-X.

Marshall J. S. & Palmer W. M. (1948). The distribution of raindrops with size. *Journal of Meteorology*, **5**(4), 165–166. doi: 10.1175/1520-0469(1948)005<0165:TDORWS>2.0.CO;2.

McKee J. L. & Binns A. D. (2016). A review of gauge–radar merging methods for quantitative precipitation estimation in hydrology. *Canadian Water Resources Journal*, **41**(1–2), 186–203. doi: 10.1080/07011784.2015.1064786.

Messer H. A., Zinevich A. & Alpert P. (2006). Environmental monitoring by wireless communication networks. *Science*, **312**(5774), 713. doi: 10.1126/science.1120034.

Michelson D., Einfalt T., Holleman I., Gjertsen U., Friedrich K., Haase G., Lindskog M. & Jurczyk A. (2005). *Weather Radar Data Quality in Europe: Quality Control and Characterisation*. Review, COST Action 717 - Use of radar observations in hydrological and NWP models, Luxembourg.

Miriovsky B. J., Bradley A. A., Eichinger W. E., Krajewski W. F., Kruger A., Nelson B. R., Creutin J.-D., Lapetite J.-M., Lee G. W., Zawadzki I. & Ogden F. L. (2004). An experimental study of small-scale variability of radar reflectivity using disdrometer observations. *Journal of Applied Meteorology and Climatology*, **43**(1), 106–118. doi: 10.1175/1520-0450(2004)043<0106:AESOSV>2.0.CO;2.

Mishra K. V., Krajewski W. F., Goska R., Ceynar D., Seo B.-C., Kruger A., Niemeier J. J., Galvez M. B., Thurai M., Bringi V. N., Tolstoy L., Kucera P. A., Petersen W. A., Grazioli J. & Pazmany A. L. (2016). Deployment and performance analyses of high-resolution Iowa XPOL radar system during the NASA IFloodS campaign. *Journal of Hydrometeorology*, **17**(2), 455–479. doi: 10.1175/JHM-D-15-0029.1.

Neff E. L. (1977). How much rain does a rain gage gauge? *Journal of Hydrology*, **35**(3–4), 213–220. doi: 10.1016/0022-1694(77)90001-4.

Nielsen J. E., Thorndahl S. & Rasmussen M. R. (2014). A numerical method to generate high temporal resolution precipitation time series by combining weather radar measurements with a nowcast model. *Atmospheric Research*, **138**, 1–12. doi: 10.1016/j.atmosres.2013.10.015.

Niemczynowicz J. (1986). The dynamic calibration of tipping-bucket raingauges. *Nordic Hydrology*, **17**(3), 203–214. doi: 10.2166/nh.1986.0013.

Ochoa-Rodriguez S. (2017). *Rainfall Estimates for Urban Drainage Modelling: An Investigation into Resolution Requirements and Radar-Rain Gauge Data Merging at the Required Resolutions*. PhD Thesis, Imperial College London, London, UK.

Ochoa-Rodriguez S., Wang L.-P., Gires A., Pina L. R., Reinoso Rondinel R., Bruni G., Ichiba A., Gaitan S., Cristiano E., van Assel J., Kroll S., Murla-Tuyls D., Tisserand B., Schertzer D., Tchiguirinskaia I., Onof C., Willems P. & ten Veldhuis J. A. E. M.-C. (2015). Impact of spatial and temporal resolution of rainfall inputs on urban hydrodynamic modelling: a multi-catchment investigation. *Journal of Hydrology*, **531**(2), 389–407. doi: 10.1016/j.jhydrol.2015.05.035.

Ochoa-Rodriguez S., Wang L.-P., Willems P. & Onof C. (2019). A review of radar-rain gauge data merging methods and their potential for urban hydrological applications. *Water Resources Research*, **55**(8), 6356–6391. doi: 10.1029/2018WR023332.

Otto T. & Russchenberg H. W. J. (2011). Estimation of specific differential phase and differential backscatter phase from polarimetric weather radar measurements of rain. *IEEE Geoscience and Remote Sensing Letters*, **8**(5), 988–992, doi:10.1109/LGRS.2011.2145354.

Overeem A., Leijnse H. & Uijlenhoet R. (2011). Measuring urban rainfall using microwave links from commercial cellular communication networks. *Water Resources Research*, **47**(12), W12505, 16 p. doi: 10.1029/2010WR010350.

Pedersen L., Jensen N. E. & Madsen H. (2010). Calibration of local area weather radar - Identifying significant factors affecting the calibration. *Atmospheric Research*, **97**(1–2), 129–143. doi: 10.1016/j.atmosres.2010.03.016.

Rahimi A. R., Holt A. R., Upton G. J. G. & Cummings R. J. (2003). The use of dual-frequency microwave links for measuring path-averaged rainfall. *Journal of Geophysical Research Atmospheres*, **108**(15), 4467, 12 p. doi: 10.1029/2002JD003202.

Rahimi A. R., Upton G. J. G. & Holt A. R. (2004). Dual-frequency links - a complement to gauges and radar for the measurement of rain. *Journal of Hydrology*, **288**(1–2), 3–12. doi: 10.1016/j.jhydrol.2003.11.008.

Rahimi A. R., Holt A. R., Upton G. J. G., Krämer S., Redder A. & Verworn H.-R. (2006). Attenuation calibration of an X-band weather radar using a microwave link. *Journal of Atmospheric and Oceanic Technology*, **23**(3), 395–405. doi: 10.1175/JTECH1855.1.

Rico-Ramirez M. A., Liguori S. & Schellart A. N. A. (2015). Quantifying radar-rainfall uncertainties in urban drainage flow modelling. *Journal of Hydrology*, **528**, 17–28. doi: 10.1016/j.jhydrol.2015.05.057.

Ruf C. S., Aydin K., Mathur S. & Bobak J. P. (1996). 35-GHz dual-polarization propagation link for rain-rate estimation. *Journal of Atmospheric and Oceanic Technology*, **13**(2), 419–425. doi: 10.1175/1520-0426(1996)013<0419:GDPPLF>2.0.CO;2.

Scarchilli G., Goroucci E., Chandrasekar V. & Seliga T. A. (1993). Rainfall estimation using polarimetric techniques at C-band frequencies. *Journal of Applied Meteorology and Climatology*, **32**(6), 1150–1160. doi: 10.1175/1520-0450(1993)032<1150:REUPTA>2.0.CO;2.

Seo B.-C. & Krajewski W. F. (2015). Correcting temporal sampling error in radar-rainfall: effect of advection parameters and rain storm characteristics on the correction accuracy. *Journal of Hydrology*, **531**(2), 272–283. doi: 10.1016/j.jhydrol.2015.04.018.

Sevruk B. (1996). Adjustment of tipping-bucket precipitation gauge measurements. *Atmospheric Research*, **42**(1–4), 237–246. doi: 10.1016/0169-8095(95)00066-6.

Simpson M. J. & Fox N. I. (2018). Dual-polarized quantitative precipitation estimation as a function of range. *Hydrology and Earth System Sciences*, **22**(6), 3375–3389. doi: 10.5194/hess-22-3375-2018.

Sinclair S. & Pegram G. (2005). Combining radar and rain gauge rainfall estimates using conditional merging. *Atmospheric Science Letters*, **6**(1), 19–22. doi: 10.1002/asl.85.

Smith J. A. & Krajewski W. F. (1991). Estimation of the mean field bias of radar rainfall estimates. *Journal of Applied Meteorology and Climatology*, **30**(4), 397–412. doi: 10.1175/1520-0450(1991)030<0397:EOTMFB>2.0.CO;2.

Strangeways I. C. (2007). *Precipitation – Theory, Measurement and Distribution*. Cambridge University Press, Cambridge (UK), 302 p. ISBN 978-0521851176.

Tapiador F. J., Checa R. & de Castro M. (2010). An experiment to measure the spatial variability of rain drop size distribution using sixteen laser disdrometers. *Geophysical Research Letters*, **37**(16), L16803, 6 p. doi: 10.1029/2010GL044120.

Thorndahl S. & Rasmussen M. R. (2012). Marine X-band weather radar data calibration. *Atmospheric Research*, **103**, 33–44. doi: 10.1016/j.atmosres.2011.04.023.

Thorndahl S., Nielsen J. E. & Rasmussen M. R. (2014). Bias adjustment and advection interpolation of long-term high resolution radar rainfall series. *Journal of Hydrology*, **508**, 214–226. doi: 10.1016/j.jhydrol.2013.10.056.

Thorndahl S., Einfalt T., Willems P., Nielsen J. E., ten Veldhuis M.-C., Arnbjerg-Nielsen K., Rasmussen M. R. & Molnar P. (2017). Weather radar rainfall data in urban hydrology. *Hydrology and Earth System Sciences Discussions*, **21**(3), 1359–1380. doi: 10.5194/hess-21-1359-2017.

Thurai M., Petersen W. A., Tokay A., Schultz C. & Gatlin P. (2011). Drop size distribution comparisons between Parsivel and 2-D video disdrometers. *Advances in Geosciences*, **30**, 3–9. doi: 10.5194/adgeo-30-3-2011.

Todini E. (2001). A Bayesian technique for conditioning radar precipitation estimates to rain-gauge measurements. *Hydrology and Earth System Sciences*, **5**, 187–199. doi: 10.5194/hess-5-187-2001.

Tokay A., Kruger A. & Krajewski W. F. (2001). Comparison of drop size distribution measurements by impact and optical disdrometers. *Journal of Applied Meteorology and Climatology*, **40**(11), 2083–2097. doi: 10.1175/1520-0450(2001)040<2083:CODSDM>2.0.CO;2.

Uijlenhoet R. (2001). Raindrop size distributions and radar reflectivity–rain rate relationships for radar hydrology. *Hydrology and Earth System Sciences*, **5**(4), 615–628. doi: 10.5194/hess-5-615-2001.

Upton G. J. G., Holt A. R., Cummings R. J., Rahimi A. R. & Goddard J. W. F. (2005). Microwave links: The future for urban rainfall measurement? *Atmospheric Research*, **77**(1–4), 300–312. doi: 10.1016/j.atmosres.2004.10.009.

van de Beek C. Z., Leijnse H., Stricker J. N. M., Uijlenhoet R. & Russchenberg H. W. J. (2010). Performance of high-resolution X-band radar for rainfall measurement in The Netherlands. *Hydrology and Earth System Sciences Discussions*, **14**(2), 205–221. doi:10.5194/hess-14-205-2010.

Villarini G. & Krajewski W. F. (2010). Review of the different sources of uncertainty in single polarization radar-based estimates of rainfall. *Surveys in Geophysics*, **31**(1), 107–129. doi: 10.1007/s10712-009-9079-x.

Villarini G., Mandapaka P. V., Krajewski W. F. & Moore R. J. (2008). Rainfall and sampling uncertainties: A rain gauge perspective. *Journal of Geophysical Research Atmospheres*, **113**(11), D11102, 12 p. doi: 10.1029/2007JD009214.

Wagner A. (2009). *Literature Study on the Correction of Precipitation Measurements*. Bavarian State Institute of Forestry, Freising (Germany), Deliverable C1-Met-29 of the Life+ FutMon project, 32 p.

Wang L.-P., Ochoa-Rodríguez S., Simões N. E., Onof C. & Maksimović C. (2013). Radar-raingauge data combination techniques: a revision and analysis of their suitability for urban hydrology. *Water Science and Technology*, **68**(4), 737–747. doi: 10.2166/wst.2013.300.

Wang L.-P., Ochoa-Rodriguez S., van Assel J., Pina R. D., Pessemier M., Kroll S., Willems P. & Onof C. (2015). Enhancement of radar rainfall estimates for urban hydrology through optical flow temporal interpolation and Bayesian gauge-based adjustment. *Journal of Hydrology*, **531**(2), 408–426. doi: 10.1016/j.jhydrol.2015.05.049.

Wauben W. (2006). *KNMI contribution to the WMO laboratory intercomparison of rainfall intensity gauges*. De Bilt (The Netherlands): KNMI, Technical report TR-287, 174 p. Available at http://bibliotheek.knmi.nl/knmipubTR/TR287.pdf (accessed 06 Oct. 2020).

Willems P. (2001). Stochastic description of the rainfall input errors in lumped hydrological models. *Stochastic Environmental Research and Risk Assessment*, **15**, 132–152. doi: 10.1007/s004770000063.

WMO (2018a). *Guide to Instruments and Methods of Observation – Volume I: Measurement of Meteorological Variables*. World Meteorological Organization, Geneva (Switzerland), 573 p. ISBN 978-92-63-10008-5. Available at https://library.wmo.int/doc_num.php?explnum_id=10179 (accessed 18 June 2020).

WMO (2018b). *Guide to Instruments and Methods of Observation – Volume III: Observing Systems*. World Meteorological Organization, Geneva (Switzerland), 573 p. ISBN 978-92-63-10008-5. Available at https://library.wmo.int/doc_num.php?explnum_id=9872 (accessed 04 Oct. 2020).

Yang D., Goodison B. E., Metcalfe J. R., Louie P., Leavesley G. H., Emerson D. G., Hanson C. L., Golubev V. S., Elomaa E., Gunther T., Pangburn T., Kang E. & Milkovic J. (1999). Quantification of precipitation measurement discontinuity induced by wind shields on national gauge. *Water Resources Research*, **35**(2), 491–508. doi: 10.1029/1998WR900042.

Zinevich A., Alpert P. & Messer H. (2008). Estimation of rainfall fields using commercial microwave communication networks of variable density. *Advances in Water Resources*, **31**(11), 1470–1480. doi: 10.1016/j.advwatres.2008.03.003.

Chapter 3

Water level and discharge measurements

Frédérique Larrarte[1,2], Mathieu Lepot[3,4], Francois H. L. R. Clemens-Meyer[3,5,6], Jean-Luc Bertrand-Krajewski[7], Damjan Ivetić[8], Dusan Prodanović[8] and Bram Stegeman[3]

[1]*Université Gustave Eiffel, Department for Geotechnics, Environment, Natural hazards and Earth sciences, Marne-la-Vallée, France*
[2]*Laboratoire d'Hydraulique Saint-Venant, Chatou, France*
[3]*Delft University of Technology, Department Water Management, Delft, The Netherlands*
[4]*Un Poids Une Mesure, Lyon, France*
[5]*Deltares, Unit Hydraulic Engineering, Department of Experimental Facility Support, Delft, The Netherlands*
[6]*Norwegian University of Science & Technology, Faculty of Engineering, Dept. Civil & Environmental Engineering, Trondheim, Norway*
[7]*University of Lyon, INSA Lyon, Laboratory DEEP, Villeurbanne, France*
[8]*University of Belgrade, Faculty of Civil Engineering, Department of Hydraulic and Environmental Engineering, Belgrade, Serbia*

ABSTRACT

The knowledge of water levels and discharges in urban drainage and stormwater management (UDSM) systems is of key importance to understand their functioning and processes, to evaluate their performance, and to provide data for modelling. In this chapter, devoted mainly to underground combined and separate sewer pipe systems, various methods and technologies are described and discussed. After an introduction to important aspects to deal with when measuring discharges in sewer systems, the following parts are presented successively: (i) measurement of water level with rulers, and pressure, ultrasonic and radar sensors, (ii) measurement of flow velocity with ultrasonic, Doppler, velocity profiler, free surface, and electromagnetic sensors, (iii) direct measurement of discharge with

doi: 10.2166/9781789060102_0035

pre-calibrated devices, physical scale models, computational fluid dynamics modelling and use of pumping stations, and (iv) detection and/or measurement of infiltration into and exfiltration from sewers, with flow or pressure measurements, tracer experiments, distributed temperature sensing and geophysical methods.

Keywords: Discharge, in- and exfiltration, measuring principles, pumping stations, sensor, technology, tracers, velocity, water level.

SYMBOLS

(Some symbols are used for different parameters; it should be clear from the context what is meant in a specific case.)

a	numerical coefficient (–)
A_r	aspect ratio B/h (–)
b	numerical coefficient (–)
$\vec{B}$	magnetic induction field (T)
B_{fs}	width of the free surface (m)
c_{air}	celerity of sound in the air (typically ~340 m/s)
c_{water}	celerity of sound in the water (typically ~1480 m/s)
C	tracer concentration (kg/m^3 or mol/m^3)
C_{BG}	tracer background concentration (kg/m^3 or mol/m^3)
C_{in}	concentration of tracer injected (kg/m^3)
C_{IND}	concentration of indicator tracer (kg/m^3 or mol/m^3)
C_{INJ}	concentration of tracer injected during a tracer experiment (kg/m^3)
C_{MEAS}	concentration of tracer measured during a tracer experiment (kg/m^3)
C_{REF}	concentration of reference tracer (kg/m^3 or mol/m^3)
d_i	distance between the Doppler sensor and the particle i (m)
D	diameter (m)
E	exfiltration fraction (–)
E_m	voltage output of an electromagnetic sensor (V)
f_D	difference between transmission and reception frequencies (Hz) of a Doppler sensor
f_R	frequency of reception of a Doppler sensor (Hz)
f_S	frequency of emission of a Doppler sensor (Hz)
Fr	Froude number (–)
g	gravity acceleration (m/s^2)
h	water level (m)
h_0	vertical distance between a water level ultrasonic sensor and the pipe invert level (m)
h_{max}	maximum water height at a point during a measurement period (m)
h_s	vertical distance from top of an underwater acoustic level sensor to pipe invert (m)
H	characteristic dimension of a physical scale model (m)
i	index
I	electrical current (A)
$\vec{j}$	virtual current vector (A)
k	numerical coefficient to estimate U_m from $\hat{U}$ (–)
K	Manning-Strickler coefficient (m$^{1/3}$/s)
K_x	dispersion coefficient (m^2/s)

L	length (m)
m	index for physical scale model
M_{in}	mass of tracer injected in the investigated reach (kg)
$M_{IND,in}$	mass of indicator tracer injected in the investigated reach (kg)
$M_{IND,out}$	mass of indicator tracer downstream of the investigated reach (kg)
M_{INJ}	mass of tracer injected during a tracer experiment (kg)
M_{MEAS}	mass of tracer measured during a tracer experiment (kg)
M_{out}	mass of tracer measured downstream of the investigated reach (kg)
$M_{REF,in}$	mass of reference tracer injected in the investigated reach (kg)
$M_{REF,out}$	mass of reference tracer downstream of the investigated reach (kg)
n	geometrical scale factor (–)
N	number of particles observed by a Doppler sensor (–)
p	index for full-scale structure
P	pressure (Pa)
P_{elec}	electrical power (W)
P_{hydr}	hydraulic power (W)
q_{IND}	dosing rate of indicator tracer (m^3/s)
q_{INJ}	tracer injection discharge (m^3/s)
q_{REF}	dosing rate of reference tracer (m^3/s)
Q	discharge (m^3/s)
Q_{MEAS}	discharge measured by a tracer experiment (m^3/s)
Re	Reynolds number (–)
s	pipe invert slope (m/m)
S_i	elementary surface (m^2)
S_m	wet section (m^2)
t	time (s)
T_E	end time of the tracer spike at the downstream measurement location
T_r	return travelling time (s)
T_S	start time of the tracer spike at the downstream measurement location
U	voltage (V)
$\hat{U}$	mean velocity measured by a sensor in a fraction of the wet section (m/s)
U'	flow velocity along the measuring line for ultrasonic travel time sensors (m/s)
$U_{e\text{-}x}$	estimated longitudinal velocity (m/s)
U_{fs}	water velocity at the free surface (m/s)
U_i	flow velocity across the elementary area S_i (m/s)
U_m	mean flow velocity across a wet section (m/s)
U_{max}	maximum flow velocity across a wet section (m/s)
U_{ri}	local radial velocity of a particle i (m/s)
$U_{s\text{-}x}$	estimated longitudinal velocity by a Doppler sensor for a particle i (m/s)
U_x	longitudinal component of the flow velocity (m/s)
U_y	transverse component of the flow velocity (m/s)
U_z	vertical component of the flow velocity (m/s)
V	water velocity (m/s)
$\vec{v}$	streamwise velocity field (m/s)
$V(x, y, z)$	water velocity at the given point of coordinates (x, y, z) (m/s)

V_{in}	volume of tracer injected (m^3)
V_r	flow velocity measured by a Doppler sensor in its sampling volume (m/s)
$w(z)$	weighting function (–)
$\vec{W}$	Bevir's weight vector (–)
We	Weber number (–)
x	longitudinal coordinate (m)
y	transverse coordinate (m)
z	vertical coordinate (m), related to water height
Z_L	lower limit of integration of an electromagnetic sensor (m)
Z_{surf}	height of an electromagnetic sensor (m)
Z_U	upper limit of integration of an electromagnetic sensor (m)
ΔH	total pump head (m)
ΔH_{static}	static pump head (m)
η_{elec}	electromotor efficiency (–)
η_{pump}	pump efficiency (–)
ρ	density of water (kg/m^3)
Δt	measurement time step in tracer experiments (s)
α	angle of measurement for ultrasonic travel time sensors (rad or °)
β_i	angle between the direction of the movement of the particle *i* and the Doppler sensor (rad or °)
δ	depth of sediment (m)
λ	friction coefficient (–)
ν	kinematic viscosity of water (m^2/s)
θ	angle of emission of a Doppler sensor (rad or °)
θ_w	opening angle of the emission cone of a Doppler sensor (rad or °)
σ	surface tension of water (N/m)
τ	control volume of an electromagnetic sensor (m^3)
ξ	local head loss coefficient (–)

3.1 INTRODUCTION

Several types of data can be recorded to better understand and manage urban drainage and stormwater management (UDSM) systems. However, discharge is probably the most common one, especially to operate facilities and evaluate pollutants loads (Joannis, 2001). Therefore, methods to measure discharges are widely used in sewer systems. The discharge can barely be measured with a single sensor and its calculation is mostly based on combining data from two measurements. It can be estimated by:

(a) Measuring a volume (m^3) and measuring a duration (s): this method is limited to small channels and low flows as the key challenge is to be able to capture all the volume flowing through the channel (Figure 3.1). Obviously, this method cannot be used for continuous monitoring, and is hard to implement in urban drainage systems. In addition, such a volumetric method is invasive, with possible risks for the operators (see Section 7.2).
(b) Combining two measurements: typically, a water level measurement to calculate the wet section (m^2) and a velocity measurement (m/s).
(c) Using water level measurement(s) and a relation between the water level and the discharge.
(d) Electromagnetic induction measurement.

Figure 3.1 Implementation of a device to capture all the volume flowing through a weir. *Source*: Université Gustave Eiffel.

Even if the measuring device delivers results as discharge values, some calculation is always carried out, sometimes implicitly, using the readings of the sensor(s) (often two or more) to estimate the discharge. The following sections of this chapter are devoted to the (b) and (c) estimations listed above. Two cases are distinguished:

- Open channels, where the flow is driven by gravity, in Sections 3.3.1 to 3.4.5.
- Pressurized pipes, in Section 3.4.6.

A free surface flow is presented in Figure 3.2 with solid walls, and water level h (m) from the invert to the free surface where the width is B_{fs} (m). The discharge calculation usually requires knowledge of:

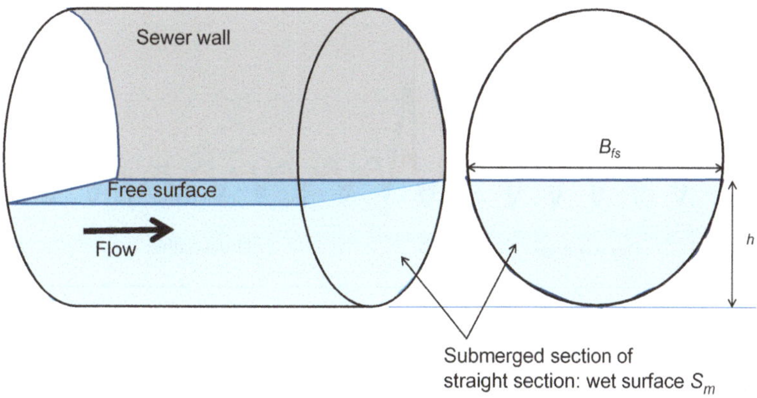

Figure 3.2 Definitions of main quantities for a free surface flow. *Source*: adapted from Larrarte *et al.* (2016a).

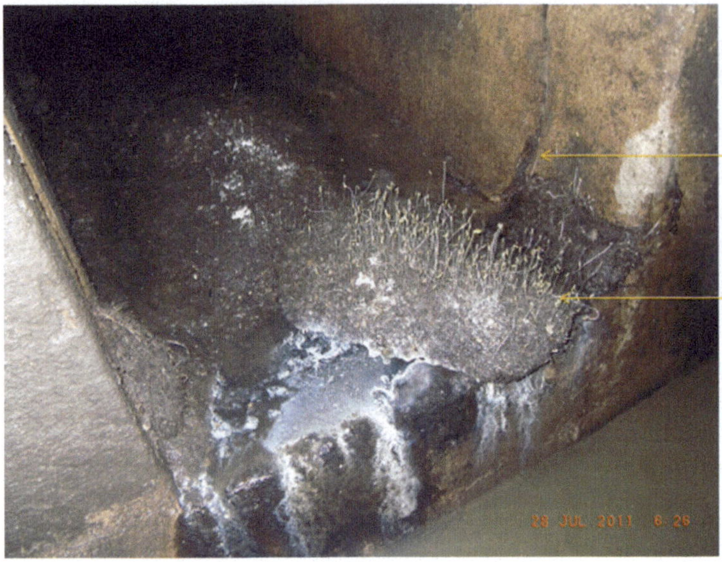

Figure 3.3 A sewer section where sediment deposit and damaged wall hinder installation of sensors for discharge measurement. *Source*: Université Gustave Eiffel.

- Geometric quantities: cross section, slope, and roughness.
- Hydraulic quantities: water level (and, therefore, wet section or surface) and flow velocity.

The geometric quantities of the channel are usually considered as constant. However, this may not be true after important works, when sediment deposit occurs into the channel modifying the cross section

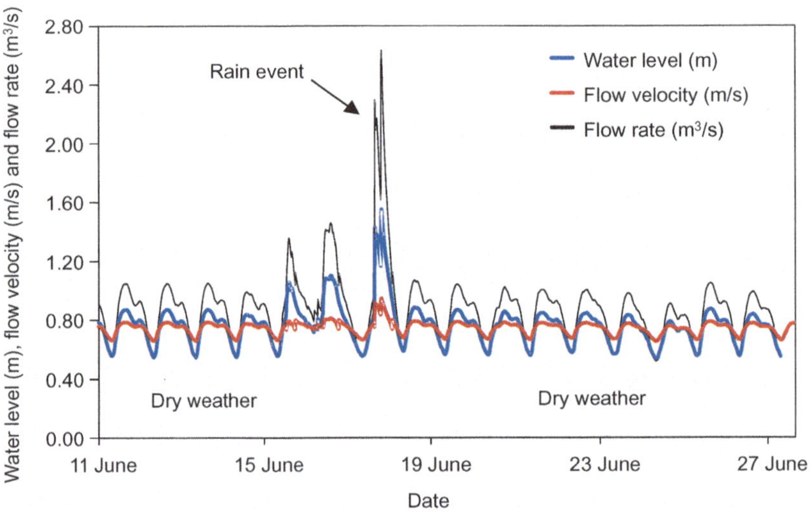

Figure 3.4 Typical flowrate patterns in a combined sewer in Nantes, France. *Source*: Université Gustave Eiffel.

(Figure 3.3), or when corrosion of the walls modifies the roughness (e.g. Stanić *et al.*, 2016). In addition, the hydraulic quantities change with time as shown in Figure 3.4 for a combined sewer pipe.

Figure 3.5(a) shows that water level and velocity do not vary with the same dynamics during a dry weather day. If the velocity is presented as a function of the water level (at the same time), the hysteresis becomes visible (Figure 3.5(b)).

Figure 3.6 shows a more complex example. During dry weather days from 22 to 31 December, the hysteresis is negligible but, during the rainy period (approx. 16–17 Dec.), an important downstream influence (backwater effect) occurs, causing the velocity to reduce when the water level rises.

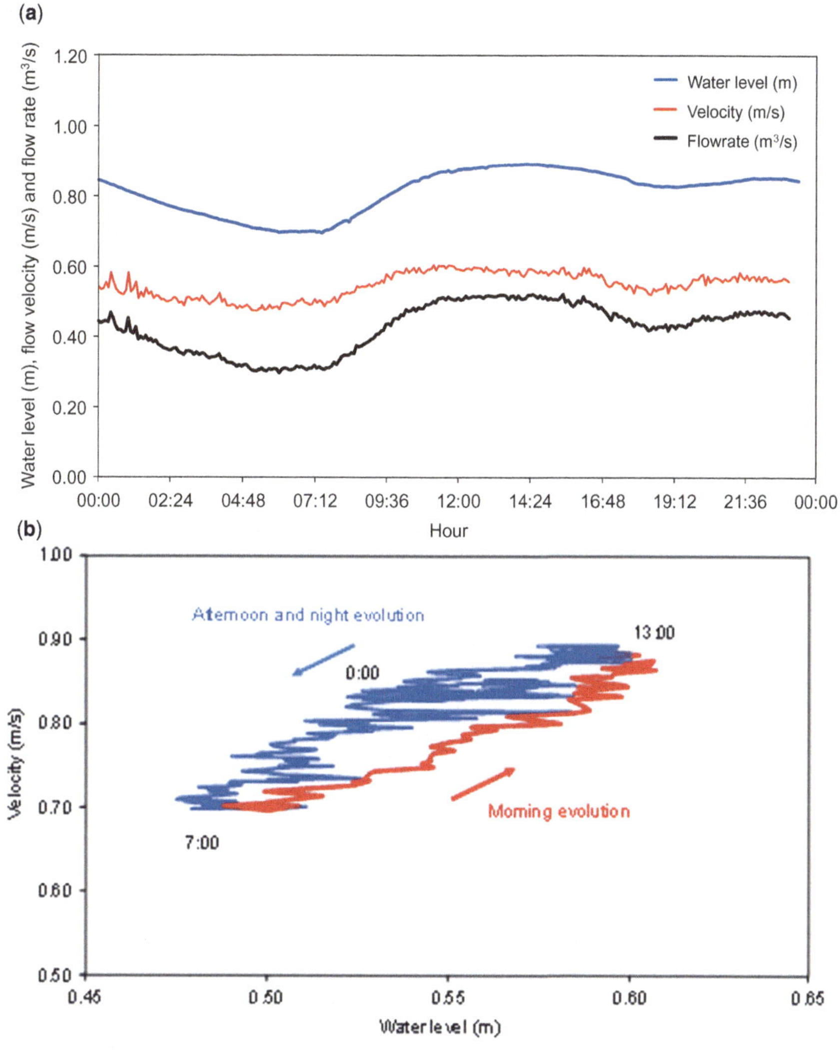

Figures 3.5 Typical flowrate patterns in a combined sewer during dry weather conditions (a. time series, b. hysteresis). (a) temporal evolution of hydraulic parameters. (b) velocity as a function of water level: hysteresis. *Source*: Université Gustave Eiffel.

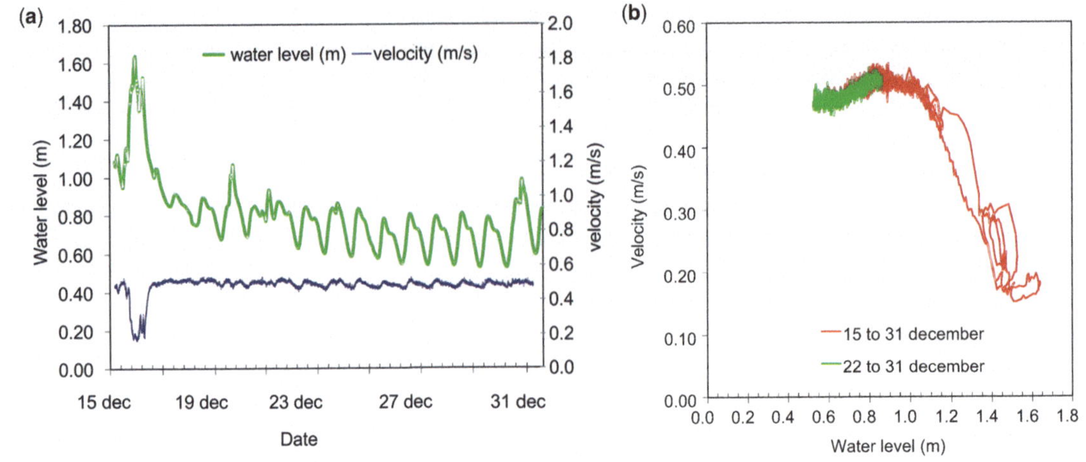

Figure 3.6 Flowrate patterns in a combined sewer with backwater effect (downstream influence). (a) temporal evolution of hydraulic parameters. (b) velocity as a function of water level: a hysteresis. *Source*: Université Gustave Eiffel.

In conclusion:

- When a univocal relation giving the velocity as a function of the water level exists (it happens either for artificially controlled sections (combined sewer overflows, weirs, Venturis, orifices, etc.) or for uniform steady flows, measuring solely the water level h is sufficient to estimate the discharge.
- When a univocal relation giving the velocity as a function of the water level does not exist, both the water level and the velocity have to be measured to estimate the discharge.

The first case is rare in sewer systems. Therefore, it is highly recommended to carefully select the measurement site before implementing any sensor (see Sections 6.2 and 6.3). Sewer operators may be good advisors as, by working daily in the network, they know areas affected by deposits, pipes with downstream influence, and other aspects that may affect measurements. Consulting maps of the network allows singularities such as confluences (Figure 3.7), bends, and changes in channel section or slope, that may have influence on the flow, to be located. Hydraulic simulations or short-term measuring campaigns may be useful to ensure the location is exempt of such effects. If no previous information exists, it is highly recommended to plan a measurement campaign to measure velocity and water level during several weeks with a few rain events to include diverse hydraulic conditions. Last but not least, the existence of a univocal relation has to be systematically validated before being used.

Another point needs to be highlighted. This book deals with metrology in urban drainage and stormwater management, the technologies presented are therefore those used in this context. It means that any part of the metrological chain implemented in sewer networks must be able to be operated in a corrosive and confined environment, sometimes complying with ATEX directives 1999/92/EC and 94/9/EC (applicable to explosive atmospheres including the electrical directives regarding Safety Extra-Low Voltage (SELV)). Additionally, of course, national/local sanitation regulations must be respected (see Section 7.2 for more information).

Figure 3.7 The water flowing from the left side channel influences the flow coming from the bottom. *Source*: Université Gustave Eiffel.

Key messages on sensors, measuring devices and sites

- KM 3.1: *Hydraulic context* – The hydraulic context of each site has to be taken into account at the early stage of the instrumentation process, especially velocity distribution and deposits
- KM 3.2: *Technologies* – For most data, there are various technologies available, with their respective pros and cons.
- KM 3.3: *Corrosion* – Sewer networks have a corrosive and confined atmosphere.
- KM 3.4: *Regulations* – National regulations must be respected.
- KM 3.5: *Staff* – Staff is a key element for professional metrology.

3.2 WATER LEVEL MEASUREMENT

The technologies commonly used for continuous monitoring of water level (Bertrand Krajewski *et al.*, 2000; Colin *et al.*, 2016) are presented in Sections 3.2.1 to 3.2.4. However, it is important to mention that in several cases direct measurement using a ruler may be necessary.

Water level measurement allows calculation of the wet section when the geometry of the section is known. The wet section is correct only if there are no deposits in the cross section. This point should be checked systematically. In the case where deposits are present, the sediment height has to be measured as well, in order to calculate the real wet section. To the authors' knowledge, the only existing device able to continuously measure sediment height is a research prototype (Larrarte *et al.*, 2016b). When applying a free surface Venturi flume (e.g. a Diskin venturi device in Diskin, 1963), the risk of deposits in the measuring section is reduced thanks to the locally increased flow velocity and shear stress.

3.2.1 The simplest sensor: a ruler

A ruler can be considered as the simplest sensor because: (i) it does not require specific skills, (ii) there is no energy requirement and (iii) it is the cheapest one. However, this device has two main drawbacks: it is (i) costly in terms of staff costs and (ii) potentially dangerous since it requires direct access to the water in the sewer pipe. Of course, it cannot be used for continuous monitoring without image acquisition and image processing tools (e.g. Jeanbourquin *et al.*, 2011). When applied in flowing water, a ruler is an invasive measuring method (Figure 3.8): a 'bow wave' appears making the measuring result ambiguous. Even in motionless water, the reading accuracy cannot be better than 2–3 mm, which can be relatively inaccurate for small water depths.

3.2.2 Pressure sensor

The measuring principle of pressure sensors is based on the Bernoulli relation (Equation (3.1)): along a given stream line, the water height z (m) at the given point (x,y) is related to the pressure P (Pa) and to the velocity $V(x,y)$ (m/s), with conservation of the sum:

$$\frac{P}{\rho g} + z(x, y) + \frac{V(x, y)^2}{2g} \tag{3.1}$$

Several types of pressure sensors are available on the market. All of them require an atmospheric pressure compensation, a low velocity around the sensor and must be placed at the invert level. As an order of magnitude, a velocity of 1 m/s leads to an overestimation of the water level of 5 cm. For 2 m/s, the overestimation will be 20 cm (Equation (3.1)).

The first kind of sensor is named a bubbler (Figure 3.9(a)): it is a bubble generator, with the outlet at the invert level, that measures the required pressure to release a bubble. This pressure is equal to the hydraulic

Figure 3.8 Not so easy to read a ruler! Water level of 36 cm upstream, 35 cm downstream. *Source*: Université Gustave Eiffel.

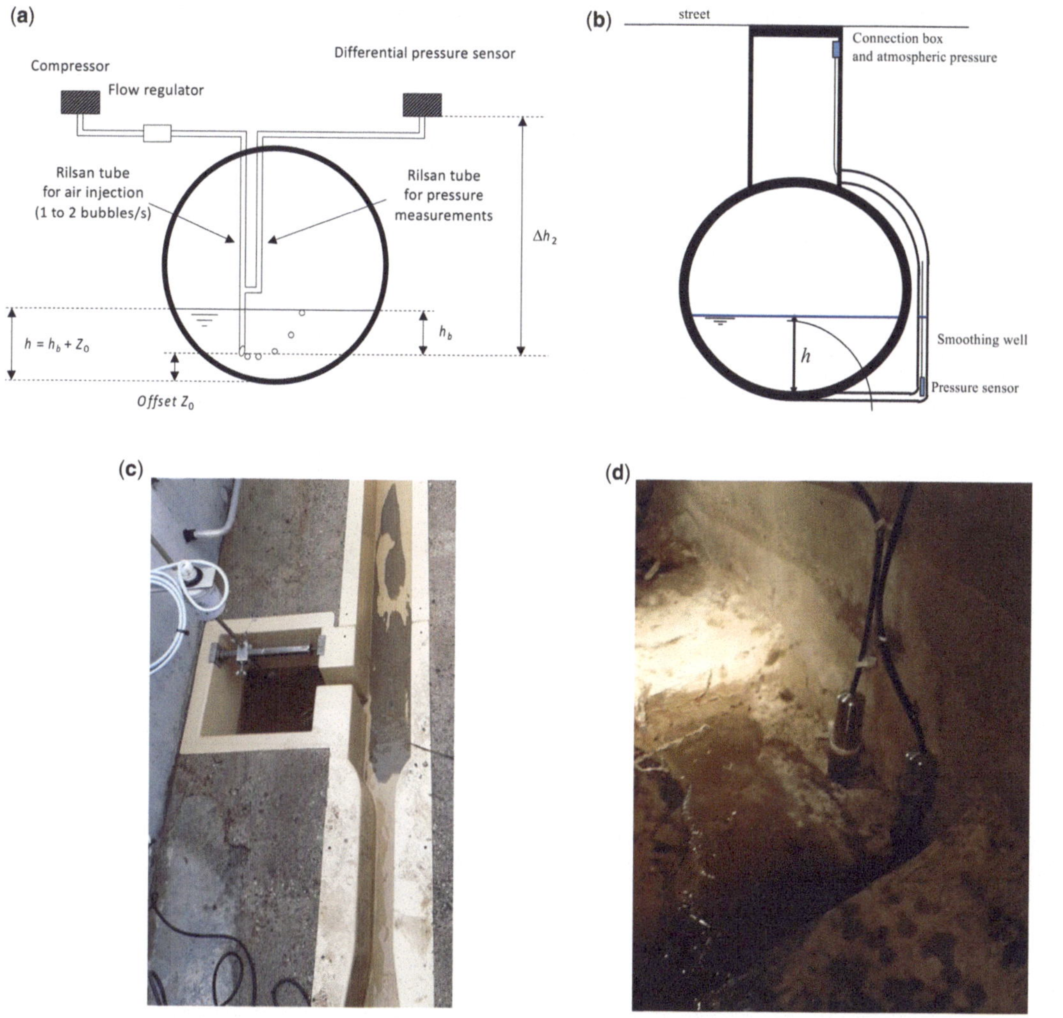

Figure 3.9 Pressure sensors: (a) installation of a bubbler, (b) installation of a piezometric sensor, (c) bubbler installed at the upstream end of a free surface Venturi channel, (d) piezometric sensor. *Sources*: (a) and (b) adapted from Bertrand Krajewski *et al.* (2000); (c) courtesy of Paul Verkroost (Efcon/A.V.M.) and (d) courtesy of Nicolas Walcker (INSA lyon).

head and can be converted into the water level (Equation (3.1)). This device is obviously sensitive to frost for its off-sewer part. These sensors require a gas source (e.g. bottle of nitrogen, atmosphere) and a compressor.

An alternative to the bubbler is the piezometric sensor (Figure 3.9(b)) that uses the piezoelectric properties of some materials. When the piezometric crystal is submitted to a pressure, electrical charges appear on the faces opposed to the constraint exerted on the sensor membrane. The intensity of the electrical signal is proportional to the pressure. Piezometric sensors are of particular interest in pipes where dimensions are too small to install an ultrasonic sensor (which has a dead zone of typically a few tens of centimetres – see Section 3.2.3) or when the flow is or can be pressurized.

Pressure sensors are necessarily submerged and therefore sensitive to clogging and deposits.

3.2.3 Ultrasonic sensor

Ultrasonic sensors are widely used for long term monitoring stations. The water level is calculated through the measurement of the travel time of acoustic waves emitted by the sensor and reflected by the free surface, i.e. the wave travel from the sensor towards the free surface and back to the sensor.

For emerged situations (aerial sensor) the water level is calculated using Equation (3.2a):

$$h = h_0 - \frac{c_{air} T_r}{2} \tag{3.2a}$$

where c_{air} is the celerity of sound in the air (typically 340 m/s), T_r (s) is the return travelling time, and h_0 (m) is the vertical distance between the sensor membrane and the sewer invert level, as shown in Figure 3.10(a).

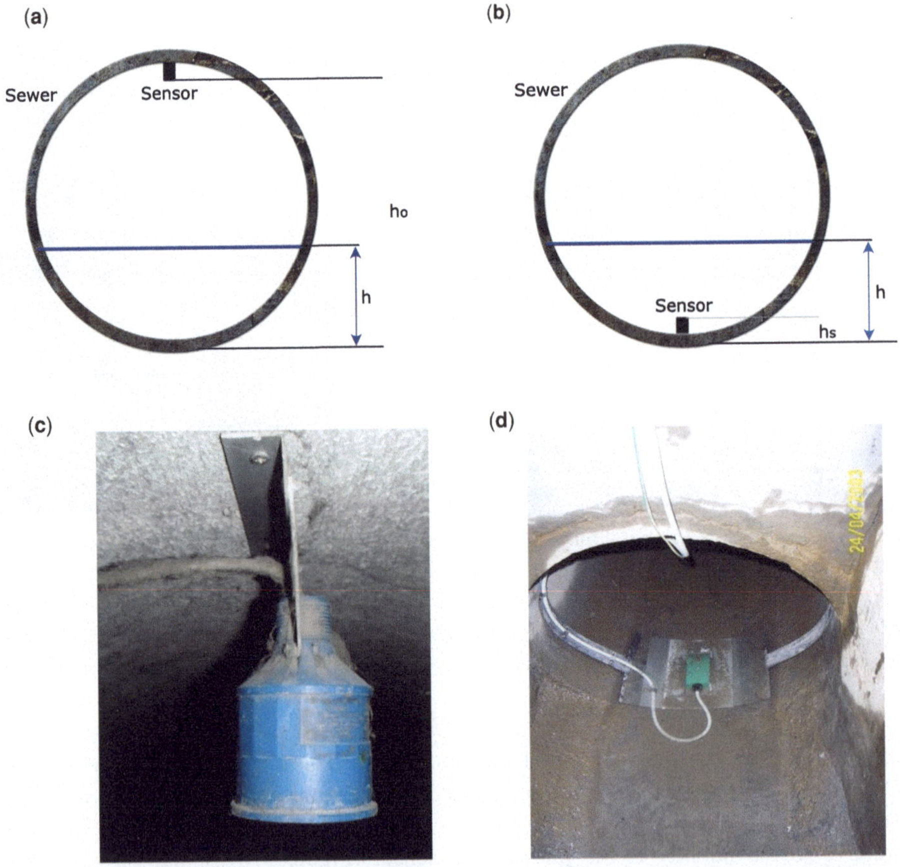

Figure 3.10 Ultrasonic sensors: installation principles at the crown (a) and invert (b) and photos of installation at the crown (c) and at the invert (d). *Sources*: (a) and (b) adapted from Bertrand Krajewski *et al.* (2000); (c) courtesy of Nicolas Walcker (INSA Lyon) and (d) courtesy of Paul Verkroost (Efcon/A.V.M.).

Aerial ultrasonic sensors present some advantages: they are small, cheap, not really prone to drift and require less maintenance than the submerged ones. Given the measurement principle, the results are unreliable when:

- Foam or floating debris is present at the free surface and in the measuring area.
- The composition, temperature, pressure and/or moisture of the atmosphere significantly influences the celerity of the sound.

While aerial sensors are the most widespread, there are also submerged ultrasonic sensors (Figure 3.10(b)). Equation (3.2a) then becomes Equation (3.2b):

$$h = h_s + \frac{c_{water}\,T_r}{2} \tag{3.2b}$$

where c_{water} is the celerity of sound in water (typically 1480 m/s), T_r (s) is the return travelling time, and h_s (m) is the vertical distance between the sensor membrane and the sewer invert level.

For aerial systems, the implicit assumption that the velocity of sound is known is often too strong due to variations in temperature, pressure and/or composition of the atmosphere. Additional measurement of these parameters, in particular temperature and air humidity, can partially correct those variations. However, this

Table 3.1 Advantages and disadvantages of various technologies for water level measurement.

Technology	Advantages	Disadvantages
Ruler	• Cheap • Almost always available • Non-sensitive to humidity	• No continuous measurements without image recording device • Not so easy to read
Bubbler	• Continuous measurements • Hydrostatic pressure measurement • Easy to install	• Requires regular maintenance as it is sensitive to fouling and clogging • Slow response time
Piezometric sensor	• Continuous measurements • No dead zone • Average investment cost • Low power consumption • Works also for pressurized flow • Easy to calibrate	• Contact with the water • Requires regular maintenance as it is sensitive to fouling • Drifts easily
Ultrasonic sensor	• Continuous measurements • Easy to install and maintain • No contact with the effluent • Low drift over time • Rather low cost	• Presence of a dead zone • Does not measure when the water level goes up to the sensor • Several disturbance factors (foams, floats, temperature gradients, haze, etc.)
Radar sensor	• Continuous measurements • Undisturbed by variations in temperature, mist, wind, foam and floatings • Easy to install and maintain • No contact with the effluent • Low drift over time • Low cost	• Slightly more expensive than the ultrasonic sensor • Energy consumption is higher than for the ultrasonic sensor • Does not measure when the water level goes up to the sensor

requires additional sensors (i.e. other additional potential problems and added uncertainties in the obtained measured values for the water level). The assumptions on the geometry (h_0 or h_S – Figure 3.10) need to be carefully checked. 'False' echoes may occur due to e.g. the shape of the manhole construction, or due to the presence of objects like spiderwebs. One main advantage with submerged systems is that the water temperature can be considered as constant along the distance h_s, the main disadvantage is that the sensor is very sensitive to clogging and deposits.

3.2.4 Radar sensor

Water level measurement with radar sensors is also based on a measurement of the travel time, but with electromagnetic waves instead of ultrasound waves. Radar waves do not need a support to propagate and they are not disturbed by variations in temperature, mists, wind, foams nor floating materials. Radar sensors are therefore preferred to ultrasound sensors if one of these constraints is present on the site. They are less subject to disturbances and produce non-ambiguous information if they cannot measure, which makes it easier to sort out false values in the case of loose free surface or dysfunction. With significant diminution of their costs during recent decades, radar sensors are being used more and more.

3.2.5 Summary

The main advantages and disadvantages of the above technologies most used in UDSM are summarized in Table 3.1.

3.3 VELOCITY MEASUREMENTS

As explained in the introduction of Section 3.2, when a univocal relation giving the velocity as a function of the water level does not exist, the velocity has to be measured in conjunction (space) and in synchronization (time) with the water level. In most cases, the geometry of the pipe or channel is known or can be assumed to be known. However, corrosion can generate a significant deviation from the initial cross section, in

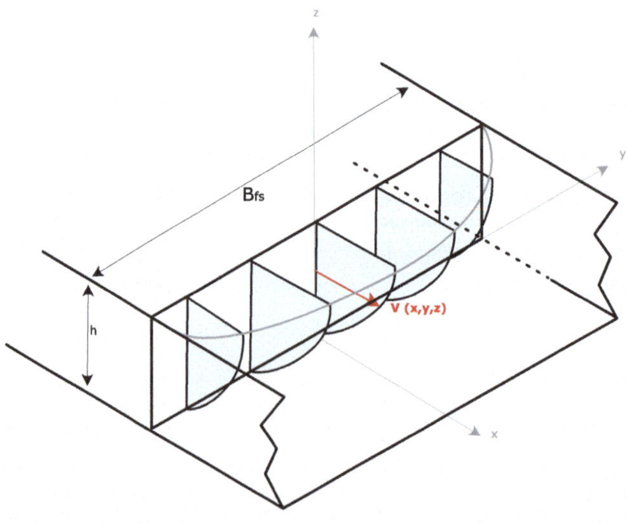

Figure 3.11 Streamwise velocity distribution in a wide channel section. *Source*: adapted from Larrarte *et al.* (2016a).

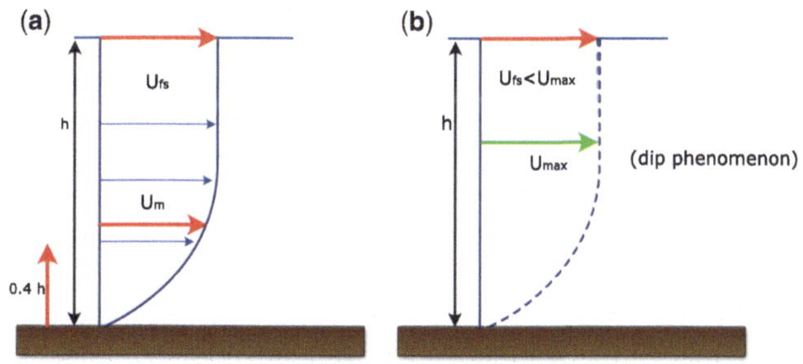

Figure 3.12 Vertical velocity profile in a channel central plane far from any singularities: large (a) and narrow (b) channels, i.e. respectively $A_r > 5$ and $A_r < 5$. *Source*: adapted from Larrarte *et al.* (2016a).

particular for concrete sewer pipes (see e.g. Clemens *et al.*, 2015 or Stanic, 2016). The presence of deposits is also a source of biases and difficulties in measurements.

Longitudinal velocities (Figure 3.11) increase from the invert level to the surface and from the walls towards the median plane of the flow. This schematization corresponds to the case of wide channels characterized by an aspect factor $A_r > 5$, with $A_r = B_{fs}/h$. In this case, the maximum velocity is observed at the free surface (Figure 3.12(a)). For narrower channels ($A_r < 5$), the velocity distribution shows a 'dip-phenomenon' effect: the maximum velocity is below the free surface (Figure 3.12(b)).

Circular or egg-shaped (ovoid) section pipes have an aspect ratio lower than 5 (Figure 3.13). Therefore, the maximum velocity is usually below the free surface (Figure 3.15), except for low filling rates. Consequently, their velocity profiles mismatch with the specifications of the international standard ISO 748 (ISO, 2007). The height of the measured average velocity may be different from the standard, as illustrated in Figure 3.14 for two combined egg-shaped sewers in Nantes, France (Larrarte, 2006): in the

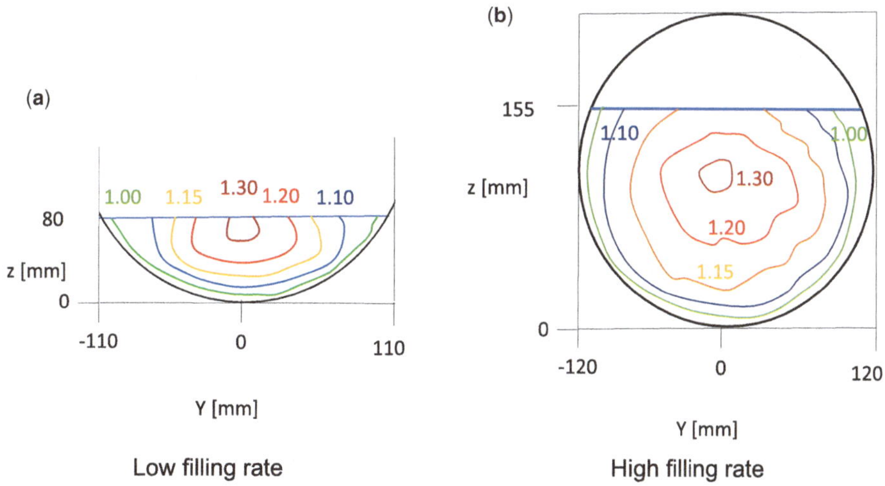

Figure 3.13 Velocity field (the isolines reflect the normalized values with respect to the mean velocity) in a circular section for low (a) and high (b) filling rates. *Source*: Experimental results adapted from Knight & Sterling (2000).

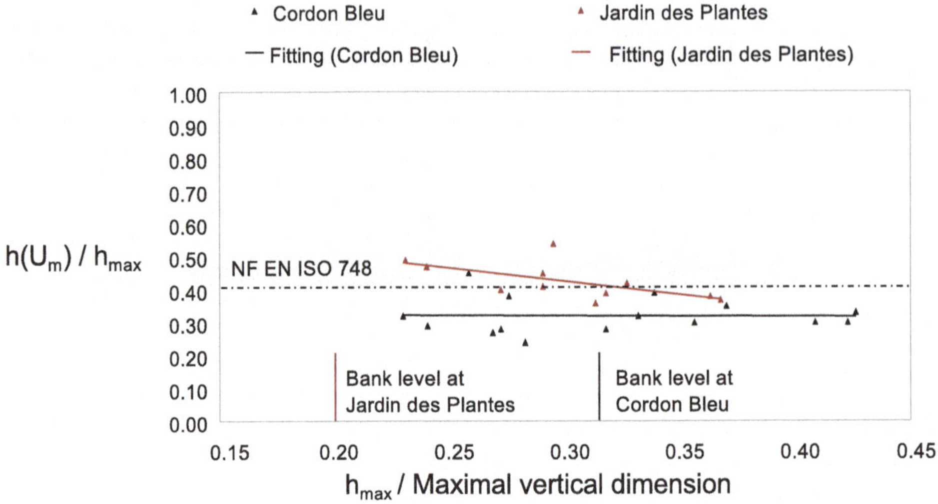

Figure 3.14 Vertical position of the average velocity for two combined egg-shaped channels in Nantes, France. *Source*: Larrarte (2006).

Jardin des Plantes sewer, the position of the mean velocity changes and moves towards the invert level with increasing filling rate, whereas it remains almost constant in the Cordon Bleu sewer.

Very large sewers are sometimes equipped with banks or sidewalks that increase the flow velocity for low discharges and thus reduce the risk of sedimentation. At low flow rates, the velocity field is distributed like in a single section (Figure 3.15(a)) with maximum velocity below the free surface and transverse and vertical velocity gradients. At higher flow rates, the section becomes compound and the velocity field is more heterogeneous: one observes in particular the presence of a local minimum velocity above the sidewalk (Figure 3.15(b)) and a very strong transversal velocity gradient in the vicinity of the vertical face of the bench. Such a velocity field cannot be expected to be stable and may vary over time, even when the flow rate remains stationary.

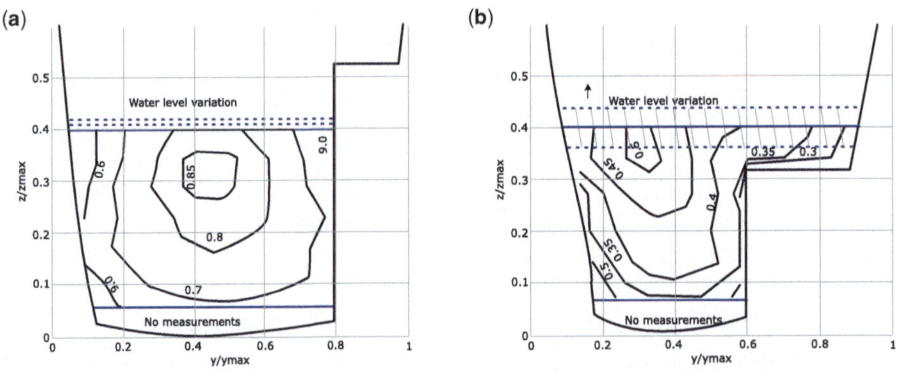

Figure 3.15 Velocity fields in a large combined sewer: low (a) and high (b) filling rates. *Source*: adapted from Larrarte (2006).

The mean velocity U_m (m/s) through a cross section is the flow divided by the wet section: it is the average of the longitudinal component (i.e. perpendicular to the section) of the velocities of all fluid particles which pass through the section at a given time. U_m can be approximated by sampling n points distributed across the section S_m (ISO, 2007):

$$U_m = \frac{1}{S_m}\sum U_i S_i \tag{3.3}$$

where U_i (m/s) is the local velocity at point i, perpendicular to the section, and S_i (m^2) is the surface element associated with this velocity (Figure 3.16).

An ideal velocity sensor should perform a complete sampling of the wet section to account for the entire velocity field. However, a real sensor usually does not deliver the true mean velocity U_m across the wet section but a mean velocity $\hat{U}$ representative of only a fraction of the wet section. This implies that a function f should be determined to estimate U_m from $\hat{U}$ (Equation (3.4)):

$$U_m = f(\hat{U}) \tag{3.4}$$

This function f is preferably reduced to a single numerical coefficient k over the whole range of velocities and filling rates (Equation (3.5)):

$$U_m = k\hat{U} \tag{3.5}$$

The two most used velocity sensors (transit time sensor and Doppler sensor) and some other emerging sensors or technologies are briefly described in the following sections.

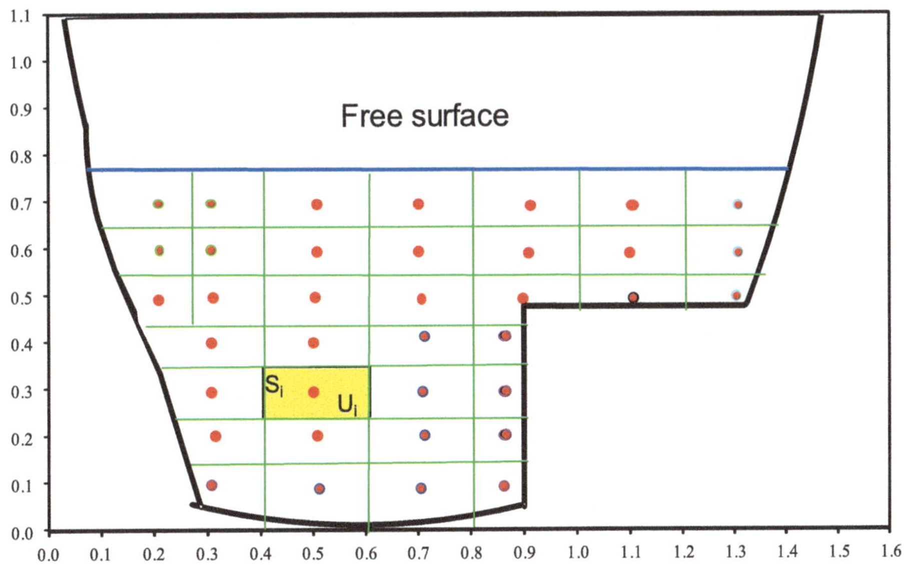

Figure 3.16 Example of discretization of a velocity field in surface elements S_i, each element being associated with a local velocity U_i. *Source*: Larrarte *et al.* (2016a).

Velocities

Sewer channels differ from rivers; the maximum velocity is below the free surface.
The height of the mean velocity may change with the filling rate.

3.3.1 Ultrasonic travel time

The principle is based on the measurement of the travel time of ultrasonic waves between two sensors A and B (emitters and receivers), which are positioned at the same elevation, either on both sides of the pipe (Figure 3.17(a)) or on the same side (Figure 3.17(b)). The sensors are installed with an angle α specified by the manufacturer. L is the wave travel distance between A and B.

In practice, A and B are alternately transmitters and receivers. Let t_1 be the time taken by an acoustic signal emitted from A to reach B and vice versa t_2 from B to A. The transit time measurement considers the flow velocity by its projection U' along the line AB.

Considering, at the height z at which the sensors A and B are installed above the pipe invert (Figure 3.18), that the velocity component $U_x(z)$ parallel to the pipe axis is predominant compared to the respectively transverse and vertical components $U_y(z)$ and $U_z(z)$ ($U_z(z)$ is perpendicular to the plane of Figure 3.18 and is therefore not visible on the figure), the velocity $U'(z)$ measured along the line AB is converted into the longitudinal velocity $U_{e\text{-}x}(z)$ by application of the coefficient $1/\cos(\alpha)$. The mean velocity $\hat{U}(z)$ measured by the sensor along the line AB is given by Equation (3.6):

$$\hat{U}(z) = \frac{L}{2\cos(\alpha)}\left(\frac{t_2 - t_1}{t_1 t_2}\right) \tag{3.6}$$

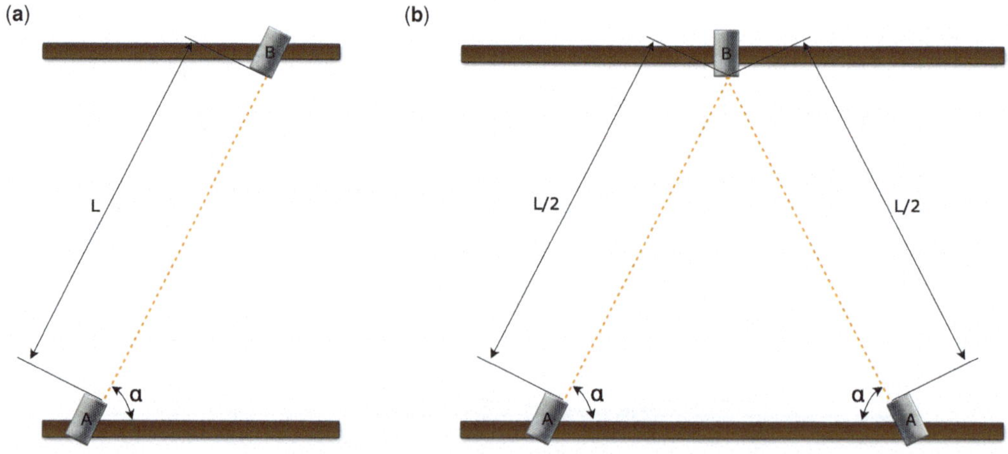

Figure 3.17 Schematic diagram of a transit time measurement system: sensors on both sides (a) or same side (b) of the channel. *Source*: adapted from Larrarte *et al.* (2016a).

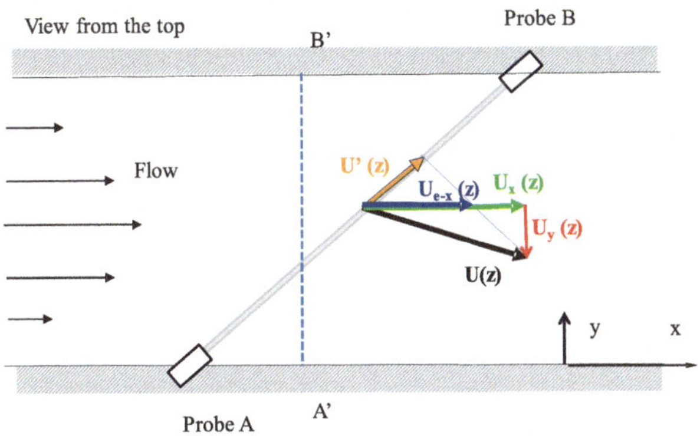

Figure 3.18 Influence of a strong transverse component of the velocity on the difference between the velocity $U_{e\text{-}x}$ measured according to the probes A and B and the longitudinal velocity U_x. *Source*: adapted from Larrarte *et al.* (2016a).

$\hat{U}(z)$ is equal to the mean flow velocity $U_m(z)$ at the level z of the line AB (Equation (3.7)):

$$\hat{U}(z) \cong \frac{1}{L\cos(\alpha)} \int_0^L U'.dl = \frac{1}{L} \int_0^L U.dl = U_m(z) \tag{3.7}$$

Figure 3.18 shows that if the transverse component $U_y(z)$ of the velocity is significant, the measured longitudinal velocity $U_{e\text{-}x}(z)$ is very different from the real longitudinal velocity $U_x(z)$ (in green). The presence of non-negligible transverse velocity components is therefore a major source of error.

The line AB is called a path in the geometric sense of the term. This line allows homogeneous horizontal sampling of the transverse velocity profile (Figure 3.19(a)), but only at a given height. This measurement is representative of the flow at the height of the path. To obtain a better mean velocity estimate through the entire wet section, it is common to use several paths at different heights, by installing several couples of sensors (Figure 3.19(b)).

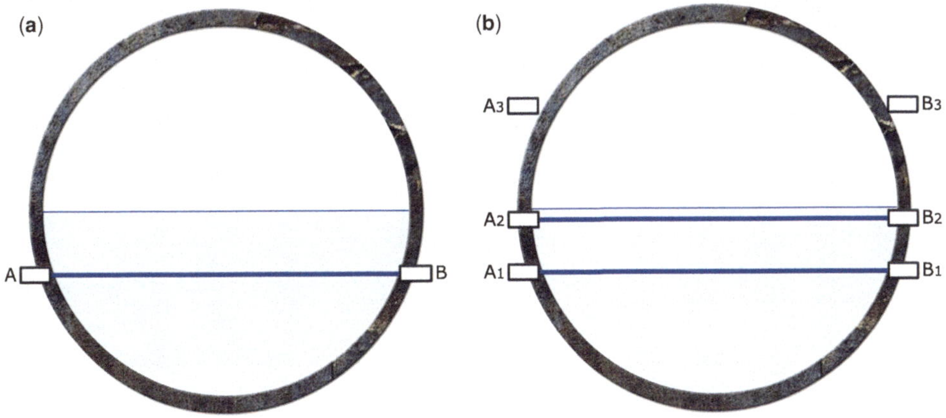

Figure 3.19 Scheme of installation of one couple (a) or several couples (b) of sensors. *Source*: adapted from Larrarte *et al.* (2016a).

The discharge is calculated by assigning a defined surface to each path and applying, where appropriate, corrective coefficients k_i to the velocities measured by the different paths (Equation (3.8)):

$$Q = \sum_{i=1}^{n} k_i \hat{U}_i S_i \tag{3.8}$$

where $\hat{U}_i$ and S_i are respectively the velocity measured by the i-th path and the corresponding wet surface. The coefficients k_i are different from 1 only for the lowest (near the pipe invert) and highest (close to the free surface) surface elements. The standard ISO 6416 (ISO, 2017) provides two methods for assigning surfaces to paths (not detailed here).

3.3.2 Acoustic Doppler flowmeter

An acoustic Doppler flowmeter (Figure 3.20) measures the velocity in a sampled volume by measuring the difference between the emitted (f_S) and received ($f_S + f_D$) frequencies (Hz) of acoustic waves. The frequency f_S is in the range 0.5 to 1 MHz. When acoustic waves encounter an acoustic impedance contrast (i.e. a density, e.g. particles or gas bubbles), the waves are reflected with a frequency f_R. The velocity V_R measured in the sample volume is directly proportional to the difference of frequencies $f_D = f_R - f_S$ (Equation (3.9)):

$$V_r = \frac{c_{water} f_D}{2 f_s} \tag{3.9}$$

where c_{water} (m/s) is the celerity of ultrasound in water, f_S (Hz) is the frequency of emission, and f_D (Hz) is the difference between the transmission and reception frequencies.

In theory, Doppler sensors can be placed anywhere in the flow (Figure 3.20). In practice, the location is of key importance to obtain accurate results. Indeed, the velocity V_r measured by the Doppler sensor is converted to the estimated mean velocity U_m according to hypotheses about the expected velocity field and profile (see below). The most common specifications correspond to a location of the sensor at the bottom of the pipe, in a central position (Figure 3.20(a), Figure 3.22). If the sensor is in a different

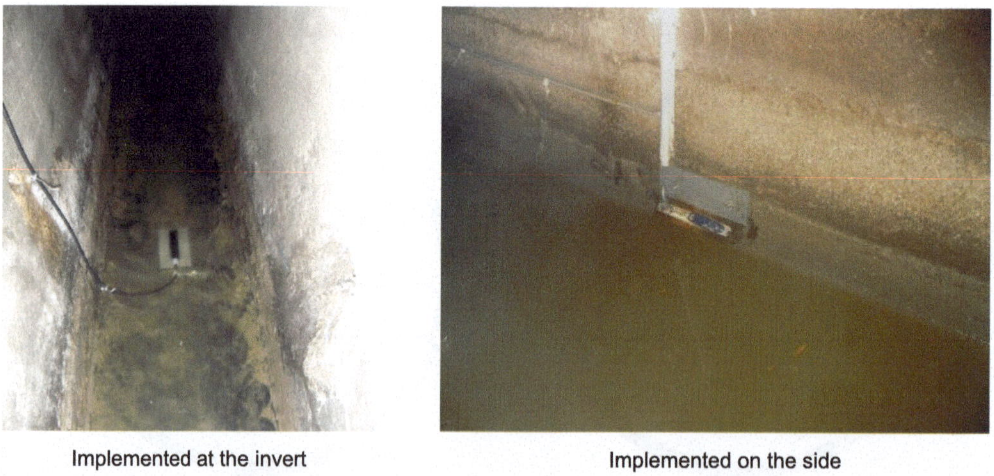
(a) Implemented at the invert

(b) Implemented on the side

Figure 3.20 Examples of acoustic Doppler sensor locations: on the invert (a) or on the side wall (b). *Source*: Université Gustave Eiffel.

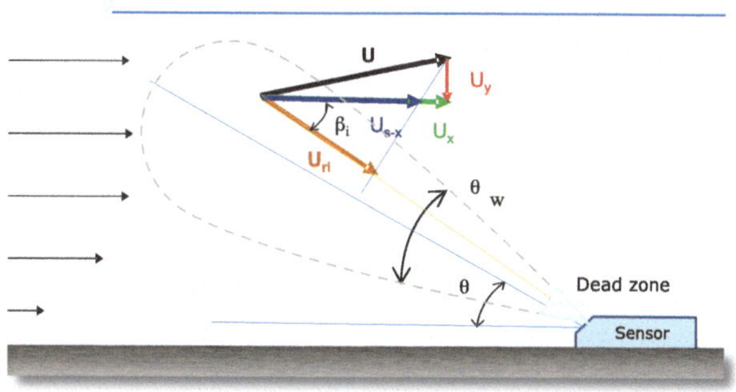

Figure 3.21 Principle of velocity measurement by a Doppler sensor. *Source*: adapted from Larrarte *et al.* (2016a).

location (e.g. on the wall, as shown on Figure 3.20(b)), an appropriate correction to estimate U_m from V_r is necessary to avoid biases which may be very significant (Lepot *et al.*, 2014). In addition, sensor locations with deposits or prone to clogging should be avoided.

For a given particle *i* within the flow (Figure 3.21), the velocity $U_{s\text{-}x}$ along the flow axis is calculated from the velocity U_{ri} measured by the Doppler sensor by Equation (3.10):

$$U_{s-x} = \frac{U_{ri}}{\cos(\beta_i)} \tag{3.10}$$

where β_i is the angle between the direction of the movement of the particle *i* and the emission axis of acoustic waves.

For typical sensor location conditions (i.e. on the invert), the angle β_i is assumed to be equal to the angle θ_i, located in the vertical plane of symmetry of the sensor, which itself is assimilated to the emission angle θ of the sensor, even if strictly speaking this is only true for the particles located on the axis of the emission cone (Figure 3.21). The measurement volume explored by the Doppler sensor is a 3D cone as shown in Figure 3.22.

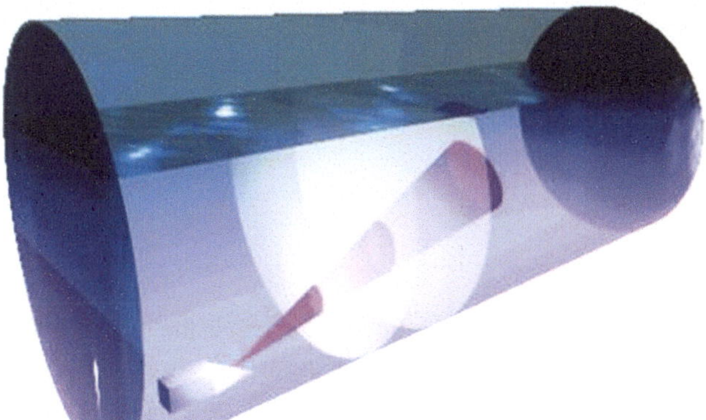

Figure 3.22 Scheme of the three-dimensional measurement cone (in pink) of a Doppler sensor and two sampled cross sections (in white). *Source*: courtesy Claude Joannis.

Table 3.2 Characteristics of three different Doppler sensors noted A, B and C (Larrarte *et al.*, 2008).

Sensor	Sensor A	Sensor B	Sensor C
Frequency (MHz)	1.0	0.5	1.0
Emission angle (degree)	15	31	14
Opening angle (degree)	17	10	24
Range (m)	>3.5	0.8	1.2

The international standard ISO 15769 (ISO, 2010) recommends some specifications for Doppler sensors:

- Emission angle θ.
- Opening angle θ_w of the ultrasonic beam.
- Range of the ultrasonic beam (i.e. the maximum distance along the measurement cone axis at which measurements are effective).
- Emission frequency f_S.
- Signal analysis.

As for ultrasonic travel time sensors (Figure 3.18), transverse or vertical velocity components tend to distort the measurement (Figure 3.21). Indeed, the true longitudinal velocity U_x (in green) may differ from the longitudinal velocity $U_{s\text{-}x}$ estimated by the sensor (in blue) because the true velocity vector U (in black) at the location of the particle i is not necessarily parallel to the main axis of the pipe.

To estimate some specifications of Doppler sensors, Larrarte *et al.* (2008) developed a test bench and showed that these quantities vary considerably for different sensors (Table 3.2). During these tests, it was assumed that the maximum range is reached when the sensor is no longer able to read with an acceptable accuracy (i.e. less than 20%) the velocity of a controlled flow, located in a tube transparent to ultrasound, and immersed in a still water tank at a given distance from the Doppler sensor. It is therefore an arbitrary definition, which however partially reflects the actual *in situ* measurement capabilities and allows objective comparisons between different instruments.

The range is limited due to the attenuation of the received signals, whose intensity decreases when the distance to the sensor increases. This attenuation is a geometrical effect due to the distribution of the ultrasound energy on a spherical cap surface proportional to the square of the distance between the sensor transducer and the reflector (particle or bubble), along the return trip of the acoustic wave. Thus, the average velocity $\hat{U}$ given by the sensor is estimated from Equation (3.11):

$$\hat{U} = \frac{1}{\cos(\theta)} \frac{\sum_1^N \frac{U_{ri}}{d_i^4}}{\sum_1^N \frac{1}{d_i^4}} \tag{3.11}$$

where U_{ri} (m/s) is the local radial velocity of a reflector i, d_i (m) is the distance between the sensor and the reflector i, and N is the number of reflectors. When the term $1/d^4$ becomes too small, the corresponding signal becomes too weak with respect to noise, or, if it is not, the weight of the corresponding velocity becomes negligible.

Laboratory experiments have shown that wastewater, at usual suspended solids concentrations observed in sewers (below 1 g/L), does not attenuate ultrasound (Larrarte & Francois, 2012) and therefore has no influence on the range.

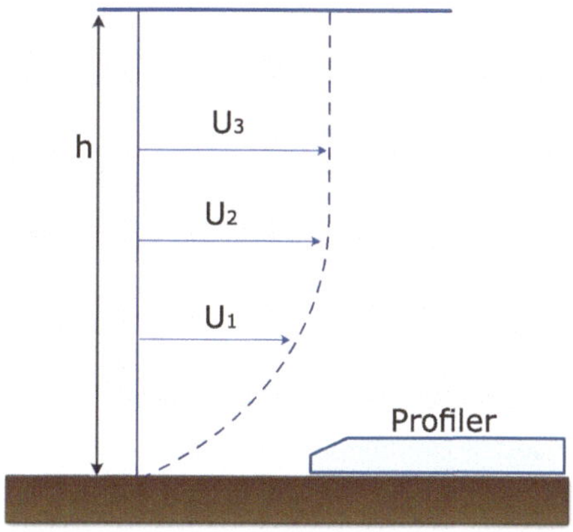

Figure 3.23 Principle of a Doppler profiler. *Source*: adapted from Larrarte *et al.* (2016a).

3.3.3 Velocity profilers

Aiming to correct the geometric attenuation effect affecting previously described standard Doppler sensors, new devices, named Doppler velocity profilers, have been available since the late 2000s (Figure 3.24 and Figure 3.25). They record both the scattered frequencies and the travel time of the emitted waves, which allows the determination of complete vertical velocity profiles, giving the velocities U_i, at different depths z_i (Figure 3.23). The discharge is then calculated using an area method (see Equation (3.3)), either assuming that the velocity is the same across the width of the wet section at each depth z_i or applying a weighting factor to account for slowing near the walls.

Figure 3.24 An ordinary Doppler sensor (left side, black sensor) and a Doppler profiler (right side, blue-grey sensor) during a comparison campaign. Note the tissues clogging on the connecting wires. *Source*: Université Gustave Eiffel.

Figure 3.25 The acoustic profiler used by Hemmerle *et al.* (2014) and Cedillo *et al.* (2016). *Source*: Hemmerle *et al.* (2014).

3.3.4 Free surface velocity measurements

Velocity measurement at the free surface is possible with sensors attached to the crown of the pipe (Figure 3.26), i.e. without contact with the effluent, except in case of exceptionally high flows. This technique offers decisive advantages in terms of maintenance, but the conversion of the velocity U_{fs} measured at the free surface to the mean velocity U_m through the wet section is more challenging (see introduction of Section 3.3.2). Proprietary (and often blind) data processing and algorithms provided by manufacturers for this conversion have to be carefully checked. Indeed, such a conversion is site specific and generic calculations cannot be assumed to be valid.

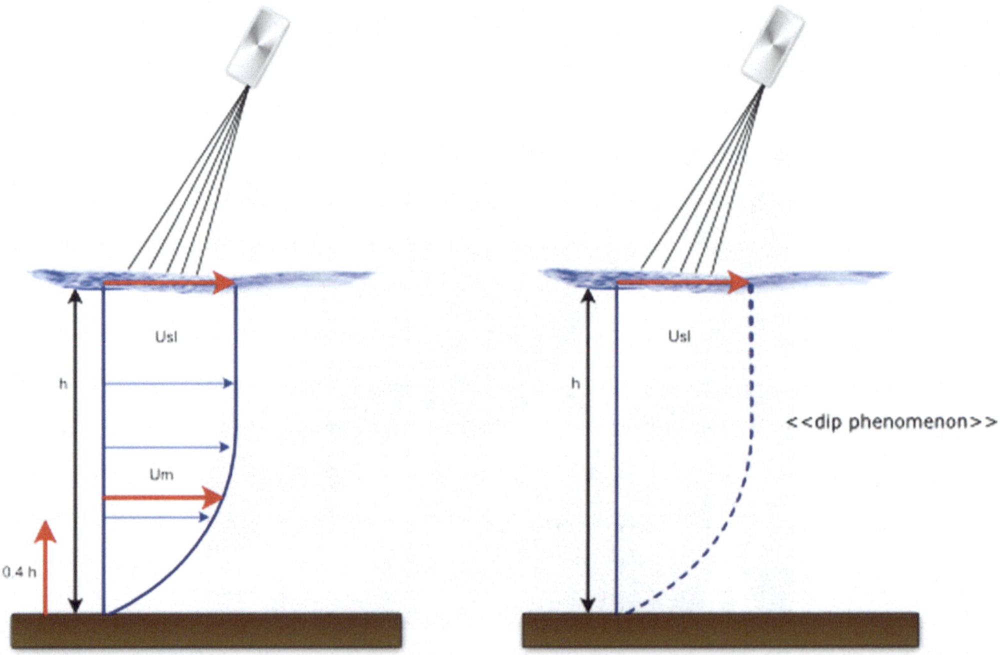

Figure 3.26 Measurement of free surface velocity U_{fs} and estimation of average velocity U_m. *Source*: adapted from Larrarte *et al.* (2016a).

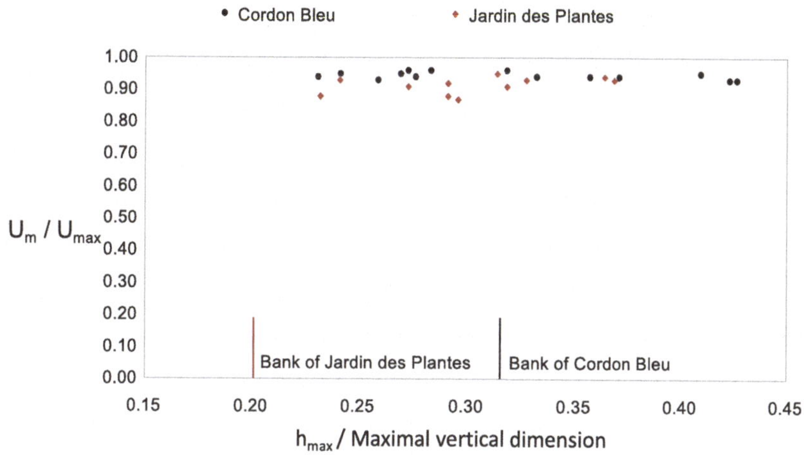

Figure 3.27 Evolution of the ratio of the average velocity U_m with the maximum velocity U_{max} for various filling rates for two egg-shaped sewers in Nantes, France. *Source*: adapted from Larrarte (2006).

The ISO 748 standard (ISO, 2007) indicates that the numerical coefficients to convert the free surface velocity U_{sl}, assumed to be the maximum velocity U_{max} along the vertical profile, to the mean velocity U_m, assumed to be observed at 60% of the total depth (i.e. $0.4 \times h$ from the pipe invert), vary from 0.84 to 0.90 depending on the pipe roughness. However, in two egg-shaped sewers, Larrarte (2006) found that empirical values of these numerical coefficients were between 0.90 and 0.96 (Figure 3.27).

Velocity measurements at the free surface typically use radar waves, but video imaging techniques are also under development. These techniques are developing rapidly for both sewers and rivers (Nguyen *et al.*, 2009).

Ideas on ultrasonic sensors

- I 3.1: Ultrasonic transit time flowmeter to measure flow velocity.
- I 3.2: Ultrasonic Doppler flowmeter to measure the velocity of particles and bubbles, assuming they are the same as the water velocity.

3.3.5 Electromagnetic sensor

The operating principle of electromagnetic (EM) flow/velocity sensors is based on Faraday's law of induction. The motion of the conductive fluid through a transversal magnetic field generates a voltage (Shercliff, 1962). To allow for the stationary analysis of the electromagnetic induction phenomenon, some electric and magnetic properties of the environment are assumed (Michalski *et al.*, 2001). Originally, under these assumptions, Kolin (1936) has given the basic relationship for the EM theory (Equation (3.12)):

$$\nabla^2 E = \text{div}(\vec{V} \times \vec{B}) \tag{3.12}$$

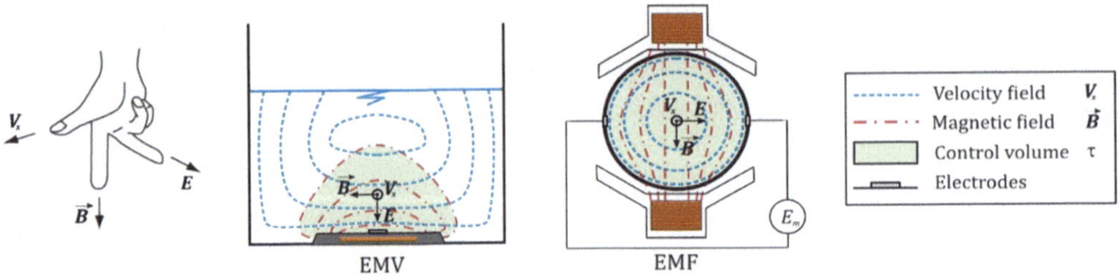

Figure 3.28 Right-hand rule governing Faraday's law of induction with cross-sectional illustrations of the EMV (bed mounted EM velocity meter) vs. the EMF (closed pipe EM flowmeter) sensors, and the reach of the respective control volumes. *Source*: Damjan Ivetić.

where $\vec{V}$ is the streamwise velocity field, $\vec{B}$ is the magnetic induction and $\mathrm{div}(\vec{V} \times \vec{B})$ is treated as a charge distribution. The raw output signal is the voltage $E_m = E_m^1 - E_m^2$, induced between the electrodes of the EM sensor. The relations used in the electrical networks motivated an idea to describe how each part of the flow contributes to the voltage E_m (Equation (3.12)) through the weight function w (Shercliff, 1962) or, in a more rigorous formulation, through the weight vector $\vec{W}$ (Bevir, 1970):

$$E_m = -\int_0^\tau (\vec{V} \times \vec{B}) \cdot \vec{j} d\tau = \int_0^\tau \vec{V} \cdot (\vec{B} \times \vec{j}) d\tau = \int_0^\tau \vec{V} \cdot \vec{W} d\tau \tag{3.13}$$

where the cross product $\vec{B} \times \vec{j}$ defines Bevir's weight vector $\vec{W}$, τ is the control (sampling, or integrating) volume of the EM sensor (Figure 3.28) and $\vec{j}$ is the virtual current vector (i.e. the current density set up in the liquid by driving an imaginary unit current between a pair of electrodes). Since Faraday's law of induction is governed by the right-hand rule, the dominant contributor to the output E_m is the longitudinal component of the velocity vector, V_x, which is needed for flow measurement.

If the sampling volume τ envelops the whole cross section, such an EM device can be classified as a closed-pipe EM flowmeter (EMF). For EMFs, the output E_m is directly proportional to the average cross-sectional velocity $\bar{U}$. Conventional EMFs have high accuracy and precision and are common in pressurized flow application, where for axisymmetric flows, errors lower than 0.1% have been reported (Leeungculsatien & Lucas, 2013). In sewer systems, the EMF is mostly used downstream of pumping stations or on inclined reaches where aerated full-pipe conditions can be easily met, for diameters up to 0.6 m. Similar devices, but with lower accuracy, are available for application in pipes with varying flow depth, for diameters up to 0.8 m.

However, in sewers, the bed mounted EM velocity (EMV) sensors are more commonly used (Figures 3.29 and 3.30). In the bed mounted EMV application, τ is variable and depends on several factors: excitation current, coil design, conduit geometry, and water depth (for low depths). Since the excitation coil of the bed mounted EMV sensors is relatively small, the reach of the produced magnetic field is limited to the relative vicinity of the EMV (Figure 3.28). Consequently, the output voltage E_m is proportional to 'some' local velocity $\hat{U}$. The small control volume τ is considered as the biggest drawback of these sensors, as usually it is significantly smaller than in the case of the acoustic Doppler flowmeters. Therefore, additional care is necessary when considering the functional relationship needed to calculate $\bar{U}$ from the measured velocity $\hat{U}$ (Equation (3.13)). Laboratory tests as reported by Ivetić

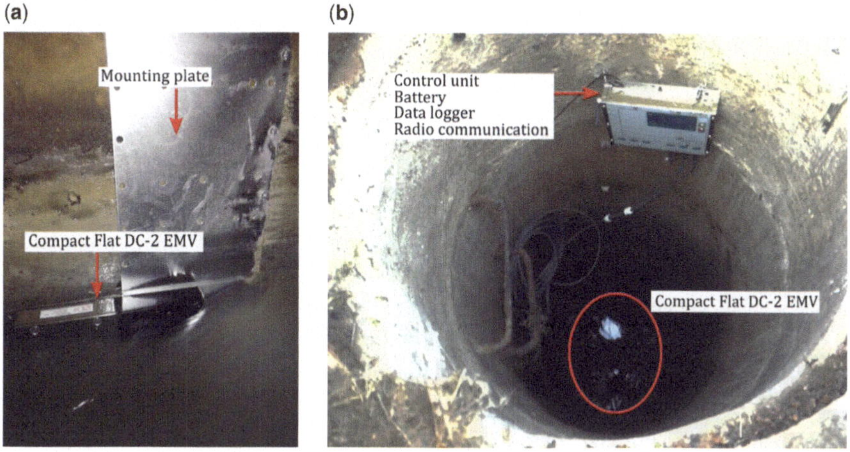

Figure 3.29 Application of the compact flat DC-2 EMV in the combined sewer system of Belgrade: (a) wall mounted; (b) bed mounted. *Source*: courtesy of Svet Instrumenata Ltd.

et al. (2018) imply that the constant coefficient k (Equation (3.5)) can be used to a certain extent, but it is deemed that in larger sewer pipes a function f (Equation (3.4)) is more suitable. If the variation of the longitudinal velocity distribution V_x is negligible across the width and length of the sensor control volume, a simplified mathematical model of the bed-mounted EMV operating principle can be used to define k and f (Ivetić *et al.*, 2019). The simplified model describes the EMV operating principle with

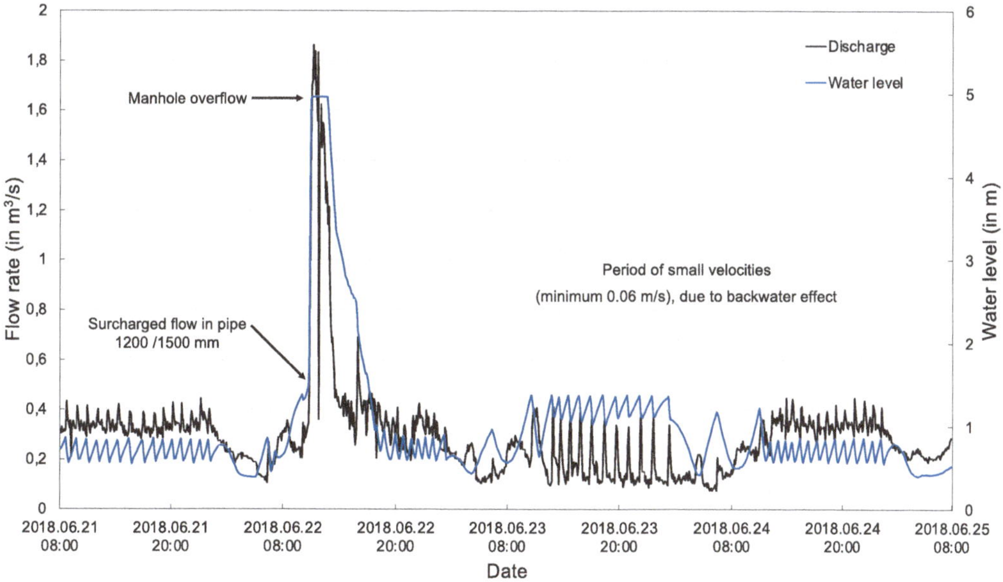

Figure 3.30 Flow rate and depth (pressure head) measured with compact flat DC-2 EMV, in the combined sewer collector of Belgrade. *Source*: adapted from Prodanović & Ivetić (2019).

only two technical parameters, one-dimensional weighting function $w(z)$ (Equation (3.14)) and the reach of the control volume noted as τ_{max}:

$$\hat{U} = \int_{Z_L}^{Z_U} w(z) \cdot V_x(z) dz \tag{3.14}$$

where z is the direction perpendicular to the sensor electrodes, and $Z_L = \max\{Z_{surf}, \delta\}$ and $Z_U = \min\{Z_{surf} + \tau_{max}, h\}$ define respectively the lower and upper limits of integration. If some porous sediment deposit of depth δ is present above the sensor body, the lower integration limit is shifted upwards. Similarly, if the water depth h is less than the sum of the sensor height Z_{surf} and reach of the control volume τ_{max}, the upper integration limit is shifted downwards.

Direct laboratory comparisons between EMVs and acoustic Doppler flowmeters (Aguilar *et al.*, 2016; Ivetić *et al.*, 2018) revealed that, although the control volume of the EMV is close to the probe, due to the operating principle, it is more precise and robust in the assessment of the average cross-sectional velocity $\bar{U}$. Also, the EMV can operate in full-bidirectional flows with velocities lower than a few centimetres per second, so they are convenient to use in flows influenced by backwater effects (Figure 3.30).

Potentially, the most interesting benefit of EMV use in urban drainage, experimentally examined in Ivetić *et al.* (2018), is the ability for velocity measurements in the case of sedimentation over the sensor housing (Figure 3.31). This is a common situation in sewers and even a small layer of sediment (or plastic bags, rags, toilet paper, etc.) may prevent the correct functioning of some other types of velocity sensors (mostly ultrasonic ones). However, if the sediment is porous and not affecting the EM properties of the device, EMVs continue to operate. For small sediment depths (i.e. a few millimetres), errors are negligible. For larger sediment depths δ, the output is biased. Preliminary laboratory results imply that a correction function model (CFM) can be experimentally determined for the particular sediment composition and EMV sensor model, to minimize the resulting bias. To apply the CFM at the measuring site, continuous sediment depth measurements are needed. The depth of the sediment above the EMV sensor is directly

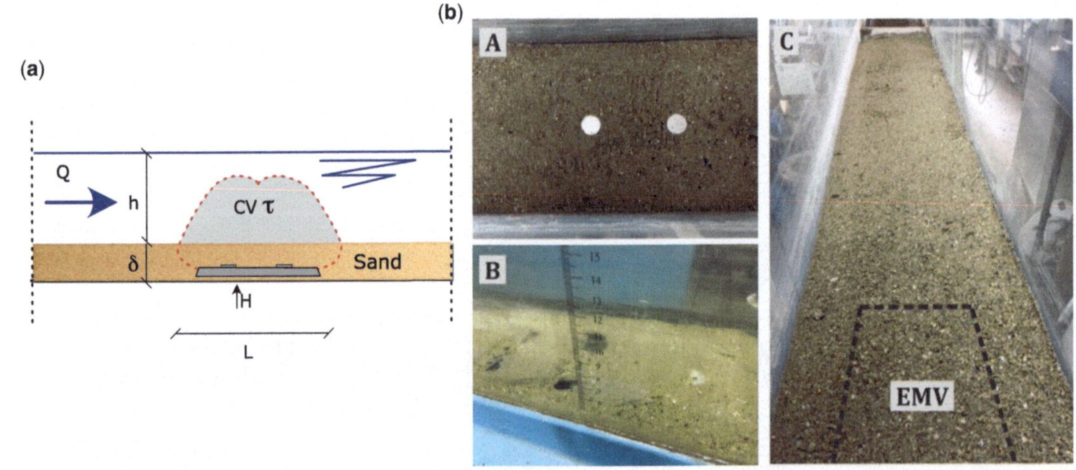

Figure 3.31 (a) schematic illustration of the EMV operation under sand sediment; (b) laboratory experiments: A) sand sediment depth of $\delta = 23$ mm, B) linearity loss with $\delta = 23$ mm due to dune formation, C) $\delta = 80$ mm. *Source*: adapted from Ivetić *et al.* (2018).

proportional to the slope and intercept values of the correction functions. It should be highlighted that in the case of sedimentation, the user should also compensate the water level measurements (reduced wet section area) for the presence of sediment.

On the other hand, failure mechanisms encountered with these devices are related to disruptions in the contact between the electrodes and the conductor (i.e. water). When the electrodes of the EM sensor are covered with an impervious material, such as cemented sediment or plastic bags, the current $\vec{j}$ path is disrupted or elongated and the resistance of the formed 'conductor' (water path between electrodes) is increased, resulting in a significant or complete loss of the output signal. Another failure mechanism is related to the hardness of water. If an insulating layer of limescale, resulting from the presence of calcium and magnesium carbonates, is formed on the electrodes of the EM sensor, again the resistance of the formed conductor (between electrodes) is dramatically increased, resulting in a decrease of the output voltage E_m. Highly aerated water can also present a problem, since air pockets can easily cover the electrodes and increase the uncertainty of the velocity measurements.

Ideas on EMV

- I 3.3: EMV meters can measure both pressure and velocity.
- I 3.4: EMV can operate under the porous sediment layer of few centimeters – but the output will be biased proportionally to the depth of the sediment layer.
- I 3.5: EMV measures the voltage difference generated by the motion of the conductive liquid through the sensor's magnetic field.

3.3.6 Manning-Strickler relation

The Manning-Strickler relation is just one of the most famous relations giving the flowrate without measuring the velocity. It is theoretically valid only for uniform permanent flows and it gives the mean velocity U_m as a function of the geometry of the site and the water level (Equation (3.15)):

$$U_m = K R_h^{\frac{2}{3}} s^{\frac{1}{2}} \tag{3.15}$$

where s (m/m) is the invert slope and K ($m^{1/3}/s$) is the Manning-Strickler coefficient that characterizes the roughness of the sewer walls. This coefficient should be validated on each site as shown by Jaumouillié *et al.* (2002). Although the Manning-Strickler relation is seemingly straightforward and easy to apply, it is not recommended for use because

- Backwater effects or hydraulic jumps can occur unnoticed and introduce unseen and significant systematic errors.
- The value of K cannot be assumed to be a constant but has rather to be named the Manning-Strickler variable quantity as it varies with the water depth (Bertrand Krajewski *et al.*, 2000).
- A fully developed velocity profile is implicitly assumed to be present, which is hard to guarantee or to check in practice.

Table 3.3 Advantages and disadvantages of various technologies for velocity measurements.

Technology	Advantages	Disadvantages
Transit time	• Accurate measurement • Less sensitive to fouling than Doppler sensors • Approved in ATEX zones • Continuous measurements • Can operate in mixed flow regimes (pressurized and free surface flows)	• Requires important civil engineering work • The sensors are fixed and cannot be removed • Delicate installation for alignment of the sensors • Relatively expensive equipment • Few manufacturers • Not suitable for small diameters (<1 m) • Not suitable for use in the presence of air bubbles
Acoustic Doppler flowmeter	• Little civil engineering work and quick installation • Prices and large number of suppliers • Possibility to associate velocity and water level measurements on the same device • Possibility of temporary use and autonomous power supply • Continuous measurements	• Range poorly defined, can be poorly adapted to channels greater than 1.5 m wide • Sensitive to clogging and sediment deposits • May be poorly adapted to slow flows (velocity <0.1 m/s), and low water levels (<0.1 m) • Sensitive to air bubbles
Doppler velocity profiler	• Rather easy to install • Various suppliers • Continuous measurements	• May be not so cheap • Range may be poorly defined
Free surface velocity measurements (radar)	• Reduced maintenance • Suitable for all types of water where the free surface is disturbed • Large velocity range • Quick installation • Suitable for high flow velocities (up to 10 m/s) • Continuous measurements	• Large device: minimum 40 to 90 cm wide • Conversion of free surface velocity to mean velocity to be calibrated
Electromagnetic flowmeter (EMF)	• Very good performance in closed pressurized pipes • Low maintenance • Does not disturb the flow • Suitable for highly contaminated media • Continuous measurements	• Sensitive to air bubbles • Civil engineering work constraints • Significant investment • No measurement below 10% pipe filling rate • Not easy to apply for free surface flows (need to envelop the whole wet perimeter)
Electromagnetic velocity meter (EMV)	• Reduced maintenance • Does not disturb the flow • Suitable for all diameters • Suitable for highly contaminated media • Possibility to associate velocity and water level measurements on the same device • Possibility of temporary use and an autonomous power supply • Continuous measurements • Operation under porous sediment	• Insulation of electrodes by limescale or plastic bags • Small control volume • Sensitive to air bubbles
Manning-Strickler relation	• Basically, requires only a water level measurement	• Limited to uniform steady flows, inaccurate, site specific calibration needed

Manning-Strickler

The Manning-Strickler relation is usually not applicable as a reliable method to estimate discharges in sewers.

3.3.7 Summary

The main advantages and disadvantages of the technologies that are most commonly used to measure the flow velocity in urban sewers are summarized in Table 3.3.

3.4 DIRECT DISCHARGE MEASUREMENTS

3.4.1 Pre-calibrated devices

A wide range of pre-calibrated devices exist for open channel flow measurement. The principle is to have the water level rising due to a contraction of the section and then, using a calibrated relation $Q(h)$, to calculate the flowrate. There are many different devices such as rectangular broad-crested weirs (ISO, 2008), triangular profile weirs (ISO, 2020), trapezoidal broad-crested weirs (ISO, 1999), round-nose horizontal broad-crested weirs (ISO, 1990), flat-V weirs (ISO, 2012) and calibrated flumes such as Venturi flumes (ISO, 2013), each with a number of advantages and disadvantages of varying importance depending on the characteristics of the sites to be monitored.

One common problem with those devices is their sensitivity to clogging due to debris in the flow or due to biofouling, which is certainly the case in wastewater. One has to be aware that when implementing these so-called pre-calibrated devices, the $Q(h)$ relations given in the standards or textbooks are valid only for a set of defined conditions (temperature range, certain water level thresholds due to surface tension effects, minimum water level to apply the relation, requirements on upstream and downstream flow conditions). Unless incorporated during the design, such conditions are difficult to apply in a practice situation at reasonable costs. A further disadvantage is that the implementation of any weir, flume or Venturi in an existing system will cause a backwater effect not taken into account in the system design, with possible risk of upstream flooding. Therefore, it is suggested to avoid using such constructions for discharge measurements as much as possible, with the exception of existing weirs or construction taken into account in the original design. It is stressed that for such constructions an *in situ* calibration is mandatory when data with a known uncertainty are to be obtained. Section 3.4.2 illustrates a comparison between $Q(h)$ relations resulting from a physical scale model and a computational model.

Pre-calibrated devices vs. conditions

Pre-calibrated devices are sensitive to clogging and sediment deposits.

3.4.2 *Q*(*h*) relation using laboratory physical scale models

An alternative to pre-calibrated devices consists of using existing structures in sewer systems as measuring devices, with case-by-case determined $Q(h)$ relations. Such $Q(h)$ relations can be determined by means of laboratory physical scale models, CFD (computational fluid dynamics) simulations or *in situ* measurements.

In complex geometries, like junctions, confluent channels or other locations with a strongly varying 3D velocity flow field (Figure 3.32), it is often not easy to identify a location for water level measurement that can be used as the input for a $Q(h)$ relation. Although a CFD numerical model can be used to simulate the 3D velocity flow field, a laboratory physical scale model can be chosen as an alternative.

There are some reports in which the results of the comparison of CFD and physical models are reported. For example, van Daal-Rombouts (2017) and Dufresne *et al.* (2018) conclude that 'fair' agreement is found, with deviation in velocities up to 20–30% and water levels that are in good agreement between the CFD model and physical model. Both approaches have their strong and weak points. Especially in multi-phase problems (sediments, air entrainment), CFD models sometimes may show flows that are difficult to recognize yet represent 'unphysical' results. In such cases, an independent verification is needed, e.g. by means of a physical scale model. On the other hand, physical models are normally built to some geometric scale. A geometrical scale between 5 and 10 is generally accepted to result in manageable models while not suffering too much from scale effects. In a sense, physical scale models and CFD are used as complementary means. By using the results of a physical scale model to validate a CFD model, the latter can be used to answer 'what if' questions in a design process or scale up to the prototype scale to eliminate the scale effects of the physical scale model. For application of either CFD or a physical scale model, specialist knowledge and professional means are needed. With respect to CFD modelling, the calculation effort can be considerable, thus limiting the number of simulations that can be made within a reasonable timeframe. On the other hand, in a physical scale model, geometrical changes are not easily made and making a 'model run' requires time from specialized personnel. There is no general preference for either CFD or a physical scale model, both have strong and weak points. Making a choice is largely a matter of carefully considering the dominant processes involved. For example, when designing a pump sump, it is of importance to know the pre-rotation of the flow approaching the pump to avoid cavitation, further the occurrence of unstable vortices and air intake should be avoided. These

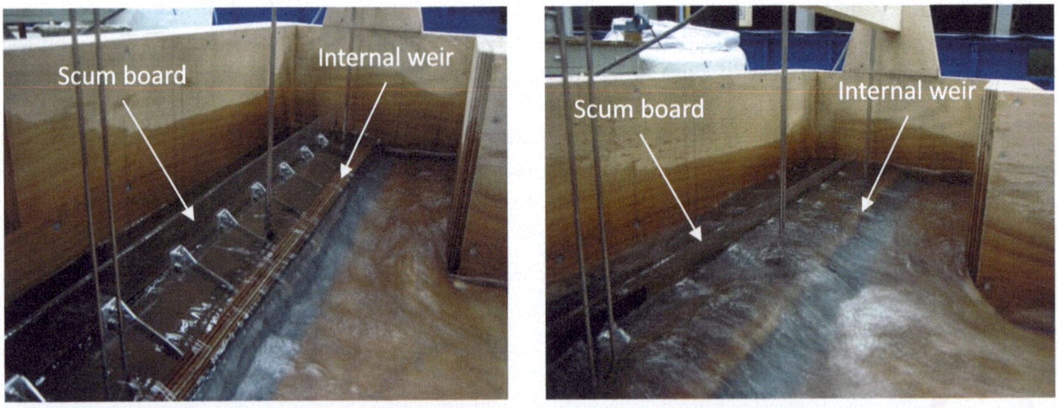

Figure 3.32 Example of a combined sewer overflow (CSO) laboratory physical scale model. *Source*: van Daal-Rombouts (2017).

phenomena are not considered to be correctly reproduced by CFD models. For this reason, application of physical scale models is recommended or even prescribed in some standards (see Hydraulic Institute, 2018; Verhaart *et al.*, 2016).

A key issue in the application of physical scale models is the choice of the geometrical scale while keeping in mind that the physical properties (viscosity, density) of the fluid used (virtually always water) cannot, or can to a limited extent only, be scaled. The viscosity is temperature dependent, so by varying the temperature the viscosity can be changed. As for surface tension, this parameter can be influenced by using a (preferably non-foaming) detergent, a reduction with a factor of ~2 is feasible in this manner. Scaling effects can be reduced as much as possible. However, in many situations, it is impossible to apply geometric scaling and maintain a dynamic similarity between the full-scale structure and the physical scale model. In the case of a free surface flow, the scale is mostly chosen such that the Froude number Fr in both the physical scale model and full-scale structure are equal. This is the Froude similarity, i.e. the ratio between kinetic and potential energy is conserved as it is the dominant energy ratio in gravity driven flows. This poses some more or less severe restrictions on the range of water levels that can be tested effectively due to the non-scale-ability of surface tension and viscosity. In this respect the deviation between the physical scale model and the real structure for the Weber and Reynolds numbers (resp. We and Re) should be accounted for. In most practical situations in urban drainage, the flow is turbulent (i.e. $Re > 10^4$–10^6). As a rule of thumb, the Reynolds number in the physical scale model should not drop below ~10^3–10^4. With respect to the Weber number, there is no real consensus on the lower critical limit (opinions vary between 10 and 100). This implies, that when taking a geometrical scale n, the following limitations are to be respected (Equations (3.16)):

$$Fr_p = \frac{V_p}{\sqrt{gh_p}} = Fr_m = \frac{V_m}{\sqrt{gh_m}} \tag{3.16a}$$

$$Re_m = \frac{H_m V_m}{\upsilon} > 10^4 \tag{3.16b}$$

$$We_m = \frac{\rho V_m^2 L}{\sigma} > 10 - 100 \tag{3.16c}$$

where the indexes p and m refer respectively to the full-scale structure and the physical scale model, V (m/s) is the flow velocity, g (m/s^2) is the gravity acceleration, h (m) is the water level, H (m) is the characteristic dimension of the physical scale model, v (m^2/s) is the kinematic viscosity of water, ρ (kg/m^3) is the density of the water, L (m) is the characteristic length, and σ (N/m) is the surface tension.

The Weber number W_e (ratio between inertial forces and surface tension forces) is normally in the order of magnitude of 10^2–10^4. As with the Reynolds number, the Weber number cannot be assumed identical in the full-scale structure and scale model. Added surfactants can reduce the surface tension by approximately 30–50%, this may be utilized in a scale model when the Weber number becomes too small. When deciding on a geometrical scale n, this implies that:

- All geometrical measures scale with this factor n.
- When striving for Froude similarity, the velocity scales with $n^{0.5}$.
- The Reynolds number scales (in Froude similarity) with $n^{1.5}$.
- The Weber number scales with n^2.
- The discharge scales with $n^{2.5}$.

The impact of scaling is illustrated by an example taken from van Daal-Rombouts (2017) and van Daal-Rombouts *et al.* (2017). In this example, a scale model was used to find a $Q(h)$ relation and a

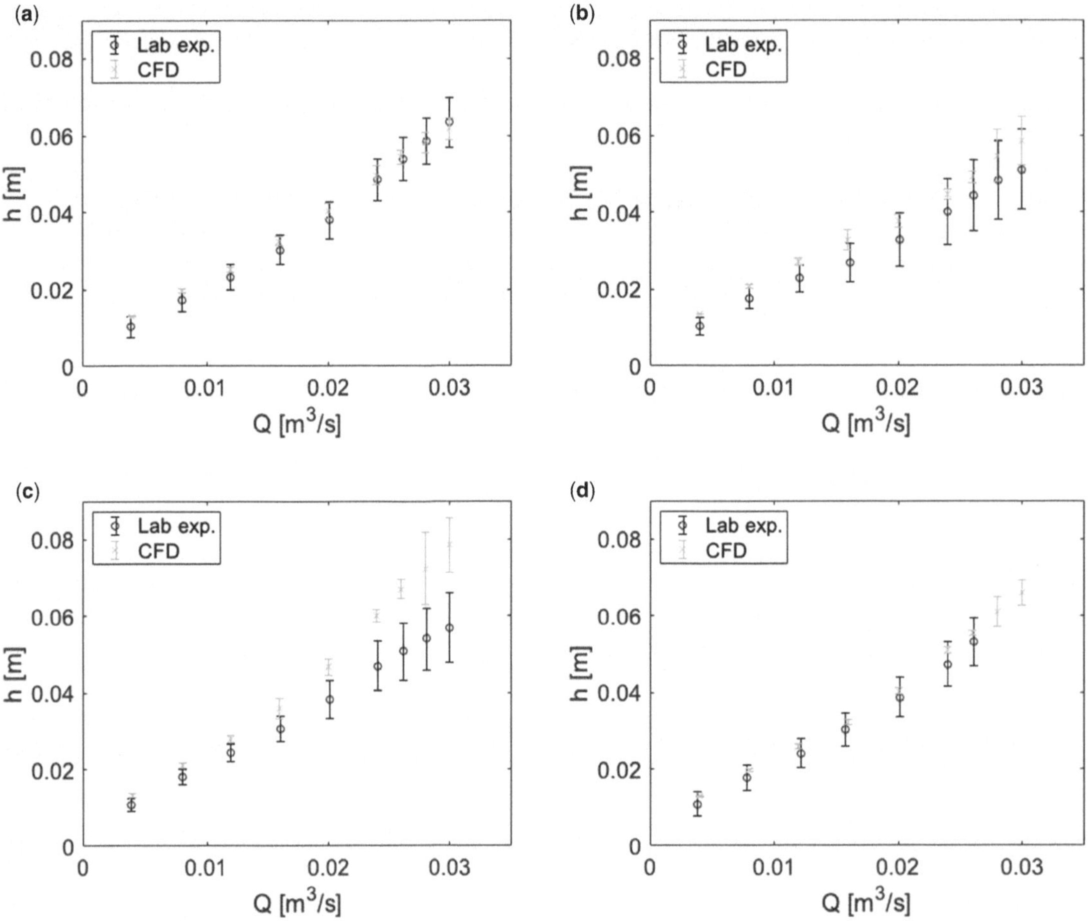

Figure 3.33 Results from physical scale model experiments and CFD modelling: (a), (b), (c) and (d) refer to four different locations along the length of the weir. As can be seen, there is a reasonable match between physical scale model and CFD model (after validation) apart from location (c) where significant deviations occur. *Source*: van Daal-Rombouts (2017).

suitable monitoring location for placing a water level sensor for a CSO construction as shown in Figure 3.32. In this case, $n = 8$. For the prototype a discharge of ~5.5 m^3/s was expected. The discharge in the scale model was reduced to a maximum of ~0.03 m^3/s. Given the choice for the characteristic length scale, the Reynolds number in the prototype was ~5×10^4 which is reduced by a factor of 81.5 to *ca.* 2000 in the physical scale model, which is just in the turbulent region. The Weber number, given the choice of the characteristic length in the order of magnitude of 100, is reduced by a factor of 64. As the Weber number reflects the ratio between inertial forces and surface tension forces, this indicates that in the full-scale structure the effect of surface tension (curvature of streamlines) is much less explicit than in the model. This directly translates back into a lower imitation of the range of discharges and water depths that can be explored in the physical scale model while assuming a classic $Q(h)$ relation in the form of $Q = ah^b$ in which the effect of surface tension is omitted. For all practical purposes, when measuring a water depth over the

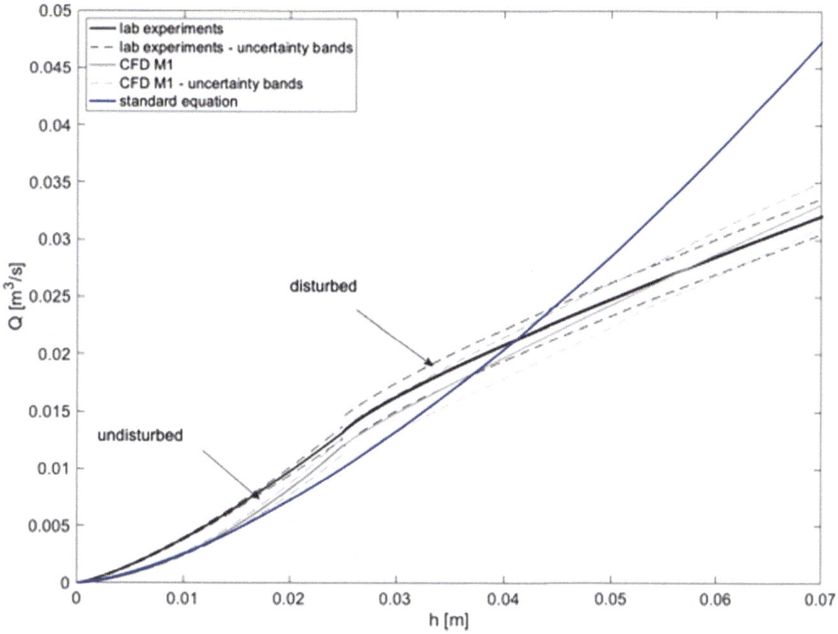

Figure 3.34 $Q(h)$ relations found for two locations as derived from physical scale and CFD models. 'Disturbed' refers to the situation in which backwater effects occurred. *Source*: van Daal-Rombouts (2017).

crest of a weir that is less than ~6 mm, the usually applied $Q(h)$ relations (e.g. the Kindsvater & Carter (1957) formula) are, strictly speaking, not valid, nor in the full-scale structure, nor in the scale model.

A comparison between the results obtained from the physical scale model and results from a CFD simulation are shown in Figure 3.33.

In Figure 3.34 the resulting $Q(h)$ relations for two locations based on the physical scale model and the CFD model are shown. It can be observed that:

- The relations for the two locations are mutually different.
- There is no significant difference between physical scale and (validated) CFD models.
- The 'standard' equation shows significant deviations relative to the model-based relations.

The added value of the physical scale model in this case was found in the identification of suitable locations for the water level sensor and the validation of the CFD model. For a more in-depth treatise on hydraulic scale models and associated scale effects, the reader is referred to the literature, e.g. Ettema *et al.* (2000) and Heller (2011).

3.4.3 Chemical tracing

3.4.3.1 Principle

Tracer experiments offer a method to measure discharges, with a simple and single assumption: the mass of tracer is constant along the reach, i.e. the mass measured downstream is equal (Equation (3.17)) to the mass

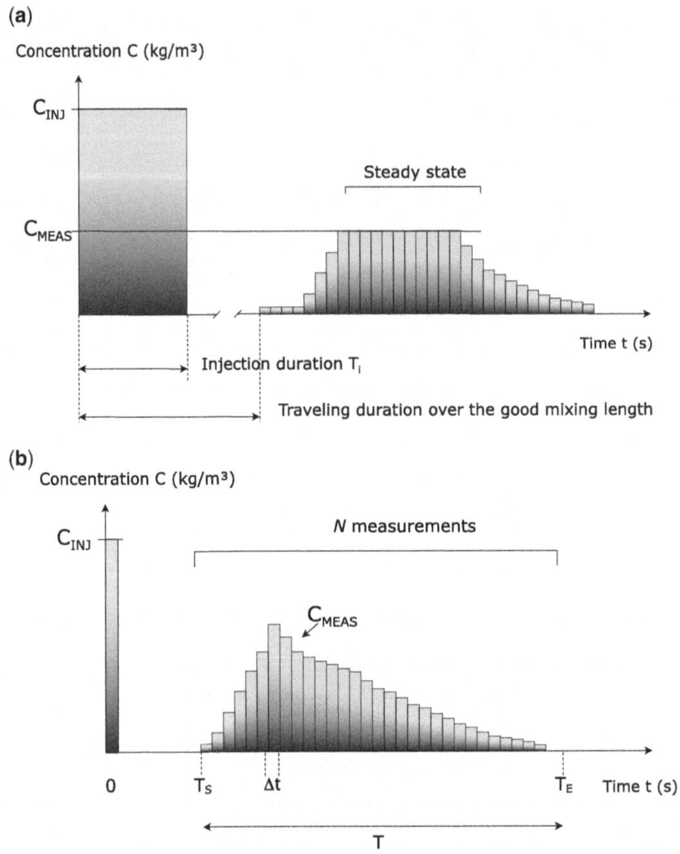

Figure 3.35 Injection and measured time series for both continuous (a) and spike (b) injections. *Source*: adapted from Bertrand Krajewski *et al.* (2000).

injected upstream:

$$M_{INJ} = M_{MEAS} \tag{3.17}$$

where M_{INJ} and M_{MEAS} are respectively the injected and measured masses of tracer (kg). There are two kinds of tracer experiments: continuous or spike injection (Figure 3.35).

The average discharge Q_{MEAS} (m^3/s) along the reach can be calculated from Equations (3.18): Equation (3.18a) for continuous injection, Equation (3.18b) for spike injection.

$$Q_{MEAS} = q_{INJ} \times \frac{C_{INJ}}{C_{MEAS}} \tag{3.18a}$$

$$Q_{MEAS} = \frac{M_{INJ}}{\sum_{t=T_S}^{t=T_E} C_{MEAS} \times \Delta t} \tag{3.18b}$$

where q_{INJ} (m^3/s) is the tracer injection discharge, C_{INJ} (kg/m^3) is the concentration of the injected tracer, C_{MEAS} (kg/m^3) is the tracer concentration measured downstream, T_S and T_E are respectively the start and end

Table 3.4 Available tracers and their pros and cons.

Tracers	Advantages	Disadvantages
Lithium	• –	• Not measurable continuously
Heavy water	• –	• Not measurable continuously
De-icing salts	• Low price • Availability • Measurable continuously with a conductivity meter	• Not constant background signal • Sometimes requires large volume for injection
Rhodamine B	• Measurable continuously with a fluorometer • Usually constant background signal	• Sensitive to TSS (total suspended solids)
Rhodamine WT	• Measurable continuously with a fluorometer • Usually constant background signal • Insensitive to TSS	• High price • Toxicity

times of the tracer spike at the downstream measurement location, and Δt (s) is the time step between two measurements.

For both methods, the estimation of Q_{MEAS} directly depends on the measurement of C_{MEAS}, which must absolutely be uniformly distributed across the measurement section. To ensure this uniform distribution, three conditions must be carefully checked along the reach: (i) there should be no active lateral house connection, (ii) the flow should not be split in two or more parts, and (iii) the length of the reach should be 75 times (respectively 150 times) longer than the largest hydraulic dimension of the wet section (water depth, width, etc.) if the injection is done in the middle (respectively on the side) of the flow. Several tracers can be used in urban drainage. Table 3.4 presents their pros and cons.

The continuous injection requires a larger quantity of tracer and more sophisticated hardware equipment, but the data processing is easier. On the contrary, spike injection experiments are easier to conduct, but require more elaborate data processing: to facilitate its application, the reader is invited to use the Excel® file and the Matlab® code available for download at https://doi.org/10.2166/9781789060102. For the tracer, de-icing salts are recommended for situations with a small discharge and short reach. When the injection volume becomes too large (a few tens of litres or more), Rhodamine WT is recommended.

3.4.3.2 A nine step operation protocol

This section gives a summary (Figure 3.36) of the spike injection method presented in detail in Lepot *et al.* (2014).

After the reach has been identified and appears to be suitable for such experiments (good mixing conditions), the first three steps deal with the preparation in the office or in the laboratory: (i) calibration of the measuring device (fluorometer for fluorescent tracer or conductivity sensor for de-icing salt), (ii) study of the injection device, i.e. study of the true volume injected by pipettes in the case where fluorescent tracer is used, and (iii) preparation of the solutions to be injected (de-icing salt at 180 g/L or dilution of commercial solution of fluorescent tracer).

The following steps can be done *in situ*. The fourth step consists of estimating the mass of tracer to inject, given the sewer reach and its hydraulic conditions. The basic equation of longitudinal dispersion (which describes the evolution of a tracer concentration C along a reach and over time) is required to understand

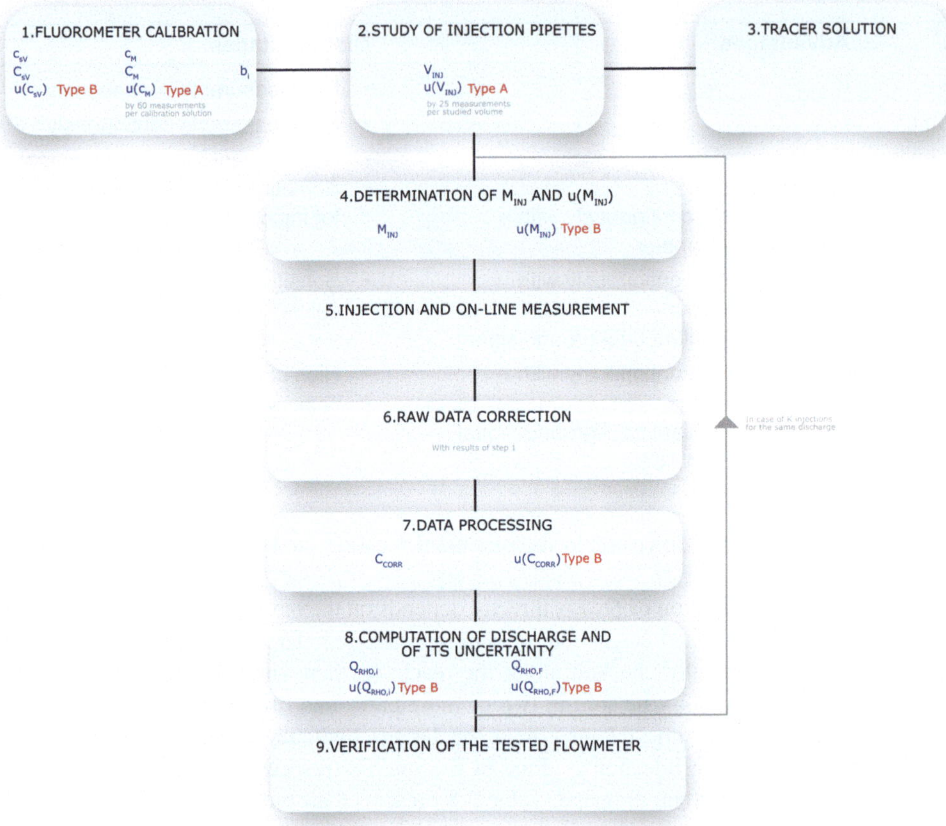

Figure 3.36 Sketch of the nine-step protocol, given here for Rhodamine WT. For details, please refer to Lepot *et al.* (2014). *Source*: adapted from Lepot *et al.* (2014).

this calculation (Equation (3.19)).

$$\frac{\partial C}{\partial t} + U_m \frac{\partial C}{\partial x} = K_X \frac{\partial^2 C}{\partial x^2} \tag{3.19}$$

where C is the tracer concentration (kg/m^3), x (m) is the distance from the upstream injection location, t (s) is the time, U_m (m/s) is the average flow velocity along the reach, and K_X (m^2/s) is the dispersion coefficient.

There are various possibilities to set the value of K_X (Rieckermann *et al.*, 2005). For most sewer reaches, straight and without back flow, the dispersion coefficient given in Equation (3.20) is recommended.

$$K_X = 6h\sqrt{ghs} \tag{3.20}$$

where h (m) is the water level, and s (m/m) is the slope of the reach.

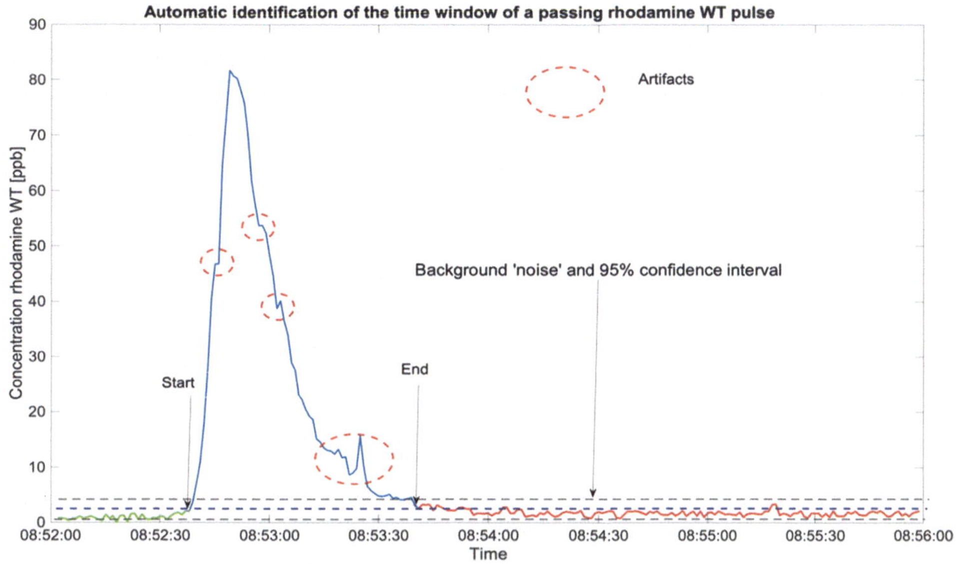

Figure 3.37 Data processing for the transit of a spike injection of Rhodamine WT. *Source*: adapted from Lepot *et al.* (2014).

Ideas on the M_{INJ} estimation

- I 3.6: *Sensors in place* – Since M_{INJ} should be only roughly estimated, data coming from even doubtful flowmeters could be used for this calculation (water level, velocity).
- I 3.7: *Water level* – Estimate the water level with a ruler.
- I 3.8: *Velocity* – By throwing floating materials (e.g. plant, balls), the free surface flow velocity U_{fs} can be estimated from their travel time along the reach. The mean flow velocity U_m can then be estimated by multiplying U_{fs} by a numerical coefficient approximately equal to 0.85 to account for the velocity profile (see Figure 3.12).

Equation (3.21) presents a possible analytical solution to the partial differential Equation (3.19):

$$C_{MEAS}(x,t) = \frac{M_{INJ}}{S_m \times \sqrt{4\pi K_X t}} e^{-\frac{(x-ut)^2}{4K_X t}} + C_{BG} \tag{3.21}$$

where S_m (m^2) is the average wet section along the reach. M_{INJ} should be chosen to ensure that the maximal value of C_{MEAS} at the measurement location is: (i) 3 to 4 times larger than the background concentration, C_{BG}, and (ii) within the linear range of the sensor calibration function. These maximal values are typically 2 g/L for salts and 80 ppb for Rhodamine WT.

Once M_{INJ} has been injected and C_{MEAS} is recorded (step 5), raw data need to be corrected (see Section 7.6 on sensor calibration) and processed (step 6, Figure 3.37). The data processing includes two steps: automatic identification of the start and end of the spike injection transit and removal of artefact and outlier values.

Once data have been cleaned, the discharge can be calculated (step 8, Equation (3.18b)). Its value might be used to check an existing flowmeter. Lepot *et al.* (2014) give all details to consider several injections and assess the uncertainties in all intermediate quantities and in the value of Q_{MEAS}.

No connection nor divergence along the reach

The tracer should be uniformly distributed across the downstream measuring section. Attention should be paid to the absence of or negligible inflows upstream of the measuring section, or to any divergence (e.g. bypass of the flow) downstream of the injection point as it engenders a loss of tracer resulting in an incorrect mass balance. In such cases, the locations of both injection and measuring points should be adapted to ensure appropriate conditions for the tracer experiment.

3.4.3.3 Examples of application

3.4.3.3.1 Rhodamine WT and de-icing salt tracings vs. flowmeter measurement of small discharges in a large sewer

Rhodamine WT and de-icing salt tracer experiments have been conducted at the inlet of a stormwater retention tank in Chassieu, France, in a large (diameter of 1.6 m) and steep (slope of 0.01 m/m) sewer. Results show that the flowmeter in place underestimates the flow and that Rhodamine WT and salt experiments deliver consistent results (Figure 3.38).

3.4.3.3.2 *Rhodamine WT tracing for two monitoring stations: large pipe, large discharges*

In a catchment of Greater Lyon, France, two measuring locations gave inconsistent discharge values: daily cumulated volumes in both locations were similar despite the fact that between the upstream and downstream locations, the population connected to the sewer system in the catchment increases by 20%. Rhodamine WT tracer experiments have been conducted for discharges varying from 250 to 550 L/s.

Figure 3.39 shows two important facts. The first being the difficulty in conducting tracer experiments immediately downstream of a pumping station: Figure 3.39(a): starts and stops of the pump generate mean values of the flow which are almost constant over pumping cycles due to the averaging by the *in situ* flowmeter, while the more instantaneous tracer experiment reveals that the flow varies significantly during pumping cycles. The second is the underestimation of the flow by the *in situ* flowmeter at the downstream location: all tracing experimental data are located above the expected ideal case (dashed red line) where the discharges given by the tracer experiments are supposed to be equal to the ones given by the flowmeter.

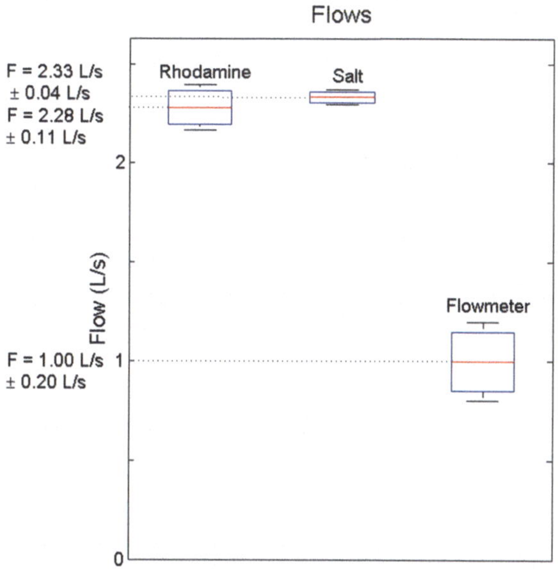

Figure 3.38 Comparative flow measurements with Rhodamine WT tracer experiment (left box plot), de-icing salt tracer experiment (central box plot) and in place flowmeter (right box plot) in the inlet sewer of a stormwater retention tank in Chassieu, France. *Source*: Lepot *et al.* (2014).

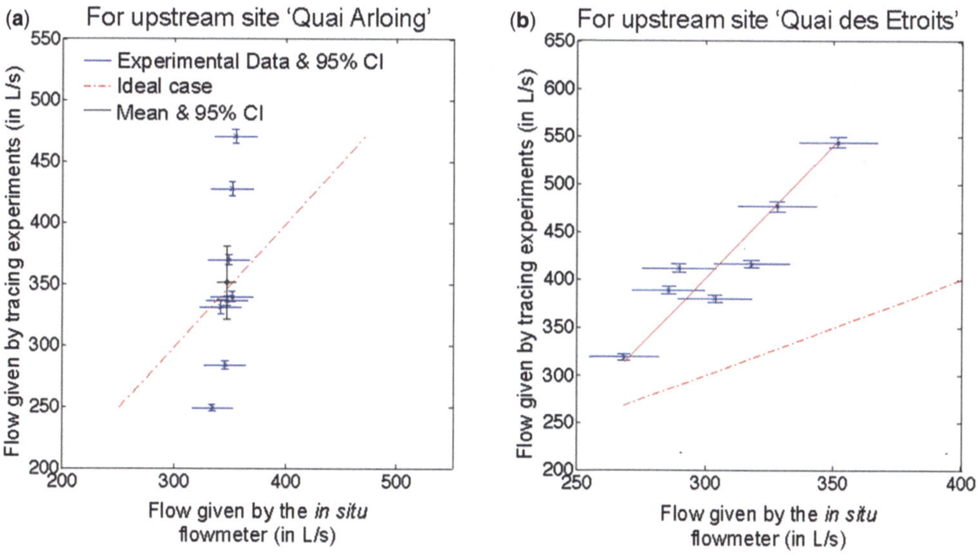

Figure 3.39 Flow measurements by Rhodamine WT tracer experiments and by in place flowmeters upstream (a) and downstream (b) of a pumping station in Greater Lyon, France. Blue horizontal bars represent the flow and its uncertainty given by the *in situ* flowmeter, and the blue vertical bars represent the corresponding flow and uncertainty given by tracer experiments carried out at the same moment. *Source*: adapted from Lepot *et al.* (2014).

Ideas on tips in the field

- I 3.9: *Control the flow* – Check if the flow can be increased, decreased or controlled in order to ensure stable conditions during the tracer experiments, or cover a wide range of hydraulic conditions as shown in the example of Figure 3.39.
- I 3.10: *Repeat* – Given the low marginal cost of additional spike injections, it is often worth performing several injections to get more accurate results (repeated measurements) when staff and all equipment are on site.
- I 3.11: *Be patient* – Be patient during two consecutive injections in the field, to ensure you can record the background signal between two peaks of tracer.
- I 3.12: *Always check immediately* the recorded data (presence of values, clean peaks) in the field to ensure that the data can be processed later: do not wait to be back in the office to discover possible problems, errors, failures, etc.

3.4.4 Pumping stations

3.4.4.1 Introduction

Pumping stations play a key role in many urban drainage systems, especially in wastewater and combined systems in flat areas. The importance of monitoring pumping stations in urban drainage has been highlighted in some publications, e.g. Korving *et al.* (2006) or Kooij *et al.* (2015). Monitoring their performance is important for a range of reasons:

- Qualification/testing of newly built installations.
- In applications of Real Time Control, variation of pump discharge is often applied as one of the methods to adapt a system's behaviour.
- For asset management: decreasing performance can be detected, and corrective actions can be induced.
- Controlling energy consumption.
- Early warning of ageing processes.

The reasons mentioned necessitate permanent monitoring. However, this is not always applied due to financial reasons and/or a lack of interest. Next to a permanent monitoring set-up, monitoring campaigns with a short duration are applied as well, e.g. for commissioning tests or for troubleshooting reasons. Both applications require a different approach and will be briefly discussed hereafter.

The performance of a pumping station can only be judged by considering the influence of the pressure main (including appendages like (air)valves, check valves, Venturis, etc.). Further, the presence of siphons is known to be a source of malfunctioning due to the risk of the accumulation of air/gas pockets (Pothof & Clemens, 2010, 2011). In severe cases, this may even cause a reduction to a capacity of zero while a high energy consumption is sustained.

Most significant pumping stations that have been (re)constructed over the past few decades are equipped with at least the possibility of performing monitoring activities. This is mainly motivated by asset managerial considerations like the planning of maintenance or rehabilitation/replacement.

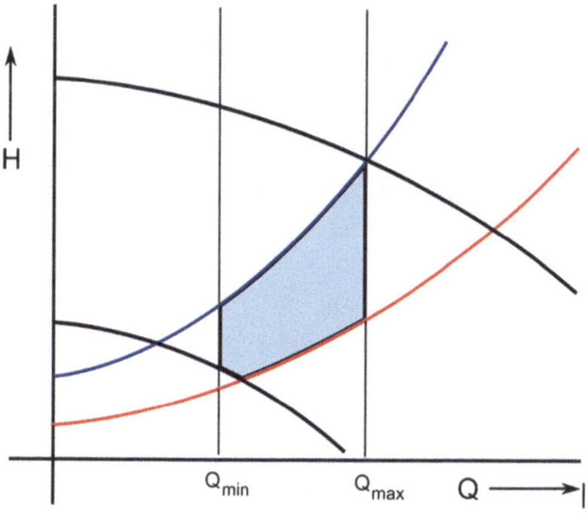

Figure 3.40 Characteristics of a variable speed pump. The black lines describe the pump characteristics, the blue and red lines indicate the characteristic of the pressure main. The blue line has a larger value for the static head and a higher friction. The shaded blue/grey area in the figure indicates the design window in which the system is supposed to operate under satisfactory conditions (among others, no occurrence of cavitation). *Source*: adapted from Tukker *et al.* (2016).

3.4.4.2 Theoretical considerations

A pump characteristic describes the relation between discharge and head, in conjunction with the hydraulic characteristics of the (system of) pressure main(s) to which the pump is connected. In particular, it defines the so-called operation area (Figure 3.40).

In the design phase, the area in which the pump is supposed to operate is defined (Figure 3.40). This area is chosen in such a manner that:

- Stable operation of the pump is guaranteed.
- The occurrence of cavitation is avoided as much as possible.
- The energy consumption is kept as low as possible.

The characteristic of the pressure main (Equation (3.22)) is defined by the static head difference in water head between pump sump and the level at which the water is discharged from the pressure main and the dynamic head that is defined by the friction losses in the system. The latter depends on the dimensions of the pipe, the number and type of bends, valves, etc.

$$\Delta H = \Delta H_{static} + \lambda \frac{LV^2}{2gD} + \sum \xi_i \frac{V^2}{2g} \tag{3.22}$$

where ΔH (m) is the total head, ΔH_{static} (m) is the static head, λ (−) is the friction coefficient, L (m) is the pipe length, D (m) is the pipe diameter, V (m/s) is the flow velocity, g (m/s^2) is the gravitational acceleration and ξ_i are local head loss coefficients.

Given the hydraulic characteristics of the pump and the pressure main, the pump must be supplied with enough power to operate in the desired work area. The power consumption is defined by Equation (3.23):

$$P_{elec} = \frac{P_{hydr}}{\eta_{pump} \times \eta_{elec}} = \frac{\rho g Q \Delta H}{\eta_{pump} \times \eta_{elec}} \tag{3.23}$$

where P_{elec} (W) is the electrical power, P_{hydr} (W) is the hydraulic power, η_{pump} (−) is the pump efficiency, η_{elec} (−) is the electromotor efficiency, ρ (kg/m^3) is the water density, Q (m^3/s) is the discharge, and ΔH (m) is the pump head.

The electrical power can be calculated directly from the power consumption and the voltage. Often, only the power consumption (Equation (3.24)) of the pump is registered because the voltage is assumed to be constant.

$$P_{elec} = UI \tag{3.24}$$

where U (V) is the voltage, and I (A) is the current.

For more information, Tukker *et al.* (2016) offer an in-depth treatise on design, operation and troubleshooting of (waste)water pumping stations.

It is known that, due to ageing, the characteristic of the pressure main changes over time. This is due to biofilm growth, sedimentation and/or the accumulation of gas/air bubbles in the system. In addition, the pump characteristic will also vary over time due to mechanical wear of bearings and propeller blades. These ageing processes result in an increased energy consumption and, in some extreme cases, in a total loss of capacity mainly caused by gas accumulation.

A routine activity when managing a pumping station is to regularly check for dynamic losses (last two terms in Equation (3.21)), especially when pressure difference and discharge are monitored for operational purposes. It provides a powerful and relatively cheap instrument to check on the performance of a pumping station/pressure main. The preferred situation is that a first reading is taken directly after putting the pumping station in operation and this reading is then used as a reference. Over time, an increase of the friction will occur, resulting in an increase in the power consumption (Equation (3.24)), as this is a slow process when caused by scaling of the pressure main or wear of the pump. Notwithstanding this, a regular check (e.g. once every month) is advised as the formation of biogas and/or the air entrainment in the pumping station may cause gas pockets that accumulate in the pressure mains and can cause a very significant increase in resistance and hence power consumption. Pothof *et al.* (2009) estimate that, due to the presence of air/gas pockets, a yearly excess amount of approx. 10,000 tons of CO_2 is emitted in the Netherlands by malfunctioning wastewater pumping stations.

In general, three main monitoring objectives can be distinguished:

- Check on the hydraulic characteristic of the pressure main.
- Check on the hydraulic characteristic of the pump.
- Pump performance.

3.4.4.3 Quantities to be measured

Obviously, one needs to simultaneously measure the head difference over the pressure main and the discharge for a first global view of the work point of the system. This can be done in a relatively simple manner when the pumping station has a built-in discharge sensor, which is the case in most large pumping stations.

When pump behaviour and energy consumption have to be known as well, one needs to measure:

- The discharge.
- The hydraulic head difference over the whole system.
- The hydraulic head difference over the pump.
- The energy consumption (electrical potential and electrical current, in the case of an electro-powered pump, which is the predominant type in practice).
- The rotation per minute (RPM) of the pump.

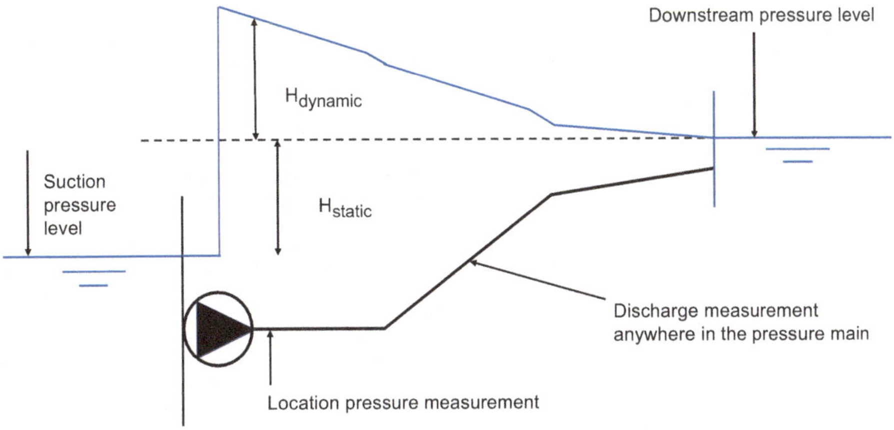

Figure 3.41 Generic scheme of a single pump/pressure main system. *Source*: Deltares.

It has to be emphasized that a stationary situation is assumed. One should therefore observe the monitoring data and choose a time window in which the system is stable. When it takes a long time (order of minutes or longer), depending on the size of the system and the presence of wind vessels to stabilize, this may hint at the presence of gas pockets in the system (see e.g. Lubbers, 2007).

When only interested in the hydraulic characteristic of the pressure main, only the following quantities need to be monitored (Figure 3.41):

- The discharge.
- The pressure directly downstream of the pump.

The downstream pressure level is normally known and to a good approximation stationary. If not, this water level needs to be monitored as well.

When doing so, the combination of discharge and total head is measured, allowing determination of the dynamic losses due to wall friction and local losses. If local losses need to be quantified individually, a pressure difference over the local part of the main (be it a bend, a valve or air vent) has to be measured.

When the pump characteristic needs to be measured (which, for practical reasons, is not recommended in an *in situ* set-up!), the following quantities need to be measured:

- The discharge.
- The suction pressure level.
- The pressure directly downstream of the pump.
- The downstream level only when not stationary.
- The rotations per minute (RPM) of the pump.
- The electrical potential and current.

When only interested in the behaviour at the working point, one measurement is sufficient. However, in most cases, then validating the pump characteristic has to be done over a wide range of discharges and total head. This can be tedious to achieve *in situ* as each measurement takes a minimum amount of time to obtain a stable situation, which puts demands on the availability of enough water to perform the test. Amongst other reasons, this is why the validation of a pump characteristic is preferably done in a laboratory under known conditions, which also allows comparison of the pump characteristic with the original characteristic provided by the manufacturer.

3.4.4.4 Permanent monitoring set-up

In a permanent monitoring set-up, all quantities mentioned in Section 3.4.4.3 are measured simultaneously and allow generation of on-line information on the performance of the system, which can be used for planning maintenance or to detect defects at an early stage (see e.g. Kooij *et al.*, 2015). In such permanent monitoring set-ups, one is well advised to mount vibration sensors on the pump axis as well, as wear or damage on the propeller blades and/or bearing can be detected at a very early stage allowing for a quick response to avoid further damage.

In a permanent monitoring set-up, in most cases an electro-magnetic discharge sensor is used, as they are known to be robust, stable and offer good accuracy. These devices can only be implemented when this is foreseen during the design. Implementation in a later phase is often difficult/costly because of the demands put on the undisturbed length of pipe up and downstream of the device to ensure the correct flow conditions in the device (see e.g. ISO, 2003).

Figure 3.42 shows the specific energy consumption of a pumping station. The specific energy of a pumping station has been quantified for a range of discharges at $t = 0$, 4 and 8 weeks. The specific energy increases with time, which is an indication of either efficiency loss of the motor/pump system or an increase of the resistance of the pressure mains. As this is a two-pump system and the increase in energy consumption is present in single-pump and two-pump operation, the latter option is the most likely.

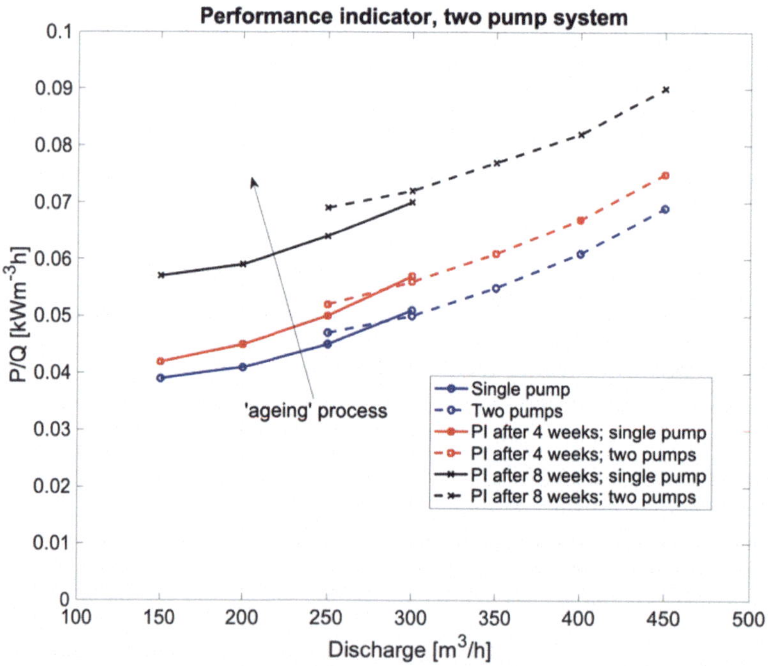

Figure 3.42 Evolution of the specific energy as a function of discharge for a dual-pump pumping station. The specific energy in this pumping station increases by a factor of almost 1.5 in the course of only 8 weeks. This may hint at the accumulation of gas pockets in the pressure main. The blue lines indicate the performance directly after cleaning at $t = 0$. *Source*: adapted from Tukker *et al.* (2016).

Figure 3.43 The cause of an increase in energy consumption became immediately clear when inspecting the pump. *Source*: courtesy Kees Kooij (Deltares).

3.4.4.5 Temporary measurements

Apart from pressure, discharge and power consumption, that can be used for defining a range of performance indicators (Figure 3.42) for daily operation and data for mid- and long-term asset management (Kooij *et al.*, 2015), the measurement of mechanical vibrations is often added as a monitoring quantity. An increase of vibrations, or changes in vibration patterns of pumps is often an indicator for either wear of the propeller and/or the bearings or the presence of some obstruction in the pump. Early registration can avoid damage and/or malfunctioning (Figure 3.43).

3.4.4.6 Incidental measurements

Many pumping stations are not equipped with permanent sensors for pressure and discharge and/or power consumption. Nevertheless, for asset managerial or legal reasons, the performance of these stations sometimes needs to be assessed. In cases where no built-in sensors are present, the main difficulty is found in getting access to locations for installing sensors. Especially, measuring discharge and pressure differences in a buried pressure main poses some challenges. As discussed in this chapter, several options for discharge measurement are available:

- Doppler acoustic (Sections 3.3.2 and 3.3.3).
- Travel time (Section 3.3.1).
- Tracer experiments (Section 3.4.3).

With respect to the latter, it has to be mentioned that depending on the choice of tracer this method can be expensive but has the advantage of offering a known and potential high degree of accuracy. Notwithstanding that this method is not often applied as it is difficult to establish during the monitoring activities, whether or not a stationary situation is obtained, tracers are mainly used to validate other discharge measuring methods.

Figure 3.44 Clamp-on discharge measuring equipment installed. *Source*: courtesy Kees Kooij (Deltares).

Depending on the accessibility of the system and the availability of sufficient undisturbed length (see ISO, 2003), clamp-on acoustic sensors based on the travel time principle can be applied relatively easily.

An alternative to flowmeters (Figure 3.44) is to isolate the pump sump from the rest of the system and record the time it takes for the pump to pump out of the pump-sump a certain amount of water. This can only result in useable data when the following conditions are met:

- The change in water level during the experiment can be neglected compared to the total head, to limit the working point (Q, h) changing too much.
- A static situation is obtained. This may take up to a few minutes, especially when gas/air pockets are present in the system. One is well advised to interview the engineers who operate the station normally to find out the characteristics of the station.
- The exact geometry of the sump is known.

By measuring the water level as a function of time, the discharge can be estimated from the results. As in many cases the water level at the suction side of the pump is lowered during pumping, the static head difference needs to be monitored (note that this implies that there is, strictly speaking, never a stationary situation). However, if the change in static head is small compared to the total head during the measurement, this effect may be ignored along with the pressure directly after the pump. The situation at the discharging end of the pressure main defines whether or not one needs to monitor there as well. In cases where there is free outflow with no risk of backwater effects, there is no need to monitor the downstream end.

3.4.4.6.1 Measuring set-up for determining the hydraulic characteristic of the pressure main

In most cases, the pressure level at the downstream end of the main is known to a fairly good approximation and is determined by the local geometry of the construction. When striving for a stationary situation, there is

Figure 3.45 *In situ* constructed pressure measuring mount. *Source*: courtesy Kees Kooij (Deltares).

no real need for the two measurements to be perfectly synchronized with each other. The overall resistance factor can be determined from the stabilized signals obtained from the reading.

The construction of the measuring point for pressure requires some attention. Figure 3.45 shows an *in situ* constructed monitoring point. The most important issues to consider when constructing a measuring location for pressure in a pressure main are the following ones:

- Preferably mount the entrance at the side of the pipe, as the risk of clogging due to fat, etc. at the ceiling or bottom is substantial, as is the risk of gas or air bubbles disturbing the measurement at the ceiling of the pipe.
- Install a valve, this allows for mounting a compressor for cleaning out the pressure mount when it gets clogged.
- Avoid installing a pressure measuring point directly after a bend, a valve or any other appendage, as the pressure at such a location may show substantial variation in time due to locally disturbed pressure and velocity profiles and may prove to yield unusable results. As a rule of thumb, a distance longer than 5 times the pipe diameter is regarded as a minimum, although in many practical situations this is hard to achieve due to the existing design.

3.4.4.6.2 Measuring set-up for determining the hydraulic characteristic of the pump

In situ determination of the characteristics of a pump is normally only carried out for 'dry' installed set-ups (normally the bigger pumping stations). For most submersible pumps, the best option is to transport them to a test-stand in a laboratory or workshop. For *in situ* testing, the pump has to be operated at a range of values for discharge and hydraulic head. When the effect of mechanical settings (RPM) and efficiency has to be verified as well, power consumption has to be monitored. To effectuate a full *in situ* verification of the pump characteristic requires detailed planning, and the exact measuring set-up strongly depends on the local conditions, the engineering of pipes and appendages, and the type of pump. It is therefore not possible to go into further detail here.

3.4.5 Use of computational fluid dynamics

As the Hydrometry Charter (Banque Hydro, 1998) points out, hydrometry should not be a routine, but a profession open to technology, intelligence and questioning. It is the same state of mind that prompted many teams to develop computational fluid dynamics (CFD) skills and then studies to understand the hydraulics of sewers or to manage combined sewer overflows (Fach *et al.*, 2008, 2009; Jarman *et al.*, 2008) or to improve the implementation of sensors (Bonakdari, 2006; Bonakdari *et al.*, 2008; El Bahlouli and Larrarte, 2018; Larrarte *et al.*, 2004, 2010, 2017; Lipeme Kouyi, 2004; Lipeme Kouyi *et al.*, 2003, 2005; Vazquez *et al.*, 2005, 2006). Those numerical tools have also been used for flowmeter calibration by Hrabak *et al.* (1998).

In this context, Larrarte *et al.* (2004) have drafted a methodology for qualifying measurement sites in sewer systems. This approach has been refined within the framework of the MENTOR project (MEasurement sites conception method for sewer NeTwORks), funded by the French National Research Agency and showing how numerical modelling can contribute to measurements in sewer systems (MENTOR, 2016). In brief, CFD allows the simulation of flows at a potential measurement site described by its geometry, by the range of flows to be measured and by the range of variation of the different boundary conditions (other than the flow to be measured, and often confined to water levels). Two approaches can be defined. The first one is a generic approach for studying the sensitivity of the measured value and its associated error to measurement conditions. The second approach is a modelling of the combination "flows + sensors" specific to a given measurement site. A digital calibration can then be performed, i.e. a relationship between the measured quantities (height and velocity) and the flow rate that one wants to know. It is not easy to judge the quality of the results from a CFD simulation as the colourful pictures can look very convincing, while in reality the relation with the real world can be weak. It requires either the judgement of a very experienced hydraulic engineer and/or a well-documented validation of the CFD results. In very critical cases or when large investments are to be based on knowledge of the local hydraulics, the reader is well advised to ask for a validation the CFD results against measured data obtained e.g. from a physical scale model.

3.4.6 Summary

The main advantages and disadvantages of the methods that are most commonly used to measure discharges in sewer systems are summarized in Table 3.5.

3.5 INFILTRATION AND EXFILTRATION

3.5.1 Introduction

In- and exfiltration of water into and out of urban sewer systems is known to be an issue from many perspectives. These processes are unwanted as they negatively influence the system performance and increase the operational costs. In addition, local infiltration of groundwater may easily cause the occurrence of sinkholes in urban areas posing a risk of material damage and/or unsafe situations for the public. On the other hand, the on-purpose infiltration processes occurring in stormwater source control measures or sustainable urban drainage systems (SUDS) are essential for the performance of such infrastructures. In both cases, it is importance to estimate these quantities when managing UDSM systems.

This section focuses on underground sewer systems, infiltration measurements in SUDS are discussed in Chapter 4.

Table 3.5 Advantages and disadvantages of various methods for direct discharge measurements.

Method	Advantages	Disadvantages
Pre-calibrated devices	• Wide range of systems	• Require stable hydraulic conditions • Sensitive to clogging and sediment deposits
Physical scale model	• Allows real physical study at lower scale	• Requires the construction of the model • Similarity issues
Tracer experiments	• Measurement fully independent of the *in situ* sensors (water level, velocity) • Independent of hydraulic conditions	• Non-continuous measurements
Pumping station	• Already built, no need for additional sensors	• Effect of ageing on measured values • Requires *in situ* verification
CFD modelling	• Numerical method without measurement	• Requires expertise and computational power • Careful attention to hydraulics is mandatory

Leakage (in- or exfiltration) in underground sewer systems is notoriously tedious to detect/quantify. This especially holds true for exfiltration out of non-pressurized systems like gravity sewers. The main questions to be answered when monitoring in- and exfiltration are:

- Is there a leak?
- How much is leaking?
- Where is the leak located?

These questions can be answered at different spatial and time scales:

- At catchment scale (long term ~ years).
- At pipe scale (short term ~ days).
- At the pipe section scale (short term ~ days).

3.5.2 Large scale measurement of infiltration

Infiltration of groundwater into gravity sewers poses a serious problem. In delta areas for example, the relative contribution of groundwater to the daily water volume treated in wastewater treatment plants (WWTPs) can mount up to 70% for ageing systems. In such a situation, it is of importance to find out which urban areas are responsible for this. A variety of relatively simple methods, known as triangle methods, have been proposed and applied to estimate the amount of parasitic water being collected and transported to the WWTP for treatment. All these methods follow more or less the approach outlined by Weiß *et al.* (2002).

The basic required input data are:

- The catchment.
- The daily amount of water discharged from the catchment for a given period of time (typically months to years).

- The amount of wastewater produced in the catchment as a daily average over a long period (typically months to years).
- Daily information on rainfall occurrence over the catchment.

From these data, a graph as shown in Figure 3.46 is produced.

The procedure is simple:

- Plot the days in order of ascending daily volume.
- Label each day with a precipitation larger than a certain threshold (e.g. 2–3 mm/day) as 'wet', and the other days as 'dry'.
- Determine the theoretical discharge of wastewater in the catchment (water supply data, possible groundwater extraction by industries).
- Construct a graph as shown in Figure 3.46.
- Determine the area labelled as 'parasitic water' in Figure 3.46. Note, the term 'parasitic water' is preferred over 'infiltrating water' as the excess volume may be caused by other mechanisms than infiltration alone (e.g. illegal discharges, leaking return valves, surface water pouring in via weirs that have a too low crest level etc.).

Obviously, this method cannot be considered very accurate, as many implicit assumptions are made. Nevertheless, it offers a relatively quick and cheap method to find out whether or not infiltration is a significant issue in a given catchment. Schilperoort (2004) and Schilperoort *et al.* (2007) report systematic deviations up to 50% when compared to results obtained with methods based on natural water isotopes. An elaborate description of the natural water isotope method can be found in e.g. De Bénédittis and Bertrand-Krajewski (2005).

Once a catchment is found to contribute significantly to infiltration, a certain refinement as to which locations are most likely to contribute to the infiltration can be made, using readily available data (in conjunction with GIS applications). In most cases, data on each conduit regarding dimensions, material, year of construction and geotechnical conditions are known.

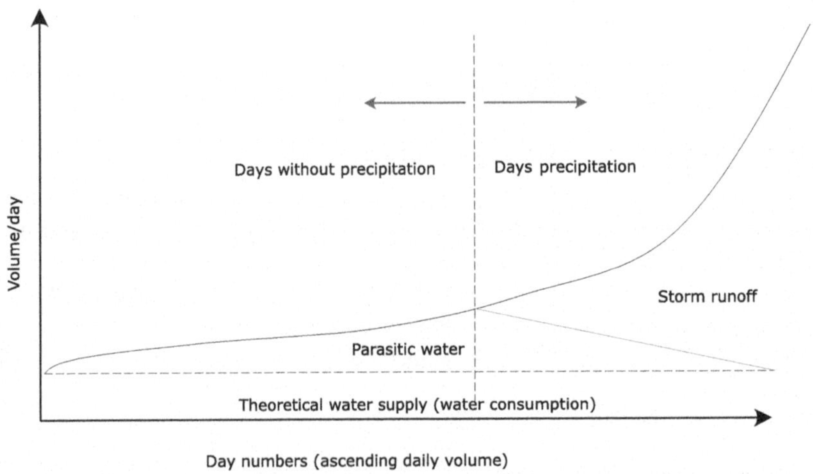

Figure 3.46 Scheme of the 'triangle method' as proposed by Weiß *et al.* (2002). *Source*: Francois Clemens-Meyer (Deltares/TU Delft/NTNU).

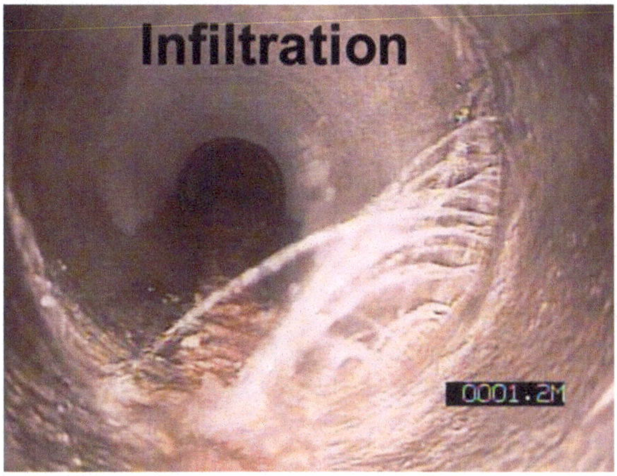

Figure 3.47 Clear example of infiltration of groundwater into a sewer pipe. *Source*: Deltares.

The first conduits to scrutinize are pipes situated below groundwater level, being of a minimum age (e.g. >30 years of operational service), known to have short pipe section lengths (this implies more joints), and being the suspected cause of reports on the occurrence of sinkholes and other citizen complaints. This selection of conduits may be subject to further investigation using more advanced technologies to detect leakage (e.g. closed circuit television (CCTV) inspections).

3.5.3 Detailed monitoring of in- or exfiltration

3.5.3.1 General

A first and obvious method to consider for detailed inspection of in- or exfiltration is CCTV inspection. This method, however, has some severe limitations:

- Quantification of the amount of leaking water is not possible.
- Exfiltration can only be detected indirectly (presence of displaced joints, cracks or root intrusion only hint at the possibility of exfiltration).
- Infiltration can be observed directly in a limited number of cases only, when occurring above water level and in a quantity that can be described as 'pouring' (Figure 3.47).
- CCTV inspection is known to suffer from serious quality flaws (Dirksen *et al.*, 2013).

CCTV can provide a first view of the overall status of the assets but is not a guarantee of obtaining high-quality information on where leakages occur or on their quantity. A range of alternative methods have thus been developed to detect and quantify in- and exfiltration from underground urban water systems. In the course of time, many technologies have been developed and applied in practice with various levels of success.

It is not intended to supply here a comprehensive and detailed overview of all technologies. A brief description along with some literature entries is given. It has to be stressed that, as not all technologies presented here have been subjected to rigorous scientific scrutiny yet, not all claims on the quality of information obtained can be taken to be correct.

3.5.3.1.1 Listening stick

The easiest method to use is the listening stick. This stick has an earpiece and is used to listen to the sound generated by leaks on e.g. pipe fittings (e.g. Glisic, 2014). An obvious drawback of this method is the fact that a human is used as an 'organic sensor', which is known to suffer from subjectivity and sometimes even bias (see e.g. Miller, 1956; Macmillan & Creelman, 2005). Apart from that, it is unlikely that all leakages can be detected in this manner as access to underground systems to apply the method is a serious obstacle.

3.5.3.1.2 Smoke testing

A relatively old but simple method is the smoke test. Smoke is injected into the sewer at a manhole. If there is a crack or a leak above the waterline the smoke is likely to show up at the surface (Figure 3.48).

3.5.3.1.3 Dye testing

Another simple method is the dye tracer test. A dye (e.g. Rhodamine WT) is injected into the sewer and the dilution of the dye is measured (Gokhale & Graham, 2004).

3.5.3.1.4 Flow monitoring

The simplest flow monitoring method is manual survey. During night-time, manholes are inspected in the presence of a significant flow (Gokhale & Graham, 2004). Alternatively, this flow could also be measured with two or more flowmeters to estimate the difference. These are however prone to unreliable results under specific circumstances (Smits *et al.*, 2008). Discharge measurements could however also be applied at a larger (e.g. network) scale to set up a water balance. To this end, discharge measurements from pumping stations could be compared to drinking water consumption, the water balance deficiency is a measure for in- or exfiltration (e.g. Korving *et al.*, 2012). Using this method one should be aware that not all water supplied into a household or an industrial process is discharged into the wastewater system, evaporation,

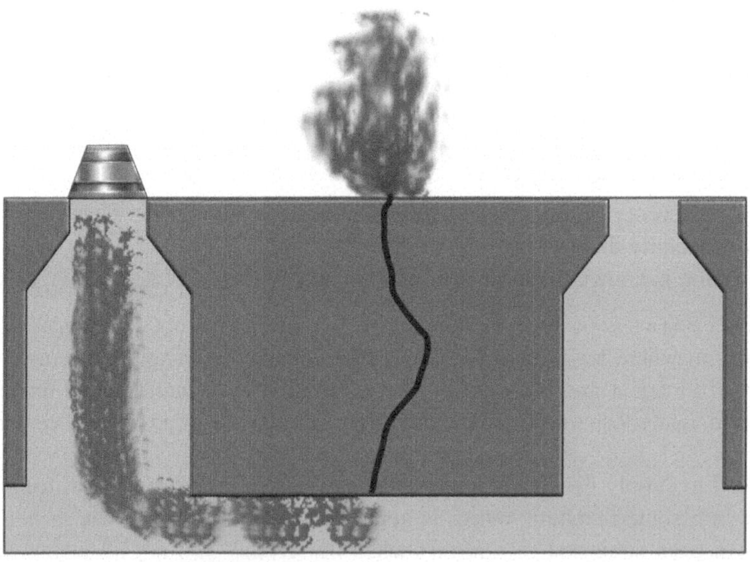

Figure 3.48 Principle of the smoke test method. *Source*: adapted from Gokhale & Graham (2004).

water being an ingredient of products made and garden watering cause a systematic deficiency. The same holds for water discharged into the wastewater systems that came from another source than the water supply system: private wells, companies having their own groundwater source. A comprehensive survey of water sources and water users in a given catchment is a mandatory action when considering flow monitoring as a means to detect in- or exfiltration.

3.5.3.1.5 Pressure test

For a pressure test, a (part of the) pipeline is isolated, and the internal pressure is set at a specific value. This pressure has to remain constant for a certain period (e.g. 30 minutes). During this period, the variation of the volume of water in the pipe is measured (EN, 2015). As an alternative, a prototype named the double packer consists of two inflatable discs to seal a pipe section of 80 cm. Freshwater is subsequently added to the interspace. Exfiltration is determined by monitoring the water volume in the interspace (Wolf, 2006).

3.5.3.1.6 Tracing methods

In the EU Project APUSS (Ellis & Bertrand-Krajewski, 2010), the QUEST (Rieckermann *et al.*, 2005) and QUEST-C (QUantification of Exfiltration from Sewers using artificial Tracers – with Continuous dosing) (Rieckermann *et al.*, 2007) methods have been developed. Artificial tracers are added to the wastewater flow to quantify exfiltration in a sewer reach by means of the quantification of the loss of tracer mass. The APUSS project has also developed various methods to quantify infiltration. One of these methods is the stable isotope method (Kracht *et al.*, 2007), which compares the isotopic compositions of the foul sewage and the infiltrating water.

3.5.3.1.7 Sampling and modelling

Sampling can also be used to trace leakage. Soil samples near the pipeline are collected to be analysed for parameters related to deterioration, e.g. contaminants or moisture content (Liu & Kleiner, 2013). Samples of wastewater and groundwater are analysed for drug remains to calibrate a groundwater flow model to estimate exfiltration (Fenz *et al.*, 2005). More recent research (Guérineau *et al.*, 2014) combined surface water quality modelling with analysis of surface water and sediment samples for *E. Coli* and wastewater micropollutants to estimate the amount of sewer exfiltration into a surface water canal.

3.5.3.1.8 Distributed temperature sensing

Distributed temperature sensing (DTS) is a technology developed in the 1980s (e.g. Dakin *et al.*, 1985) and introduced in hydrology and urban drainage in the past decade for, amongst other applications, detecting and locating infiltration (Hoes *et al.*, 2009; Panasiuk *et al.*, 2017; Schilperoort *et al.*, 2013). In this method, an in-sewer fibre optic cable is installed to conduct high-frequency temperature measurements over a long stretch of sewer pipes. This detects and locates temperature anomalies due to exchange of groundwater with wastewater. To detect and locate exfiltration and leakage from a pressure main, a fibre optic cable could be placed in the pipeline bed to monitor the temperature differences (Nikles *et al.*, 2004).

The measuring principle is based on the shift in wavelength of scattered laser light sent down a glass fibre. In Figure 3.49(a), a schematic representation of the measuring set-up is shown (Lopez-Higuera, 2002). A laser sends a light pulse through a directional coupler into a glass fibre. In the glass fibre, two types of scattering occur: Raman scattering (induced by molecular vibrations) and Brillouin scattering (bulk vibrations). The light is scattered back and recombined in the directional coupler for processing in an electro-optical device to quantify the shift(s) in wavelength. Figure 3.49(b) shows a typical result of the shifts occurring in the wavelength observed. The travelling time is measured as well and is used (Rayleigh scattering, assuming the speed of light to be known and constant) to identify the distance from

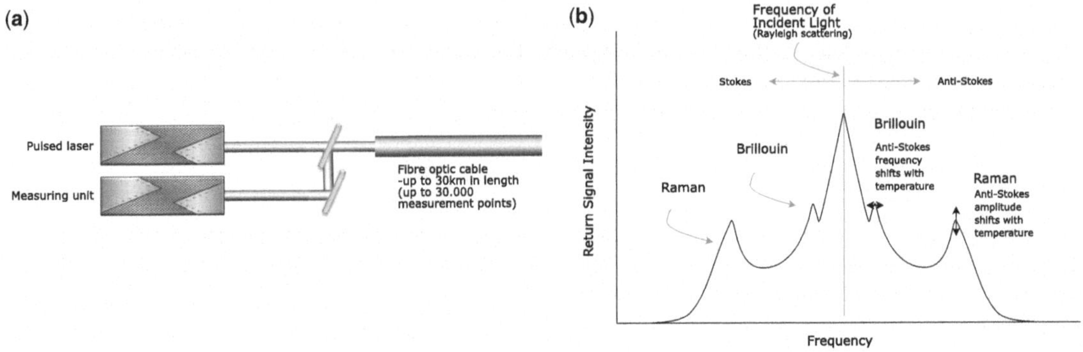

Figure 3.49 (a) DTS measuring principle, (b) Rayleigh, Stoke, anti-Stokes, Raman and Brillouin scattering. *Sources*: (a) adapted from https://roctest.com/wp-content/uploads/2017/01/SEN2-Manual1.0c.pdf, (b) adapted from https://www.bandweaver.com/technology/distributed-temperature-sensing/.

the light source where the scattering occurred. The Rayleigh backscatter has the same wavelength as the original signal and is used for distance determination. The Raman backscatter show a double shift in wavelength: (i) toward the red, with longer wavelength (named Stokes) and (ii) toward the blue, with shorter wavelength (named anti-Stokes). The ratio of the intensity peaks for the Stokes and anti-Stokes is a measurement of the temperature (for the underlying physical process details, the reader is referred to the literature, e.g. Lopez-Higuera, 2002). Brillouin scatter depends on bulk vibration, potentially the absolute difference between respectively Brillouin/Stokes-Rayleigh and Brillouin/anti-Stokes-Rayleigh can be used to detect pressure vibrations (sound) produced by small leakages in pressure mains (Sliwczynski & Krehlik, 2014).

A typical measuring result obtained from a DTS measuring campaign is shown in Figure 3.50.

When groundwater is infiltrating in a wastewater sewer and there is a temperature difference that is larger than the resolution of the measuring device (approximately 0.1 °C), the location of the infiltration can be determined. In modern systems, the uncertainty in location is in the order of 10 cm, and the uncertainty in the temperature is ~0.1 °C, depending on duration of the measurement and the feasibility given the location. In general, a trade-off has to be made between time and space resolution, uncertainty level and the effect of water velocity. On a sense it resembles the situation when taking photo: one has to find a workable combination of shutter speed, aperture and the depth of focus/sharpness of the picture sought when targeting a moving object. In practice the main limiting factor is the amount of data generated (up to several Gbytes per day) when it comes to choosing these parameters.

Prior to any monitoring campaign, the DTS cable has to be calibrated after it has been installed in the system (Figure 3.51). By inserting the cable at a number of locations (preferably at accessible locations like manholes) in a bucket filled with melting ice, the offset (deviation between the reading obtained with the cable and the known temperature of the ice (0 °C) is known and is used as a systematic correction on the raw data in the data processing. In more recent equipment, there is the option of connecting Pt100 thermometers to enable end-to-end calibration of the cable. The reader is referred to Section 6.4.2 for practicalities regarding DTS.

3.5.3.1.9 Time domain reflectometry

Time domain reflectometry (TDR) as described by Cataldo *et al.* (2014) could be applied to detect and locate leakage. In this method, a sudden voltage increase propagates along a sensor or sensing element (e.g.

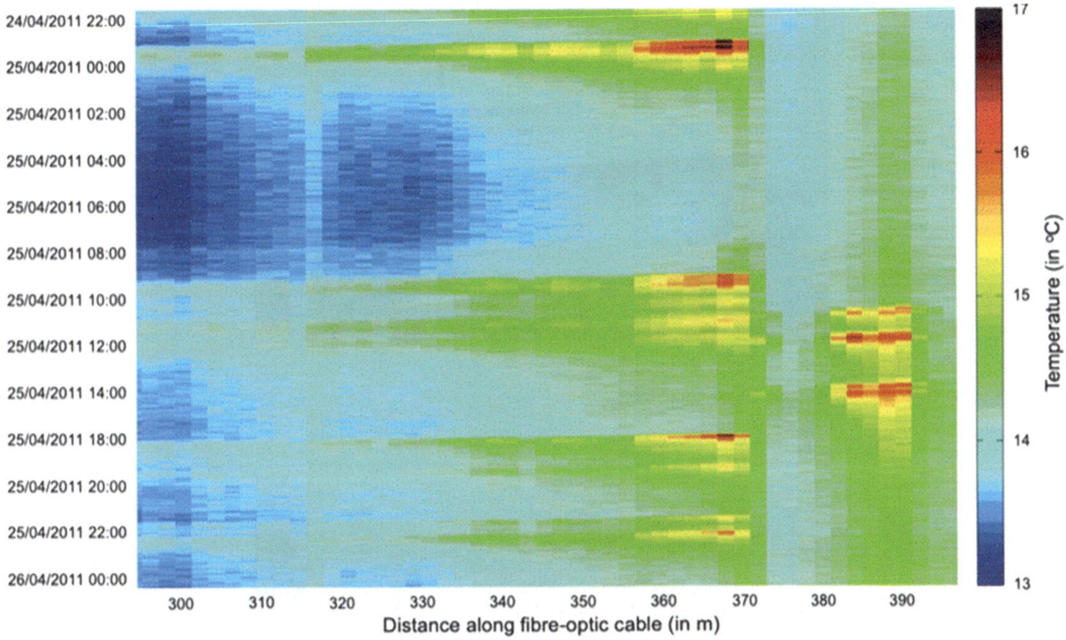

Figure 3.50 Example of a measuring result obtained with the DTS method in a storm sewer. At a distance of circa 367 m and 390 m, regular discharges of relatively warm water occur. This hints at a misconnection at those locations. *Source*: courtesy Rémy Schilperoort (Partners4UrbanWater).

two-wire transmission line) which could be installed at the pipe bedding. The variations of the electrical impedance, which is influenced by leakage, that are encountered along the way are monitored and schematized in a reflectogram.

3.5.3.1.10 Infrared thermography

Infrared thermography is a technique that can also be used to detect leaks and voids in the surrounding soil from the ground level (e.g. Wirahadikusumah *et al.*, 1998). This method detects temperature differences that occur as a result of the exchange between the pipeline and the surrounding soil. Lepot *et al.* (2017) demonstrated that infiltration through a crack can be detected using in-line infrared thermography.

3.5.3.1.11 Smartball

The Smartball is an acoustic concept. A ball equipped with acoustic sensors, an accelerometer, and a temperature and pressure sensor is inserted at an upstream part of the pipe system and flows downstream. The location of the ball and possible leaks are determined by analysing the emitted acoustic signal which is collected at the ground station (Liu & Kleiner, 2013).

3.5.3.1.12 SAHARA

Another acoustic concept is the SAHARA system. In this system, a sensor is mounted on an umbilical cable which is inserted at an upstream point of the pipe. The sensor, a hydrophone, is equipped with a small parachute which unfolds in the pipeline to let the sensor flow downstream (Figure 3.52). The

Figure 3.51 (a) DTS measuring computer, (b) Installation of a DTS cable. *Source*: courtesy Rémy Schilperoort (Partners4UrbanWater).

hydrophone can detect the sound which is generated by the leak. Subsequently, the location of the leak is recorded with a receiver at ground level (Rizzo, 2010).

3.5.3.1.13 Leak noise correlation

Leak noise correlators can also be used to detect and locate leakage. Hydrophones up and downstream of a possible leak can be used to listen to noise generated by the leak. Subsequently, the leak position is identified by the delay between the leak noise reaching each monitoring point (Davis *et al.*, 2013; Hao *et al.*, 2012).

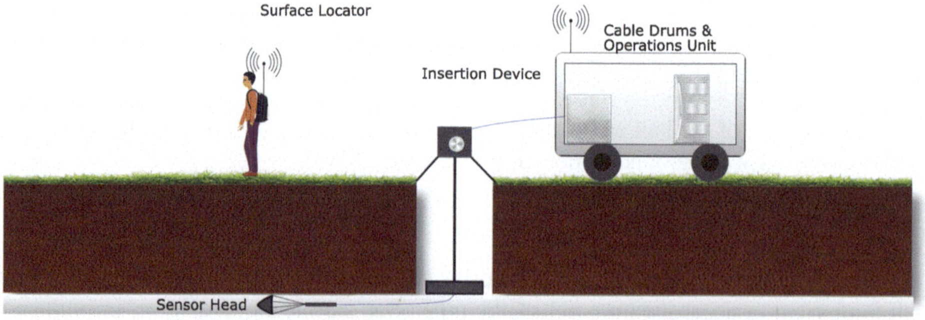

Figure 3.52 Principle of the SAHARA system. *Source*: adapted from Rizzo (2010).

3.5.3.1.14 Magnetic flux leakage

Electromagnetic methods are often used in pressurized systems. The magnetic flux leakage method uses large magnets to create a magnetic field around the pipe wall. This is applicable only to ferrous pipes. Defects are detected by measuring changes in the pipe magnetic permeability (Rizzo, 2010). A magnetic flux leakage unit is usually mounted on a pig (pipeline inspection gauges). These 'intelligent' pigs (Figure 3.53) are mainly used in oil pipelines, but there is an increasing interest in application in wastewater pressure mains (Driessen, 2016).

3.5.3.1.15 Ground penetrating radar

The ground penetrating radar (GPR) is also an electromagnetic method. GPR uses electromagnetic radiations in the microwave band. With a transmitter, microwaves are sent through materials of different dielectric constants to detect reflected signals from the subsurface. GPR can detect voids in and changes in soil saturation. Traditionally, GPR is used from the ground surface towards the soil, but in-pipe GPR systems also exist (Hao *et al.*, 2012; Liu & Kleiner, 2013).

3.5.3.1.16 Multi-sensor systems

Multi-sensor systems have also been developed over the years. In the PIRAT system (Kirkham *et al.*, 2000), CCTV is combined with sonar and laser profiling on a robot. Sonar is an acoustic technique that can be used to identify sediments and cracks below the water line. For inspection of cracks above the waterline, laser profiling can be used (Selvakumar *et al.*, 2014). In this technique a ring of light is projected on the sewer wall.

Another example of a multi-sensor system mounted on a pipe robot is SAM (Eiswirth *et al.*, 2000). This system uses, next to CCTV and a laser scanner, a microwave sensor (in-pipe GPR) and radioactive sensors (γ-g probe). These sensors are used to inspect the soil behind the pipe for changes in moisture content and voids. The γ-g probe acts as a source of gamma radiation. The backscatter of gamma rays, together with natural radiation, is recorded by the probe. The backscatter can be related to the density of the surrounding soil. Voids change the density and can therefore be detected (Heske, 2003). The rotatable microwave sensor records changes of the dielectric constant, which are different for dry and wet soils (Munser & Hartrumpf, 2003). Hydrochemical sensors (conductivity, pH and temperature) and an acoustic sensor (Herbst, 2002) are also installed on the robot to detect cracks and voids, based on impact echo. Sound waves are introduced in a concrete pipe wall with e.g. an automatic hammer.

Figure 3.53 Example of an intelligent pipeline inspection gauge. *Source*: courtesy John Driessen (Sweco).

The waves reflected by internal flaws and external surfaces are subsequently recorded. A geo-electrical sensor is installed on a cable attached to the back of the robot to detect and locate leakages (Wolf, 2006). The latter is now commonly referred to as focused electro leak location (FELL) or electro-scanning.

Tracer-based and electrical conductivity-based methods are discussed in some length hereafter.

3.5.3.2 Electrical conductivity-based methods

Electrical methods rely on the fact that when leakage occurs, be it in- or exfiltration, this implies that the electrical resistance of the pipe wall is locally reduced. In this category, the following methods are reported:

- Focused electro leak location (FELL).
- Electrical resistivity tomography (ERT).

Heterogeneity of the underground in urban areas is a challenging environment, as multiple sources of noise may be present while being unnoticed. Mainly for this reason, no reliable generic information on detection limits, uncertainty or repeatability of these methods is known. No substantiated claims on uncertainty levels in measured in- or exfiltration discharges is available either.

3.5.3.2.1 Focused electro leak location (FELL)

The focused electro leak location (FELL) system has been developed in Germany within the interdisciplinary project SAM (Sewer Assessment with Multi-sensor Systems) to detect and locate leakages (Eiswirth *et al.*, 2000). An electric potential is applied between an electrode in the pipe (the sensor), and an electrode on the surface (Figure 3.54). The sensor is located below the waterline and to complete the electrical circuit the surface electrode is often a metal stake (e.g. Wilmut *et al.*, 2005). The electrical resistance of the pipe wall is high, unless there is a defect in the pipe such as a defective joint or a crack. The resulting increase in current is registered and coupled to the location of the sensor.

The main part of the FELL method is the geo-electrical sensor. Gokhale and Graham (2004) describe a commercialized version of this technique, the FELL-41 system. Since then some alternative systems with

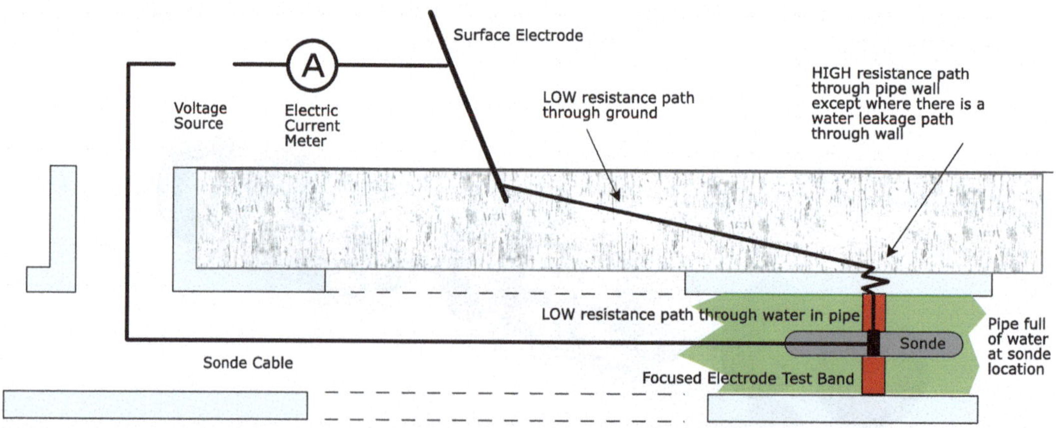

Figure 3.54 Principle of the focused electro leak location (FELL) method. *Source*: adapted from ASTM (2018).

the same measurement principle have also been developed. Next to the FELL-41 system, the MSI-1620 system is used by Tuccillo *et al.* (2011) as an alternative model.

Until now FELL has been applied in different studies to detect and locate leakage. Harris and Dobson (2006) compared FELL with joint pressure testing and CCTV. They concluded that FELL results coincide with joint pressure testing. However, the FELL results showed significantly more defective joints when compared to the CCTV, a result later confirmed by Tuccillo *et al.* (2011).

3.5.3.2.2 Electrical resistivity tomography (ERT)

Electrical resistivity tomography (ERT) is an active geo-electrical method that calculates the subsurface distribution of electrical resistivity from a large number of resistance measurements (Daily *et al.*, 2005). The subsurface resistivity is determined by applying a known electrical direct current between two electrodes. The ground resistivity is related to various geological parameters such as the mineral and fluid content, porosity, and degree of water saturation in the rock (Loke, 2020).

A range of electrode configurations is applied, also known as arrays. Different modes of deployment are available, e.g. vertical electrical sounding (VES) to determine the vertical variation in resistivity. Investigating the vertical and horizontal variation in resistivity along an investigation line is known as electrical resistivity tomography (ERT) (Figure 3.55). This 2D electrical survey assumes that resistivity does not change in the direction that is perpendicular to the survey line. To get more accurate results, a predefined grid of multiple survey lines could be combined to obtain a pseudo-3D plot (Reynolds, 2011).

Ramirez *et al.* (1996) used ERT with a combination of surface and subsurface electrodes to detect leakage from an underground tank in a test facility. The authors successfully monitored the movement of a released saline tracer from the tank in 2D, 3D and in time.

Jordana *et al.* (2001) applied ERT using surface electrodes to detect water leakages from a buried pipe in an experimental set-up and in a short pipeline buried in a farm field. Wood and Palmer (2000) investigated sewer exfiltration in Sydney with the mise-à-la-masse method in combination with ERT along a single line. They applied both methods at four different test-sites: two sewers located in open, overgrown areas allowing ease of access, and two sewers buried under roads. They concluded that in areas with large surface variations, such as might be associated with sewerage pipes laid under roads, the combination of the two methods to detect exfiltration may be ambiguous. Eiswirth *et al.* (1994) also used 2D ERT to investigate sewer exfiltration in a combination of flush experiments at a gravity sewer test-site in Rastatt, Germany. A 2D ERT profile was obtained before and after a flush experiment with a saline solution. Results of the study indicate they were able to detect exfiltration at multiple (12 out of 15) known defect locations.

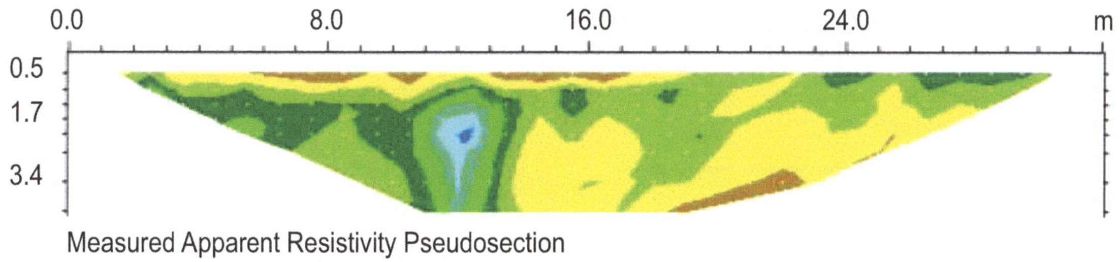

Figure 3.55 Example of apparent resistivity pseudo section. *Source*: adapted from Loke (2020).

3.5.3.3 *Tracer methods*

The basic principle of exfiltration measurements with tracers is to dose a well-known amount of tracer injected into the sewer under investigation and to apply a mass balance in the investigation reach. Given the conservative behaviour of the substance, the tracer loss can directly be related to the leakage along the reach.

3.5.3.3.1 QUEST-C method

The QUEST-C method uses two different tracers, which are continuously dosed upstream and downstream of the reach under investigation. The underlying principle is that losses of the indicator tracer are mostly identified relative to a reference tracer which is not affected by exfiltration (Figure 3.56).

The straight-forward QUEST-C method assumes steady discharge in the sewer system during the experiment. Exfiltration is computed only from the two series of tracer concentration data. The dynamic approach of the QUEST-C uses the ratio of tracer loads. To this end, simultaneous discharge measurements at the sampling point are mandatory. The exfiltration fraction E can be estimated by Equation (3.25):

$$E = 1 - \frac{M_{REF,in}}{M_{IND,in}} \cdot \frac{M_{IND,out}}{M_{REF,out}} = \frac{\int c_{REF}(t) \cdot q_{REF}(t)dt}{\int c_{IND}(t) \cdot q_{IND}(t)dt} \cdot \frac{\int C_{IND}(t) \cdot Q(t)dt}{\int C_{REF}(t) \cdot Q(t)dt} \tag{3.25}$$

where c_{REF} and c_{IND} are the respective reference and indicator tracer concentrations of the dosing solutions, q_{IND} and q_{REF} are the respective dosing rates of the tracer solutions, C_{IND} and C_{REF} are the respective tracer concentrations in the sample, and Q is the discharge at the sampling point.

However, the computed exfiltration is systematically wrong if there is significant inflow/infiltration in the investigation reach.

In the APUSS project, the QUEST-C method was solely applied in field tests. This provides no full formal validation. Different tracer combinations were used: LiCl, NaBr (Prigiobbe & Giulianelli, 2011; Rieckermann *et al.*, 2007) and Rhodamine WT (Revitt *et al.*, 2006).

3.5.3.3.2 *DEST method*

Application of the DEST (detection of exfiltration from sewers using tracers) method implies a well-known mass of a single tracer is injected at an upstream point of the investigation reach. To complete the balance, downstream of the reach or system under investigation, the remaining tracer mass is determined. To this end, discharge is measured and combined with concentration measurements. By doing so, exfiltration can still be

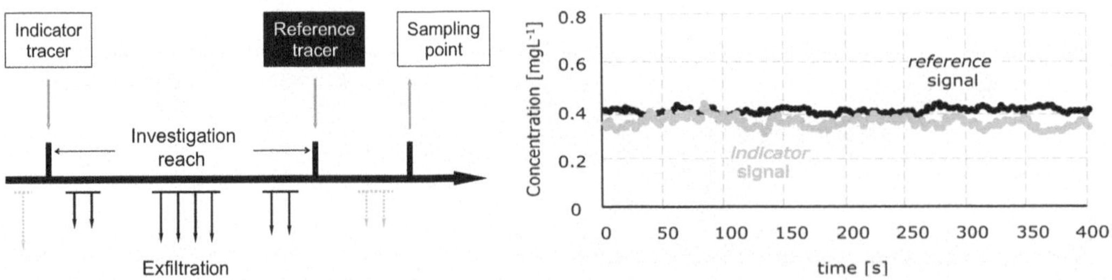

Figure 3.56 Concept scheme of the QUEST-C method. *Source*: adapted from Rieckermann *et al.* (2007).

detected if there is a significant inflow with a background tracer concentration in the sewer of interest. The exfiltration fraction E can be estimated by Equation (3.26):

$$E = \frac{M_{in} - M_{out}}{M_{in}} = 1 - \frac{\int Q(t)(C(t) - C_{BG}(t))dt}{V_{in}C_{in}} \tag{3.26}$$

where V_{in} is the dosed volume, C_{in} is the dosed tracer concentration, C is the tracer concentration in the sample, C_{BG} is the background concentration and Q is the discharge at the sampling point.

3.6 SUMMARY AND TRANSITION

This chapter introduced several sensors and methods to estimate or measure flows and validate recorded discharges data. Such measurements are conducted in many urban drainage systems in the world. They may look easy and well established, but they are often prone to error and biases that are not always obvious. Even when applying methods described in standards, one needs to be aware of the limitations posed by the conditions in field measurements on the applicability.

The combination of a water level and velocity measurement is likely the most used method in sewer pipes, but careful attention is mandatory in choosing and installing the devices. Section 6.3 of this book highlights some key points for such installations. Pre-calibrated devices are also well established. If these technologies are robust, they require detailed checks to set them up.

SUDS are slowly and surely becoming an alternative for stormwater management in urban catchment and/or remote places. Monitoring such decentralized systems poses different challenges: specific processes (evaporation, infiltration, low discharges) occur, and the constraints are slightly different (cost, number of sensors to set up, energy supply). The following chapter is devoted to those issues.

REFERENCES

Aguilar M. F., McDonald W. M. & Dymond R. L. (2016). Benchmarking laboratory observation uncertainty for in-pipe storm sewer discharge measurements. *Journal of Hydrology*, **534**, 73–86. doi: 10.1016/j.jhydrol.2015.12.052.

ASTM (2018). *ASTM F2550-13(2018) Standard Practice for Locating Leaks in Sewer Pipes by Measuring the Variation of Electric Current Flow Through the Pipe Wall.* ASTM International, West Conshohocken, PA (USA), 7 p. doi: 10.1520/F2550-13R18.

Banque Hydro (1998). *Hydrometry Quality Charter – Guide to Good Practices.* Ministry of Environment, Paris (France), 31 p. Available at https://library.wmo.int/doc_num.php?explnum_id=4709 (accessed 21 Nov. 2020).

Bertrand Krajewski J.-L., Laplace D., Joannis C. & Chebbo G. (2000). *Mesures en hydrologie urbaine et assainissement [Measurements in urban hydrology and sewer systems].* Editions Technique et Documentation, Paris (France), 794 p. ISBN 978-2-7430-0380-4. (in French).

Bevir M. K. (1970). The theory of induced voltage electromagnetic flowmeters. *Journal of Fluid Mechanics*, **43**(3), 577–590. doi: 10.1017/S0022112070002586.

Bonakdari H. (2006). *Modélisation des écoulements en collecteur d'assainissement – Application à la conception de points de mesures [Modelling of flows in sewer pipes: application to the design of measurement points].* PhD thesis, Université de Caen – Basse Normandie, Caen, France, 263 p. (in French).

Bonakdari H., Larrarte F., Joannis C. & Levacher D. (2008). Methodology for qualifying measurement sites within a drainage network: application to flow readings in a main drain. *Bulletin des Laboratoires des Ponts et Chaussées*, **272**, 9–19.

Cataldo A., Persico R., Leucci G., de Benedetto E., Cannazza G., Matera L. & de Giorgi L. (2014). Time domain reflectometry, ground penetrating radar and electrical resistivity tomography: A comparative analysis of alternative approaches for leak detection in underground pipes. *NDT and E International*, **62**, 14–28. doi: 10. 1016/j.ndteint.2013.10.007.

Cedillo S., Lepot M. & Clemens F. H. L. R. (2016). Potentiality of a velocity profiler to investigate sewers: results of laboratory experiments. *Proceedings of the 8th International Conference on Sewer Processes and Networks*, 31 Aug.–2 Sept., Rotterdam, The Netherlands, 42–43.

Clemens F. H. L. R., Stanic N., van der Schoot W., Langeveld J. G. & Lepot M. J. (2015). Uncertainties associated with laser profiling of concrete sewer pipes for the quantification of the interior geometry. *Structure and Infrastructure Engineering*, **11**(9), 1218–1239. doi: 10.1080/15732479.2014.945466.

Colin D., Herault C.-A. & Venandet N. (2016). *Guide pratique : mise en place de l'autosurveillance des réseaux d'assainissement [Practical guide: implementing self-monitoring of sewer systems]*. Agence de l'eau Rhin Meuse, Rozérieulles (France), 164 p. (in French). Available at http://cdi.eau-rhin-meuse.fr/Record.htm?idlist=4&record=19319019124911372919. (accessed 08 Dec. 2020).

Daily W., Ramirez A., Binley A. & LaBrecque D. (2005). Chapter 17 - Electrical resistance tomography - Theory and practice. In *Near-surface Geophysics (Investigations in Geophysics No. 13)*, D. K. Butler (ed.), Society of Exploration Geophysicists, Tulsa, OK (USA), pp. 525–550. ISBN 978-156080-130-6. doi: 10.1190/1. 9781560801719.ch17.

Dakin J. P., Pratt D. J., Bibby G. W. & Ross J. N. (1985). Distributed optical fibre Raman temperature sensor using a semiconductor light source and detector. *Electronics Letters*, **21**(13), 569–570. doi: 10. 1049/el:19850402.

Davis P., Sullivan E., Marlow D. & Marney D. (2013). A selection framework for infrastructure condition monitoring technologies in water and wastewater networks. *Expert Systems with Applications*, **40**(6), 1947–1958. doi: 10. 1016/j.eswa.2012.10.004.

De Bénédittis J. & Bertrand-Krajewski J.-L. (2005). Measurement of infiltration rates in urban sewer systems by use of oxygen isotopes. *Water Science and Technology*, **52**(3), 229–237. doi: 10.2166/wst.2005.0080.

Dirksen J., Clemens F. H. L. R., Korving H., Cherqui F., Le Gauffre P., Ertl T., Plihal H., Mueller K. & Snaterse C. T. M. (2013). The consistency of visual sewer inspection data. *Structure and Infrastructure Engineering*, **9**(3), 214–228. doi: 10.1080/15732479.2010.541265.

Diskin M. H. (1963). Temporary flow measurement in sewers and drains. *Journal of the Hydraulics Division*, **89**(4), 141–159.

Driessen J. (2016). Onderzoek naar intelligente pigs voor persleidingen water [Research into intelligent pigs for water pressure pipes]. *Land+Water*, **5**, 34–35. (in Dutch).

Dufresne M., Vazquez J. & Bercovitz Y. (2018). Complementarities between physical modelling and computational fluid dynamics for an ecological continuity project. *Proceedings of 7th IAHR International Symposium on Hydraulic Structures*, 15–18 May, Aachen, Germany, 8 p. doi: 10.15142/T3V07R.

Eiswirth M., Heske C., Hötzl H., Schneider T. & Burn L. S. (2000). Pipe defect characterisation by multi-sensor systems. *Proceedings of the conference No-dig 2000*, 15–19 Oct., Perth, Australia, 12 p.

Eiswirth M., Lazar C. & Merkler G. P. (1994). Sewerage leakages as source of groundwater contamination. *Proceedings of the International Hydrogeological Symposium "Impact of Industrial Activities on Groundwater"*, 23–28 May, Constantza, Romania, 175–185.

El Bahlouli A. & Larrarte F. (2018). Proposal for improving discharge quantification in urban drainage. *Flow Measurement and Instrumentation*, **60**, 51–56. doi: 10.1016/j.flowmeasinst.2018.02.014.

Ellis J. B. & Bertrand-Krajewski J.-L. (eds) (2010). *Assessing Infiltration and Exfiltration on the Performance of Urban Sewer Systems (APUSS)*. IWA Publishing, London (UK), 180 p. ISBN 9781843391494.

EN (2015). *EN 1610:2015 Construction and testing of drains and sewers*. CEN – European Committee for Standardization, Brussels (Belgium), 38 p.

Ettema R., Arndt R., Roberts P. & Wahl T. (eds) (2000). *Hydraulic Modelling: Concepts and Practice*. ASCE, Reston, VA (USA), 390 p. ISBN 978-0784404157.

Fach S., Sitzenfrei R. & Rauch W. (2008). Assessing the relationship between water level and combined sewer overflow with computational fluid dynamics. *Proceedings of 11th International Conference on Urban Drainage*, 31 Aug.–5 Sept., Edinburg, Scotland, UK, 9 p.

Fach S., Sitzenfrei R. & Rauch W. (2009). Determining the spill flow discharge of combined sewer overflows using rating curves based on computational fluid dynamics instead of the standard weir equation. *Water Science and Technology*, **60**(12), 3035–3043. doi: 10.2166/wst.2009.752.

Fenz R., Blaschke A. P., Clara M., Kroiss H., Mascher D. & Zessner M. (2005). Quantification of sewer exfiltration using the anti-epileptic drug carbamazepine as marker species for wastewater. *Water Science and Technology*, **52**(9), 209–217. doi: 10.2166/wst.2005.0321.

Glisic B. (2014). Chapter 14 – Sensing solutions for assessing and monitoring pipeline systems. In *Sensor Technologies for Civil Infrastructures – Volume 2: Applications in Structural Health Monitoring*, M. L. Wang, J. P. Lynch & H. Sohn (eds.), Woodhead Publishing Limited, Cambridge (UK), pp. 422–460. ISBN 978-1-78242-242-6. doi: 10.1533/9781782422433.2.422.

Gokhale S. & Graham J. A. (2004). A new development in locating leaks in sanitary sewers. *Tunnelling and Underground Space Technology*, **19**(1), 85–96. doi: 10.1016/j.tust.2003.08.003.

Guérineau H., Dorner S., Carrière A., McQuaid N., Sauvé S., Aboulfadl K., Hajj-Mohamad M. & Prevost M. (2014). Source tracking of leaky sewers: A novel approach combining fecal indicators in water and sediments. *Water Research*, **58**, 50–61. doi: 10.1016/j.watres.2014.03.057.

Hao T., Rogers C. D. F., Metje N., Chapman D. N., Muggleton J. M., Foo K. Y., Wang P., Pennock S. R., Atkins P. R., Swingler S. G., Parker J., Costello S. B., Burrow M. P. N., Anspach J. H., Armitage R. J., Cohn A. G., Goddard K., Lewin P. L., Orlando G., Redfern M. A., Royal A. C. D. & Saul A. J. (2012). Condition assessment of the buried utility service infrastructure. *Tunnelling and Underground Space Technology*, **28**(1), 331–344. doi: 10.1016/j.tust.2011.10.011.

Harris R. J. & Dobson C. (2006). Sewer pipe infiltration assessment: Comparison of electro-scan, joint pressure testing and CCTV inspection. *Proceedings of the Pipeline Speciality Conference 2006*, July 30–Aug. 2, Chicago, USA. doi: 10.1061/40854(211)61.

Heller V. (2011). Scale effect in physical hydraulic engineering models. *Journal of Hydraulic Research*, **49**(3), 293–306. doi: 10.1080/00221686.2011.578914.

Hemmerle N., Randrianarimanana J.-J., Joannis C. & Larrarte F. (2014). Hydraulics and deposit evolution in sewers. *Proceedings of the 9th International Symposium on Ultrasonic Doppler Methods for Fluid Mechanics and Fluid Engineering*, 27–29 Aug., Strasbourg, France, 9–12.

Herbst J. (2002). Non-destructive testing of sewer pipes by an acoustical method. *Proceedings of the 19th IEEE Instrumentation and Measurement Technology Conference*, 21–23 May, Anchorage, USA, 849–853. doi: 10.1109/IMTC.2002.1006952.

Heske C. H. (2003). Hohlraumdetektion in der Umgebung erdverlegter Abwasserkanäle mit Hilfe einer γ-γ-Sonde [Detection of cavities in the environment of burried sewer pipes using a γ-γ-probe]. *Technisches Messen*, **70**(7–8), 377–385. doi: 10.1524/teme.70.7.377.22646.

Hoes O. A. C., Schilperoort R. P. S., Luxemburg W. M. J., Clemens F. H. L. R. & van de Giessen N. C. (2009). Locating illicit connections in storm water sewers using fiber-optic distributed temperature sensing. *Water Research*, **43**(20), 5187–5197. doi: 10.1016/j.watres.2009.08.020.

Hrabak D., Pryl K., Krejcik J. & Richardson J. (1998). Calibration of flowmeters using FLOW-3D software. *Proceedings of Novatech 1998*, 4–6 May, Lyon, France, vol. **2**, 139–144.

Hydraulic Institute (2018). *Rotodynamic Pumps for Pump Intake Design – ANSI/HI 9.8-2018*. Hydraulic Institute, Parsippany, NJ (USA), 132 p. ISBN 978-1-935762-71-3.

ISO (1990). *ISO 4374:1990 Liquid flow measurement in open channels. Round-nose horizontal broad-crested weirs*. ISO – International Organization for Standardization, Geneva (Switzerland), 18 p.

ISO (1999). *ISO 4362:1999 Hydrometric determinations. Flow measurement in open channels using structures. Trapezoidal broad-crested weirs*. ISO – International Organization for Standardization, Geneva (Switzerland), 31 p.

ISO (2003). *ISO 5167-1:2003 Measurement of fluid flow by means of pressure differential devices inserted in circular cross-section conduits running full – Part 1: General principles and requirements*. ISO – International Organization for Standardization, Geneva (Switzerland), 33 p.

ISO (2007). *ISO 748:2007 Hydrometry – Measurement of liquid flow in open channels using current-meters or floats*. ISO – International Organization for Standardization, Geneva (Switzerland), 46 p.

ISO (2008). *ISO 3846:2008 Hydrometry – Open channel flow measurement using rectangular broad-crested weirs*. ISO – International Organization for Standardization, Geneva (Switzerland), 28 p.

ISO (2010). *ISO 15769:2010 Hydrometry – Guidelines for the application of acoustic velocity meters using the Doppler and echo correlation methods*. ISO – International Organization for Standardization, Geneva (Switzerland), 61 p.

ISO (2012). *ISO 4377:2012 Hydrometric determinations – Flow measurement in open channels using structures - Flat-V weirs*. ISO – International Organization for Standardization, Geneva (Switzerland), 59 p.

ISO (2013). *ISO 4359:2013 Liquid flow measurement in open channels – Rectangular, trapezoidal and U-shaped flumes*. ISO – International Organization for Standardization, Geneva (Switzerland), 75 p.

ISO (2017). *ISO 6416:2017 Hydrometry – Measurement of discharge by the ultrasonic transit time (time of flight) method*. ISO – International Organization for Standardization, Geneva (Switzerland), 58 p.

ISO (2020). *ISO 4360:2020 Hydrometry – Open channel flow measurement using triangular profile weirs*. ISO – International Organization for Standardization, Geneva (Switzerland), 30 p.

Ivetić D. (2019). *Assessment of the Liquid Flow Rate in Complex Flow Conditions with Flat Electromagnetic Sensors*. PhD thesis, University of Belgrade – Faculty of Civil Engineering, Belgrade, Serbia (in Serbian).

Ivetić D., Prodanović D. & Stojadinović L. (2018). Bed-mounted electromagnetic meters: Implications for robust velocity measurement in urban drainage systems. *Journal of Hydrology*, **566**, 455–469. doi: 10.1016/j.jhydrol. 2018.08.068.

Ivetić D., Prodanović D., Stojadinović L. & Pavlović D. (2019). Bed-mounted electromagnetic meters: Assessment of the (missing) technical parameters. *Flow Measurement and Instrumentation*, **68**, 101588. doi: 10.1016/j. flowmeasinst.2019.101588.

Jarman D. S., Faram M. G., Butler D., Tabor G., Stovin V. R., Burt D. & Throp E. (2008). Computational fluid dynamics as a tool for urban drainage system analysis: a review of applications and best practice. *Proceedings of the 11th International Conference on Urban Drainage*, 31 Aug.–5 Sept., Edinburg, Scotland, UK, 10 p.

Jaumouillié P., Larrarte F. & Milisic V. (2002). New devices for 2D sampling of velocities and pollutant concentrations in sewers. *Proceedings of the 3rd International Conference on Sewer Process and Networks*, 15–17 April, Paris, France, 171–178.

Jeanbourquin D., Sage D., Nguyen L., Schaeli B., Kayal S., Barry D. A. & Rossi L. (2011). Flow measurements in sewer based on image analysis: automatic flow velocity algorithm. *Water Science and Technology*, **64**(5), 1108–1114. doi: 10.2166/wst.2011.176.

Joannis C. (2001). La mesure de débits en assainissement [Discharge measurement in sewers]. *La Houille Blanche*, **5**, 58–62. doi: 10.1051/lhb/2001057. (in French).

Jordana J., Gasulla M. & Pallàs-Areny R. (2001). Electrical resistance tomography to detect leaks from buried pipes. *Measurement Science and Technology*, **12**(8), 1061–1068. doi: 10.1088/0957-0233/12/8/311.

Kindsvater C. E. & Carter R. W. C. (1957). Discharge characteristics of rectangular thin plate weirs. *Journal of the Hydraulics Division*, **83**(6), 1–36.

Kirkham R., Kearney P. D., Rogers K. J. & Mashford J. (2000). PIRAT – A system for quantitative sewer assessment. *The International Journal of Robotics Research*, **19**(11), 1033–1053. doi: 10. 1177/02783640022067959.

Knight D. & Sterling M. (2000). Boundary shear in circular pipes running partially full. *Journal of Hydraulic Engineering*, **126**(4), 263–275. doi: 10.1061/(ASCE)0733-9429(2000)126:4(263).

Kolin A. (1936). An electromagnetic flowmeter – Principle of the method and its application to blood flow measurements. *Proceedings of the Society for Experimental Biology and Medicine*, **35**(1), 53–56. doi: 10. 3181/00379727-35-8854P.

Kooij C., Mühle S., Clemens F. H. L. R., Pothof I. W. M. & Blokzijl F. H. (2015). Performance indicators for complex wastewater pumping stations and pressure mains. *Proceedings of the 1st International Conference on Industrial Networks and Intelligent Systems (INISCom)*, 2–4 March, Tokyo, Japan, 94–99. doi: 10.4108/icst.iniscom. 2015.25838.

Korving J. L., Admiraal N., Veurink J. & van Bijnen M. (2012). Rioolvreemd water efficiënt opsporen en effectief aanpakken [Find and cure parasitic water]. *H twee O*, **45**(2), 35–37. (in Dutch).

Korving J. L., Clemens F. H. L. R. & van Noortwijk J. M. (2006). Statistical modeling of the serviceability of sewage pumps. *Journal of Hydraulic Engineering*, **132**(10), 1076–1085. doi: 10.1061/(ASCE)0733-9429(2006)132:10(1076).

Kracht O., Gresch M. & Gujer W. (2007). A stable isotope approach for the quantification of sewer infiltration. *Environmental Science and Technology*, **41**(16), 5839–5845. doi: 10.1021/es062960c.

Larrarte F. (2006). *Contributions à la métrologie en réseau d'assainissement [Contributions to metrology in sewer systems]*. Habilitation thesis (HDR), Université de Caen – Basse Normandie, Caen, France, 104 p. + annexes. (in French).

Larrarte F., Bardiaux J.-B., Battaglia P. & Joannis C. (2008). Acoustic Doppler flow-meters: a proposal to characterize their technical parameters. *Flow Measurement and Instrumentation*, **19**(5), 261–267. doi: 10.1016/j.flowmeasinst.2008.01.001.

Larrarte F., Dufresne M., Mignot E., Lipeme Kouyi G., Rivière N., Vazquez J. & Joannis C. (2017). Débitmètrie et mécanique des fluides numérique : contribution à l'évaluation et à la réduction des incertitudes des mesures de vitesse moyenne [Flow measurement and computational fluid dynamics: contribution to the assessment and control of uncertainties on mean velocity measurement]. *La Houille Blanche*, **6**, 67–72. doi: 10.1051/lhb/2017060. (in French).

Larrarte F. & Francois P. (2012). Attenuation of an ultrasonic beam by suspended particles and range of acoustic flow meters in sewer networks. *Water Science and Technology*, **65**(3), 478–483. doi: 10.2166/wst.2012.873.

Larrarte F., Jaumouillié P. & Joannis C. (2004). Computational fluid dynamics: an aid for designing the instrumentation of sewer sections. *Proceedings of Novatech 2004*, 6–10 June, Lyon, France, vol. **1**, 729–736.

Larrarte F., Joannis C. & Bonakdari H. (2010). Qualification and design of flow meter measurement sites within sewer networks. *Bulletin des Laboratoires des Ponts et Chaussées*, **277**, 31–41.

Larrarte F., Joannis C., Mignot E., Rivière N. & Lipeme Kouyi G. (2016a). *Guide Technique – Fiabiliser les mesures de vitesse – Représentativité spatiale des mesures en continu de vitesse et incertitudes sur les mesures de débit – Guides techniques du projet MENTOR [Technical guide – Making flow velocity measurements reliable – Spatial representativeness of continuous flow velocity measurements and uncertainties on flow measurements – Technical guides of the MENTOR project]*, 28 p. Available at http://wikhydro.developpement-durable.gouv.fr/index.php/Fichier:MENTOR_GT_Fiabilisation_debits-vfinale.pdf (accessed 21 Nov. 2020). (in French)>.

Larrarte F., Sturycz E., Lebouc L. & Riochet B. (2016b). New technique for continuous monitoring of sediment height. *Flow Measurement and Instrumentation*, **49**, 40–45. doi: 10.1016/j.flowmeasinst.2016.04.005.

Leeungculsatien T. & Lucas G. P. (2013). Measurement of velocity profiles in multiphase flow using a multi-electrode electromagnetic flow meter. *Flow Measurement and Instrumentation*, **31**, 86–95. doi: 10.1016/j.flowmeasinst.2012.09.002.

Lepot M., Makris K. F. & Clemens F. H. L. R. (2017). Detection and quantification of lateral, illicit connections and infiltration in sewers with infra-red camera: Conclusions after a wide experimental plan. *Water Research*, **122**, 678–691. doi: 10.1016/j.watres.2017.06.030.

Lepot M., Momplot A., Lipeme Kouyi G. & Bertrand-Krajewski J.-L. (2014). Rhodamine WT tracer experiments to check flow measurements in sewers. *Flow Measurement and Instrumentation*, **40**, 28–38. doi: 10.1016/j.flowmeasinst.2014.08.010.

Lipeme Kouyi G. (2004). *Expérimentations et modélisations tridimensionnelles de l'hydrodynamique et de la séparation particulaire dans les déversoirs d'orage [Experimentations and three-dimensional modeling of hydrodynamics and particle separation in storm overflows]*. PhD thesis, Université Louis Pasteur, Strasbourg, France, 274 p. (in French).

Lipeme Kouyi G., Vazquez J. & Poulet J.-B. (2003). 3D free surface measurement and numerical modelling of flows in storm overflows. *Flow Measurement and Instrumentation*, **14**(3), 79–87. doi: 10.1016/S0955-5986(03)00011-6.

Lipeme Kouyi G., Vazquez J., Rollet D., Gallin Y. & Sadowski A.-G. (2005). Use of 3D modelling and several ultrasound sensors to assess overflow rate. *Water Science and Technology*, **51**(2), 187–194. doi: 10.2166/wst.2005.0047.

Liu Z. & Kleiner Y. (2013). State of the art review of inspection technologies for condition assessment of water pipes. *Measurement*, **46**(1), 1–15. doi: 10.1016/j.measurement.2012.05.032.

Loke M. H. (2020). *Tutorial: 2-D and 3-D Electrical Imaging Surveys*. Revised version 17 Aug. 2020, 231 p. Available at https://www.geotomosoft.com/downloads.php (accessed 22 Nov. 2020).

Lopez-Higuera J. M. (ed.). (2002). *Handbook of Optical Fibre Sensing Technology*. John Wiley & Sons Ltd., Chichester (UK), 828 p. ISBN 978-0-471-82053-6.

Lubbers C. L. (2007). *On Gas Pockets in Wastewater Pressure Mains and their Effect on Hydraulic Performance*. PhD thesis, TU Delft, The Netherlands, 304 p. ISBN 978-1-58603-789-5.

Macmillan N. A. & Creelman C. D. (2005). *Detection Theory – A User's Guide, 2nd edn*. Lawrence Erlbaum Associates, Mahwah, NJ (USA), 512 p. ISBN 978-0-8058-4230-6.

MENTOR (2016). *MENTOR Project Final Deliverables*. Available at http://wikhydro.developpement-durable.gouv.fr/index.php/Autosurveillance_et_diagnostic_en_r%C3%A9seau_d%27assainissement_-_Projet_MENTOR (accessed 20 Nov. 2020). (in French).

Michalski A., Starzynski J. & Wincenciak S. (2001). Electromagnetic flowmeters for open channels - Two-dimensional approach to design procedures. *IEEE Sensors Journal*, **1**(1), 52–61. doi: 10.1109/JSEN.2001.923587.

Miller G. A. (1956). The magical number seven, plus or minus two: some limits on our capacity for information processing. *Psychological Review*, **63**(2), 81–97. doi: 10.1037/h0043158.

Munser R. & Hartrumpf M. (2003). Detektion verdeckter Anomalien im Bettungsbereich von Abwasserrohren mit einem Mikrowellen-Rückstreusensor [Detection of hidden anomalies around sewer pipes by means of a microwave back-scattering sensor]. *Technisches Messen*, **70**(7–8), 359–369. doi: 10.1524/teme.70.7.359.22643. (in German).

Nguyen L. S., Schaeli B., Sage D., Kayal S., Jeanbourquin D., Barry D. A. & Rossi L. (2009). Vision-based system for the control and measurement of wastewater flow rate in sewer systems. *Water Science and Technology*, **60**(9), 2281–2289. doi: 10.2166/wst.2009.659.

Nikles M., Vogel B. H., Briffod F., Grosswig S., Sauser F., Luebbecke S., Bals A. & Pfeiffer T. (2004). Leakage detection using fiber optics distributed temperature monitoring. *Proceedings SPIE 5384, Smart Structures and Materials 2004: Smart Sensor Technology and Measurement Systems*, 27 July, San Diego, USA, 8 p. doi: 10.1117/12.540270.

Panasiuk O., Hedström A., Langeveld J. G., Liefting E., Schilperoort R. P. S., de Haan C. & Viklander M. (2017). Methods for localization and volume estimation of the infiltration and inflow: Comparative study. *Proceedings of the 14th International Conference on Urban Drainage*, Prague, Czech Republic, 392–395.

Pothof I. W. M. & Clemens F. H. L. R. (2010). On elongated air pockets in downward sloping pipes. *Journal of Hydraulic Research*, **48**(4), 499–503. doi: 10.1080/00221686.2010.491651.

Pothof I. W. M. & Clemens F. H. L. R. (2011). Experimental study of air–water flow in downward sloping pipes. *International Journal of Multiphase Flow*, **37**(3), 278–292. doi: 10.1016/j.ijmultiphaseflow.2010.10.006.

Pothof I. W. M., Kooij C., Clemens F. H. L. R. & Schuit A. D. (2009). Gasbellen in persleidingen: beter voorkomen dan genezen [Gas pockets in pressure mains, prevention is better than recovery]. *Land+Water*, **49**(6/7), 30–31. (in Dutch).

Prigiobbe V. & Giulianelli M. (2011). Quantification of sewer leakage by a continuous tracer method. *Water Science and Technology*, **64**(1), 132–138. doi: 10.2166/wst.2011.639.

Prodanović D. & Ivetić D. (2019). Primeri primene ravnih elektromagnetnih sondi za merenje protoka u Kolektorima [Examples of flat EM sensor application for flow measurements in collectors]. *Vodoprivreda*, **51** (300–302), 197–209. Available at http://www.vodoprivreda.net/wp-content/uploads/2020/01/5-Dusan-Prodanovic-i-Damjan-Ivetic-Redigovano.pdf (accessed 30 Nov. 2020). (in Serbian).

Ramirez A., Daily W., Binley A., LaBrecque D. & Roelant D. (1996). Detection of leaks in underground storage tanks using electrical resistance methods. *Journal of Environmental and Engineering Geophysics*, **1**(3), 189–203. doi: 10.4133/JEEG1.3.189.

Revitt D. M., Ellis J. B. & Paterakis N. (2006). Comparison of tracer techniques for monitoring sewer losses. *Journal of Environmental Monitoring*, **8**(5), 564–571. doi: 10.1039/B600522E.

Reynolds J. M. (2011). *An Introduction to Applied and Environmental Geophysics, 2nd edn*. Wiley-Blackwell, Chichester (UK), 710 p. ISBN 978-0-471-48535-3.

Rieckermann J., Neumann M., Ort C., Huisman J. L. & Gujer W. (2005). Dispersion coefficients of sewers from tracer experiments. *Water Science and Technology*, **52**(5), 123–133. doi: 10.2166/wst.2005.0124.

Rieckermann J., Vojtěch B., Kracht O., Braun D. & Gujer W. (2007). Estimating sewer leakage from continuous tracer experiments. *Water Research*, **41**(9), 1960–1972. doi: 10.1016/j.watres.2007.01.024.

Rizzo P. (2010). Water and wastewater pipe nondestructive evaluation and health monitoring: A review. *Advances in Civil Engineering*, paper ID **818597**, 13 p. doi: 10.1155/2010/818597.

Schilperoort R. P. S. (2004). *Natural Water Isotopes for the Quantification of Infiltration and Inflow in Sewer Systems*. MSc thesis, Delft University of Technology, Delft, The Netherlands, 151 p.

Schilperoort R. P. S., Hoppe H., de Haan C. & Langeveld J. G. (2013). Searching for storm water inflows in foul sewers using fibre-optic distributed temperature sensing. *Water Science and Technology*, **68**(8), 1723–1730. doi: 10.2166/wst.2013.419.

Schilperoort R. P. S., Meijer H. A. J., Flamink C. M. L. & Clemens F. H. L. R. (2007). Changes in isotope ratios during domestic wastewater production. *Water Science and Technology*, **55**(4), 93–101. doi: 10.2166/wst.2007.099.

Selvakumar A., Tuccillo M. E., Martel K. D., Matthews J. C. & Feeney C. (2014). Demonstration and evaluation of state-of-the-art wastewater collection systems condition assessment technologies. *Journal of Pipeline Systems Engineering and Practice*, **5**(2), paper ID 04013018, 11 p. doi: 10.1061/(ASCE)PS.1949-1204.0000161.

Shercliff J. A. (1962). *The Theory of Electromagnetic Flow-Measurement*. Cambridge University Press, Cambridge (UK), 164 p. ISBN 978-0521335546.

Sliwczynski L. & Krehlik P. (2014). Measurement of acoustic noise in field-deployed fiber optic cables. *Proceedings of the Conference European Frequency and Time Forum (EFTF)*, 23–26 June, Neuchatel, Switzerland, 339–342. doi: 10.1109/EFTF.2014.7331504.

Smits J., Moens M. R., Klootwijk M. & van Vliet H. (2008). Testing flow-meters using a field laboratory. *Proceedings of the 11th International Conference on Urban Drainage*, 31 Aug.–5 Sept., Edinburgh, Scotland, UK, 12 p.

Stanic N. (2016). *Assessment Methods for Structural and Hydraulic Properties of Concrete Sewer Pipes*. PhD thesis, Delft University of Technology, The Netherlands, 162 p. ISBN 978-94-6233-2058. doi: 10.4233/uuid:3ea2146b-21cb-49b0-9aca-40ae8338e4d3.

Stanic N., Clemens F. H. L. R. & Langeveld J. G. (2016). Estimation of the hydraulic roughness of concrete sewer pipes by laser scanning. *Journal of Hydraulic Engineering*, **143**(2), paper 04016079. doi: 10.1061/(ASCE)HY.1943-7900.0001223.

Tuccillo M. E., Wilmut C., Feeney C., Martel K. & Selvakumar A. (2011). Field demonstration of electro-scan defect location technology for condition assessment of wastewater collection systems. *Proceedings of the Water Environment Federation*, **5**, 265–281. doi: 10.2175/193864711802837309.

Tukker M., Kooij K. & Pothof I. (2016). *Hydraulic Design and Management of Wastewater Transport Systems – CAPWAT Manual*. IWA Publishing, London (UK), 152 p. ISBN 9781780407821. Available at https://www.iwapublishing.com/books/9781780407821/hydraulic-design-and-management-wastewater-transport-systems (accessed 20 Nov. 2020).

van Daal-Rombouts P. M. M. (2017). *Performance Evaluation of Real Time Control in Urban Wastewater Systems*. PhD thesis, TU Delft, Deft, The Netherlands, 256 p. ISBN 978-94-6233-707-7. doi: 10.4233/uuid:b1d5d733-b271-474f-ad32-b5fdde257161.

van Daal-Rombouts P. M. M., Tralli A., Verhaart F. I. H., Langeveld J. G. & Clemens F. H. L. R. (2017). Validation of computational fluid dynamics for deriving weir discharge relationships with scale model experiments and prototype measurements. *Flow Measurement and Instrumentation*, **58**, 52–61. doi: 10.1016/j.flowmeasinst.2017.09.011.

Vazquez J., Lipeme Kouyi G. & Zug M. (2006). Modelling and instrumentation of the storm overflows of the combined sewer system of sélestat. *Urban Water Journal*, **3**(2), 91–110. doi: 10.1080/15730620600855936.

Vazquez J., Zug M., Buyer M. & Lipeme Kouyi G. (2005). CSOs: Tools for assessing their operation in our systems. *Water Science and Technology*, **51**(2), 179–185. doi: 10.2166/wst.2005.0046.

Verhaart F. I. H., de Fockert A. & Zwanenburg S. A. A. (2016). Velocity profiles in the bell mouth throat of vertically submersible pumps. *Journal of Applied Water Engineering and Research*, **4**(2), 102–111. doi: 10.1080/23249676.2015.1090350.

Weiß G., Brombach H. & Haller B. (2002). Infiltration and inflow in combined sewer systems: long-term analysis. *Water Science and Technology*, **45**(7), 11–19. doi: 10.2166/wst.2002.0112.

Wilmut C. G., Dsouza B. J. & Guidry B. A. (2005). Electro-scan technology - Is it the 'silver bullet' for rainfall dependent infiltration (RDI) and exfiltration. *Proceedings of the Water Environment Federation*, **2005(4)**, 697–706. doi: 10.2175/193864705784291529.

Wirahadikusumah H., Abraham D. M., Iseley T. & Prasanth R. K. (1998). Assessment technologies for sewer system rehabilitation. *Automation in Construction*, **7**(4), 259–270. doi: 10.1016/S0926-5805(97)00071-X.

Wolf L. (2006). *Influence of Leaky Sewer Systems on Groundwater Resources Beneath the City of Rastatt, Germany*. PhD thesis, Universität Karlsruhe, Karlsruhe, Germany, 231 p.

Wood W. & Palmer D. (2000). The application of mise-à-la-masse and resistivity surveys to the detection of pollution from leaking sewers. *Exploration Geophysics*, **31**(3), 515–519. doi: 10.1071/EG00515.

Chapter 4

Measuring the water balance in stormwater control measures

Tim D. Fletcher[1], Jean-Luc Bertrand-Krajewski[2], Jérémie Bonneau[3], Matthew J. Burns[1], Peter J. Poelsma[1] and Jasmine K. Thom[1]

[1]The University of Melbourne, Waterway Ecosystem Research Group, School of Ecosystem and Forest Sciences, Melbourne, Australia
[2]University of Lyon, INSA Lyon, Laboratory DEEP, Villeurbanne, France
[3]INRAE, UR RiverLy, Villeurbanne, France

ABSTRACT

Stormwater control measures (SCMs), also frequently referred to as sustainable urban drainage systems (SUDS), are of growing importance in cities, as part of a global move towards mitigating the impacts of stormwater on receiving environments. They need to be monitored as parts of UDSM systems but require specific and sometimes innovative methods and sensors. This is particularly the case for SCMs such as swales, rain-gardens, bioretention filters, infiltration trenches, green roofs, etc., which have complex and varied configurations and hydrologic behaviour. This chapter deals with measuring the water balance in SCMs by accounting for its various components: inflows, outflows, overflows, storage, infiltration, exfiltration, intrusion, evaporation, and evapotranspiration. It presents a range of suitable methods and tools, indicates key points to consider, and discusses possible difficulties in obtaining accurate monitoring data. Routine monitoring of decentralized and diversified SCMs is still an emerging field for both researchers and practitioners. A significant evolution is therefore expected with its generalization in the next years.

Keywords: Evaporation, evapotranspiration, exfiltration, infiltration, inflow, intrusion, outflow, overflow, water content.

doi: 10.2166/9781789060119_0105

SYMBOLS

A	area (m^2)
AET	actual evapotranspiration (m^3s^{-1})
c_p	specific heat of the air
D	drainage
$e_s{-}e_a$	vapour pressure deficit of the air (Nm^{-2})
exf	exfiltration (m^3s^{-1})
E	transpiration
ET	evapotranspiration (m^3s^{-1})
ET_0	reference evapotranspiration (m^3s^{-1})
f	function (–)
G	soil heat flux (Wm^{-2})
h	water level (m)
i	time step index (–)
I	irrigation
K_c	crop factor (–)
K_{sat}	hydraulic conductivity at saturation
P	rainfall (m^3s^{-1})
PET	potential evapotranspiration (m^3s^{-1})
Q	discharge (m^3s^{-1})
Q_{in}	inflow (m^3s^{-1})
Q_{out}	outflow (m^3s^{-1})
Q_{over}	overflow (m^3s^{-1})
Q_{sap}	sapflow (m^3s^{-1})
r_a	aerodynamic resistance
r_s	surface resistance
R	rainfall depth (m)
R_a	extraterrestrial radiation (also known as solar constant) (Wm^{-2})
R_n	net solar radiation (Wm^{-2})
sub	as index: refers to the subsurface of an SCM (–)
$surf$	as index: refers to the surface of an SCM (–)
T_{max}	maximum temperature (K)
T_{mean}	mean temperature (K)
T_{min}	minimum temperature (K)
Vol	volume of water in a lysimeter per unit lysimeter area (m^3m^{-2})
Δ	slope of the saturation vapour pressure temperature relationship
ΔS	variation of water storage within an SCM during one step (m^3)
ΔV	change in the storage volume in a porosity measurement (m^3)
γ	psychrometric constant ($Nm^{-2}\,K^{-1}$)
ρ_a	mean air density at constant pressure (kgm^{-3})

4.1 INTRODUCTION

Fundamental to understanding the performance of a stormwater control measure (SCM) is the quantification of its water balance – inflows, storage and outflows (Figure 4.1) – or at least of the components of the water balance that are of interest. While conceptually simple, the reality is quite complex, particularly for systems where infiltration and evapotranspiration fluxes are present. In this chapter we describe methods and tools that can be used to measure or estimate the various fluxes and storages, drawing on illustrative examples.

Key messages on measuring the water balance in stormwater control measures

KM 4.1: Ensuring reliable flow measurements usually requires that (i) either new SCM facilities must be designed in order to allow and facilitate measurements by means of dedicated arrangements and structures which are usually ignored in most SCM facilities, or (ii) existing SCM facilities must be adapted and/or retrofitted.

KM 4.2: Understanding the water balance of SCMs is vital to understanding their performance, not only for hydrological aspects, but for pollution reduction also. It also provides important information on long-term maintenance and performance, which are vital for overall sustainability and asset management considerations.

KM 4.3: The specificities of SCMs require specifically adapted monitoring techniques and equipment, which are outlined by this chapter. As this is an evolving area, however, readers are advised to look for up-to-date information before designing or embarking on a monitoring campaign for SCMs.

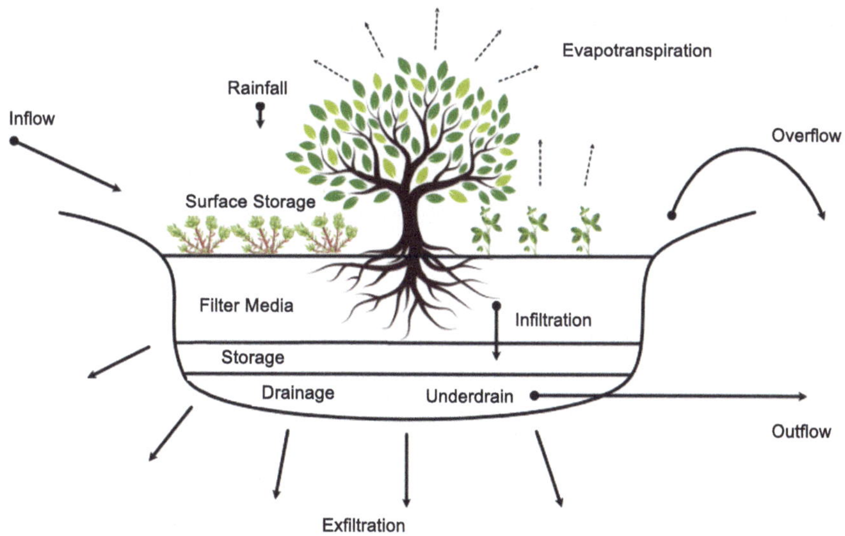

Figure 4.1 Schematic representation of the water balance of a stormwater control measure. *Source*: adapted from Jérémie Bonneau (INRAE).

4.2 DESCRIPTION OF THE WATER BALANCE

The water balance of an SCM is given, in its most general form, by the sum of the inflows and outflows (Equation (4.1)):

$$\Delta S = \sum in - \sum out \tag{4.1}$$

where ΔS is the change in storage over the time-step being considered.

The inflows and outflows depend on the configuration of the SCM of interest. For example, Equation (4.2) shows the case for a vegetated system with infiltration components, and both surface (ponding) storage and internal water storage (i.e. water stored in the substrate). In this example the possibility of groundwater intrusion into the SCM is ignored.

$$\Delta S_{surf+sub} = Q_{in} + P - Q_{out} - Q_{over} - exf - ET \tag{4.2}$$

where $\Delta S_{surf+sub}$ is the combined surface (ponding) and subsurface (substrate water content) storage volume, Q_{in} is the inflow (for example from a pipe or other inlet), P is the rainfall, Q_{out} and Q_{over} are the outflow and overflow respectively, and exf and ET are the exfiltration and evapotranspiration fluxes, respectively.

In determining the monitoring strategy for a given SCM, consideration should be given to which fluxes are important to properly quantify, and which can be estimated. For example, if the only performance metric of interest for an infiltration basin is the reduction in peak flows, measurements of infiltration or evapotranspiration may not be necessary. Conversely, if a vegetated bioretention system is being monitored for its contribution to reducing overall flow (both through surface runoff and groundwater recharge via infiltration), evapotranspiration will need to either be measured directly, or indirectly estimated through quantification of all other elements of the water balance. An appropriate monitoring strategy can only be developed once agreement on the important fluxes has been reached.

Compared to traditional centralized underground stormwater collection systems, where monitoring some key points (e.g. overflow structures, outlets) may be sufficient to get data and information about the corresponding entire catchment, SCMs are usually numerous, decentralized and spatially distributed over the entire catchment: getting data and information at this spatial scale necessitates the monitoring of several SCMs or getting appropriate and representative information about a subset of them, for extrapolation to others having similar properties and/or behaviour. Moreover, monitoring traditional underground collection systems, even if it is never obvious, benefits from well-established and various experiences, whereas monitoring SCMs is still emerging and prone to significant evolution in the coming years with e.g. the use of low-costs sensors, new data transmission technologies, the development of IoT (Internet of things), etc. (Cherqui *et al.*, 2019).

4.3 INFLOW, BYPASS, OUTFLOW AND OVERFLOW

Inflow, bypass (water diverted around the system and thus bypassing all treatment), outflow and overflow (excess water spilling from within the system) can be measured using a range of instruments, depending on the configuration of each element.

Wherever possible, design of the SCM should consider proposed or potential monitoring. Ideally, the system should be constructed with dedicated monitoring points, allowing monitoring weirs, flumes or other proposed measurement apparatus to be installed easily. For example, a bypass channel around a wetland can only be accurately measured if there is a stable cross section with easily characterized geometry, and it is not affected by backwater.

Similarly, when choosing between multiple SCMs as potential monitoring sites, the complexity of the system will be an important selection criterion. A system with multiple inlets will require either:

- Separate monitoring infrastructure at each inlet (thus substantially increasing both the capital and operating costs of monitoring), or
- Estimation of flows from additional inlets, using a rainfall-runoff model, or simply a catchment area *pro rata* estimation. Regardless of the approach, estimation will introduce substantial errors, which may undermine the objectives of the monitoring and bias results and conclusions.

4.3.1 Inflows

Inflows through a pipe or constructed channel can theoretically be measured using depth measurement (if a suitable rating relationship can be achieved, for example using a V-notch or other calibrated weir, or Venturi or Parshall flume – see Sections 3.4.1 and 3.4.2), although attention needs to be paid to obstruction and sediment accumulation and other limitations mentioned in Chapter 3. In such situations, flow calculations based on depth-area-velocity measurement, using technology such as acoustic Doppler, ultrasonic, laser or radar sensors are likely to be more effective. Piped inflows can be challenging to measure, due to factors described in the following three subsections: sediments accumulation, flow regime transition and backwater effect.

Low inflows or flows in small pipes, commonly encountered in SCMs (particularly at the outlet), are not easy to measure with traditional sensors due to low water levels and high relative uncertainties, low flow velocities, influence of immersed sensors themselves on water levels and flow velocity. Methods used for large sewers as presented in Chapter 3 may thus be affected by very high uncertainties and bias, and sometimes may not even be applicable. They may require other methods and devices to be measured accurately, e.g. electromagnetic flow meters in siphons, or tipping bucket gauges, as in the case of SCM outflows (see Section 4.3.2).

4.3.1.1 Obstruction of the flow sensor caused through the accumulation of sediment or debris

Weirs, flumes, or other structures used to measure inflows can accumulate sediment and debris usually present in untreated inflows (Figure 4.2). Unless inflows contain very little sediment (e.g. pre-treated by

(a) (b)

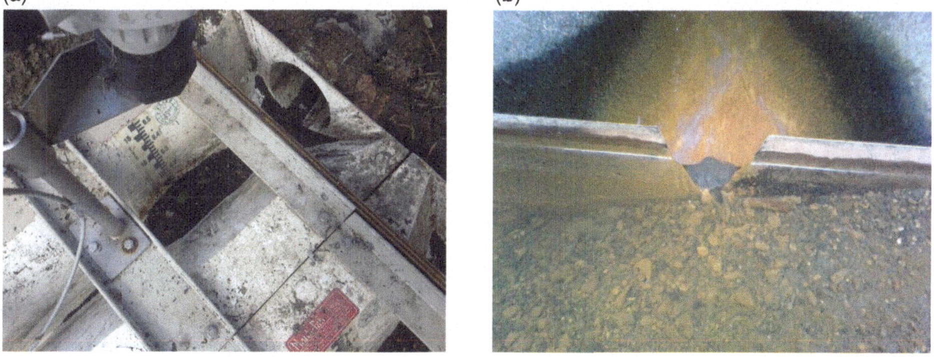

Figure 4.2 Parshall flume (a) and V-notch weir (b) rendered unusable due to sediment accumulation. *Source*: Peter Poelsma (University of Melbourne).

a gross pollutant trap or sediment basin), using such structures is not suitable. Even apparently 'clean' inflows can contain some sediment which then accumulates and requires regular removal.

Submerged sensors can also be frequently covered with sediment or debris (Figure 4.3). Large pieces of debris and litter can get caught on the sensor (ragging), or sediment can accumulate during low velocities at the end of events or in pipes with low slopes. Debris can stop the sensor from working by blocking the path of beams (e.g. Doppler sensor), or sediment can enter openings in the sensor, e.g. to measure pressure, resulting in inaccurate measurements.

Sensors covered by sediments may lose their ability to measure water levels accurately. For example, the 'weight' of sediment on a pressure sensor can lead to an inaccurate level measurement (Figure 4.4).

Sensors mounted above the water surface can also result in inaccurate water depths if sediment accumulates in the bottom pipe or channel. The water surface/level can increase because of sediment accumulation, and thus artificially increase the recorded flow rate.

If possible, measurements should be made downstream of a primary treatment which removes the litter and sediment, thus making flow measurement feasible and reliable. In some cases, it may even be

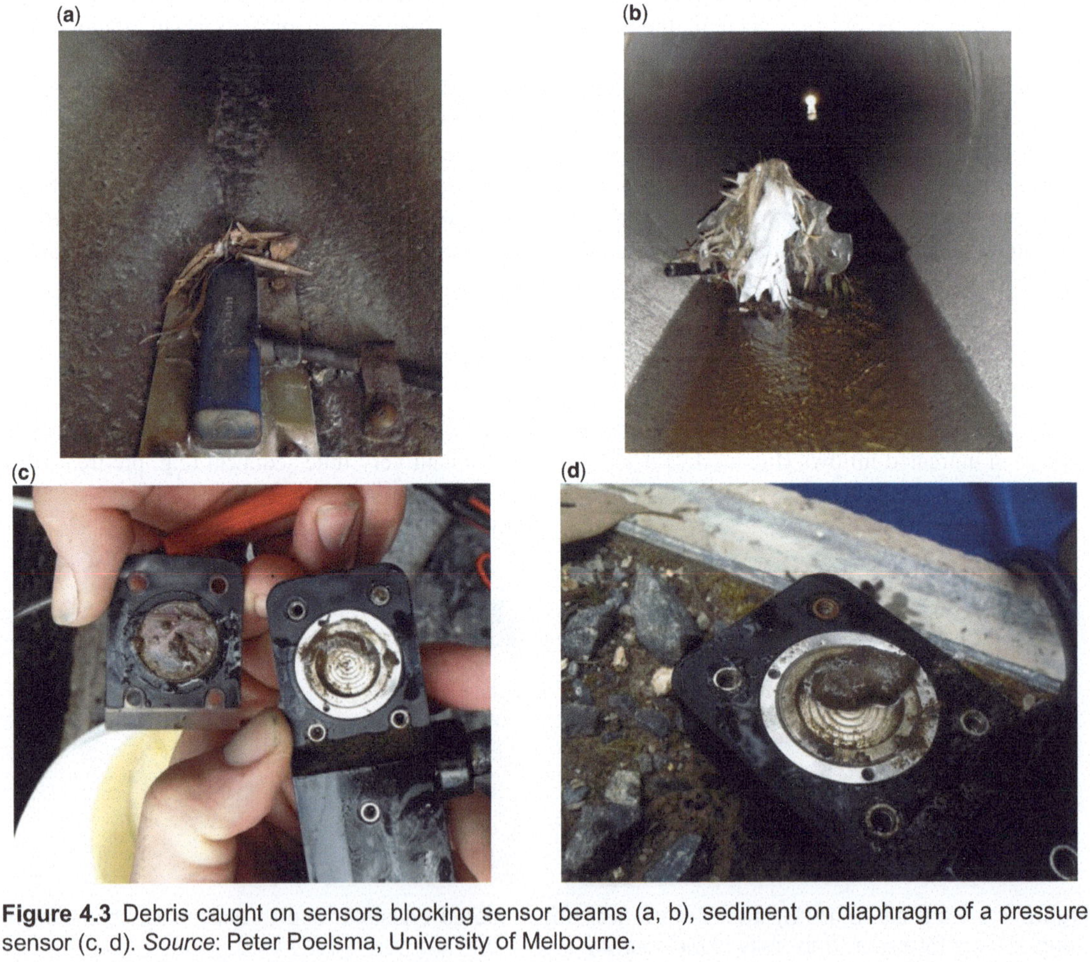

Figure 4.3 Debris caught on sensors blocking sensor beams (a, b), sediment on diaphragm of a pressure sensor (c, d). *Source*: Peter Poelsma, University of Melbourne.

Figure 4.4 Example of a flowmeter being covered by sediments after a storm event, leading to loss of reliable data. In this case sediment accumulation was due to the construction of a small weir downslope of the probe. *Source*: Peter Poelsma (University of Melbourne).

appropriate to install a dedicated sediment trap. Figure 4.5 shows a litter fence built to prevent the blockage of downstream flow diversion and monitoring infrastructure.

The maintenance of sensors is very important (see Section 7.4), especially in the challenging conditions at inlets – high sediment loads, debris, long periods of no flow, etc. Regular removal of debris or sediment may be required, as well regular cleaning and calibration of the sensors.

Steeper pipes or channels can also reduce the accumulation of sediment which causes the measurement problems mentioned above, but this introduces other issues, in terms of transition from subcritical to supercritical flows.

4.3.1.2 Transition of flows from subcritical to supercritical

A hydraulic jump happens in a pipe or a channel when the flow of water transitions from supercritical (typically in steep pipes) to subcritical (e.g. under the influence of a break in slope, an access hole, or a weir), as illustrated in Figure 4.6. As a consequence, waves and turbulence can develop on top of the probes, creating uncertainty and variability of the water level reading. Therefore, as much as possible, measurements must be made upstream of the hydraulic jump position, which may be estimated by means of hydraulic calculations for various expected flow regimes in the SCM facility.

Figure 4.5 Example of primary treatment constructed to reduce blockage of downstream monitored pipes and diversion weir. In this case, the litter fence required monthly-cleaning to maintain satisfactory operation. *Source:* Peter Poelsma (University of Melbourne).

4.3.1.3 Backwater influence from within the SCM itself during certain periods

In some circumstances, backwater influences from downstream may render flow measurements inaccurate, either because the sensor is unable to cope with transition from positive to negative velocities, because the relation between water level and flow is no longer valid, or because the backwater zone leads to accumulation of sediment on the sensor. Most velocity sensors are able to cope with such influences, but they need to be installed in a position where they are protected from sediment, such as part-way up the side of a pipe. It is important to identify, as much as possible, these potential complications before monitoring equipment is installed by conducting some detailed hydraulic analysis and modelling of the flow measurement structure jointly with its upstream and downstream parts. For example, the risk of backwater influence can be easily established by survey of the inlet pipe and SCM water level range. Very frequently, an iterative approach is necessary, once the first data are collected on site and reveal its real functioning, as there are very often differences between prior information, theory and reasoning, and posterior data, practice and observations.

4.3.1.4 Rating curve

Where inflow is conveyed through a natural channel, a rating function $Q = f(h)$ needs to be established, relating the discharge Q to the water level h, based on empirical measurements of depth and velocity or tracing experiments across the channel cross section.

Water level is measured upstream of a stable cross section where a change in discharge results in a measurable change in level which is not affected by a section further downstream. Discharge is measured

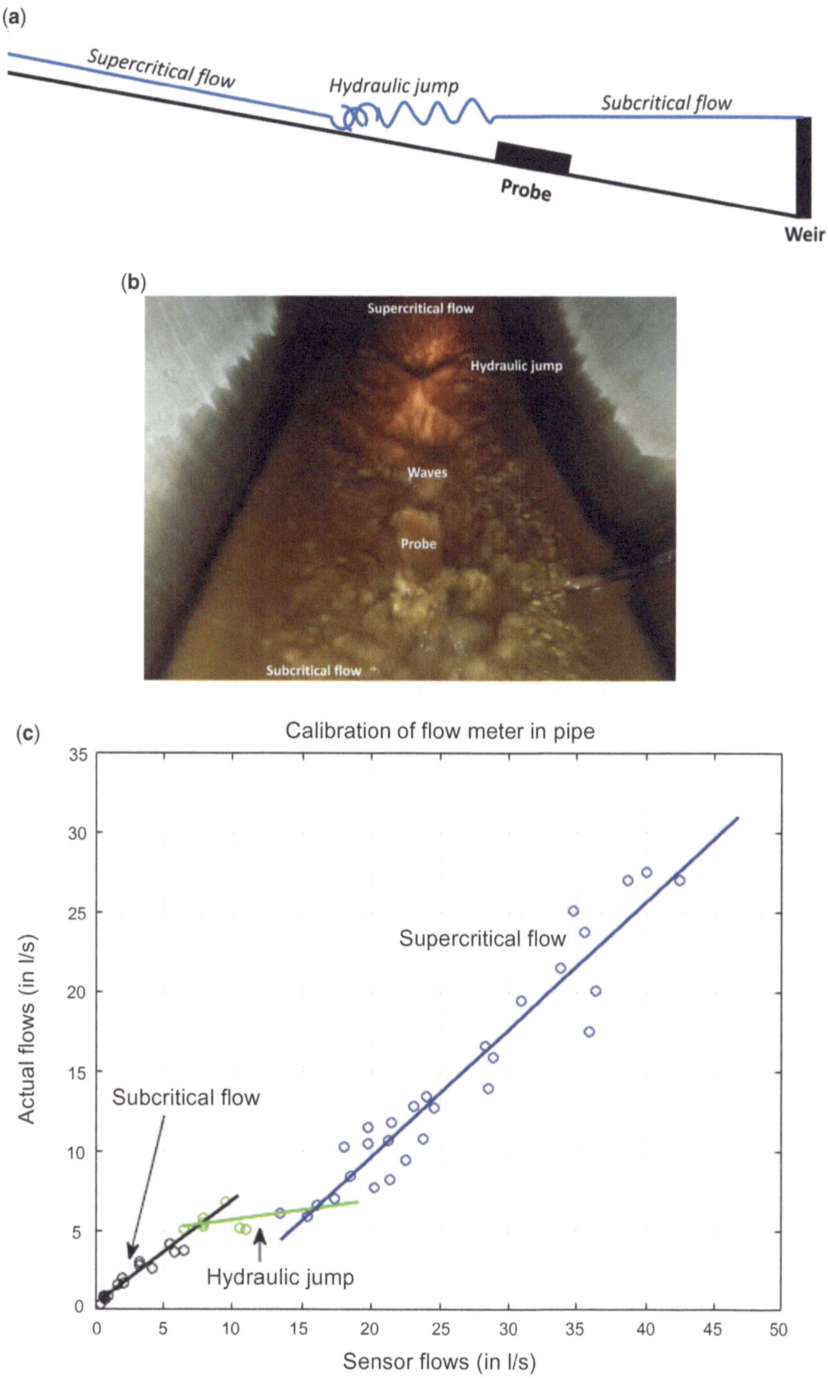

Figure 4.6 Example of transition from supercritical to subcritical flow, resulting in disturbance to flow measurement in theory (a) and in practice (b), with resultant impact on the flow rating curve (c). *Sources*: (a) and (c) adapted from Jérémie Bonneau (INRAE), (b) Jérémie Bonneau (INRAE).

at a variety of levels covering the range of levels experienced by the site. A relationship between water level and discharge is then established and used to convert level to flow. Achieving an accurate rating function is a laborious task, requiring field measurement during a wide range of conditions; sufficient accuracy will only be achieved if the necessary personnel to undertake the rating are available to go on site during times of varying flow, which may be difficult and risky during storm events (see Section 7.2). A detailed standard method for establishing rating curves is available in ISO (2013).

Chapter 3 provides details on water level sensors, flow velocity sensors, and also on velocity-area measurement methods (ISO, 2007). In selecting flow sensors, it is important to consider water quality. For example, there are occasions where the water is too clear for some sensors that rely on a signal being reflected from particles or bubbles in the water (e.g. acoustic Dopplers, laser distance measurement). This is more common in treated outflows but can also occur in the baseflow at inlets or the tail of inflow events.

4.3.2 Outflows

Outflow configurations greatly vary, ranging from a simple V-notch or square weir, to a single orifice or orifice plate, or a simple pit. In the case of a bioretention system with underdrain, outflow needs to be measured in the pipe or pit downstream, for example using depth measurement and a weir or flume or measuring depth-area-velocity. If appropriate conditions (geometry, size, flow regime, etc.) are satisfied, methods and sensors described in Section 4.3.1 for inflows and more generally in Sections 3.3 and 3.4 are possible solutions to monitor SCM outflows.

However, SCM outflows are frequently very low, for example in the range of only a few litres per hour or even less. Indeed, many SCMs are designed to maximize storage, infiltration and evapotranspiration, to delay and attenuate peak flows, resulting in outflows which are significantly lower than inflows. In such cases, most methods used to measure discharges in large pipes or even SCM inflows are no longer applicable and alternatives methods are needed.

As SCM outflows may range from $m^3\ h^{-1}$ to a few Lh^{-1} or less, no single instrument can cover such a wide range with acceptable uncertainties. Different sensors are therefore frequently combined, as shown in the examples below.

The first example deals with the measurement of the outflow from a vegetated roof, with a substrate depth from 40 to 140 mm (Figure 4.7a). Preliminary theoretical estimations indicated that the maximum outflow was expected to be between 1.8 and 9.0 $m^3\ h^{-1}$. Therefore, in a services room located under the roof, the initial vertical downspout (160 mm pipe) evacuating the roof outflow has been cut and replaced by a 25 mm pipe siphon equipped with an electromagnetic flowmeter with a measuring range of 0 to 19.8 $m^3\ h^{-1}$. In the case of blockage, a bypass above the electromagnetic flowmeter has been built (Figure 4.7b). The maximum outflow measured over the 9-month monitoring campaign Sept. 2012–May 2013 reached 2 $m^3\ h^{-1}$, i.e. just above the lowest theoretical estimation. An example of monthly data is shown in Figure 4.8 for February 2013 (Bertrand-Krajewski & Vacherie, 2014). The maximum outflow is close to 0.8 $m^3\ h^{-1}$. But the critical point was the measurement of the very low outflow. As shown in the zoom box in Figure 4.8, when the outflow is lower than approximately 55 Lh^{-1}, the velocity in the electromagnetic flowmeter is too low for the sensor and the outflow values drop abruptly to zero. Consequently, the diminishing tail of the outflow hydrograph is missed, and the water balance is biased.

In order to improve the measurement of very low outflows, a new device was created, which combines three components: a 10 mL tipping bucket (the same as in tipping bucket rain gauges – see Chapter 2), a 1 L tipping bucket and an electromagnetic flowmeter (Figure 4.9). Two control valves avoid overload and saturation of the tipping buckets. The complete device was calibrated in the laboratory with a regulated peristaltic pump associated with a scale to measure mass of water, over the range of approximately

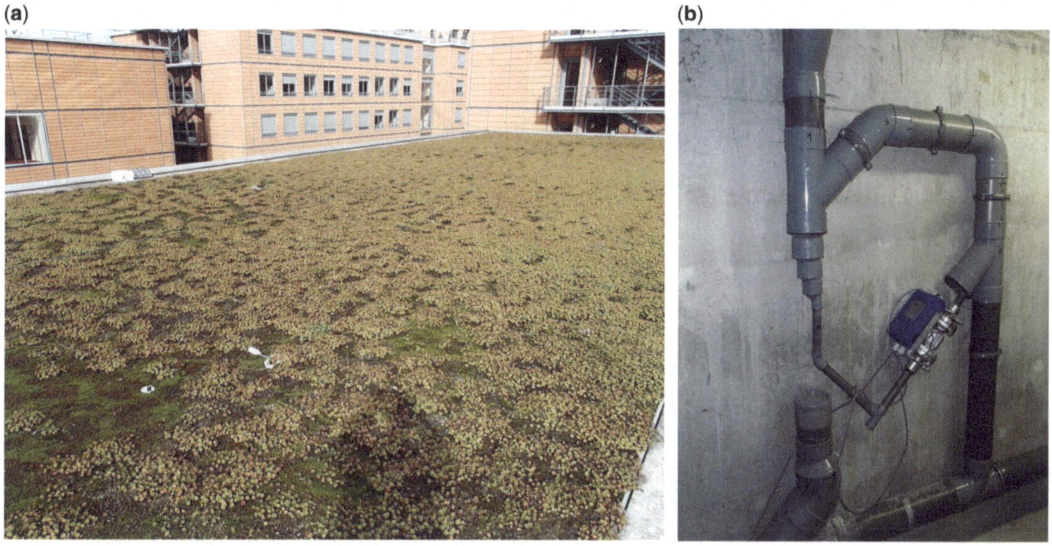

Figure 4.7 (a) Lyon Congress Centre 282 m^2 vegetated roof; (b) modified downspout equipped with an electromagnetic flowmeter and an upper bypass. *Source*: Laboratory DEEP, INSA Lyon.

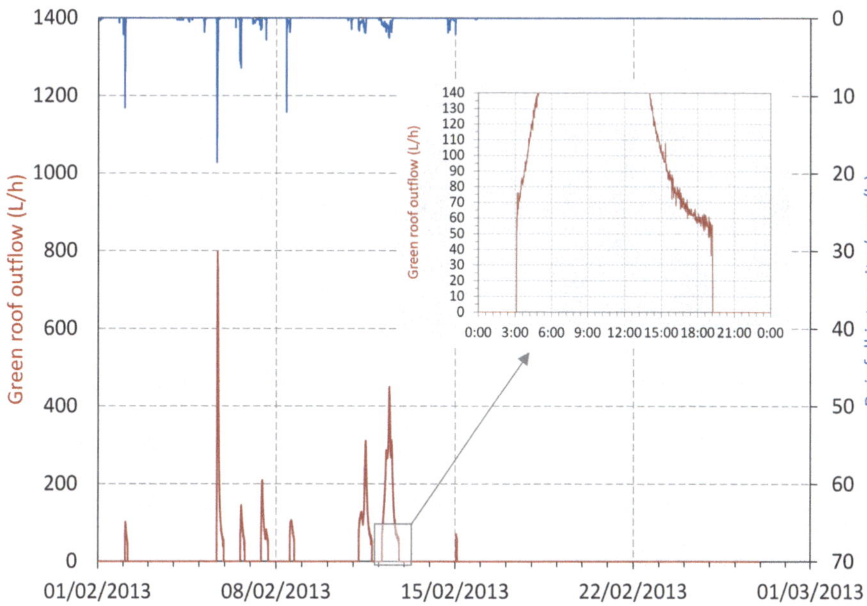

Figure 4.8 Lyon Congress Centre rainfall intensity and vegetated roof outflow measured in February 2013. *Source*: Jean-Luc Bertrand-Krajewski (INSA Lyon).

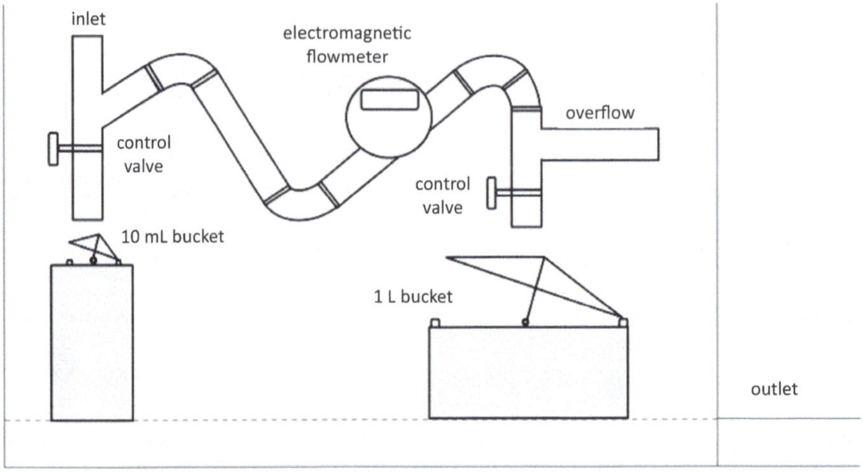

Figure 4.9 Scheme of a flowmeter combining a 10 mL tipping bucket, a 1 L tipping bucket and an electromagnetic flowmeter (manufactured by Précis-Mécanique as the Trio flowmeter and as the Duo flowmeter without the electromagnetic flowmeter). *Source*: Laboratory DEEP, INSA Lyon.

0.0035 to 0.55 $m^3 h^{-1}$ (Arias *et al.*, 2016b). The measuring range of the 1 L tipping bucket has some overlap with both the 10 mL tipping bucket and the electromagnetic flowmeter, to ensure continuity between the components and avoid gaps or jumps in the measured values. This equipment was first implemented (Figure 4.10) to monitor the outflows of three experimental green roofs. An example of data (Arias *et al.*, 2016a) is given in Figure 4.11 for two green roofs: compared to Figure 4.8, the diminishing tail of the outflow hydrograph is not interrupted, and the water balance can be calculated accurately.

Figure 4.10 (a) installation of a Duo flowmeter to measure the outflow of a green roof; (b) view of a Trio flowmeter. *Sources*: (a) Rémy Bournique (INSA Lyon & Le Prieuré-Vegetal ID), (b) Le Prieuré-Vegetal ID, Moisy.

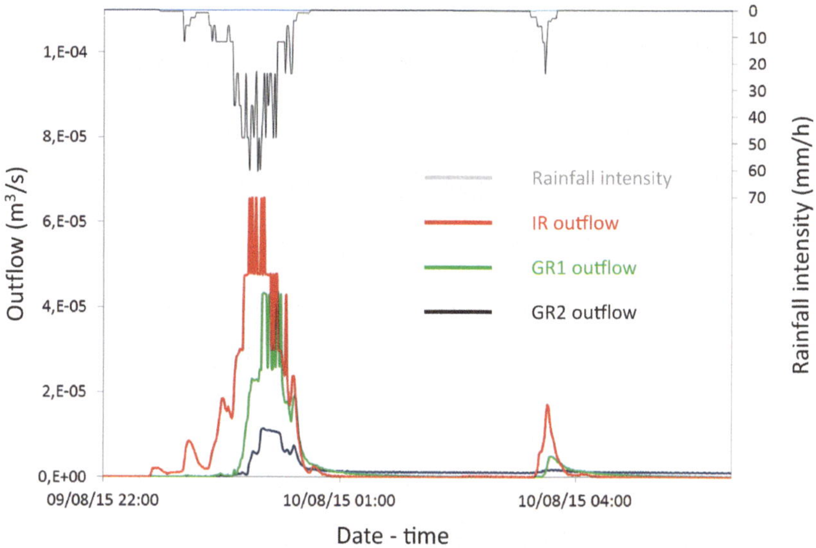

Figure 4.11 Gepeto experimental green roof outflows: example of data on 09–10 Aug. 2015. IR: impervious reference green roof; GR1: basic green roof with 60 mm thick substrate and GR2: GR1 with a 95 mm additional underlying storage reservoir. *Source*: Luis Arias (INSA Lyon).

Similar devices, with local adaptations, have also been used more recently to monitor the outflows from a permeable parking lot and an infiltration trench (Figure 4.12), where space is much more limited (Garnier *et al.*, 2017).

A critical consideration in the monitoring of SCM outlets is ensuring that they are regularly inspected to minimize the effect of blockage. Partial blockages are of particular concern, because they may not be apparent in looking at the measured water level/discharge data. Figure 4.13 shows an example of a partially blocked orifice place, where the discharge is greatly reduced, meaning that the measured water level behind the plate no longer reflects the observed discharge.

4.3.3 Bypass

Bypass occurs when the flow is bypassed around the stormwater control measure, usually through means such as a diversion weir into a pipe or channel. Measuring the flow over a bypass can be difficult, particularly where the diversion weir is subject to highly turbulent flows. A common approach is to measure both the upstream and downstream flow, with downstream flow being measured either in the system or the bypass, or both. In the case where only one of the two is measured, simple subtraction can be used to infer one from the other. Subtraction is simple but may lead to high uncertainties.

4.4 STORAGE VOLUMES

Storages within SCMs may be either surface storage (i.e. ponding), or subsurface (water contained within the substrate, for example in an infiltration, bioretention system or green roof).

Measurement of storage volumes in a stormwater control measure is conceptually simple. Surface ponding volume can be calculated from water depth (which can be obtained using a range of sensors,

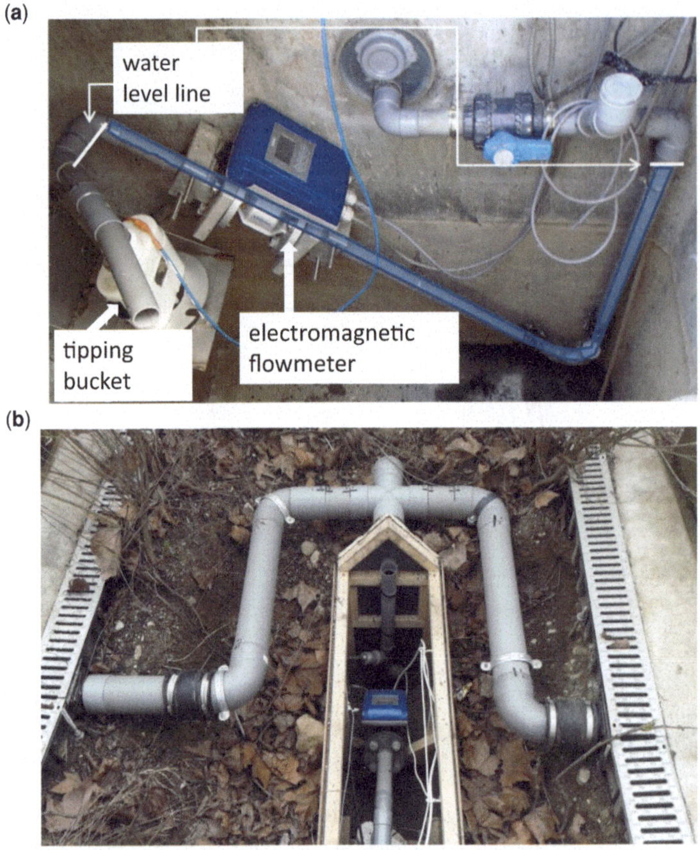

Figure 4.12 Installation of a two-component (a 20 mL tipping bucket and an electromagnetic flowmeter) flowmeter to measure the outflow of an infiltration trench. (a) permeable parking; (b) infiltration trench. *Source*: courtesy Robin Garnier (INSA Lyon).

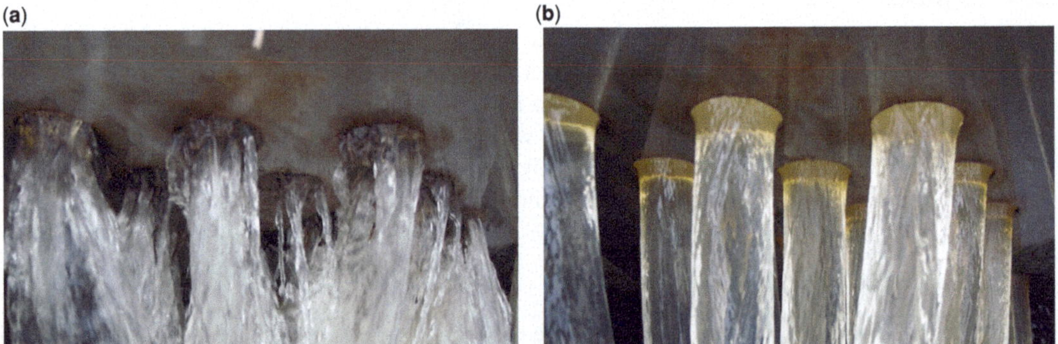

Figure 4.13 Blocked (a) and unblocked (b) orifices in wetland outlet demonstrating the need to ensure regular maintenance to obtain accurate outflow estimates. *Source*: Peter Poelsma (University of Melbourne).

such as pressure, capacitance, ultrasonic or radar sensors (see Section 3.2 for more information on water level sensors) matched to an accurate survey of the bathymetry of the system. ISO (2010) provides useful guidance on volume measurement in this manner (cubature method). It is important to note that the bathymetric survey should be as-constructed and not as-designed, as the difference can be quite large and introduce a substantial error into estimated storage volumes. There may also be the need to account for changes in storage volume over time, such as the decrease due to sediment accretion in the inlet zone. Rapid accurate survey is now readily accessible using drones equipped with LiDAR.

Many SCMs also contain subsurface storage. Examples include infiltration systems, bioretention systems, sand filters, green roofs, etc. Storage volume estimation in the substrate(s) of such systems can be undertaken using depth sensors as listed above, coupled with bathymetry of the subsurface component, and a reliable measurement of the porosity of the substrate. In many cases this requires measurement of the porosity of each of the substrate layers, such as in a bioretention system, which typically has a loamy sand substrate sitting over a gravel drainage layer, with one or several transition layers (Figure 4.14). In such systems, the water balance may require porosity to be taken into account, by one of the following methods:

- Direct measurement of porosity, typically taken by coring the substrate and undertaking a laboratory analysis. Standard methods for such analysis are widely published (see for example ISO, 2017). Ideally, analysis of the substrate porosity should be undertaken close to the period for which the water balance analysis is being undertaken, as the substrate properties will change over time due to the growth and death of plant roots and other biomass.

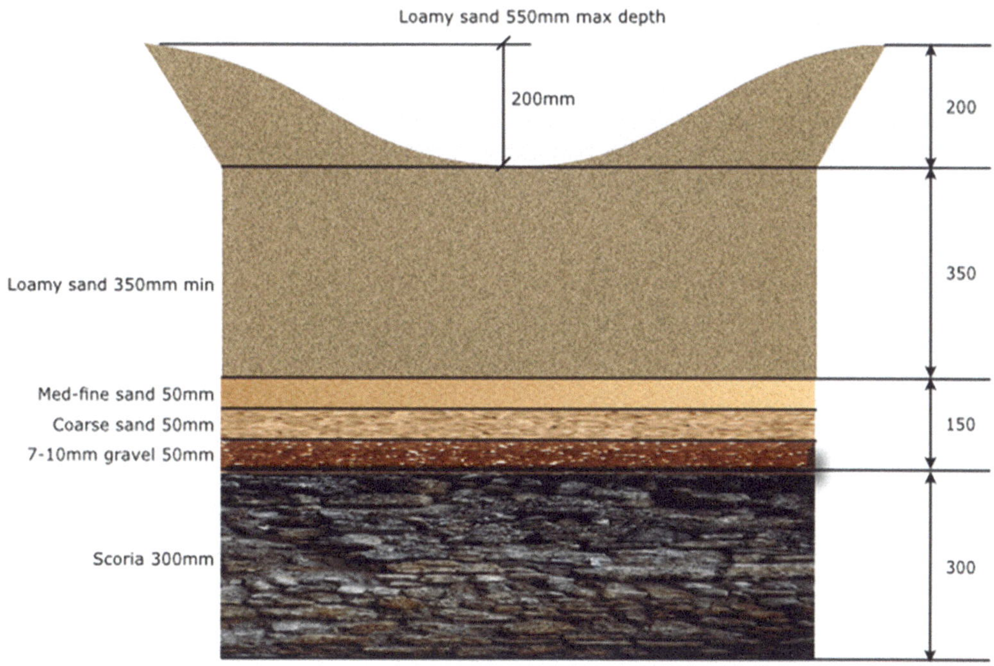

Figure 4.14 Example of filter media profile of a bioretention system. *Source*: Tim Fletcher (University of Melbourne).

- Estimation of porosity from known properties of the substrate (provided for example during the construction). This method is not as accurate as direct measurement, because it does not account for changes which occur to the material *in situ*.
- Direct measurement of water volume, by closing the water balance such that the change in subsurface water level can be calculated from a known inflow volume. For example, observing the change in water level within a system closed off from any input and output during a rainfall event of known depth allows an *in situ* estimation of the substrate porosity by means of Equation (4.3).

$$\text{Porosity} = R \times A/\Delta V \tag{4.3}$$

where R is the observed rainfall depth, A is the area of the system considered and ΔV is the change in storage volume.

Sensors used to measure the subsurface water level should be installed in such a way that water from the surface storage above does not leak down along the sensor housing in such a way that it gives a false water level reading in the substrate.

4.5 INFILTRATION AND EXFILTRATION

For many SCMs such as infiltration basins and bioretention systems (often called biofiltration systems or rain-gardens), the movement of water from the ponding zone through the filter media substrate (referred to herein as infiltration), or from the filter media substrate to the underlying soil (referred to herein as exfiltration) are important fluxes to quantify (Figure 4.15). In some situations, there may also be an intrusion of groundwater into the SCM from adjacent soil, and it may be necessary to quantify this, for example to analyse seasonal variation in performance.

Infiltration can be measured either using the water level in the entire system or measured at specific locations using experimental apparatus. The former method is preferable wherever possible, because it gives an integrated measurement of the performance of the whole system. Conversely, where the interest is whether there is clogging in a particular part of the system (typically this more likely occurs around the inlet), location-specific infiltrometer testing will be most helpful. The reader should refer to

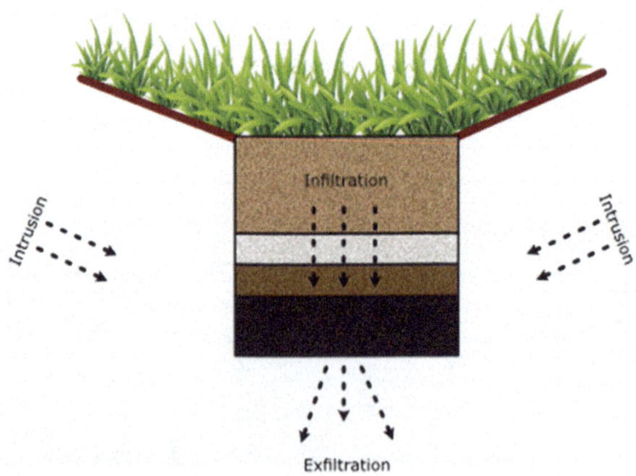

Figure 4.15 Conceptual definition of infiltration, exfiltration, and intrusion fluxes. *Source*: Tim Fletcher (University of Melbourne).

Bertrand-Krajewski *et al.* (2007) for guidance on methods for determining the spatial distribution of such infiltration testing within a large infiltration system.

4.5.1 Measuring infiltration

4.5.1.1 Whole-of-system measurement

Infiltration from the ponding zone into the filter media can be most simply determined by measuring the depth of ponded water, using water level sensors (see Chapter 3). The change in storage depth (or, more precisely, the change over time in storage volume/infiltration area, to account for changes in the surface and subsurface bathymetry) gives the infiltration rate. Calculation of the infiltration rate depends on the configuration of the system. For example, where infiltration then passes to an underdrain, which is considered to be non-limiting, the infiltration rate can be calculated using Darcy's equation, with the hydraulic gradient given as the ratio of the total depth (ponding depth + substrate depth) to the substrate depth. Where it is suspected that the pipe may be restricting outflows, measurement of the pipe outflow should be made (see Section 4.3.2) and compared to the infiltration rate.

4.5.1.2 Infiltrometer-based measurement

Infiltration rates can be measured using infiltration rings (often called infiltrometers), where water is applied to a ring inserted into the media, either a single- or double-ring, with either a constant head or falling head. Other commonly used methods include an air-entry permeameter or using boreholes. The commonly used methods have been reviewed by several authors (e.g. in Angulo-Jaramillo *et al.*, 2000; Philips & Kitch, 2011). The constant head test is simple to apply, is generally considered to be reliable and aims to measure infiltration rate after a steady state has been achieved. In theory, it measures the infiltration rate under saturated conditions (K_{sat}), allowing measurements at different locations or times to be compared.

One recent innovation in the measurement of *in situ* infiltration rate is that of Di Prima *et al.* (2016), using an automated single-ring infiltrometer operating with a quasi-constant head to measure steady state infiltration, and connected to an Arduino-controlled pressure transducer, thus recording infiltration rate to a logger (Figure 4.16). This dramatically reduces the personnel hours required to complete such *in situ* testing. In addition, Di Prima (2015) provides (in Appendix B) a free software code to automatically process the data.

4.5.2 Measuring exfiltration

Exfiltration from a stormwater control measure is of particular interest for determining, for example, its contribution to groundwater recharge. The suitable method for its measurement depends on the SCM configuration. For a system with temporary ponding, and without any infiltration medium, an infiltrometer approach is feasible, but for all other systems, the only suitable approach is to derive infiltration from change in water storage, taking into account both the surface water and subsurface storages, as described in Section 4.3.2. Again, it is noted that this approach requires that other components of the water balance (other inflows and outflows) can either be accurately quantified or can be ignored (for example, only data from periods with zero inflows and during the night, when evapotranspiration is minimal, may be considered).

In an SCM containing both surface and subsurface storages, a separate water level sensor is needed for each component, along with details of porosity and subsurface bathymetry.

Figure 4.16 (a) example application of single-ring infiltration testing using laborious manual measurement or, (b) automated systems. *Sources*: (a) Peter Poelsma (University of Melbourne); (b) Simone Di Prima (Università degli Studi di Palermo).

4.5.3 Measuring groundwater intrusion

Intrusion of groundwater into an SCM is perhaps the most difficult flux to quantify, given that its source will be diffuse. The approach to estimating this flux is the inverse of the whole-of-system infiltration measurement described above. For example, intrusion could be estimated at times when all other fluxes are either zero or known, so that the water balance can be closed. In systems with a measurable outlet (e.g. an outlet pipe or underdrain), long periods of zero inflow could thus be used to directly measure intrusion, assuming that discharge remaining after the normal system draining period is due to intrusion. However, in systems where there is no outlet, estimation is very difficult, as the water level in the system likely equilibrates with surrounding groundwater. It is thus recommended that a series of piezometer wells be installed around the SCM to measure groundwater level, so that it can be matched to water level measured within the SCM itself. Continuously measured groundwater levels can thus be used to estimate groundwater intrusion into the SCM.

4.6 EVAPOTRANSPIRATION

The importance of measuring evaporation and transpiration depends on (i) the aim of the monitoring and (ii) the configuration of the SCM being investigated. In an infiltration system with high infiltration rates, where water remains ponding for a matter of only a few hours after rain ceases, evaporation will not play a major role in the water balance. Conversely, in a vegetated bioretention system without an underdrain, and sitting on heavy clay soils, both evapotranspiration and transpiration could be important. Green roofs are also characterized by high transpiration and evapotranspiration rates. Table 4.1 gives the definitions of various quantities related to both transpiration and evapotranspiration.

Table 4.1 List of definitions.

Names	Definitions
Transpiration (*E*)	Transpiration (mm) refers to the movement of water through a plant from roots, to leaf surfaces, where it is released by stomata (small pores at the leaf surface) and is transferred to the atmosphere as vapour.
Evapotranspiration (*ET*)	Evapotranspiration (mm) refers to the combined movement of water from transpiration and evaporation of water (from vegetation or other surfaces) across a given area.
Reference Evapotranspiration (ET_0)	Reference evapotranspiration (mm) refers to the evapotranspiration derived from meteorological measurements for a given reference crop (e.g. grass).
Potential Evapotranspiration (*PET*)	Potential evapotranspiration (mm) refers to the maximum evapotranspiration estimated for a given surface area (or SCM), derived from reference evapotranspiration equations or pan evaporation.
Actual Evapotranspiration (*AET*)	Actual evapotranspiration (mm) refers to *PET* that has been modified by soil or crop coefficients to estimate evapotranspiration for given vegetation or soil conditions.
Crop factor (K_c)	A unitless coefficient that modified ET_0 according to the maximum transpiration of a given plant under well-watered conditions, relative to well-watered grass.
Sap flux density	Sap flux density ($cm^3\ cm^{-2}$) refers to the movement of sap through the xylem of woody vegetation, it can be positive, (representing movement from the roots), or negative (movement toward the roots) and is comprised of both transpiration, and stem refilling
Sapflow (Q)	Sapflow (L) refers to volumetric whole-plant water use estimated from sap flux density and the sapwood area of woody vegetation.

In non-vegetated systems, where transpiration can be ignored, evaporation can be readily estimated from local pan evaporation data (Locatelli *et al.*, 2017), noting that local microclimate factors may influence this estimate. In particular, daily shading of the SCM (affecting solar exposure) and local site factors such as surrounding buildings or vegetation (affecting relative humidity, air temperature, and wind speed) will strongly influence evaporation rates, as these are key microclimate drivers of evaporation.

For vegetated systems, evapotranspiration can be derived using several possible approaches:

- Calculation from meteorological data.
- Direct measurement of evapotranspiration or transpiration.
- Indirect estimation from the water balance.

4.6.1 Calculation of PET from meteorological data

Potential evapotranspiration can be derived from meteorological data, obtained either from an onsite or nearby weather station. Several equations are commonly used for this purpose, drawing on methods developed by Penman, Monteith and others (see Table 4.2). These equations estimate the reference

Table 4.2 Commonly used methods for calculation of *PET* from meteorological data.

Method	Equation	Crop/Conditions	Source & Comment
FAO Penman-Monteith (FAO56-PM)	$ET_0 = \frac{\Delta(R_n - G) + \rho_a c_p((e_s - e_a)/r_a)}{\Delta + \gamma(1 + r_s/r_a)}$ where R_n is the net solar radiation, G is the soil heat flux, $(e_s - e_a)$ represents the vapour pressure deficit of the air, ρ_a is the mean air density at constant pressure, c_p is the specific heat of the air, Δ represents the slope of the saturation vapour pressure temperature relationship, γ is the psychrometric constant, and r_s and r_a are the (bulk) surface and aerodynamic resistances.	Reference crop (alfalfa & grass)	(Monteith, 1965) This is the standard method recommended by the FAO (Food and Agriculture Organization).
Hargreaves	$ET_0 = 0.0023(T_{mean} + 17.8)(T_{max} - T_{min})^{0.5} R_a$ where T_{mean} is the mean air temperature, T_{max} is the maximum air temperature, T_{min} is the minimum air temperature, and R_a is the extraterrestrial radiation, also known as the solar constant, which varies geographically.	Well-watered short green crop	(Hargreaves *et al.*, 1985) This is a simplified version of the FAO56-PM equation used when solar radiation, relative humidity and wind speed data are not available. This approach tends to underestimate *ET* during high wind speeds and overestimate *ET* during high relative humidity. Considered as a good substitute for the Penman-Monteith equation at a monthly timestep.
Oudin	$ET_0 = 0.408 \cdot R_a \cdot [0.01 \cdot (T_{mean} + 5)]$ where the variables have already been defined above		(Oudin *et al.*, 2010) Similar to Hargreaves, a simplified version of the FAO56-PM equation utilized when climate data availability is limited. Suitable for coarse temporal scales such as monthly or annual.

evapotranspiration (ET_0) based on a well-watered reference crop (typically grass), which assumes that soil water content is not a limiting factor. Use of these equations without consideration of soil moisture limitations may result in significant errors during dry periods, especially for vegetation that quickly respond to drought by down-regulating transpiration. For example, Johannessen *et al.* (2019) demonstrated that the use of a simplified calculation of reference *ET* (Hargreaves equation in Table 4.2) combined with a single coefficient for soil recovery does not adequately estimate the volume of *ET* under soil drying.

The models for deriving *ET* from meteorological data presented in Table 4.2 were developed by their authors for a given reference crop, ranging from grass (for example for the Penman and FAO-24 Penman equations) to alfalfa for the Kimberly-Penman equation. There is thus a need to apply an adjustment – a crop factor noted K_c – to determine the potential evapotranspiration by accounting for different rates of transpiration likely from the vegetation in the SCM of interest (Grey *et al.*, 2018; Talebi *et al.*, 2019), such that:

$$PET = ET_0 \cdot K_c \tag{4.4}$$

where *PET* is the potential evapotranspiration from an SCM, ET_0 is the reference evapotranspiration, and K_c is the crop factor.

Finding values of crop factors can be difficult. A starting point could be to consider the table of crop factors provided by the Food and Agriculture Organization (FAO, 2019) and consider species which could be used as appropriate analogues for the species being considered (Grey *et al.*, 2018; Kristvik *et al.*, 2019). Alternatively, a literature review could be conducted for water use studies of the species of interest. A combined approach can derive crop factors from direct measurements (see below) of species of interest (Jahanfar *et al.*, 2018; Szota *et al.*, 2017).

Adjusting reference evapotranspiration (ET_0) to estimate the potential evapotranspiration (*PET*) by a crop factor is an important first step. However, Szota *et al.* (2018) suggest that including additional species-specific coefficients that capture sensitivity to soil drying represented by drought stress (leaf water potential) and transpiration rates will not only provide a more accurate estimation of *ET* in the water balance, but critically, give insight to the likely exposure and response of vegetation in SCMs (e.g. rain-gardens, green roofs, street-tree rain-gardens) to drought.

The most commonly used equations for calculating PET from meteorological data are summarized in Table 4.2. It is important to note, however, that a wide range of methods exist. Guo *et al.* (2019), for example, provide an R package 'Evapotranspiration' which calculates the potential or actual *ET* using up to 17 different methods.

4.6.2 Direct measurement of evapotranspiration, transpiration and evaporation

Evaporation can be simply measured using a pan (Figure 4.17), using standard methods (World Meteorological Organization [WMO], 2019). Where measurement of actual evapotranspiration is important, it can be directly measured using: (i) flux chambers or (ii) lysimeters, while transpiration alone can be estimated from measurements of (iii) sap flux or (iv) leaf stomatal conductance. A summary of these methods is given in Table 4.3.

4.6.2.1 Flux chambers

One direct way of measuring evapotranspiration is through the use of flux chambers, designed to quantify the flux of water vapour (and potentially of other gases, such as carbon dioxide), as shown in Figure 4.18.

Figure 4.17 Direct measurement of evaporation in a pan by means of a water level radar sensor installed above the pan. *Source*: courtesy Adama Kone (INSA Lyon).

Table 4.3 Summary of methods for direct measurement of evapotranspiration.

Method	Description	Advantages	Limitations	Application
Lysimetry	Measures total evapotranspiration from changes in weight of vegetation planted in a container.	Continuous data Can separate transpiration from evaporation	Measure between watering events	Glasshouse
Sapflow	Estimates transpiration from temperature changes in the xylem of woody vegetation following the application of heat.	Continuous data Measure during watering events	Damage to plants Requires scaling Requires sapwood properties Measures transpiration only	Field Glasshouse
Stomatal conductance	Estimates transpiration from the measurement of the stomatal conductance of a representative leaf (sunlit and shaded)	Measure transpiration during watering events	Point data Requires scaling Requires leaf area and fraction of shaded: sunlit leaves Measures transpiration only	Field Glasshouse
Flux chamber	Estimates transpiration from changes in the relative humidity in a closed chamber within a given time period quantified	Continuous measurements	Measures total evapotranspiration only Not suitable for large vegetation, Complex to establish	Field Glasshouse

Figure 4.18 Example of application of flux chambers to measure evapotranspiration in a vegetated infiltration system ("rain-garden") and in surrounding grass. *Source*: adapted from Hamel *et al.* (2014).

Flux chambers work by measuring changes in the relative humidity within a given time period, allowing the amount of water vapour emitted from the soil and vegetation surface to be quantified. Such a system has practical limitations in terms of its size, with chambers typically being limited to measurement of fluxes in grasses through to small shrubs (Figure 4.18). A review of chamber design options is outlined by Hamel *et al.* (2014), and includes open-flow, closed chamber and dynamic closed chamber. The reader is referred to papers by Deguchi *et al.* (2008) and McLeod *et al.* (2004) for a full description of the theory, design, application and validation of flux chambers. It should be noted that these systems can be quite complex to construct, and that their accuracy depends on system-specific calibration, as outlined by Hamel *et al.* (2014).

4.6.2.2 Lysimetry

A lysimeter refers to a device that measures actual plant-soil evapotranspiration from a bounded soil volume in a tank or container. It provides a way of estimating *ET* of experimental SCMs from the change in water fluxes of mass weight to allow the measurement of the overall water balance. There are two types of lysimeter (i) drainage and (ii) weighing. Drainage lysimeters can calculate *ET* using a water balance approach, whereby drainage is subtracted from a known quantity of water applied to the soil volume (either precipitation or irrigation) to determine *ET*. Drainage can be measured using tipping-bucket devices or weighing collected drainage on a separate load cell to the lysimeter. Weighing lysimeters measure *ET* through changes in mass between two time periods where no irrigation or precipitation occurs. The water balance of a lysimeter is therefore given as (Howell, 2005):

$$Vol_i = Vol_{i-1} + P_i + I_i - D_i - ET \tag{4.5}$$

where Vol_i is the volume of water in the lysimeter per unit lysimeter area (mm) at time i, Vol_{i-1} is the volume of water in the lysimeter per unit lysimeter area (mm) at time $i-1$, P_i is precipitation (mm) at time i, I_i is irrigation (mm) at time i, D_i is drainage from the lysimeter (mm) at time i and ET is evapotranspiration (mm).

Lysimeter systems are well equipped to determine both infiltration during a rainfall event and the evapotranspiration that occurs between events. However, they are limited in that they cannot measure *ET* during a rainfall event. Fortunately, *ET* during a rainfall event is normally minimal due to high relative humidity and low solar radiation, which are primary drivers for transpiration.

Lysimeters are highly diverse in form and design (Figure 4.19), ranging from large-scale in-ground tanks on a large load cell, to small potting containers on a small load cell. They have been used to determine the impact of *ET* on the water balance for experimental SCMs such as green roofs (Kemp *et al.*, 2019; Wadzuk *et al.*, 2013), or biofilters (Hess *et al.*, 2017; Szota *et al.*, 2018). This approach is widely used to measure transpiration of plants in a glasshouse setting and to calibrate other methods of measuring evapotranspiration from vegetated modules (Bleby *et al.*, 2004; Forster, 2017; Wadzuk *et al.*, 2013), as it is the most accurate measure of whole-plant transpiration. For weighing lysimeters, the accuracy of this approach is directly related to the accuracy of the load cell. Despite its accuracy, the reliance of this method on bounded soil volume means it is rarely applicable for SCMs in the field. In addition, the applicability of this approach to large SCMs is limited, as increasing plant or system size requires larger load cells at the cost of decreased accuracy and increased expense. Indeed, for full *in situ* systems, lysimetry is not possible.

Figure 4.19 Examples of direct measurement of evapotranspiration from changes in weight measured by load cells (lysimeter approach): (a) experimental green roof using mobile load cell above the module to collect periodic weight measurements, to determine evapotranspiration from change in module weights between simulated rainfall events. (b) advanced trees in potting containers atop load cells continuously logging weight to a central datalogger, to determine daily evapotranspiration as the change in container weight from morning (pre-dawn) to evening (dusk). *Sources*: (a) Zhang *et al.*, 2018; (b) Jasmine Thom (University of Melbourne).

4.6.2.3 Sapflow

In practice, vegetation planted in SCMs can rarely be weighed to determine the contribution of evapotranspiration to the water balance in the field. It is therefore necessary to utilize alternative methods. Sapflow sensors can estimate transpiration from SCMs, but are mostly limited to measurements of woody vegetation, such as trees, shrubs, or vines. Sapflow sensors measure the movement of water (sap flux density) through a plant, from temperature changes in the xylem (conducting wood), following the application of heat (Forster, 2017). The movement of heat through the plant is corrected for specific wood properties (to calculate sap flux density) and scaled up to estimate whole-plant transpiration from sapwood area.

There are several types of sapflow sensors that utilize three main approaches: thermal dissipation, heat velocity (Burgess *et al.*, 2001), or heat balance, each with their limitations and advantages (see Forster, 2017 for further detail). The advantage of using sapflow sensors is their discrete application in the field (Figure 4.20), continuous data, and ability to measure *ET* during stormwater events. However, despite careful installation and the application of correction factors for specific wood properties, accurately scaling sap flux density to whole-tree transpiration remains a challenge (Looker *et al.*, 2016; Wullschleger *et al.*, 2011).

While there are a number of studies that utilize sapflow measurements to describe the water use of crops (Silva *et al.*, 2008), forests (Pfautsch *et al.*, 2010), and urban trees (Litvak *et al.*, 2017), their application for estimating transpiration from SCMs has, so far, been limited. The only relevant study to date, Tirpak *et al.* (2018), describes patterns of sap flux density for trees planted in bioretention systems (suspended

Figure 4.20 Sapflow probes used to directly measure transpiration from SCMs with establishing trees. *Source*: Jasmine Thom (University of Melbourne).

pavements), but stops short of estimating the contribution of these trees to the overall water balance. Both researchers and practitioners are increasingly advocating for the inclusion of trees in SCMs to bolster the benefits of SCMs and urban forests concurrently (Berland *et al.*, 2017). Urban trees can transpire large quantities of water daily (up to 260 Lday^{-1}; Pataki *et al.*, 2011), so their contribution to the water balance of tree-based SCMs is substantial. Therefore, this method is likely to become an important part of measuring the water balance of tree-based SCMs in the field.

4.6.3 Stomatal conductance

Leaf stomatal conductance measurements are an alternative approach for measuring transpiration in field applications using a leaf porometer. These instruments measure the amount of water vapour passing through stomata at the surface of a leaf (stomatal conductance) by determining the difference in vapour pressure between the leaf and the air in a porometer chamber. Measurements are conducted on both sunlit and shaded leaves throughout the day to account for temporal and spatial differences in transpiration for a plant of interest. Values are then scaled for all leaves according to total leaf area and proportioned for sunlit or shaded leaves (Konarska *et al.*, 2016; Scharenbroch *et al.*, 2016). Using this method, Scharenbroch *et al.* (2016) was the first study to estimate the contribution of transpiration from tree-based SCMs to the water balance in the field. They suggested transpiration from trees in a 'green parking lot' in Illinois (that included stormwater control measures such as permeable pavements and bioswales) represents a substantial proportion of the water balance (46–72%). The challenge in using this method, as with sapflow measurements, is accurately scaling flow rates to the whole tree canopy. Unlike sapflow measurements, stomatal conductance also needs to be temporally scaled from point measurements to estimate continuous or daily data. This requires an understanding of tree characteristics (e.g. total leaf area, proportion of sunlit to shaded leaves) as well as the temporal variability of stomatal conductance across a range of climatic conditions. Given the limitations of using lysimetry in field applications, both stomatal conductance and sapflow measurements are the best approximation we have for estimating the contribution of transpiration from large woody vegetation, such as trees, to the water balance of SCMs.

4.6.4 Estimation of *ET* from the water balance

Evapotranspiration can be estimated from the water balance, provided that other parts of the balance can be estimated to give a closed balance. In this approach, the data are separated into periods where the *ET* flux can be separated from the exfiltration flux (if present). To do so, changes in storage are calculated when there is no inflow or outflow. The data are then separated into 'day' and 'night', on the basis that during the night there will be no *ET*, or that night-time *ET* is a negligible fraction of daily *ET*; any change in water level will be a result of exfiltration. This allows the exfiltration flux to be estimated, meaning that it can be subtracted from daytime changes in water storage, to derive net *ET* (Figure 4.21).

Inflow and outflow data are used to isolate periods when exfiltration and evapotranspiration are the only fluxes influencing the water level (i.e. stored volume) within the filter media of the bioretention basin. By isolating night time (with no *ET*), the infiltration rate can be estimated, and then subtracted from the rate of storage change within the filter media during the day (Bonneau *et al.*, 2018). It should be acknowledged that the assumption of no night-time transpiration may result in overestimation of exfiltration at night and subsequent underestimation of transpiration, since night-time transpiration has been shown to occur across a range of species, biomes, and seasons (Forster, 2014). For example, night-time transpiration rate of street trees in Sweden was measured as 11% of daytime transpiration for sun-exposed leaves and 23% of that for shaded leaves (Konarska *et al.*, 2016). How much this influences the overall water balance

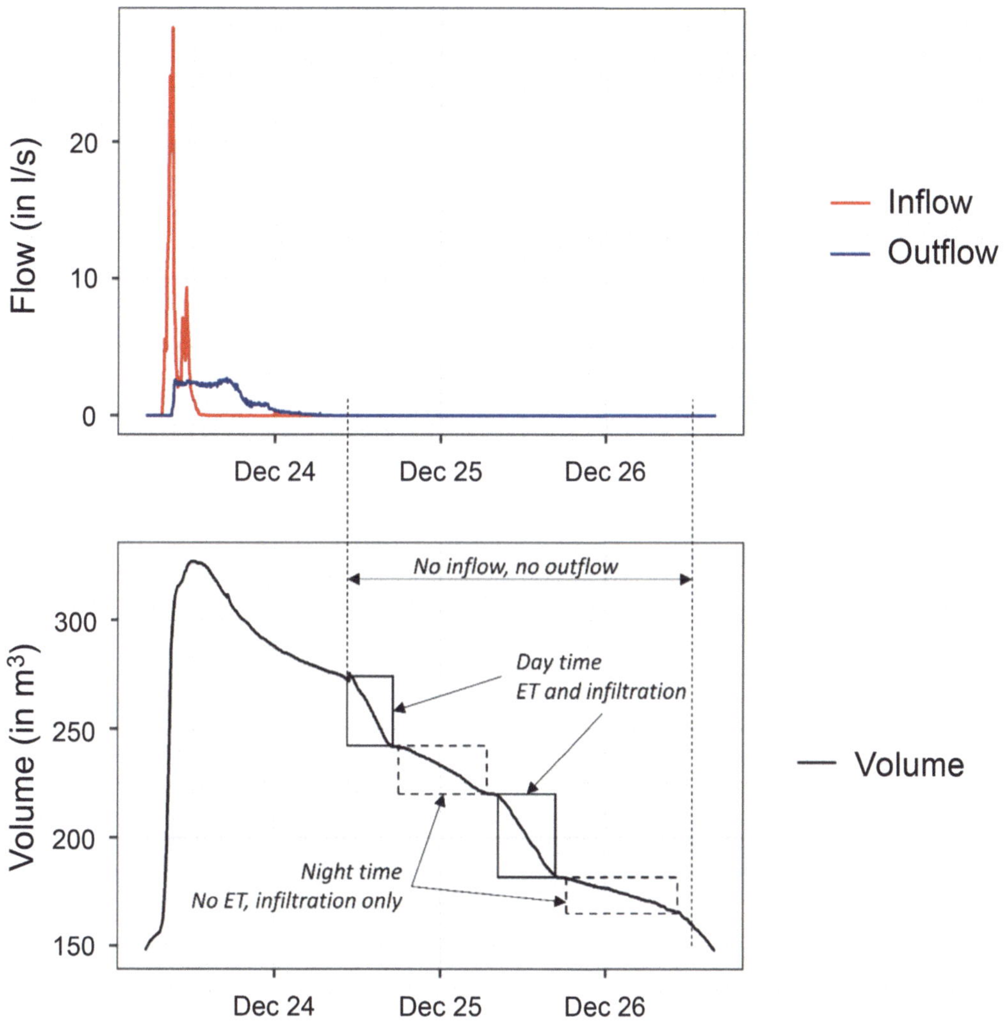

Figure 4.21 Example of using water volume in an SCM during periods free of other inflows or outflow, to derive evapotranspiration rate. *Source*: Jérémie Bonneau (INRAE).

will largely depend on the size, type, and distribution of vegetation in the SCM. In most cases, this error is likely to be within the uncertainty of other aspects of the water balance, and thus should not be an impediment to use of this approach to *ET* estimation.

4.7 SUMMARY AND TRANSITION

This chapter detailed the latest research experiences for SCM (or sustainable urban drainage systems [SUDS]) monitoring, with a focus on the water balance which is considered an important driver of urban hydrology. This new topic raised some questions/challenges, but due to the development of measuring technologies most of them have been overcome or will be overcome in the future. The progress on those systems is being updated regularly and a regular literature review is strongly advised to be aware of the latest findings, before embarking on any given monitoring programme or design.

Various processes occur in SCMs (infiltration, evaporation, etc.) and are central to the way in which SCMs operate and perform their design functions: special monitoring set-ups and technologies are required to collect data in these stormwater infrastructures. The often dispersed or decentralized nature of SCMs must also be taken into account in the design of SCM monitoring.

REFERENCES

Angulo-Jaramillo R., Vandervaere J.-P., Roulier S., Thony J.-L., Gaudet J.-P. & Vauclin M. (2000). Field measurement of soil surface hydraulic properties by disc and ring infiltrometers: A review and recent developments. *Soil and Tillage Research*, **55**(1–2), 1–29. doi: 10.1016/S0167-1987(00)00098-2.

Arias L., Grimard J.-C. & Bertrand-Krajewski J.-L. (2016a). First results of hydrological performances of three different green roofs. *Proceedings of Novatech 2016*, 28 June-1 July, Lyon, France, 4 p. Available at http://documents.irevues.inist.fr/bitstream/handle/2042/60504/1B24-131ARI.pdf (accessed 20 September 2020).

Arias L., Vacherie S. & Bertrand-Krajewski J.-L. (2016b). *Rapport du projet Gepeto – Performances hydrauliques et thermiques de trois toitures végétalisées expérimentales [Gepeto project report – Hydraulic and thermic performances of three experimental vegetated roofs]*. INSA Lyon, Villeurbanne (France), DEEP, 51 p. (in French).

Berland A., Shiflett S. A., Shuster W. D., Garmestani A. S., Goddard H. C., Herrmann D. L. & Hopton M. E. (2017). The role of trees in urban stormwater management. *Landscape and Urban Planning*, **162**, 167–177. doi: 10.1016/j.landurbplan.2017.02.017.

Bertrand-Krajewski J.-L. & Vacherie S. (2014). *Projet ECCLAIRA – Rapport de fin de contrat – Partie INSA Lyon LGCIE-DEEP: suivi expérimental de la toiture végétalisée du Centre des Congrès de Lyon [ECCLAIRA Project – Final report – INSA Lyon LGCIE-DEEP part: experimental monitoring of the Lyon Congress Centre vegetated roof]*. INSA Lyon, Villeurbanne (France), 24 p. (in French).

Bertrand-Krajewski J.-L., Fletcher T. D. & Mitchell V. G. (2007). Chapter 5 – Temporal and spatial scale considerations. In *Data Requirements for Integrated Urban Water Management*, T. D. Fletcher & A. Deletic (eds), Taylor and Francis, London (UK), Urban Water series – UNESCO IHP, pp. 45–64. ISBN 9780415453455.

Bleby T. M., Burgess S. S. O. & Adams M. A. (2004). A validation, comparison and error analysis of two heat-pulse methods for measuring sap flow in *Eucalyptus marginata* saplings. *Functional Plant Biology*, **31**(6), 645–658. doi: 10.1071/FP04013.

Bonneau J., Fletcher T. D., Costelloe J. F., Poelsma P. J., James R. B. & Burns M. J. (2018). Where does infiltrated stormwater go? Interactions with vegetation and subsurface anthropogenic features. *Journal of Hydrology*, **567**, 121–132. doi: 10.1016/j.jhydrol.2018.10.006.

Burgess S. S. O., Adams M. A., Turner N. C., Beverly C. R., Ong C. K., Khan A. A. H. & Bleby T. M. (2001). An improved heat pulse method to measure low and reverse rates of sap flow in woody plants. *Tree Physiology*, **21** (9), 589–598. doi: 10.1093/treephys/21.9.589.

Cherqui F., Szota C., James R., Poelsma P., Perigaud T., Burns M. J., Fletcher T. D. & Bertrand-Krajewski J.-L. (2019). Toward Proactive Management of Stormwater Control Measures Using Low-Cost Technology. *Proceeding of Novatech 2019*, 01–05 July, Lyon, France, 4 p.

Deguchi A., Hattori S., Daikoku K. & Park H. T. (2008). Measurement of evaporation from the forest floor in a deciduous forest throughout the year using microlysimeter and closed-chamber systems. *Hydrological Processes*, **22**(18), 3712–3723. doi: 10.1002/hyp.6974.

Di Prima S. (2015). Automated single ring infiltrometer with a low-cost microcontroller circuit. *Computers and Electronics in Agriculture*, **118**, 390–395. doi: 10.1016/j.compag.2015.09.022.

Di Prima S., Lassabatère L., Bagarello V., Iovino M. & Angulo-Jaramillo R. (2016). Testing a new automated single ring infiltrometer for Beerkan infiltration experiments. *Geoderma*, **262**, 20–34. doi: 10.1016/j.geoderma.2015.08.006.

FAO (2019). *Chapter 6 – ETc – Single Crop Coefficient (Kc)*. Food and Agriculture Organization, Rome (Italy). Available at http://www.fao.org/3/X0490E/x0490e0b.htm (accessed 12 September 2020).

Forster M. A. (2014). How significant is nocturnal sap flow? *Tree Physiology*, **34**(7), 757–765. doi: 10.1093/treephys/tpu051.

Forster M. A. (2017). How reliable are heat pulse velocity methods for estimating tree transpiration? *Forests*, **8**(9), 350, 11 p. doi: 10.3390/f8090350.

Garnier R., Castebrunet H., Cherqui F., Winiarski T., Barraud S. & Vacherie S. (2017). Micropollutant Removal Efficiency: Hydraulic Monitoring and Sampling Strategy for Source Control Stormwater Measures. *Proceedings of the 14th ICUD – International Conference on Urban Drainage*, 10–15 September, Prague, Czech Republic, pp. 1553–1561.

Grey V., Livesley S. J., Fletcher T. D. & Szota C. (2018). Tree pits to help mitigate runoff in dense urban areas. *Journal of Hydrology*, **565**, 400–410. doi: 10.1016/j.jhydrol.2018.08.038.

Guo D., Westra S. & Peterson T. (2019). *Modelling actual, potential and reference crop evapotranspiration – R Package 'Evapotranspiration'*. Adelaide University, Adelaide (Australia). Available at https://cran.r-project.org/web/packages/Evapotranspiration/Evapotranspiration.pdf (accessed 12 September 2020).

Hamel P., McHugh I., Coutts A., Daly E., Beringer J. & Fletcher T. D. (2014). An automated chamber system to measure field evapotranspiration rates. *Journal of Hydrologic Engineering*, **20**(2), 1–7. doi: 10.1061/(ASCE)HE.1943-5584.0001006.

Hargreaves G. L., Hargreaves G. H. & Riley J. P. (1985). Irrigation water requirements for Senegal River basin. *Journal of Irrigation and Drainage Engineering*, **111**, 265–275. doi: 10.1061/(asce)0733-9437(1985)111:3(265).

Hess A., Wadzuk B. & Welker A. (2017). Evapotranspiration in rain gardens using weighing lysimeters. *Journal of Irrigation and Drainage Engineering*, **143**(6), 04017004. doi: 10.1061/(ASCE)IR.1943-4774.0001157.

Howell T. A. (2005). Lysimetry. In *Encyclopedia of Soils in the Environment*, D. Hillel (ed.), Elsevier, pp. 379–386. doi: 10.1016/B0-12-348530-4/00391-X.

ISO (2007). *ISO 748:2007: Hydrometry – Measurement of liquid flow in open channels using current-meters or floats*. International Standards Organisation, Geneva (Switzerland), 46 p.

ISO (2010). *ISO 2425:2010 Hydrometry – Measurement of liquid flow in open channels under tidal conditions*. International Standards Organization, Geneva (Switzerland), 27 p.

ISO (2013). *ISO 18365:2013: Hydrometry – Selection, establishment and operation of a gauging station*. International Standards Organization, Geneva (Switzerland), 17 p.

ISO (2017). *ISO 11508:2017 Soil quality – Determination of particle density*. International Standards Organization, Geneva (Switzerland), 9 p.

Jahanfar A., Drake J., Sleep B. & Gharabaghi B. (2018). A modified FAO evapotranspiration model for refined water budget analysis for green roof systems. *Ecological Engineering*, **119**, 45–53. doi: 10.1016/j.ecoleng.2018.04.021.

Johannessen B. G., Hamouz V., Gragne S. & Muthanna T. M. (2019). The transferability of SWMM model parameters between green roofs with similar build-up. *Journal of Hydrology*, **569**, 816–828. doi: 10.1016/j.jhydrol.2019.01.004.

Kemp S., Hadley P. & Blanusa T. (2019). The influence of plant type on green roof rainfall retention. *Urban Ecosystems*, **22**, 355–366. doi: 10.1007/s11252-018-0822-2.

Konarska J., Uddling J., Holmer B., Lutz M., Lindberg F., Pleijel H. & Thorsson S. (2016). Transpiration of urban trees and its cooling effect in a high latitude city. *International Journal of Biometeorology*, **60**(1), 159–172. doi: 10.1007/s00484-015-1014-x.

Kristvik E., Johannessen B. G. & Muthanna T. M. (2019). Temporal downscaling of IDF curves applied to future performance of local stormwater measures. *Sustainability*, **11**, 24 p. doi: 10.3390/su11051231.

Litvak E., Manago K. F., Hogue T. S. & Pataki D. E. (2017). Evapotranspiration of urban landscapes in Los Angeles, California at the municipal scale. *Water Resources Research*, **53**(5), 4236–4252. doi: 10.1002/2016WR020254.

Locatelli L., Mark O., Mikkelsen P. S., Arnbjerg-Nielsen K., Deletic A., Roldin M. & Binning J. P. (2017). Hydrologic impact of urbanization with extensive stormwater infiltration. *Journal of Hydrology*, **544**, 524–537. doi: 10.1016/j.jhydrol.2016.11.030.

Looker N., Martin J., Jencso K. & Hu J. (2016). Contribution of sapwood traits to uncertainty in conifer sap flow as estimated with the heat-ratio method. *Agricultural and Forest Meteorology*, **223**, 60–71. doi: 10.1016/j.agrformet.2016.03.014.

McLeod M. K., Daniel H., Faulkner R. & Murison R. (2004). Evaluation of an enclosed portable chamber to measure crop and pasture actual evapotranspiration at small scale. *Agricultural Water Management*, **67**(1), 15–34. doi: 10.1016/j.agwat.2003.12.006.

Monteith J. L. (1965). Evaporation and environment. *Symposia of the Society for Experimental Biology*, **19**, 205–234.

Oudin L., Moulin L., Bendjoudi H. & Ribstein P. (2010). Estimating potential evapotranspiration without continuous daily data: possible errors and impact on water balance simulations. *Hydrological Sciences Journal*, **55**(2), 209–222. doi: 10.1080/02626660903546118.

Pataki D. E., McCarthy H. R., Litvak E. & Pincetl S. (2011). Transpiration of urban forests in the Los Angeles metropolitan area. *Ecological Applications*, **21**(3), 661–677. doi: 10.1890/09-1717.1.

Pfautsch S., Bleby T. M., Rennenberg H. & Adams M. A. (2010). Sap flow measurements reveal influence of temperature and stand structure on water use of *Eucalyptus regnans* forests. *Forest Ecology and Management*, **259**(6), 1190–1199. doi: 10.1016/j.foreco.2010.01.006.

Philips E. C. & Kitch W. (2011). A review of methods for characterization of site infiltration with design recommendations. *Journal of the Nevada Water Resources Association*, **6**, 29–46.

Scharenbroch B. C., Morgenroth J. & Maule B. (2016). Tree species suitability to bioswales and impact on the urban water budget. *Journal of Environment Quality*, **45**(1), 199–206. doi: 10.2134/jeq2015.01.0060.

Silva R. M., Paço T. A., Ferreira M. I. & Oliveira M. (2008). Transpiration of a kiwifruit orchard estimated using the granier sap flow method calibrated under field conditions. *Proceedings of the 5th International Symposium on Irrrigation of Horticultural Crops*, **792**, 593–600. doi: 10.17660/ActaHortic.2008.792.70.

Szota C., Farrell C., Williams N. S. G., Arndt S. K. & Fletcher T. D. (2017). Drought-avoiding plants with low water use can achieve high rainfall retention without jeopardising survival on green roofs. *Science of the Total Environment*, **603–604**, 340–351. doi: 10.1016/j.scitotenv.2017.06.061.

Szota C., McCarthy M. J., Sanders G. J., Farrell C., Fletcher T. D., Arndt S. K. & Livesley S. J. (2018). Tree water-use strategies to improve stormwater retention performance of bio filtration systems. *Water Research*, **144**, 285–295. doi: 10.1016/j.watres.2018.07.044.

Talebi A., Bagg S., Sleep B. E. & Carroll D. M. O. (2019). Water retention performance of green roof technology: A comparison of Canadian climates. *Ecological Engineering*, **126**, 1–15. doi: 10.1016/j.ecoleng.2018.10.006.

Tirpak R. A., Hathaway J. M. & Franklin J. A. (2018). Evaluating the influence of design strategies and meteorological factors on tree transpiration in bioretention suspended pavement practices. *Ecohydrology*, **11**(8), e2037. doi: 10.1002/eco.2037.

Wadzuk B. M., Schneider D., Feller M. & Traver R. G. (2013). Evapotranspiration from a green-roof storm-water control measure. *Journal of Irrigation and Drainage Engineering*, **139**(12), 995–1003. doi: 10.1061/(ASCE)IR.1943-4774.0000643.

WMO (2019). *Chapter 10 – Measuring Pan Evaporation.* World Meteorological Organization, Geneva (Switzerland), 13 p. Available at https://library.wmo.int/doc_num.php?explnum_id=3156 (accessed 12 September 2020).

Wullschleger S. D., Childs K. W., King A. W. & Hanson P. J. (2011). A model of heat transfer in sapwood and implications for sap flux density measurements using thermal dissipation probes. *Tree Physiology*, **31**(6), 669–679. doi: 10.1093/treephys/tpr051.

Zhang Z., Szota C., Fletcher T. D., Williams N. S. G., Werdin J. & Farrell C. (2018). Influence of plant composition and water use strategies on green roof stormwater retention. *Science of the Total Environment*, **625**, 775–781. doi: 10.1016/j.scitotenv.2017.12.231.

Chapter 5

Data communication and storage

Flora Branger[1], Simon Tait[2], Véronique Chaffard[3], Elodie Brelot[4], Vivien Lecomte[4], Isabelle Mallet[5] and Peter Skipworth[6]

[1]*INRAE, UR Riverly, Villeurbanne, France*
[2]*University of Sheffield, Civil and Structural Engineering, Sheffield, UK*
[3]*IRD – Institut de Recherche pour le Développement, IGE, Grenoble, France*
[4]*GRAIE, Villeurbanne, France*
[5]*Métropole de Lyon, Lyon, France*
[6]*Environmental Monitoring Solutions Ltd., Sheffield, UK*

ABSTRACT

Monitoring programs in urban drainage systems generate, potentially, a huge amount of data from sources distributed in the urban environment, working at relatively high sampling rates for extended periods of time. Collecting data using adaptable and reliable communication systems is the first challenge. Then structuring the collected data is a first requisite for effectively managing the quality and accessibility of the data. In adjacent fields of research, the topic of managing huge collections of data has resulted in several (open) standards and protocols for database structure, transfer and storage to ensure unambiguous definitions on which parties can build their workflows/software. This chapter describes relevant approaches for urban drainage and stormwater management systems, and appropriate standards along with examples from case studies.

Keywords: Data accessibility, database, data communication, data standards.

doi: 10.2166/9781789060119_0135

5.1 INTRODUCTION

The cost of reliable data collection is becoming a significant burden for many water utilities and a major constraint for researchers in the field of urban drainage and stormwater management (UDSM). For water utilities there is often a need to demonstrate to a government regulator that a particular level of performance has been achieved. This need is likely to increase as monitoring is a key aspect of the Water Framework Directive (European Union [EU], 2000) and requires demonstration of compliance with earlier legislation such as the Urban WasteWater Treatment Directive (EU, 1991, 1998). An example of this need for enhanced monitoring is the instruction sent to UK water companies in September 2013 requiring that the majority of combined sewer overflows (CSOs) must be monitored for their frequency and duration of operation. Whilst the measurement itself may have its own complexity (see Chapters 3 and 6), the need to transfer the data securely, the number of sites involved and the need to demonstrate the quality of the collected data to the environmental regulator made this a technically challenging task. Currently the regulator, the UK's Environment Agency, has taken a risk-based approach to implement this monitoring program. It is expected that by 2020 over 30,000 CSOs in the UK will have event-based monitoring in which the data are telemetered back to central sites or that data are collected and stored securely before being collected manually. This example is a good illustration of the costs of data storage and communication for system monitoring for environmental purposes.

With various quantities measured at many locations, with, in general, consistent frequency data collection needed, for example every minute or so (see Section 6.2.5) to be able to capture the dynamic behaviour of an urban drainage system, and over long periods of time of the order of many months, data sets can quickly become overwhelmingly large. The challenges of data communication and storage are therefore not losing any information and ensuring that the data can be efficiently accessed and used, whatever that use might be (data analysis, calibration of numerical models, publication, etc.). This requires a carefully thought workflow from the sensor to the database, and appropriate strategies for data structuration and storage. Another challenge is that there is also increasing pressure from public authorities to make data available for the public. For example, the new European Open Data Directive (EU, 2019) requires from the Member States that all data produced by public sector activities (including research, ministries, state agencies and municipalities) should be available with open access or at marginal costs, and should be reused as much as possible, including for economic purposes. This goes

Key messages on data communication and storage and workflow in urban drainage systems

- KM 5.1: *Data Communication* – Two key issues need to be considered: reliability of data transfer and data latency. Acceptable threshold of reliability and latency are determined by the intended use of the data, for example if the data is to be used for monitoring or warning, each use requires different minimum levels of reliability and latency. Cost and power usage are secondary concerns.
- KM 5.2: *Data storage* – The use of data base management systems requires an important initial investment but guarantees data security, fast and efficient access and opportunities for data sharing and interoperability in the long-term. However tempting, spreadsheets are no DBMS (Data Base Management System) and are not appropriate for data storage.
- KM 5.3: *Workflow* – The workflow process for supplying data to the database and updating it regularly is as important as the design of the database structure itself.

along with the concept of FAIR data that is now becoming a major objective of data management in research and many public administrations (Wilkinson *et al.*, 2016). FAIR means Findable, Accessible, Interoperable and Reusable. These are the conditions that must be met in order to make data available in a long-term perspective and taking advantage of the current advanced technologies (cloud storage, webservices, big data). The development of a communication and storage strategy should therefore also take that into account.

This chapter presents the current state of the art and guidelines for building a strategy and choosing the appropriate tools for data storage and communication, from *in situ* sensors to interoperable databases. It also provides practical case studies with recommendations from practitioners.

5.2 FROM *IN SITU* SENSORS TO DATA FILES – DATA TRANSFER METHODS

Collecting spatially distributed real time data in urban drainage and stormwater management systems is challenging. The traditional way in which researchers and water authorities have transferred data is using SCADA (supervisory control and data acquisition) systems. This approach uses technology that was developed in the 1980s. This traditional approach requires significant investment and delivers systems with inflexibility and significant ongoing costs and is often unsuitable for research studies. More recently many commercial sensor units have coupled to GSM (global system for mobile communication) communication modules, but these often have issues with data reliability, mainly caused by the intermittence of the GSM network and poor power reliability. Wireless communication between sensors and local hubs via radio telemetry is becoming more popular due to its lower costs and reduced latency and reliability issues compared with cellular communication. Radio telemetry is a low power technology. Transmitters and receivers can thus be developed with long battery lives of up to 5 years. Some radio devices can be set to transmit at pre-set intervals, increasing the transmit frequency if certain urban drainage conditions are met, such as high-water levels.

The industrial, scientific and medical (ISM) radio band is a portion of the radio spectrum reserved internationally for the use of radio frequency (RF) energy for industrial, scientific and medical purposes. The frequencies and power of these bands vary from country to country. The most common frequencies encountered are the 868 MHz band – Europe, the 915 MHz band – North America, South America and some other countries, and the 2.4 GHz band – nearly worldwide. In Europe, the use of the ISM band is covered by Short Range Device regulations issued by the European Commission. In the United States, uses of the ISM bands are governed by the Federal Communications Commission (FCC). Short Range Devices (SRD) is a general term, applied to various radio devices designed to operate on a licence-exempt basis, over short range and at low power levels. This includes alarms, telemetry and tele-command devices with maximum power of up to 500 mW at VHF/UHF (very high frequency/ultra high frequency). Frequencies are made available for licence-free operation in the bands of 433 MHz and 868 MHz in Europe and 915 MHz in the USA, Australia and Canada. Modules on different frequencies have different performance characteristics. The 433 MHz frequency has good penetration through structures but low transmit power (10 mW) and hence a limited range. The higher frequency 868 MHz gives improved transmit range thanks to 500 mW power, and can provide 10 km line of sight, and up to 1.3 km transmission lengths through urban environments. It must be noted that both radio frequencies are lower than those used by 2G and 3G services, so they perform better in below ground structures but have far less range than cellular systems due to the limited power. The ISM frequencies are used in networks often referred to as low power wide area networks (LPWAN) used by WiFi (802.11), LoRa, Bluetooth and ZigBee devices (LoRaWan, 2015). However, recommending a particular technology for a specific field application is difficult as the different technologies are developing rapidly in a highly competitive market and have specific relative strengths and weaknesses. This frequency band has a good range if unobstructed, but limited penetration through structures. One of the most interesting aspects of these technologies is the straightforward ability to form local networks as their transmission range in above ground urban areas is a few hundred metres, but their range can be extended indefinitely by organizing them into mesh networks (See *et al.*, 2012). The disadvantage of this type of network is the cost of the additional communication/repeater hubs required to obtain the desired transmission range. For underground applications there is significant attenuation loss and transmission ranges are consequently reduced. The key aspects for this type of mesh-based networks are required range, battery life, robustness to interference and data transfer capacity. Such communication networks have begun to be deployed for research studies associated with urban drainage systems. Ebi *et al.* (2019) reported on the development of a LoRaWAN based arrangement of underground sensor nodes, synchronized to close by over ground repeater nodes which can then transmit data to more remote gateways. At two sites with test periods up to two months, they demonstrated high reliability of data transfer, although data transfer rates were very limited. These studies have demonstrated the flexibility, low cost and adaptability of such radio telemetry-based systems.

5.3 FROM DATA FILES TO STRUCTURED DATABASE

5.3.1 Principles and advantages of relational databases

Data are usually retrieved from the field as unstructured files, either text or binary files. They can be stored as such on a personal computer, or in a shared repository of the organization or company intranet, if they are to be shared between several users. It is common that data are imported in Microsoft Excel: it is a widely available software that most know how to use. Embedded calculation formulas, programming functionalities and automated graph generation allow the user to get a first insight into the data without too much effort. However, these solutions, even if they are certainly widely used, present major drawbacks:

- Apart from very small datasets, the data have to be structured manually: the files are organized in several folders, and/or in worksheets in the case of a workbook. One must be very rigorous to keep the structure as the dataset grows and becomes more complex. Eventually it can become impossible to maintain.
- In some cases, the data files do not contain all the information required to use the data, such as variable names, units, time zones, contact of data producer, types of sensors and so on. These metadata must be added to the dataset, thus adding complexity for manual maintenance, particularly when several people are involved in data collection.
- Access to the data themselves is difficult and slow: one has to find the right file in the right folder, open it or import the data into a processing software without making errors with the format and so on.
- There is no possibility of sharing the data while ensuring its integrity and consistency: copies of the file can be multiplied in several places, files can be modified, updates are not necessarily transmitted.

These limitations can be overcome using relational databases. In a relational database, information is organized in bidimensional tables (or relations, with columns and rows) that are related to each other. Each record in a table (i.e. each row) can be identified individually using a unique identifier (usually an integer value) that is called a primary key. The connection between tables is achieved using foreign keys. It consists of adding a column to a table containing the identifier for each row of the related row in the other table. All complex relations between tables can usually be decomposed into simpler ones.

The main advantages of relational databases compared to unstructured files are:

- Information is not often duplicated, thanks to the use of primary and foreign keys. Thus, there is no information redundancy. The data volume is lower, which optimizes storage.
- The structure of the database is fixed at once, which ensures a better consistency of data and a better sustainability of the database, as the dataset grows over the years. There can be constraints on the data (e.g. one value per date for a given time series, to ensure that there are no duplications).
- Relational databases use a specific programming language for accessing the data: SQL (Structured Query Language). It is a standardized language that allows quick and efficient access to the data, whatever the size of the dataset, which can be unlimited. Complex manipulations of data can also be carried out with SQL.

All these functionalities are implemented in dedicated software tools called Data Base Management Systems (DBMS). Most of these tools come also with additional functionalities such as:

- Data centralization (on a server) and management of access rights (data integrity/security).
- Automatic backups.
- Spatial extension/compatibility with GIS software tools.

5.3.2 Existing DBMS and software solutions

Several Data Base Management Systems are available on the market. A short list of the five 'best known' DBMS is presented in Table 5.1. What distinguishes the software tools are mostly the licence conditions (proprietary vs. free and open-source), the maximum size allowed for the database, and the possibility of adding spatial data into the database. This spatial extension feature makes databases compatible with GIS (geographical information system) software, such as QGIS or ArcGIS. GIS software can use a database with a spatial extension as a centralized data source. The database can then benefit from the functionalities of the GIS software (data visualization, maps and so on). Microsoft Access is distributed with the Microsoft Office Suite, and as such it is the first DBMS that one may come across. It is suitable for small datasets (due to the size limitations) that do not need to be shared between several people. It has no server service. This means that each database is one file stored on a personal computer with no access control: it can be duplicated or modified by any user of the computer. MySQL is the backbone behind most of the commercial websites in the world. PostgreSQL and OracleDB are more specialized in scientific applications. The main difference between them is the licence conditions. PostgreSQL is the open-source reference for structuring and management of environmental observations. It is used for the

Table 5.1 Main database management systems and their key characteristics.

	Max size of a table	Spatial extension (OGC standards)	Server service	Licence
Microsoft Access	2 Gb/database	No	No	Proprietary
Microsoft SQL Server	2 Gb/table	Yes (partial)	Yes	Proprietary
MySQL	Unlimited	Yes (partial)	Yes	Depends
PostgreSQL	Unlimited	Yes (PostGIS)	Yes	Free and open source
OracleDB	Unlimited	Yes (Oracle Spatial)	Yes	Proprietary

OGC – Open Geospatial Consortium.

storage of data from many research observatories throughout the world, as for example in BDOH that stands for Base de Données pour les Observatoires en Hydrologie (i.e. Database for Hydrologic Observatories, Branger *et al.*, 2014), AMMA-CATCH (Galle *et al.*, 2018), ILMS (Zander *et al.*, 2013), Hydroshare (Heard *et al.*, 2014; Yi *et al.*, 2018) or Drought Observatory (Magno *et al.*, 2018).

All the DBMS (even those with a server service) can be installed on a regular PC (which then becomes a 'local server'). One can then benefit from the robust data structure that they offer, the unlimited size of datasets and the efficiency of the SQL queries for data manipulation and access. However, these systems show their full potential (in particular data sharing with access control, proper handling of multiple users simultaneously and data backups) when installed on a proper server independent from the PCs of the data collectors and users.

Data users can have access to the database through specific applications installed on their own computers (heavy or rich client), or more simply with a web browser (thin client). A typical architecture is called 'three-tier' and involves the data server (that contains the DBMS and the datasets), a dedicated web application (program that accesses the server), and a web server on which the users can connect with their computer browser.

Commercial software dedicated to environmental monitoring and based on DBMS can be found on the market. Aquarius by Aquatic Informatics (https://aquaticinformatics.com/products/) and WISKI by Kusters (https://www.kisters.net/NA/products/wiski/) can be cited as examples. Both rely on Oracle DB databases. These software products provide solutions for data structuration, backups, secured access plus a wide variety of functionalities for data visualization and processing. There are also a few free and open-source alternatives made available for a wider public, such as BDOH and Hydroshare (see Section 5.5) which were initially developed for environmental research applications.

5.3.3 Typical data structuration for environmental time series

As an alternative to pre-packaged commercial or open-source solutions, it is also possible to develop custom-made databases. This requires more time and human resources for the informatic development, but it ensures that the database will be adapted to the specificities of the data and to the needs of the end-users. CUAHSI (Consortium of Universities for the Advancement for Hydrologic Science, a non-profit organization that includes many American universities) proposes guidelines about important aspects for database development, such as defining a controlled vocabulary for the names of the variables (Horsburgh *et al.*, 2014) and a database structure that can be used as a template (Horsburgh *et al.*, 2016). Since most data collected are numerical time series, they must be also supported by metadata, i.e. additional information about the measurement attributes (names, time zones, locations of sensors, physical conditions at the measurement sites, etc.). These metadata are essential for the interpretation and further use of the data. Chapters 7 and 9 highlight the type of metadata to be collected and their importance for data quality assurance, respectively.

An important step when building a database is the design of the data structure, i.e. how to dispatch the information into several tables, and how these tables are related to each other. Several options are possible, according to the specific nature of the data and the objectives of the data collectors and users. For example, if there are several measurement points scattered over a large territory, it will be particularly important to be able to locate these measurement points. In other cases, emphasis can be put on the sensors that produce the data, keeping track of information such as the nature of sensor, brand, series number, and maintenance history (sensor replacement, etc.). In other cases, the database will have to be able to deal with several possible time steps for the time series (or even variable time steps).

For example, in the BDOH database (Branger *et al.*, 2014 – see also Section 5.5.1), one of the main requirements was to be able to manage time series with variable time steps. Another important feature was that each value should be associated with an estimation of data quality and possibly a confidence interval corresponding to the data uncertainty (see Chapter 6 for uncertainty assessment). The database was designed with one single table containing all the measured values (by end of 2020, it contained over 60 million entries), that is represented in Figure 5.1. Each entry has a unique index value (id), a date-time value given at 1 millisecond discretization level, the measured value ('valeur' in Figure 5.1) itself, the min/max values of the uncertainty interval and a code corresponding to the data quality. This data quality code is linked with another table that contains all the possible quality codes for the data, hence the foreign key. There is also a reference to the id of the time series the measured values belong to ('chronique_id', linked to the table 'chronique', e.g. time series). All the metadata (description of the quantities, units, etc.) are stored in the 'chronique' table, which is linked to a table 'station' with monitoring stations localized with their XYZ coordinates (Figure 5.2).

5.3.4 Supply of information to databases

In addition to choosing the appropriate tools and building the database, one crucial point is also to define a processing workflow that ensures that the data actually get into the database and are updated regularly. It can be a fully automated process from *in situ* sensors to the database server and even to the data online publication, or it can include several manual steps such as data collection, data validation, uncertainty assessment, etc. All of these aspects must be taken into account and organized so that no data are lost and the information system is efficient (Horsburgh *et al.*, 2011). One should ensure that the time between actual measurements and the proper storage of the corresponding measured values in the database must not be too long. A delay of 3 years between data collection and storage is not acceptable, although it is unfortunately not that uncommon. In such a case, the memory of data acquisition, data validation and data processing is lost and leads to poor final quality.

5.4 DATABASE INTEROPERABILITY

5.4.1 Definition and interest

Interoperability can be defined as a set of characteristics of a system, that allows it to exchange information seamlessly with other systems. In the world of data management, interoperability means that several databases can communicate with each other and exchange information on the data they contain (metadata) and/or data itself. It is one of the pillars of the FAIR data concept.

Interoperability presents huge advantages for data management and exploitation. Instead of having to build and maintain one large and centralized database that must contain everything, it allows entities to build several smaller and specialized databases, that are easier to design and maintain, especially if the data they contain are produced by different entities. It can also bring together data of different sorts that come from various sources, thus providing additional elements for data interpretations. A typical example could be bringing together maps of the sewer system and stormwater control measures (SCMs), logs of the operations on weirs, and continuous flow measurements from *in situ* sensors. Mostly these three types of data are not available together, at least not in a straightforward way. One person has to request them from two or three different people, each time a question arises. A change of direction of flow monitored by the sensors could typically be interpreted after several weeks of investigation as due to an operation on a weir in a nearby branch of the sewer. If these data were stored in interoperable systems, they would be known to all the systems (and operators) as soon as they are uploaded in their respective databases, thus saving a lot of time and energy. Finally, interoperability is the cornerstone of public data portals, such as Data Grand Lyon (see

id	qualite_id	chronique_id	date	valeur	minimum	maximum
11076435	1	150	2006-02-12 13:30:00	7	*NULL*	*NULL*
11076436	1	150	2006-02-12 14:00:00	6.98	*NULL*	*NULL*
11076437	1	150	2006-02-12 14:30:00	6.98	*NULL*	*NULL*
11076438	1	150	2006-02-12 15:00:00	6.98	*NULL*	*NULL*
11076439	1	150	2006-02-12 15:30:00	6.98	*NULL*	*NULL*
11076440	1	150	2006-02-12 16:00:00	6.98	*NULL*	*NULL*
11076441	1	150	2006-02-12 16:30:00	6.98	*NULL*	*NULL*
11076442	1	150	2006-02-12 17:00:00	6.97	*NULL*	*NULL*
11076443	1	150	2006-02-12 17:30:00	6.97	*NULL*	*NULL*
11076444	1	150	2006-02-12 18:00:00	6.95	*NULL*	*NULL*
11076445	1	150	2006-02-12 18:30:00	6.97	*NULL*	*NULL*
11076446	1	150	2006-02-12 19:00:00	6.95	*NULL*	*NULL*
11076447	1	150	2006-02-12 19:30:00	6.97	*NULL*	*NULL*
11076448	1	150	2006-02-12 20:00:00	6.95	*NULL*	*NULL*
11076449	1	150	2006-02-12 20:30:00	6.95	*NULL*	*NULL*
11076450	1	150	2006-02-12 21:00:00	6.95	*NULL*	*NULL*
11076451	1	150	2006-02-12 21:30:00	6.93	*NULL*	*NULL*
11076452	1	150	2006-02-12 22:00:00	6.95	*NULL*	*NULL*
11076453	1	150	2006-02-12 22:30:00	6.93	*NULL*	*NULL*
11076454	1	150	2006-02-12 23:00:00	6.92	*NULL*	*NULL*
11076455	1	150	2006-02-12 23:30:00	6.92	*NULL*	*NULL*
11076456	1	150	2006-02-13 00:00:00	6.93	*NULL*	*NULL*
11076457	1	150	2006-02-13 00:30:00	6.93	*NULL*	*NULL*
11076458	1	150	2006-02-13 01:00:00	6.92	*NULL*	*NULL*
11076459	1	150	2006-02-13 01:30:00	6.91	*NULL*	*NULL*
11076460	1	150	2006-02-13 02:00:00	6.91	*NULL*	*NULL*
11076461	1	150	2006-02-13 02:30:00	6.91	*NULL*	*NULL*
11076462	1	150	2006-02-13 03:00:00	6.89	*NULL*	*NULL*
11076463	1	150	2006-02-13 03:30:00	6.91	*NULL*	*NULL*
11076464	1	150	2006-02-13 04:00:00	6.9	*NULL*	*NULL*

Figure 5.1 Extract of the main storage table in BDOH. *Source*: Flora Branger (INRAE) (Screenshot of the BDOH database).

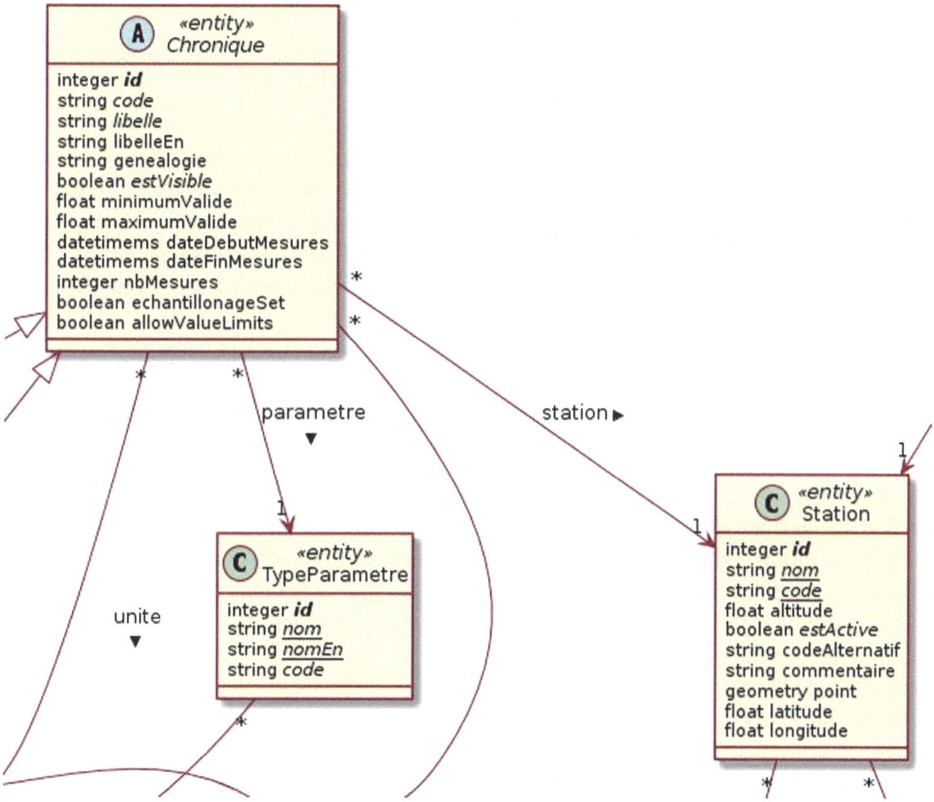

Figure 5.2 Extract of BDOH database structure: chronique and station tables. *Source*: Flora Branger (INRAE) (Screenshot of the BDOH database).

Section 5.5.3) or the Theia/OZCAR Information System (Braud *et al.*, 2020). Such data portals present and give access to datasets that come from various data producers and are stored in various databases. They are the full implementation of the FAIR data concept, and as such, can be considered as the future of data storage.

5.4.2 Interoperability standards and examples

As for sensors, interoperability means that communication protocols must be defined for databases. These communication protocols operate through the web, and as such are called webservices. In the field of environmental data, the Open Geospatial Consortium (OGC) is in charge of defining open standards for these webservices. The OGC is an international standards organization, that was created in 1994, initially to foster the development of open geospatial tools. The most well-known and widely used standard webservices proposed by the OGC are the CSW (Catalogue Service for the Web) that is behind online metadata catalogues (such as https://www.data.gouv.fr/fr/ (accessed 07 Dec 2020) in France, https://www.ukdataservice.ac.uk/ (accessed 07 Dec 2020) in the UK, etc.) and information sharing between them (called metadata harvesting: information added in one catalogue is automatically replicated in the others), and also the WMS (Web Map Service) and WFS (Web Feature Service) that are behind most displays of interactive maps on the web.

In terms of *in situ* measurements, several standards are proposed by the OGC:

- The Sensor ML standard (https://www.ogc.org/standards/sensorml (accessed 07 Dec 2020)) describes the measurement process of observation data (description of sensors and also post-measurement data processing) but does not contain the measured values themselves.
- The Observation & Measurement (O&M) standard (http://www.opengeospatial.org/standards/om (accessed 07 Dec 2020)) describes the properties of observations, including surrounding environment features (location, sampling strategy if any, organization). It is complementary to the SensorML standard and can contain the actual data.
- The Sensor Observation Service (SOS) (http://www.opengeospatial.org/standards/sos (accessed 07 Dec 2020)) defines a webservice interface for querying real-time sensor data and sensor metadata. It uses the information encoded in SensorML (description of Sensor) and O&M (description of measurements).

These three standards are the key to the implementation of data operability. They are not specialized to water-related data, and are generic enough to handle any kind of observation data. More specific to hydrological data, the WaterML standard (https://www.ogc.org/standards/waterml (accessed 07 Dec 2020)) has been defined more recently. It introduces additional concepts specific to hydrological time series, such as gauging and rating curves, or temporal data interpolation. However, it is not as widely used as the three other standards presented here.

OGC provides standards and specifications, but does not provide ready-to-use tools. The implementation of the standards has to be carried out. The standards are long and complex, they do not have to be implemented to their full extent. Several implementations of the SOS webservice are available. The most widely used implementation, recognized as reference, is the open-source German-based 52°North solution (https://52north.org/software/software-projects/sos/ (accessed 07 Dec 2020)). This solution also takes into account the WaterML standard. The 52°North SOS is currently behind many environment data portals, and in particular behind the case study Data Grand Lyon.

5.4.3 Practical recommendations

Data portals that implement interoperability are high level software constructions. As a water scientist or a practitioner, the objective is not to be able to implement them or manage them directly, but just to understand the underlying concepts enough to be able to communicate with the specialists. The development and maintenance of interoperability in databases requires skilled software developers and IT support. However, many organizations now do have the corresponding human resources, in universities or research institutes, and in local authorities' administrations (regional councils, city councils).

Although we can see interoperable data portals as the future of data storage, we must also keep in mind that the data must be properly structured and stored beforehand, following the guidelines given in the previous sections. If the data is not structured and available online, there is no way it can become interoperable.

5.5 CASE STUDIES

Four case studies are presented to illustrate emerging approaches to: (i) local wireless-based communication, and (ii) the organization of collected urban drainage system field data.

5.5.1 Case study 1: BDOH (Base de Données des Observatoires en Hydrologie), a database for the storage and publication of long-term water observation data

5.5.1.1 Context and objectives

BDOH has been developed by Irstea (now INRAE) in France, which has been responsible since 2013 for storage, management and dissemination of hydrological time series produced in its long-term environmental observatories (Figure 5.3). The main recorded quantities are rainfall, streamflow, groundwater level, soil moisture, air and water temperature, suspended solids concentration as well as concentrations of various chemical substances. At the end of 2020, BDOH contained records from 12 observatories at different locations in France, and over 55 years, which makes approximately 60 million records (https://bdoh.irstea.fr (accessed 07 Dec 2020)). Data can be browsed and visualized freely. Data download is free upon authentication for all users (scientists, management authorities, general public). BDOH now has almost 300 registered users.

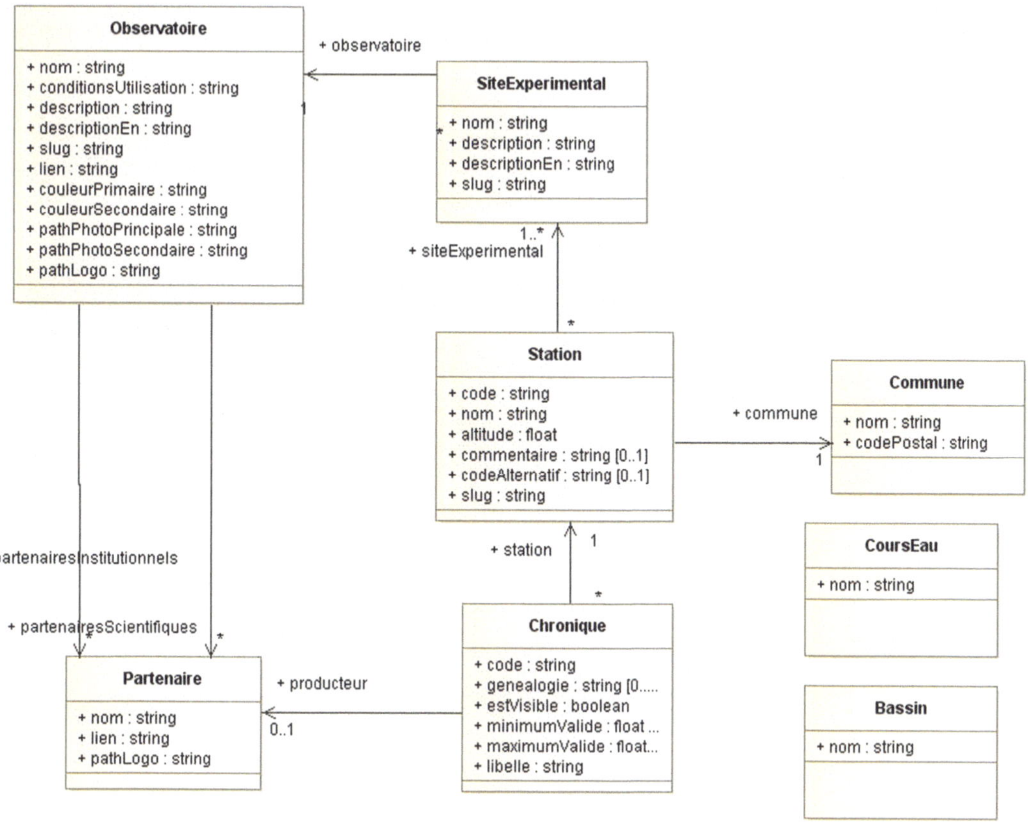

Figure 5.3 General structure of the BDOH database. *Source*: Flora Branger (INRAE) (Screenshot of the BDOH database).

5.5.1.2 Structure and main features

BDOH is based on open-source technologies such as the database management system PostgreSQL/PostGIS, and a web application developed in PHP and Javascript. The structure of the database considers stations that are points in space where physical quantities are being measured. The time series for each of these quantities are stored at variable time steps. Each value is associated with a date/time, a code appreciating the quality of the data (see also Section 9.2.2), and an estimation of the measurement uncertainty (see Section 8.4). Functionalities for data managers include data import from flat files (text files) in various formats, automatic calculation of time series derived from other time series (for example streamflow calculation from water level measurements using rating curves, and taking into account rating curve changes over time, or contaminant fluxes, see Section 9.4.3.3), and import and visualization of manual control points. An important point is that each data must be associated with a specific quality code, which means that data must be critically reviewed/validated before being imported into BDOH. Additional functionalities are also available for data users, such as search engine, graphical visualization (Figure 5.4), and options for data download (file formats, time steps).

5.5.1.3 Advantages and drawbacks

The use of a relational database management system and a web application means that the data is easy to browse and to extract, and is permanently available. It is also hosted on a server with automatic backups and security checks so that the integrity of the data can be ensured. The qualification of the data is also

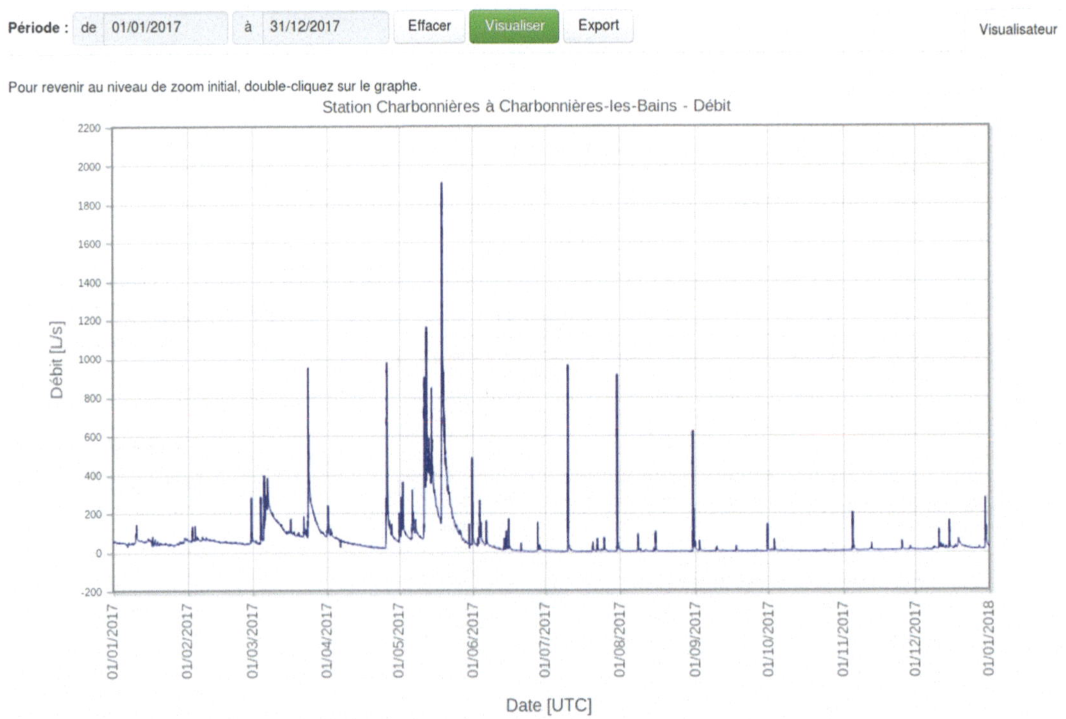

Figure 5.4 Snapshot of data visualization in BDOH. *Source*: Flora Branger (INRAE) (Screenshot of the BDOH database https://bdoh.irstea.fr/YZERON/V3015301/DEB (accessed 07 Dec 2020)).

an asset, although BDOH does not provide a method or workflow on how to assess this quality. BDOH also fostered exchanges and general improvement regarding data production practices. The main drawback is that of course it is quite some work to install and set up a BDOH instance. It requires computer resources and skilled human resources, in particular an IT system that can manage a Linux server. Otherwise, the time required for maintenance and data administration is quite low, approximately a 10% part-time job for over 60 million records.

5.5.2 Case study 2: DoMinEau, an Excel-based database for water quality monitoring

5.5.2.1 Context and objectives

The DoMinEau database (http://www.graie.org/Sipibel/projets.html#DOMINEAU (accessed 07 Dec 2020)) was developed in 2016 in response to a call by Agence Française de la Biodiversité (now Office Français de la Biodiversié, a French public national agency in charge of fostering research and expertise on biodiversity as well as helping to apply public water policies) and for the needs of the SIPIBEL observatory (Site PIlote de BELlecombe – Bellecombe Pilot Site). SIPIBEL investigates pollutants in urban and hospital wastewater and fosters research programmes in four themes (pollutant flows, treatment, risks, and sociology). The objectives of DoMinEau were to structure the datasets to (i) provide information about the pollution of wastewater and receiving water, including micropollutants, microbiology and bio-indicators, (ii) provide an estimation of the data quality and (iii) make the datasets easily accessible, searchable and interoperable by the partners of the project (scientists and water management authorities). At the end of the project, the DoMinEau database contained over 55,000 records.

5.5.2.2 Structure and main features

DoMinEau is based on Microsoft Excel. This software was chosen because of its widespread availability, because all the targeted users (scientists and water management authorities) were familiar with its use, and because there was little time awarded in the project for database development.

DoMinEau consists of four workbooks, each one organized with several worksheets:

- Sites-parameters-and-methods: contains all the metadata associated with the pollutant datasets, incl. sampling points, quantities, analytical methods. The list of quantities includes physical-chemical (temperature, water flow, pH, conductivity), micropollutants, microbiology, hydrobiology indicators and bio-essay indicators. There can be several analytical methods for each quantity.
- Campaigns-and-results: contains the sampling dates, sample identifiers, analyses results and a quality assessment for each individual result. The structure is flexible enough to record any type of sample at any frequency that was collected by the observatory. A sample is defined by a sampling point (location), the start and end dates of sampling and its duration, the type of sample (flow dependent or not) and the conditions during the sampling (river discharge, 48-hrs rainfall, etc.). The quality assessment method is shown in Figure 5.6.
- Statistics-and-graphics: calculates and displays automatically summary statistics and graphs about the data (see Figure 5.5).
- Data extraction: generates files that can be loaded into statistics and data analysis software, such as R or Matlab®. This is only an extraction utility; it does not carry out additional calculations.

A complete workflow was developed for quality assessment. Seven indicators (Figure 5.6) are used to assign a quality category (correct, dubious, incorrect) to each data. It takes into account not only the

Records	Correct data	Uncertain data	Incorrect data	Campaigns	Samples	Sampling points
55 655	25 151	27 713	2 791	236	1 007	53
	45%	50%	5%			

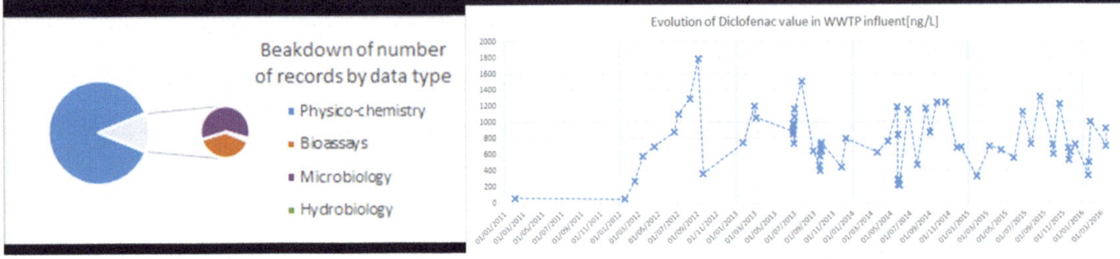

Figure 5.5 Snapshot of summary metadata statistics from the DoMinEau database. *Source*: GRAIE (http://www.graie.org/Sipibel/projets.html – DOMINEAU (accessed 07 Dec 2020)).

quality of the chemical analysis, but also the quality of the sampling itself. For each indicator there is a drop-down list with prepared values (prè-calculated). These values are set manually by the person who imports the data into the database. This step introduces some subjectivity, the reader is referred to the discussion on this important aspect in Chapter 9 on data validation. The most critical step is the coordination with the persons who were actually in the field and collected the samples. At the beginning of the DoMinEau database this was the same person as the data administrator, so that was quite easy. At present, the data administrator works with copies of the field sheets and contacts the data providers if necessary.

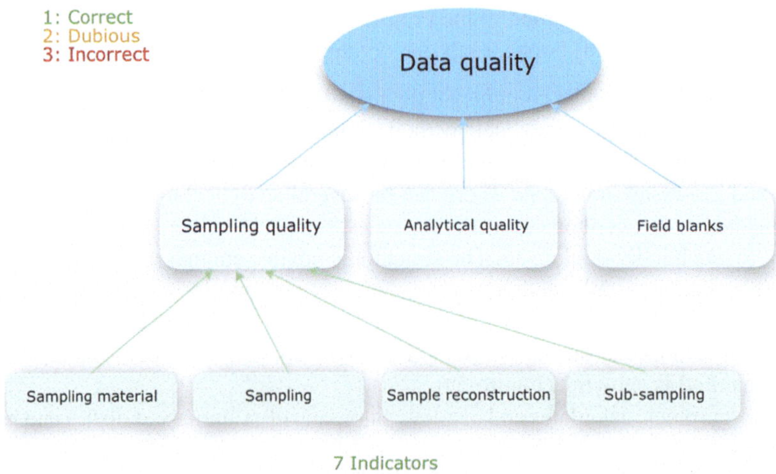

Figure 5.6 Criteria for data quality assessment in DoMinEau. For each indicator, three levels are possible. The overall data quality is assessed by the count of indicators in each colour. Overall quality is green if all indicators are green. Only one orange indicator triggers an overall orange quality. If there is one red value or more than three orange values the overall quality is red. *Source*: Flora Branger (INRAE).

DoMinEau is managed by a data administrator at GRAIE (a non-profit organization in charge of the coordination of the SIPIBEL observatory and of data management). In order to disseminate the data to the partners, copies of the Excel database are sent over on request. In addition, the data were all uploaded to the NORMAN network database system (European network for monitoring of emerging environmental substances (https://www.norman-network.com/nds/ (accessed 07 Dec 2020)).

5.5.2.3 Advantages and drawbacks

The main advantage of the DoMinEau database is the effort put on the structure and validation of the data, which results in a more rigorous approach for field work and in particular for compliance with the sampling protocols. It has contributed greatly to improve the quality of the data, which is a good example of a 'virtuous circle'. The statistics and graphic modules provide a first quick overview of the data in real time (as soon as the data is uploaded in the database) that is useful to compare campaigns. The summary dashboard is also useful for project managers and for achieving the project deliverables.

The main drawback is caused by the limitations of the selected software, i.e. Microsoft Excel. In terms of data management, new data must be added manually as new lines to the spreadsheet. With 55,000 records, this is quite cumbersome and it is very easy to make mistakes. The macros (in particular for the automatic statistics and graph generation) can be slow. The system is sensitive to successive updates and versions of Excel. Data management thus requires a lot of time from the Data Administrator (50% FTE [full time equivalent] approximately) and the database being off-line (the only way to transfer data is to send over copies), hence the data are not easily accessible. In practice DoMinEau has not been used as expected by the targeted users (scientists and water management authorities) and did not foster data sharing between project partners as expected. Its use requires significant effort and time. The data are now also stored in the online NORMAN database but they are quite difficult to find and mixed with other datasets. In conclusion, although DoMinEau proved to be useful in several ways during the SIPIBEL project, the technical choice of Excel appeared to limit its impact and the dissemination of the data during and after the project. It shows that alternative and more sustainable storage techniques would be preferable for future projects.

5.5.3 Case study 3: Data Grand Lyon – open data portal

The platform Data Grand Lyon (https://data.grandlyon.com/accueil (accessed 07 Dec 2020)) is far more ambitious than just a water data portal. It has been under development since 2011 as part of the Smart City policy of the Greater Lyon metropolis. Its ambition is to make available all types of data produced over the metropolis territory, that can be used:

- To facilitate communication and information exchange between the various management authorities.
- To encourage citizens to access the data and encourage potential participation.
- By making data freely available to foster innovation and economic initiatives in terms of data analysis and visualization.

These data contain items such as aerial photographs, street and cadastral maps, state of traffic, noise data, air temperature data, availability of Velo'v bikes (local bicycle sharing system), localization of disabled persons parking places, bus network alerts, some being available in real-time. They are produced by different stakeholders over the metropolis territory, public or private, with the ambition of data mutualization and interoperability.

Data are distributed according to three types of licence: open data, data delivered for free upon authentication, and data accessible with a licence fee. Most data are open.

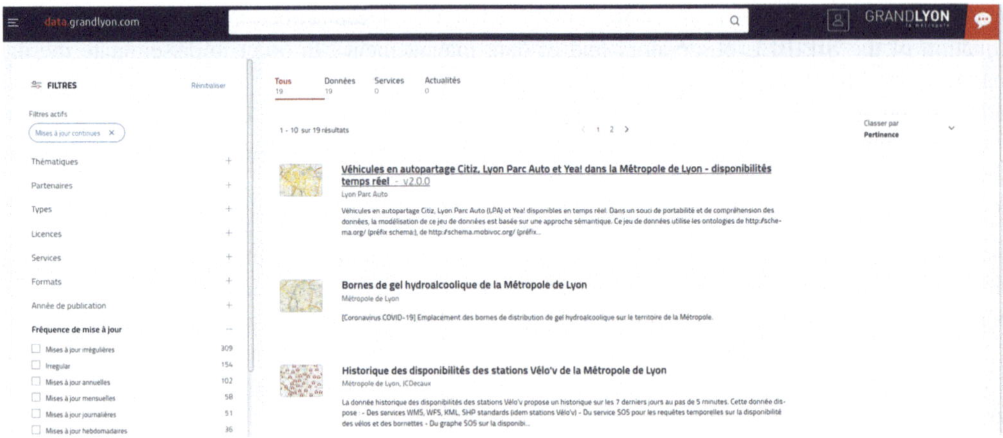

Figure 5.7 Snapshot of Data Grand Lyon metadata catalogue. *Source*: Flora Branger (INRAE) (Screenshot from Data Grand Lyon https://data.grandlyon.com/recherche (accessed 07 Dec 2020)).

Data Grand Lyon is based on a metadata catalogue (Figure 5.7) with a search engine, along with several protocols that:

- Take the data from various sources (relational databases, but also plain files).
- Format them (middle office, Figure 5.8).
- Make them available using (mostly) standard webservices (front office, Figure 5.8).

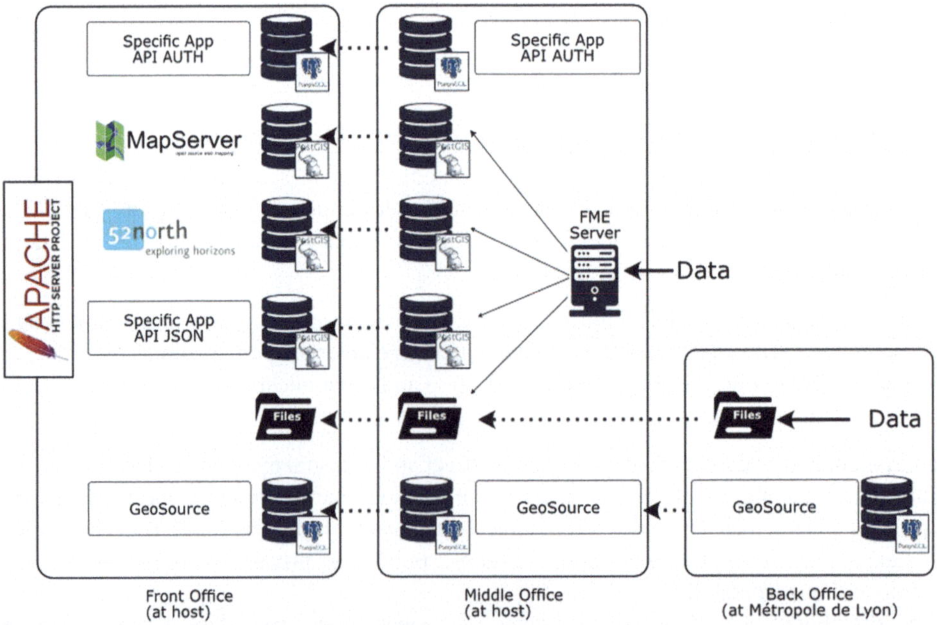

Figure 5.8 Architecture of Data Grand Lyon platform. *Source*: Grand Lyon.

Data Grand Lyon does not provide the basic structure of data such as presented in the BDOH case study (see Section 5.5.1). However, it can deal with unstructured data to some extent (plain files with stabilized formats and protocols for data updates). Data Grand Lyon relies mostly on open-source software. In terms of standards for data exchange (see Section 5.4.2), Data Grand Lyon implements the WMS/WFS standards for geographic data through MapServer. It also implements the SOS (Sensor Observation Service) for time series using the 52N open-source solution.

At the end of 2020, hydrological data were not present in the platform. Only rainfall data (from rain gauges located all over the metropolis territory) are available. They are currently being pushed on to the platform from Excel files. However, hydrological data could seamlessly fit in, using the SOS webservice that can take into account all kinds of data. Data Grand Lyon is an example of the very powerful tools that are now being developed for data storage and access. It is probably the best way to communicate data, even in real-time. However, one must keep in mind that it does not provide the basic structure of data (although it can deal with unstructured data to some extent). And it is also quite expensive to maintain as it requires specific computer resources and skilled data scientists.

5.5.4 Case study 4: local wireless based system for flood risk assessment and reduction – CENTAUR

CENTAUR (Cost-Effective Neural Technique to Alleviate Urban flood Risk, https://www.sheffield.ac.uk/centaur/home/outputs (accessed 16 Dec. 2020)) is an intelligent autonomous system for local urban flood risk reduction. It utilizes underutilized storage capacity to reduce the flow rate and water level at vulnerable potential flood locations. A gate controls flow using an intelligent algorithm which analyses local water level data, and then instructs the gate to adjust to the contemporary conditions. As this is a data driven system, reliable data transfer between the water level sensors and the gate controller is essential. CENTAUR is self-managing and easily deployed as a retrofit solution. It is far less disruptive and significantly less costly than alternative capital and space intensive solutions.

Data transfer for the system is wireless based and the communication equipment is modular and extensible; modules are lamp-post mounted or in-manhole. The types of modules are:

- Monitoring STation (MST): provides real-time monitoring of water level via radio communication between modules; it is ATEX certified for use in sewers with potentially explosive atmospheres.
- Control STation (CST): provides real time control of PLC (programmable logic controller) or analogue signal-controlled flow control devices.
- HUB (HUB): executes a fuzzy logic control algorithm based on the time-series of water level information relayed from MSTs, and relays commands to CSTs; it also communicates periodically with a web-based online dashboard, with system diagnostics transferred intermittently via GSM communications.
- RePeaTer (RPT): provides an extension of radio communication links between modules, when distances between modules are large or line-of-sight is particularly disrupted by obstacles.
- online dashboard: cloud-based dashboard which communicates with the HUB to give remote viewing of system operation, allows remote reconfiguration of modules, and flags any faults (however, the system operates autonomously and independently of the dashboard; the dashboard is in place for operational monitoring).

The technical innovation is around deploying artificial intelligence, autonomy, and tailoring different communication technologies for reliability and power efficiency. The system design exploits mainstream technologies to minimize cost and ensure reliability.

Radio communication for monitoring and issuing control instructions gives guaranteed signal over the typical distances between flow control gates and water level monitoring locations (up to several hundred metres), that characterize upstream storage availability and downstream flood risk locations. Distances can be extended indefinitely by use of a series of RPTs. The radio protocol was specially engineered for the monitoring paradigm; bi-directional, and communicating from below to above ground, and then relaying above ground. During the development of CENTAUR, the off-the-shelf protocol LoRa was unable to provide the reliability required by the system. Although line-of-sight performance was good, obstacles in built up urban areas had a significant detrimental effect consistent with noted reliability of around 90% in Ebi *et al.* (2019). This was insufficient for the CENTAUR application which featured mission-critical control as well as monitoring. An application-specific radio platform was developed based on a chip set from a major international manufacturer. This and the use of repeaters led to the 100% reliability required by the control application. Use of the GSM network was ruled out early in the design process, due to its lack of reliability and its latency. GSM communications are often intermittent and are unsuitable for mission-critical control systems. It is suitable for monitoring applications where missing data or latent communication are acceptable.

The modules can be configured using USB connection or via Bluetooth and a smartphone app. Basic reconfiguration is also achievable via the online dashboard. Bluetooth is particularly convenient for field engineers when installing the system. LED-based diagnostics on the front of each module make for convenient communication establishment and error diagnosis between modules. Such capability is important for applications in which long term, reliable performance is necessary and is often overlooked when data transfer is considered.

In terms of cyber-security, the inter-module proprietary protocol is owned by Environmental Monitoring Solutions Ltd. An attacker would require several key pieces of information to be able to interpret, modify and transmit messages to/from devices. Messages are transmitted in binary form with data embedded anonymously. Use is made of cyclic redundancy checks (CRC) and checksum functions to confirm the integrity of the data; any change to the message would require a change to the CRC and checksums before the receiving device will accept it, even if the attacker could decipher the message content. Additionally, the use of radio communications rather than the mobile phone network and internet steers firmly away from the possibility of cyber-attack. An attacker would need access to specialized listening and transmitting instrumentation rather than just access to the internet.

Communication between the HUB module and the internet-based dashboard is via JSON messages which use SSL security to encrypt messages. CENTAUR servers are virtual machines and are cloud-hosted on a cloud platform. The system therefore leverages the platform provider's cyber security strategy (based on 'assume breach'). A new server and application instance is created for each customer to ensure privacy and security. Bluetooth connections to modules are opened as required, and close automatically on a time-out.

Sensor redundancy gives reliable water level data. The system can disable itself automatically on failure of, for example, a communication link. The gate technology is easily deployed with physical fail-safes to give minimum upstream risk.

The system is self-managing; it is effectively 'fit-and-forget'. CENTAUR first deployment was in Coimbra, Portugal in 2017, where it controlled around 60 storms in its first year of operation delivering a tangible reduction in flood risk. CENTAUR can also be used for limiting CSO spills, managing flows into energy intensive assets using local water level data transferred using the radio-based communication system described above.

5.6 SUMMARY AND TRANSITION

This chapter summarized the main wired and wireless technologies that can be used to transfer data from the measurement location in UDSM systems to the location/database where data can be converted into understanding of a system state. From the raw data to databases that allow communication with other ones, this entire process requires hardware, services and IT competences to be conducted properly.

Chapters 2 to 4 list the most common sensors for monitoring and here Chapter 5 describes how to transfer, record and store data, i.e. the required hardware to transfer and then record data and make them available for any future use. Once those choices have been made, monitoring network and stations must be designed. Chapter 6 gives details on the macro- (network) and micro- (stations) design of monitoring systems whose processes for data-acquisition, -storage and -access are of key importance. There are also strong links between the current chapter and subsequent Chapters 7 to 9 in which operation & management, and data uncertainty and data validation are discussed, respectively. In these chapters the importance of employing a high-quality database will become even more clear. As each aspect of monitoring is essential in contributing to obtain information in a controlled and structured manner, a well-wrought design of the data handling allows reconstruction of the 'history' of the data offered to a range a data users. This includes data on maintenance activities, calibration results, validation results and, depending on the application, processed data.

REFERENCES

Branger F., Thollet F., Crochemore M., Poisbeau M., Raidelet N., Farissier P., Lagouy M., Dramais G., Le Coz J., Guérin A., Tallec G., Peschard J., Mathys N., Klotz S. & Tolsa M. (2014). Le projet Base de Données pour les Observatoires en Hydrologie : un outil pour la bancarisation, la gestion et la mise à disposition des données issues des observatoires hydrologiques de long terme à Irstea [Database for hydrological observatories: a tool for storage, management and access of data produced by the long-term hydrological observatories of Irstea]. *La Houille Blanche*, **1**, 33–38. doi: 10.1051/lhb/2014005.

Braud I., Chaffard V., Coussot C., Galle S., Juen P., Alexandre H., Baillion P., Battais A., Boudevillain B., Branger F., Brissebrat G., Cailletaud R., Cochonneau G., Decoupes R., Desconnets J.-C., Dubreuil A., Fabre J., Gabillard S., Gérard M.-F., Grellet S., Herrmann A., Laarman O., Lajeunesse E., Le Hénaff G., Lobry O., Mauclerc A., Paroissien J.-P., Pierret M.-C., Silvera N. & Squividant H. (2020). Building the information system of the French Critical Zone Observatories network: Theia/OZCAR-IS. *Hydrological Sciences Journal*, in press. doi: 10.1080/02626667.2020.1764568.

Ebi C., Schaltegger F., Rüst A. & Blumensaat F. (2019). Synchronous LoRA mesh network to monitor processes in underground infrastructure. *IEEE Access*, **7**, 577663–557677. doi: 10.1109/ACCESS.2019.2913985.

EU (1991). Council Directive 91/271/EEC of 21 May 1991 concerning urban waste-water treatment. *Official Journal of the European Communities*, 30 May, L135/40-L135/52. Available at https://op.europa.eu/en/publication-detail/-/publication/34860b42-95cc-4a57-b998-2e445d5befa9/language-en (accessed 15 December 2020).

EU (1998). Commission Directive 98/15/EC of 27 February 1998 amending Council Directive 91/271/EEC with respect to certain requirements established in Annex I thereof (Text with EEA relevance). *Official Journal of the European Communities*, 07 March, L67/29-L67/30. Available at https://op.europa.eu/en/publication-detail/-/publication/ff7ec087-8cc3-4619-bffc-b08ea4883d2c (accessed 15 December 2020).

EU (2000). Directive 2000/60/EC of the European Parliament and of the Council establishing a framework for Community action in the field of water policy. *Official Journal of the European Union*, 22 Dec., L327/1-L327/73. Available at https://eur-lex.europa.eu/legal-content/EN/TXT/HTML/?uri=CELEX:32000L0060&from=EN (accessed 16 December 2020).

EU (2019). Directive (EU) 2019/1024 of the European Parliament and of the Council of 20 June 2019 on open data and the re-use of public sector information. *Official Journal of the European Union*, 26 June, L172/56-L172/83.

Available at https://eur-lex.europa.eu/legal-content/EN/TXT/PDF/?uri=CELEX:32019L1024&from=EN (accessed 14 December 2020).

Galle S., Grippa M., Peugeot C., Bouzou Moussa I., Cappelaere B., Demarty J., Mougin E., Panthou G., Adjomayi P., Agbossou E. K., Ba A., Boucher M., Cohard J.-M., Descloitres M., Descroix L., Diawara M., Dossou M., Favreau G., Gangneron F., Gosset M., Hector B., Hiernaux P., Issoufou B.-A., Kergoat L., Lawin E., Lebel T., Legchenko A., Malam Abdou M., Malam-Issa O., Mamadou O., Nazoumou Y., Pellarin T., Quantin G., Sambou B., Seghieri J., Séguis L., Vandervaere J.-P., Vischel T., Vouillamoz J.-M., Zannou A., Afouda S., Alhassane A., Arjounin M., Barral H., Biron R., Cazenave F., Chaffard V., Chazarin J.-P., Guyard H., Koné A., Mainassara I., Mamane A., Oi M., Ouani T., Soumaguel N., Wubda M., Ago E. E., Alle I. C., Allies A., Arpin-Pont F., Awessou B., Cassé C., Charvet G., Dardel C., Depeyre A., Diallo F. B., Do T., Fatras C., Frappart F., Gal L., Gascon T., Gibon F., Guiro I., Ingatan A., Kempf J., Kotchoni D. O. V., Lawson F. M. A., Leauthaud C., Louvet S., Mason E., Nguyen C. C., Perrimond B., Pierre C., Richard A., Robert E., Román-Cascón C., Velluet C. & Wilcox C. (2018). AMMA-CATCH, a critical zone observatory in West Africa monitoring a region in transition. *Vadose Zone Journal*, **17**(1), 1–24. doi: 10.2136/vzj2018.03.0062.

Heard J., Tarboton D., Idaszak R., Horsburgh J., Ames D., Bedig A., Castronova A. M., Couch A., Dash P., Frisby C., Gan T., Goodall J., Jackson S., Livingston S., Maidment D., Martin N., Miles B., Mills S., Sadler J., Valentine D. & Zhao L. (2014). An Architectural Overview of Hydroshare, a Next-generation Hydrologic Information System. *Proceedings of the 11th International Conference on Hydroinformatics, HIC 2014*, New York City, USA, Paper 311, 8 p. http://academicworks.cuny.edu/cc_conf_hic/311/ (accessed 31 May 2021).

Horsburgh J. S., Tarboton D. G., Maidment D. R. & Zaslavsky I. (2011). Components of an environmental observatory information system. *Computers and Geosciences*, **37**(2), 207–218. doi: 10.1016/j.cageo.2010.07.003.

Horsburgh J. S., Tarboton D. G., Hooper R. P. & Zaslavsky I. (2014). Managing a community shared vocabulary for hydrologic observations. *Environmental Modelling and Software*, **52**, 62–73. doi: 10.1016/j.envsoft.2013.10.012.

Horsburgh J. S., Aufdenkampe A. K., Mayorga E., Lehnert K. A., Hsu L., Song L., Jones A. S., Damiano S. G., Tarboton D. G., Valentine D., Zaslavsky I. & Whitenack T. (2016). Observations Data Model 2: a community information model for spatially discrete earth observations. *Environmental Modelling and Software*, **79**, 55–74. doi: 10.1016/j.envsoft.2016.01.010.

LoRaWAN (2015). *LoRaWAN: what is it? A technical overview of LoRA and LoRaWAN*. LoRa Alliance, San Ramon, CA (USA), Technical Marketing Workgroup 1.0, 20 p. Available at https://lora-alliance.org/sites/default/files/2018-04/what-is-lorawan.pdf (accessed 7 December 2020).

Magno R., De Filippis T., Di Giuseppe E., Pasqui M., Rocchi L. & Gozzini B. (2018). Semi-automatic operational service for drought monitoring and forecasting in the Tuscany region. *Geosciences*, **8**(2), 49, 24 p. doi: 10.3390/geosciences8020049.

See C. H., Horoshenkov K. V., Abd-Alhameed R. A., Hu Y. F. & Tait S. J. (2012). A low power wireless sensor network for gully pot monitoring in urban catchments. *IEEE Sensors Journal*, **12**(5), 1545–1553. doi: 10.1109/JSEN.2011.2174354.

Wilkinson M. D., Dumontier M., Aalbersberg I. J., Appleton G., Axton M., Baak A., Blomberg N., Boiten J.-W., da Silva Santos L. B., Bourne P. E., Bouwman J., Brookes A. J., Clark T., Crosas M., Dillo I., Dumon O., Edmunds S., Evelo C. T., Finkers R., Gonzalez-Beltran A., Gray A. J. G., Groth P., Goble C., Grethe J. S., Heringa J., 't Hoen P. A. C., Hooft R., Kuhn T., Kok R., Kok J., Lusher S. J., Martone M. E., Mons A., Packer A. L., Persson B., Rocca-Serra P., Roos M., van Schaik R., Sansone S.-A., Schultes E., Sengstag T., Slater T., Strawn G., Swertz M. A., Thompson M., van der Lei J., van Mulligen E., Velterop J., Waagmeester A., Wittenburg P., Wolstencroft K., Zhao J. & Mons B. (2016). Comment: the FAIR Guiding Principles for scientific data management and stewardship. *Scientific Data*, **3**, 160018, 9 p. doi: 10.1038/sdata.2016.18.

Yi H., Idaszak R., Stealey M., Calloway C., Couch A. L. & Tarboton D. G. (2018). Advancing distributed data management for the HydroShare hydrologic information system. *Environmental Modelling and Software*, **102**, 233–240. doi: 10.1016/j.envsoft.2017.12.008.

Zander F., Kralisch S. & Flügel W.-A. (2013). Data and information management for integrated research – Requirements, experiences and solutions. *Proceedings of the 20th International Congress on Modelling and Simulation, MODSIM 2013*, 1–6 December, Adelaide, Australia, 2201–2206.

Chapter 6

Design of a monitoring network: from macro to micro design

Mathieu Lepot[1,2], Zoran Kapelan[1,3] and Francois H. L. R. Clemens-Meyer[1,4,5]

[1]*Delft University of Technology, Water Management Department, Delft, The Netherlands*
[2]*Un poids une mesure, Lyon, France*
[3]*University of Exeter, College of Engineering, Mathematics and Physical Sciences, Exeter, United Kingdom*
[4]*Deltares, Unit Hydraulic Engineering, Dept. Experimental Facility Support, Delft, The Netherlands*
[5]*Norwegian University of Science & Technology, Faculty of Engineering, Dept. Civil & Environmental Engineering, Trondheim, Norway*

ABSTRACT

Designing a monitoring network or a measuring set-up or a monitoring station is a typical (multidisciplinary) engineering enterprise: a range of potentially conflicting demands (technical, financial and managerial) and limitations (e.g. availability of resources, skilled personnel, regulations) have to be respected. This chapter addresses the design aspects on both the macro scale (a monitoring network) and on the micro scale. The macro scale addresses what to measure, where to measure, how frequently to measure and the applications of models in the design process. On the micro scale issues with safety, accessibility and practical limitations are discussed. This chapter has close links with virtually all other chapters in this book and a comprehensive set of literature references is supplied to allow the interested reader to broaden his/her knowledge on the subject.

Keywords: Micro design, monitoring, network design, practical aspects.

doi: 10.2166/9781789060119_0155

SYMBOLS

C	weir coefficient ($m^{1/2}/s$)
f_N	Nyqvist frequency (Hz)
$f(n)$	inverse Fourier transform element 'j'
$F(k)$	Fourier transform for frequency k
h	water level (m)
i	imaginary unit, integer value (-)
I	amount of information (-)
j	index (-)
J	Jacobian matrix (various)
k	integer value (-)
k_n	Nikuradse's equivalent sand roughness (m)
loc	as exponent: discrete monitoring location index
m	number of measurement locations (-)
mse	mean squared error (various)
M	selected number of monitoring locations among N possibilities (-)
n	integer value, number parameter used for evaluating the Jacobian matrix (-)
N	number of monitoring locations (-)
N_e	number of elements in a time series (-)
$p_1, p_2,.., p_n$	parameters of a model (various)
P	power of spectrum
Q	discharge (m^3/s)
$r(T)$	autocorrelation function for the time shift T (-)
SNR	signal to noise ratio (-)
t	time (s)
t	as index: discrete time index (-)
T	time shift for which the autocorrelation of the process $x(t)$ is estimated (s)
$x(t)$	a time varying process (various)
α	weighing factor (-)
β	weighing factor (-)
γ	weighing factor (-)
Δt	sampling time interval (s)
μ_x	mean value of the process $x(t)$
$\rho(T)$	normalized autocorrelation function for the time shift T (-)
ρ_x	autocorrelation of the process $x(t)$ (-)
σ_{interp}	standard deviation of an estimated interpolated value (various)
σ_m	standard deviation of measurement uncertainty (various)
σ_n	standard deviation of the noise (various)
σ_x	standard deviation of the process $x(t)$ (various)
τ	a given value of time t between two successive times t_{j-1} and t_j (s)

Motivation anecdote 'Unexpected cycles' from Walcker *et al.* (2018)

In 2016, the OTHU (Field Observatory on Urban Hydrology) in Lyon, France set up the second generation of their monitoring stations. After a decade of experience, trials and feedbacks, new monitoring stations have been designed, built, and implemented (see also Section 6.3.5). Figure 6.0 (a) shows the discharges measured at the inlet of a stormwater retention tank with the old (red) and the new (blue) data acquisition methods.

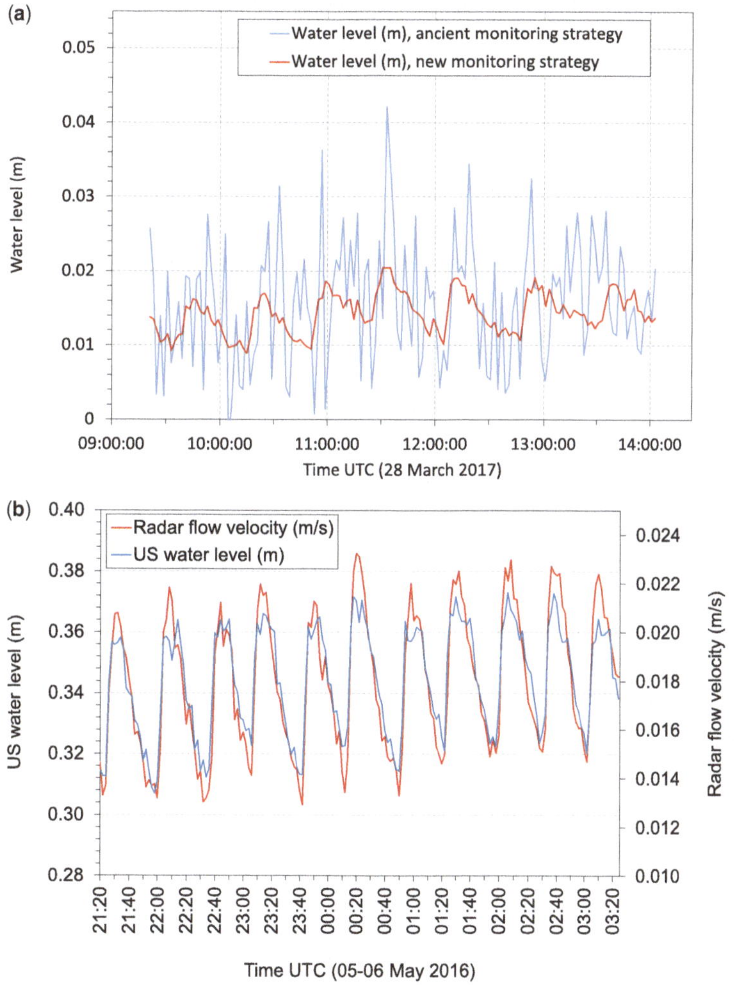

Figure 6.0 (a) water level – red line: old monitoring system, blue line: new monitoring system; (b) water level (blue line) and flow velocity (red line) measured with the new monitoring station. *Source*: Nicolas Walcker (INSA Lyon).

At first sight, the new station delivers less noisy data, mainly but not only due to the fact that the value recorded every two minutes is calculated from 120 values measured every second and no longer sampled as a single instantaneous value measured every two minutes. Looking at a few days within the time series, 30 min cycles (Figure 6.0(b)) become visible with those smooth data for different quantities: water level, flow velocity, discharge and, more surprisingly, pH and conductivity, even during rain events. Since those fluctuations occur with several sensors but not all, they cannot be explained by the data acquisition system: something happens within the sewer. Investigations have shown that water used for industrial processes is discharged into the stormwater pipe of the separate sewer system. The improved data acquisition method with on-line pre-processing of high frequency data made possible by a better monitoring system leads to new knowledge, and, in this case, to the identification of an illegal inflow within the sewer.

What can you learn if you (re)build a high-quality system? This chapter will give you few methods and tips to guide you in such projects.

Mathieu Lepot and Nicolas Walcker

6.1 INTRODUCTION

In this chapter the design of a monitoring set-up is discussed. On the one hand a distinction is made between set-ups for permanent or long-lasting use (e.g. weeks to years) and set-ups for occasional measurements. On the other hand, a distinction is made between theoretical and practical aspects of the design. Further the reader is referred to a range of matters strongly interwoven with design aspects that are discussed in other chapters. As this subject is a very comprehensive one, the reader is also supplied with entries to the literature for further study. Overall, when starting a monitoring project, a scheme as shown in Figure 6.1 is applied. A first omni-important step is to agree upon the data that needs to be gained from the monitoring set-up to fulfil the information need according to monitoring goal(s).

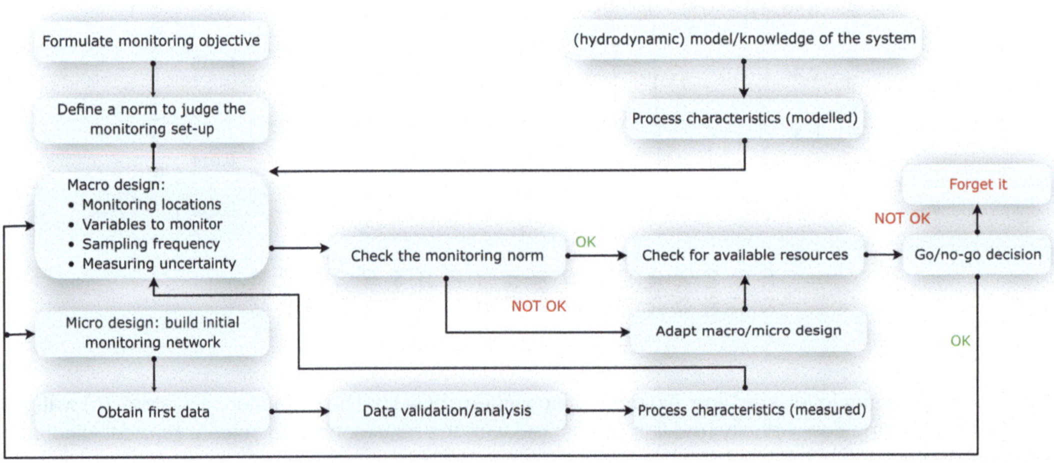

Figure 6.1 Basic flow diagram for designing monitoring networks, initially using a model or another source of *a priori* knowledge on the system to be monitored. *Source*: Francois Clemens-Meyer (Deltares/TU Delft/NTNU).

From this information need, the following questions arise:

- What quantities(s) should be monitored?
- At which location(s) should monitoring take place (where and how many)?
- At what sampling frequency?
- What is the maximum allowed uncertainty?
- What should be the duration of the monitoring campaign?

When these questions are answered a first rough budget estimate can be drafted and a managerial decision whether or not to proceed with the project has to be made. Basically, the question '*Is the information obtained worth the investment and additional costs?*' has to be answered. There is no general 'recipe' on how to handle such a question, as in many cases the possible 'gain' or 'loss' cannot be entirely expressed in monetary terms (e.g. in cases of monitoring set-ups that are mandatory for legal purposes, as frequently included in permit conditions). In the case where it is decided to proceed with the monitoring project, the information yield has to be regularly checked against the original information need in order to allow for making adaptations of the monitoring set-up (e.g. reducing the number of locations, replacing sensors, reducing the sampling frequency, etc.). Such a check, or evaluation, should be done on a regular basis with an increased frequency at the beginning of the project, and for long-term projects at least once a year. An important driver here is the regular economical re-evaluation of the project: '*Do we still conclude that the information obtained is value for money?*'

In long-term projects the rationale for checking the information yield is found in the following considerations: (i) (small) adaptations in the structure and/or geometry of the system under study and (ii) a shift in the goals strived for. These situations result in a change in information obtained and a change in information need, respectively.

This implies, especially for long-term monitoring projects, that there is a strong need to obtain:

- Information on the quality of the data obtained (see Chapter 9 on data quality and validation).
- Insight into the costs involved for keeping the monitoring project operational (that is in a physical sense as well as in the sense that data gained are being processed into information and are actually used for the defined purpose).
- Actual data on adaptations made in the system under study, so as to enable linking system behaviour as recorded to the system's status (geometry, structure, operation) over the monitoring period. This is even more important when the goal of the monitoring project is to quantify the effect of some adaptation in the system under study, or when evaluating the effect of taking large-scale measures in a system to e.g. reduce the pollutant load into the environment.

Overall, it has to be realized that the design of a monitoring set-up is, certainly for long-term monitoring, not static. Further aspects to keep in mind when developing a design are the demands put on the organization responsible for keeping the system operational and budget claims implied by design choices made during the process.

The following general cases are distinguished:

(1) Monitoring as a basic need for the (asset) management of the urban drainage system.
(2) Monitoring for legal purposes (location specific, long- and short-term depending on the legal issue).
(3) Monitoring for regulation or environmental compliance (e.g. CSO (combined sewer overflow) spills, in- and exfiltration, wrong connections).
(4) Control purposes (location specific, long term).
(5) Decision support for a redesign of an existing system.
(6) Model calibration.

It is largely stated that cases 1 to 4 (and to a lesser extent case 5) are normally straightforward in terms of what and where to monitor and for what duration, the design effort is then concentrated on the micro design.

In theory one could argue that when taking 'Model calibration' (case 6) as a goal one could, once the calibration is proven successful, use the model as a surrogate reality with which all other questions arising from 1 to 5 can be answered. This is true only up to a certain extent. The following remarks have to be made in this respect:

- Even after calibration, significant uncertainties in the model parameters (and hence the model predictions) remain, limiting its effective applicability.
- Unnoticed, significant deviations with the 'real world' may exist as not all relevant processes are correctly incorporated in the present generation of models (see e.g. Tscheikner-Gratl *et al.*, 2019), limiting again its effective use.
- Regular re-calibration is needed for the same reasons that a monitoring set-up for long term monitoring has to be evaluated on a regular basis. With respect to case 5, monitoring a system in relation to a planned redesign or enhancement of functionality, sets the same demands on the monitoring design as model calibration and is not separately discussed here any further.
- Application of the model parameters obtained beyond the domain for which they are determined (i.e. the range of loads for which the parameters are calibrated), may result in biased results.

Key messages on design of monitoring networks and stations

- KM 6.1: *Macro ~ Micro design* – 'Where and what to measure' questions are answered by the macro design approach. The micro design answers the question of 'How'.
- KM 6.2: *Mathematics vs. experts* – Expert designs are still often more reliable than the ones determined by mathematical methods.
- KM 6.3: *Teamwork* – Both designs require teamwork: never neglect the feedback from all the parties involved in the project.
- KM 6.4: *1, 2 and 3* – If you are beginner and/or too optimistic, please keep in mind the rule of '1, 2 and 3': 1 rule, the budget will be 2 times as expensive than the first expectation and the forecasted duration should be multiplied by 3.

6.2 MACRO DESIGN

Macro design of a monitoring network encompasses the choice of the number of monitoring locations, what to measure, how frequently and with what quality in terms of uncertainty in the monitoring results and data. By definition, this is a cyclic process. The initial design is based on the knowledge available on the system at hand, while the monitoring system is meant to extend and deepen the knowledge on the system. This implies that after obtaining and interpreting the data, they may hold clues for further refinement of the monitoring system. When developing the macro design it has to be realized that choices made here may affect the 'margin of freedom' in the micro design. For example, when setting up criteria for eliminating or identifying potential locations, and not taking into account the availability of certain services (data communication or power infrastructure), this limits the choice for the type of power supply and data storage and transfer methods.

6.2.1 General

For a wide range of monitoring objectives, the monitoring location(s) is(are) unambiguously defined, e.g.:

- What is the discharged CSO volume at this CSO structure?
- Does flooding occur on a specific location?
- What is the performance of this pumping station?
- What is the mass balance in a given catchment?

In other words, often the formulation of the monitoring goal explicitly defines where and, to a certain extent, what to measure. There are however cases in which this is not entirely clear, when e.g. the following monitoring goals are formulated:

- We need monitoring to calibrate a model.
- The hydraulic impact of the urban drainage system on the river must be quantified using monitoring data.

Choosing monitoring locations for such goals requires some prior information/knowledge. The system(s) should be known in some detail (structure and geometry, details on land use, connected surfaces, information on ground water levels, locations where flooding occurred, citizen's complaints). There should be means to get a preliminary impression of the system's response to loads (i.e. storm events and/or wastewater discharges). Such a preliminary impression may be supplied by using a model simulation, although simpler data can be useful as well. When translating the choice of monitoring location(s) into an engineering question, the following task is: *'Given the available budget and the monitoring goals identified, find the minimum number of monitoring locations and their actual locations in the system.'*

In some cases, the budget will be insufficient to achieve the goals set. In such a case one either needs to raise the budget (political/managerial decision) or try to find cheaper methods to achieve the goals. If none of these options is applicable, one has to abandon the idea of monitoring altogether. However obvious the latter conclusion may seem to be, in practice parties often implicitly proceed in such cases with reduced ambitions and/or poorly designed monitoring set-ups, which in most cases ends in a disappointment. It is this type of situation that erodes away the political/managerial support for monitoring campaigns.

Hereafter an approach for choosing monitoring locations based on expert judgement is discussed as well as the added value (computer) models may bring in the design process.

6.2.2 Choosing locations as a combinatorial problem

Now, let us assume that it is possible to express the monitoring goals as an 'amount of information *I*' to be obtained from *N* monitoring locations, where *N* is defined by the available budget and the choices made for sensors, data handling, etc.

Suppose that there are only *M* locations that allow for monitoring, which in many cases are in manholes, as they are the main entrances for most underground urban drainage systems. Typically, *M* is much larger than *N*. This results in a discrete optimization problem:

> *'Find the combination of N manholes, out of a population of M possibilities, that maximizes the information obtainable from these manholes and check whether this is sufficient to achieve the monitoring goal(s) set.'*

In practice, the number of possible solutions is immense: for example, even for a relatively small case with $N = 28$ and $M = 210$, the number of possible solutions is approximately 5×10^{34}. Since it is a discrete problem, popular optimization algorithms like e.g. Levenberg-Marquardt, Simplex Method, etc.

are not applicable in a straightforward manner. Apart from that, this type of optimization problem is suspected to be of NP (hard) nature (NP stands for 'Non-deterministic Polynomial-time' and refers to a class of problems for which there is no algorithm known that can solve such a problem in a predictable number of operations. A famous optimization problem in this category is the Travelling Salesman Problem and algorithms to solve it, e.g. in Diaz-Delgadillo *et al.* (2016)). This implies that there is likely no algorithm that will give the optimal solution within a limited and predictable calculational effort. Therefore, for all practical purposes, one has to settle for a 'good' solution rather than striving for the 'best'. These are the main reasons why optimizations are usually carried out with either genetic algorithms or simulated annealing algorithms (e.g. Boomgaard *et al.*, 2001; Ruiz-Cardenas *et al.*, 2010). Obviously, a very effective manner to reduce the search space is to reduce the number of possibilities to choose from, i.e. eliminate all locations that cannot be considered as a monitoring location based on practical considerations (like e.g. accessibility, safety, etc.). In the literature (e.g. Clemens, 2001; Henckens & Clemens, 2004; Thompson *et al.*, 2011) different theoretical problem formulations exist with associated sensor network design criteria and decisions to be taken. All these approaches and theoretical frameworks are ultimately based on an extended analysis of sensitivity to parameter variation of a model. Much of the theory and algorithms are developed for pressurized systems (e.g. water supply networks), and transferring these algorithms for application to urban drainage systems (in which free surface flow and pressurized flow can both occur and holding transitions between the two modes over time) is not straightforward. The methods presented so far prove to be hard to implement and are not convincingly better by any metric (costs, quality of data obtained or reliability of monitoring results) than a design obtained by expert judgement. Although the theory of optimization of a monitoring network is an interesting topic for further development, its practical applicability is at present judged to be very difficult and is therefore not discussed here in detail. References to the existing literature on the subject are provided.

6.2.3 Considerations in choosing locations

Sensor network design involves deciding on what, where and under what conditions to measure. This approach relies on some prior knowledge of the system (either expert knowledge, or existing observations or a model mimicking the system behaviour) along with related knowledge of sensors, their costs, accuracy, and potential installation, operation and maintenance issues, etc. Whilst expert knowledge can be quite valuable, especially in specific circumstances, the ultimate decisions (i.e. the sensor network design to be chosen) tend to be subjective in nature, for obvious reasons. It has to be appreciated, however, that designs based on mathematical algorithms (as briefly mentioned in the preceding paragraph), e.g. to optimize information content, do not automatically acknowledge the added value of redundant information for validation purposes or any other practical circumstance other than the elimination of locations for practical considerations. This statement should be balanced by several facts: (i) actual goals of the sensor network should be clearly defined, (ii) future goals must be considered, (iii) the data on facilities and their accessibility must be implemented correctly, and (iv) evolutions (extensions of the system, rehabilitation activities or decommissioning of elements) of the catchment must be considered.

Even though choosing the 'optimal' set of monitoring locations is, to a certain extent, a subjective process, it is still mainly based on sound engineering criteria and it involves the following practical aspects:

- Purpose of the sensor network. Experts are driven by solving a specific sensor network design problem, e.g. designing a sewer network for a specific purpose. Having said this, quite often, they tend to think beyond that purpose, i.e. tending to think about other possible future applications and, generally, what might change in the future that will impact on their decision where to locate and what sensors to choose. For example, an expert may be aware of the fact that a town and its urban drainage system will expand in the easterly direction in the near future and, as a consequence, they may decide to locate a new flowmeter and level monitoring in that part of the town even though there is no real need to do so based on the current situation.

- Accessibility. Urban water systems are often comprised of large pipe networks that stretch over significant geographical areas, both urban and rural. As a consequence, some of the potential sensor locations, e.g. remote ones or locations in parking lots or in the proximity of heavy traffic, may not be easily accessible and an expert may choose not to install a sensor in such locations for these reasons.
- Safety. Working in the urban environment can be challenging, as one needs to account for a range of hazards, e.g. traffic, working in confined spaces, risk of contamination when working with polluted (waste)water, etc. Chapter 7 touches upon these issues in more detail. It is, however, efficient to take safety aspects into consideration early in the initial design.
- Availability of power supply. Most measuring devices require power to run. Some of them can run on batteries but there is always an issue of battery life which, in turn, is linked to operation costs. An alternative is to obtain the power from a nearby property but this often leads to accessibility issues. Experts are very aware of these issues which tend to impact their decisions on where to locate sensors and what sensors to use in the first place.
- Data communication. Many modern sensors communicate with a control room in a water utility or between themselves. This is usually done nowadays by means of wireless communication. There is no point in installing such a sensor in a location where the mobile network coverage is poor when relying on GSM (global system for mobile communication).
- Security. Most modern sensors are not cheap and some may be very expensive. Consequently, they need to be secured against theft/vandalism and this is easier to do in some locations than others. Moreover, modern sensors can potentially be remotely manipulated, e.g. via internet, and this may result in triggering undesirable actions in automated systems, e.g. at treatment works or systems that utilize real time control (RTC). This requires not just ensuring the physical protection of sensors but also that cyber security is ensured.
- Budget and availability of sensors. Practitioners are very much aware of budget constraints and limited availability of sensors (within their organization or otherwise), which has a major effect on decisions made regarding the sensor network design. Over recent years, cheap DIY (Do It Yourself) electronic systems (e.g. Arduino®) have been introduced. These systems allow laymen to put together a sensor system that is cheap and easy to obtain. Although no comprehensive objective data are available to date, the authors' experience hints at issues with reliability, and sensor quality in a broad sense, along with issues related to the operational conditions in the field in general. However, for small-scale trials or experiments, the application of DIY systems may prove to be a future game changer for monitoring the urban environment, as it comes at low costs and therefore allows for making errors or misjudgements in the design without substantial (financial) consequences. Recently, a protocol for testing low-cost water level monitoring was suggested which will allow sharing of experiences with such sensors on a common basis (Cherqui *et al.*, 2020).
- Make sure there is some overlap/redundancy of the expected recorded values to allow for consistency checks when validating the data.

Figure 6.2 shows an example of two subsystems connected at some point via manhole C. Suppose this is a combined sewer system, the subsystem connected to manhole A and the subsystem connected to manhole B

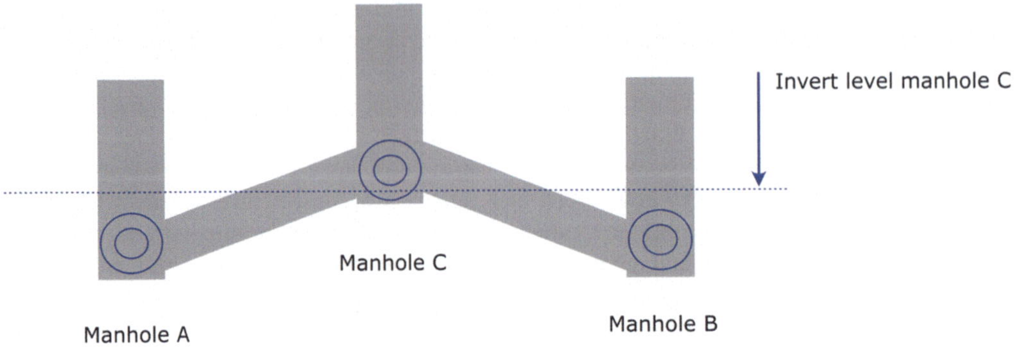

Figure 6.2 Example of a geometry that allows for overlap in readings for manholes A and B, when the water level reaches the threshold of the invert level at manhole C. *Source*: Francois Clemens-Meyer (Deltares/TU Delft/NTNU).

will show a high correlation between monitored water levels at these nodes when the water level rises above the invert level indicated in the figure. During dry weather flow and moderate storms, this connection is not present. In the case of larger storm events, an occasional overlap in water levels is present allowing for cross-checking the readings obtained in manholes A and B.

The option of changing measurement locations after a first evaluation of the network information yield should be kept open, certainly when the system is meant for long-term monitoring. Modelling the system can be informative for the expert, especially if he/she is able to perform a sensitivity analysis with respect to model parameters. The model can assist the expert in identifying locations in which two or more sensors may show some overlap in the expected data.

Expert knowledge-based sensor network design is typically carried out according to the following general principles:

- Ensure good network coverage of the analysed urban water systems. Experts tend to distribute sensors relatively evenly through the analysed pipe network/geographical area. They know, intuitively, that good network coverage is important regardless of the intended sensor network use. This ensures, among other things, an effective everyday monitoring of the system and is especially important for detecting various events in pipe networks (e.g. collapses and blockages in a sewer network). In addition to distributing the sensors evenly in space, experts know that a number of sensors need to be distributed toward the pipe network edges. Otherwise substantial parts of the pipe network may end up not being observable at all. In dendritic systems (i.e. networks with very few loops and interconnections), prior reasoning may be a very effective manner to decide where to monitor.
- Install sensors at important and key system locations. Experts know, based on experience, intuition and a 'feeling' of the system they manage, that observing some of the key urban water system elements and structures (e.g. locations where flooding occurs, pumping stations, large sewers, outfalls, CSO constructions, etc.) is critical to ensure good system observability and ensure normal system operation.
- Use good quality and reliable sensors. Experts know, based on experience, that going for cheaper (hard- and software) solutions now is likely to result in additional efforts in, amongst others, maintenance and data validation during the project.
- Calibrate (Section 7.6) and maintain (Section 7.4) the sensors regularly. This is essential for the effective operation of the designed sensor network.

6.2.4 Example of using a model as a design aid

When a model is available, it can be used in identifying locations that provide information related to certain model parameters. Basically, a sensitivity analysis is performed on the response of the model (water level and/or discharge) when changing parameter values. When this is done in a systematic way, the potential of each location with respect to each individual model parameter can be obtained.

Consider a basic example of application, with two parameters: the hydraulic roughness k_n and the weir coefficient C. Three simulations are made using a hydrodynamic model, one with the parameter vector $[k_n, C]$, and two with parameter vectors $[k_n + \Delta k_n, C]$ and $[k_n, C + \Delta C]$. Thus, for each node (manhole) of the system, three hydrographs are obtained.

Based on these model results, the Jacobian matrix is built (Figure 6.3). For two parameters p_1 and p_2, the elements in the $m \times 2$ Jacobian matrix J (Equation (6.1)) are defined as:

$$J = \begin{bmatrix} \frac{dh_{t=1}^{loc=1}}{dp_1} & \cdots & \frac{dh_{t=1}^{loc=m}}{dp_2} \\ \vdots & & \vdots \\ \frac{dh_{t=n}^{loc=1}}{dp_1} & \cdots & \frac{dh_{t=n}^{loc=m}}{dp_2} \end{bmatrix} \approx \begin{bmatrix} \frac{\Delta h_{t=1}^{loc=1}}{\Delta p_1} & \cdots & \frac{\Delta h_{t=1}^{loc=m}}{\Delta p_2} \\ \vdots & & \vdots \\ \frac{\Delta h_{t=n}^{loc=1}}{\Delta p_1} & \cdots & \frac{\Delta h_{t=n}^{loc=m}}{\Delta p_2} \end{bmatrix} \tag{6.1}$$

where m is the number of locations where the water level is calculated, and n is the number of times the water level is calculated during a simulation.

The last term in Equation (6.1) is the partial difference approximation of the Jacobian J obtained using the model results. For each node, a Jacobian is obtained showing exactly for each time step the sensitivity of the water level to a variation in the parameters, each column corresponding to one parameter.

Figure 6.4 shows a very simple, artificial network (basically a hydraulic analogon of the well-known Wheatstone bridge electrical circuit). All nodes have a free surface area of 2 m^2, the conduits all have a

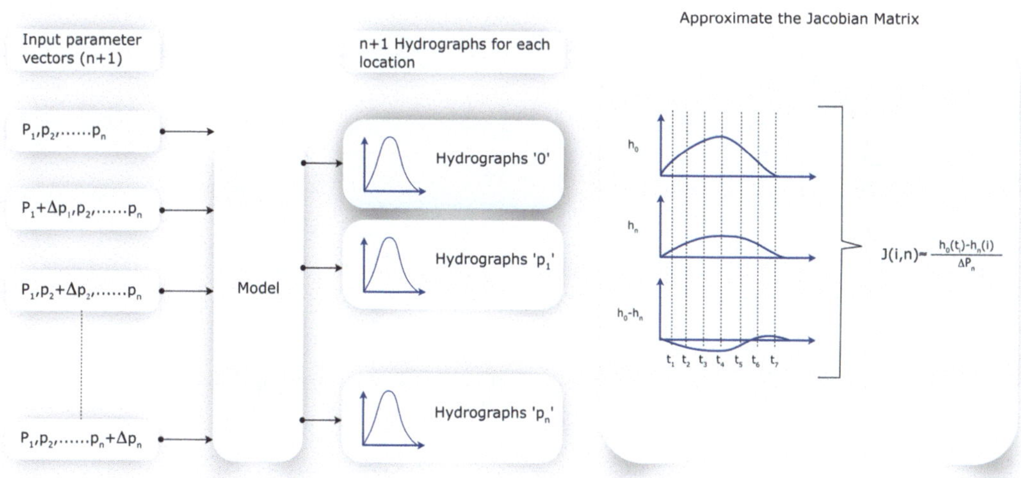

Figure 6.3 Scheme for the construction of the Jacobian matrix *J*. *Source*: Francois Clemens-Meyer (Deltares/TU Delft/NTNU).

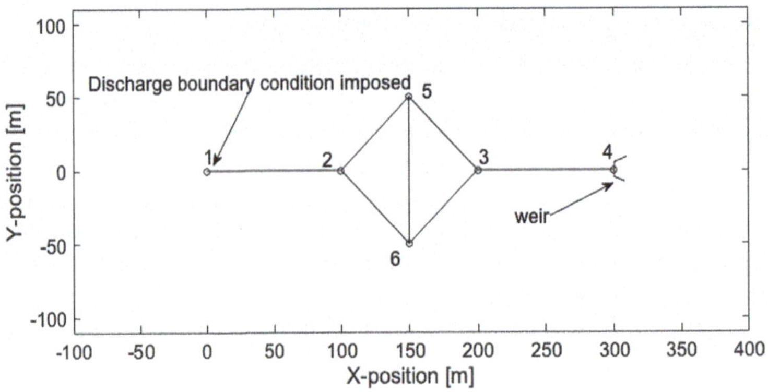

Figure 6.4 Layout of the artificial network. *Source*: Francois Clemens-Meyer (Deltares/TU Delft/NTNU).

length of 100 m (apart from the conduit between nodes 5 and 6, which is $100 \times \sqrt{2}$ m long), with a diameter of 0.5 m. The weir at node 4 has a length of 3 m and a weir level of 8 m above reference level.

At $t = 0$ the system is empty, the boundary condition at node 1 is a time varying discharge (Figure 6.5). For the sake of simplicity, only two model parameters are considered: the hydraulic roughness k_n and the weir coefficient C.

As can be seen in Figure 6.6, the water level at all nodes is more sensitive to a change in the value of the weir coefficient C (continuous line) when compared to the hydraulic roughness k_n (dashed line), i.e. larger absolute values of the elements in the Jacobian matrix. This implies that more information can be obtained related to the value of C when compared to the value of k_n. Node 4 (closest to the weir) shows a high

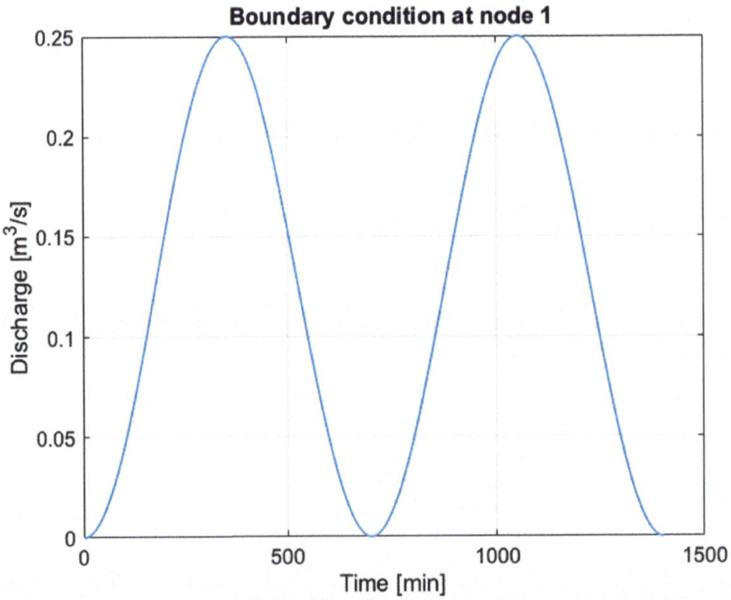

Figure 6.5 Boundary condition at node 1. *Source*: Francois Clemens-Meyer (Deltares/TU Delft/NTNU).

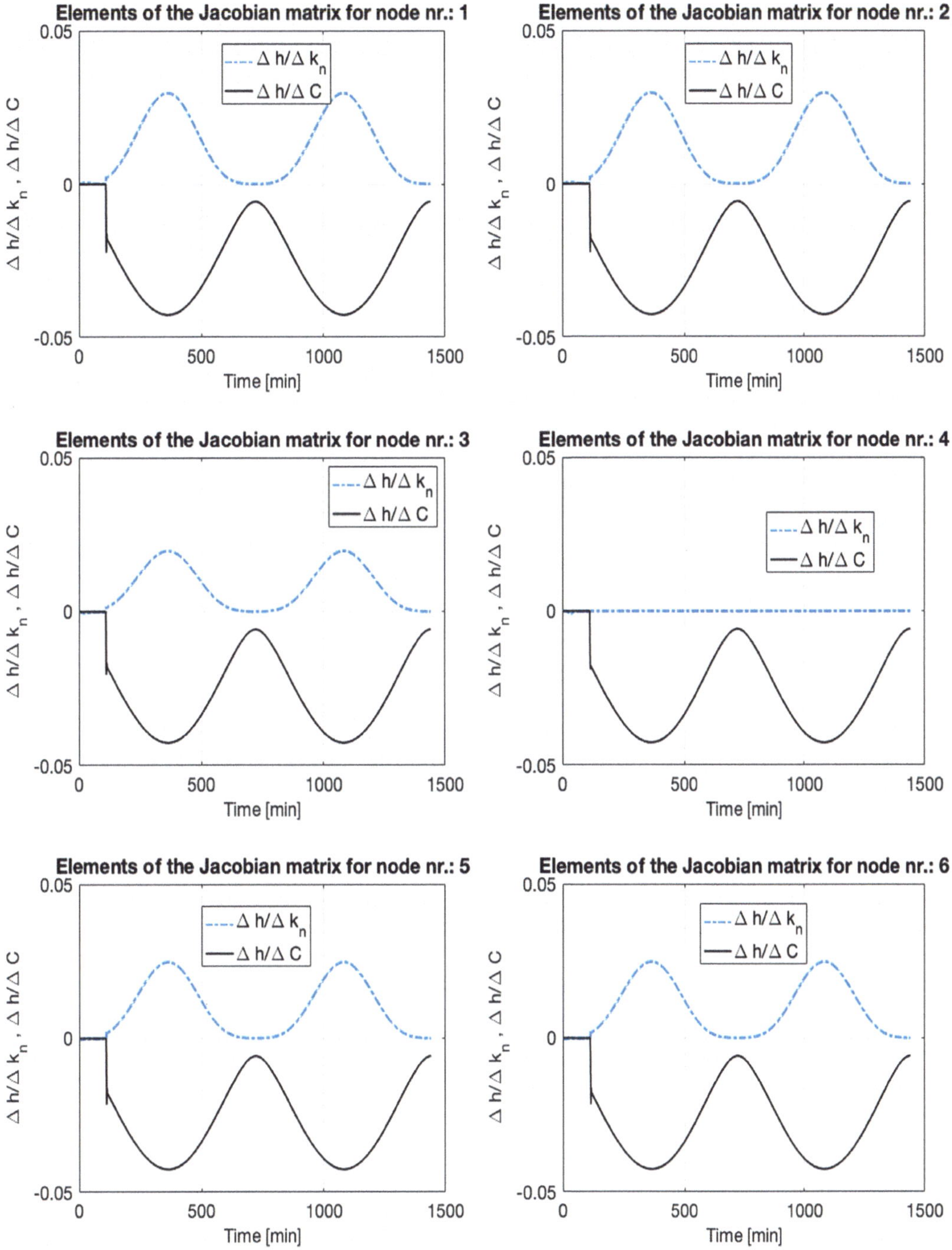

Figure 6.6 Jacobian elements for nodes 1–6. The results for node 2 are enlarged for the time interval [0,160] in Figure 6.7, as some instabilities and obviously wrong results are obtained. *Source*: Francois Clemens-Meyer (Deltares/TU Delft/NTNU).

sensitivity to variation of the value of C and no sensitivity to variation in k_n. This can be understood in the following manner: the water level response at node 4 is almost completely determined by the presence of the weir, while the water level at the other nodes is influenced by both the weir and the hydraulic losses in the conduits. So, node 4 would be an obvious choice for a measuring location as unbiased information for C is obtained (this is only true as long as the boundary condition at node 1 forces the water to flow in the direction of node 4, when flow reverses, the information obtained at node 4 is again a mix between information on C and on k_n. This illustrates the need for using several loads (storm events) when using a model to identify potential measuring locations). With respect to identifying k_n one could pick manhole 1, as this manhole is the farthest away from the weir and the water level variation is influenced by the hydraulic roughness of the conduits over the maximum length. Combined with the unambiguous information for C obtained at node 4, this allows the identification of k_n. Any combination of manholes will result in a certain covariance between the parameter values obtained for C and k_n, as the change in water level is influenced by both parameters. A further observation is that choosing manholes 5 and 6 or just one of them makes no difference in the information gained: the water level in both manholes is identical and responds identically to a variation in C and k_n. This is the same as stating that the water levels at both locations are one-to-one correlated. This implies that on the one hand acquiring information from manholes that are in a hydraulic sense 'neighbours' does not add substantial amounts of information. On the other hand, it allows for consistency checks between the two sensors. An expert will look for manholes that on the one hand provide information but on the other hand, preferably, also show some (limited) overlap in the expected validity range of the parameter which enhances the data validation options. Such an overlap in water levels can easily be checked using a model.

There are some other observations that can be made from the model results:

- The absolute value of the elements in the Jacobian depends strongly on the discharge (Figure 6.6).
- In the time window between $t = 100$ and $t = 120$ minutes, 'artifacts' show up (Figure 6.7): when the water level at node 4 reaches the weir level, a brief time window follows in which some small oscillations in discharge occur resulting in unrealistic values for the Jacobian elements.

These issues are due to the limitations of the software, and some incorrect modelling (in this example this was done on purpose to illustrate the effect): settings of the numerical solver, choice of time steps, etc. Further it has to be realized that in the determination of the Jacobian elements, a finite difference approach is used for quantifying a gradient: when small gradients occur, even round off errors may induce unrealistic values for the gradients.

Overall, a model can be a very valuable aid when choosing monitoring locations, although when doing so the following issues are to be kept in mind:

- Be aware of the occurrence of numerical 'artifacts' (even in very simple networks, as shown previously).
- Make sure to use a range of loads (storm events), as the sensitivity of model results (e.g. water levels) to a variation of model parameters related to any $Q(h)$ relation strongly depends on the load, therefore one is well advised to use a series of loads to judge the potential of all relevant nodes to be a monitoring location.
- Be sure all processes related to the calibration parameters chosen do indeed occur in the model simulation.
- The results of the sensitivity analysis (the resulting Jacobian matrix) can, and in the general case very likely will, vary with varying initial parameters.

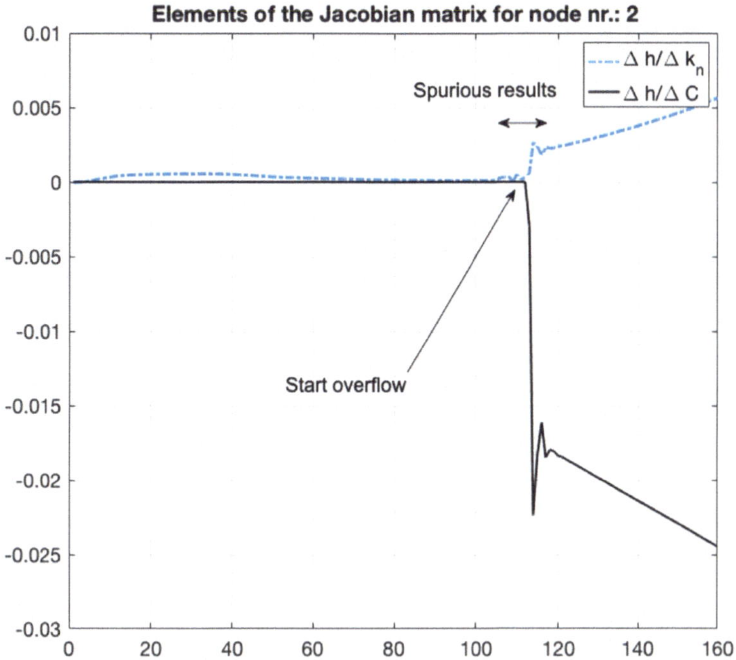

Figure 6.7 Anomalies in the time window between $t = 100$ and $t = 120$ minutes (onset of the overflow). *Source*: Francois Clemens-Meyer (Deltares/TU Delft/NTNU).

The information obtained by estimating the elements of the Jacobian matrix can be used in different ways to determine the optimal locations of sensors. In addition to a simple approach as explained above, more sophisticated methods could be employed to use this related information. These include statistical theory and information theory-based methods. Examples of statistical theory methods include:

- *Alphabetic design methods* that are based on conventional statistical theory and linearization of the system models in question. Among the most used alphabetic metrics are the D-optimality, A-optimality and V-optimality metrics (e.g. Kapelan *et al.*, 2003, 2005).
- *Bayesian theory-based methods* that seek to maximize the gain in information between the prior and posterior distributions of parameters, inputs, or outputs.

In addition to statistical theory methods, information theory-based methods can be used to locate sensors. These methods work by maximizing the information content that sensors can provide. Examples include:

- *Entropy based method* based on the principle of maximum entropy, i.e. a selection of measurement locations and other sampling variables that result in best current knowledge about the observed phenomena.
- *Mutual information-based methods* that use the amount of information that one variable of interest contains about another variable of interest. An optimal sensor network design should avoid collecting repetitive or redundant information, i.e. it should reduce the mutual (shared) information between sensors in the network.

- Other methods such as the *value of information based method* (based on the value a decision-maker is willing to pay for extra information before making a decision where to locate sensors), *fractal-based method* (that utilizes the concept of Gaussian self-affinity to measure the dimensional deficit between the observations of the analysed system and its real domain) and *network theory based method* (makes use of three variables, average clustering coefficient, average path length, and degree distribution to distinguish between different sensor network designs).

More details about the methods mentioned above can be found in Chacon-Hurtado *et al.* (2017).

6.2.5 Timescales, sampling frequency and measuring uncertainty

Assessing the timescales at which some of the processes show significant changes in time provides a first indication of the order of magnitude of the sampling frequency needed.

Table 6.1 provides a first indication of the timescales of major processes encountered in urban drainage systems (Schilperoort *et al.*, 2012). These values, of course, deviate in specific cases but they can serve as a rule of thumb in the preliminary stages of the design.

In most cases it is not just one process that is of importance to the monitoring goal. In these cases, theoretically, one could decide on a varying sampling frequency (e.g. switching from one sample per hour under dry weather flow conditions to once every 5 minutes when a storm event is detected). This approach was applied in the past (1980s–1990s) when data transmission and storage capacity were expensive and frequently unreliable. The reasons for doing so have become less and less valid as data handling has shown considerable improvement in recent decades. Therefore, one is well advised to pick a constant and uniform sampling frequency throughout the monitoring period for all sensors involved, defined by the process with the smallest expected characteristic timescale.

Table 6.1 Range of characteristic timescales of some relevant hydraulic and hydrological processes in urban drainage and stormwater management.

Process	Timescale
Dry weather flow in foul- and combined sewer systems	1–4 hours. For larger catchments (i.e. >100 000 inhabitants) the timescales tend to be at the higher end of the range, and for smaller catchments towards the lower end.
Storm induced CSO events	1–15 minutes, strongly depending on the system under study and the storm event
Infiltration	Minutes-hours, depending on the soil type, initial conditions, climate
Rain	Minutes-hours, depending on the type of storm event (intense summer convective showers to frontal storm events)
Evaporation	Hours-weeks-months
Flow in storm sewers	Minutes-hours, strongly depending on the forcing storm event
Emptying of system using pumps or outlet	1–10 hours
Run-off	See storm events (minutes-hours)
Clogging of filters (SUDS – sustainable urban drainage systems)	Months-years

There are other advantages of using a constant time step for monitoring. Most methods for analysing time series assume equidistance (fixed time interval) between data points, as time series analysis is an important tool in the assessment of data quality and data validation (see Sections 9.3 and 9.4). This important aspect of monitoring becomes easier and more straightforward to implement when choosing a constant and uniform sampling frequency. It has been shown in practice that implementing rules for switching sampling frequencies adds significantly to a decrease in the reliability of the monitoring system reflected in the data-yield (after data validation, see Chapter 9) of the monitoring networks. For example, the case where the detection of rain triggers a switch in sampling frequency of e.g. a water level sensor results in loss of data for water level measurement when the rain sensor is not correctly functioning. In other words, adding complexity to the monitoring system should be considered only when the consequences with respect to data- and information yield strived for are considered. Another general issue to reckon with, when choosing a sampling frequency, is that reducing the sampling frequency is a decision that can be made on evaluation of historical results from a given monitoring station, increasing the sampling frequency can only be made based on 'gut' feeling. Therefore, it is suggested to start a monitoring project with a sampling frequency as high as practically feasible and decide on reducing that frequency after a first evaluation of the data obtained from the system. For more information on reasons for data/information loss, the reader is referred to literature (e.g. Schilperoort *et al.*, 2008 and Chapters 2, 3, 4, 5, 9 and 11 of this book).

6.2.5.1 Application of a model to quantify timescales of a system

In cases where a model of the system that mimics the process(es) one is interested in is available, such a model can be used as a means for obtaining information for picking a sampling frequency. It has to be realized however that such a model is not (yet) calibrated and may contain serious deviations from the 'real' world due to missing data, wrong data, etc. Further all methods described in the following sections are more or less sensitive in their outcomes to the quality of the model, the loads chosen (storm events) and model parameter settings (see Section 6.2.4). The implicit underlying assumption here is that, in close resemblance to the famous story of Baron von Munchhausen (Raspe, 1895), one can bootstrap him/herself out of the swamp: i.e. in an iterative process the model is improved using monitoring data, using the improved model the monitoring design can be improved, etc. To the authors' knowledge, such a scheme has not been reported in the literature to be applied so far, and no guarantee can be given for the iterative process, as described, to converge.

6.2.5.2 Upper and lower limits of the sampling frequency related to measurement uncertainty

When taking samples from a process and the demand is that the process can be reconstructed from the data, a lower and upper limit for the sampling frequency can be defined in relation to the uncertainty of the sampled parameter values. Loosely defined the lower limit is that sampling frequency below which information on the process is lost, while the upper limit is defined by that sampling frequency beyond which 'noise' is being monitored. Both limits depend on the process characteristic timescale(s) and the uncertainty of the monitoring data. The former is to a large extent a given fact and is known in terms of order of magnitude only, while the latter is largely a matter of choice (design).

6.2.5.2.1 Lower limit

The lower limit of the sampling frequency can be quantified using a time-domain analysis on either a measured time series or a time series obtained from a model. The reasoning is as follows (for details see Lepot *et al.*, 2017).

Suppose an equidistant time series describing some (hydraulic or hydrological) process $x(t)$, a value $x(\tau)$ with $t_{j-1} < \tau < t_j$ is obtained from a simple linear interpolation (Equation (6.2)):

$$x(\tau) = \alpha x(t_{j-1}) + \beta x(t_j) \tag{6.2}$$

where α and β are weighing factors in the interpolation.

Let us assume the process has a known variance σ_x^2 and mean value μ_x, and the normalized autocorrelation function of the process $\rho_x(\tau)$ is known as well. In that case, the mean squared error introduced by the interpolation process is defined by Equation (6.3):

$$mse_{interp} = \frac{1}{2}\sigma_x^2[3 + \rho_x(t_{j-1}, t_i) - 4\rho_x(t_{j-1}, \tau)] \tag{6.3}$$

From Equation (6.3) it is easily checked that for a constant process (no change in time so $\rho_x(t_1, t_2) = 1$ for any combination of t_1 and t_2), the interpolation error is zero. In this case, theoretically, one measurement over the whole monitoring period would suffice, on the other hand, for a completely random process ($\rho_x(t_1, t_2) = 0$ for any combination of t_1 and t_2), the interpolation error (more correctly the introduced uncertainty) $\sigma_{interp} = 1.5^{0.5}\sigma_x \approx 1.22\sigma_x$. The autocorrelation function is defined as the correlation between the time series x and the same series x, but shifted over a certain time shift T (Equation (6.4)):

$$r(T) = r(t_{i-T}, t_i) = \frac{1}{n-1}\sum_{i=1}^{i=n}(x_{t-T} - \mu_x)(x_t - \mu_x) \tag{6.4}$$

The normalized autocorrelation function is defined by Equation (6.5):

$$\rho(T) = \frac{r(T)}{r(0)} \tag{6.5}$$

The demand set on the allowable interpolation error is given in Equation (6.6):

$$mse_{interp} < \gamma\sigma_m^2 \tag{6.6}$$

where σ_m is the standard deviation of the measuring uncertainty of the chosen measuring device, and γ is a multiplication factor: $\gamma = 1$ implies that the uncertainty introduced by interpolation is equal to the uncertainty of the measured data. Given a process (and hence an autocorrelation function), setting a value for γ allows the choice of a combination of sampling interval and measuring accuracy so as to comply with the demand set in Equation (6.6).

The two sources of uncertainty (interpolation and measuring) may be assumed to be independent, which implies that the uncertainty interval (95% coverage interval, see Section 8.2.3) of a value x that is based on interpolation between two data points is the sum of uncertainties due to the interpolation process, and the measuring uncertainty ($\sigma_x^2 = \sigma_m^2(1+\gamma)$) in the data itself is $[x - 1.96\sqrt{1+\gamma}\sigma_m, x + 1.96\sqrt{1+\gamma}\sigma_m]$.

One has to realize that this result has, strictly speaking, no generic validity as it is based on the characteristic of the process that is assumed in the analysis. This implies that in order to achieve a robust result, the analysis should be made for a range of mutually different events, as the dynamics of the event largely define the autocorrelation function.

6.2.5.2.2 Upper limit

Strictly speaking, the upper limit of the sampling frequency is determined by the progress of sensor innovation. However, in practice there is also a maximum sampling frequency above which no additional information is obtained. This frequency depends, likewise for the lower limit, on the process to be monitored and the uncertainty of the sensor readings.

The signal to noise ratio (*SNR*) is a key parameter when defining an upper limit for the sampling frequency. The *SNR* is defined as in Equation (6.7):

$$SNR = \frac{P_{signal}}{P_{noise}} \tag{6.7}$$

where P_{signal} is the power of the signal one wants to monitor and P_{noise} is the power of the measuring noise.

In general literature, an *SNR* of 3 is used to define the limit of detection (LoD) (e.g. Desimoni & Brunetti, 2015). The power of a signal depends on the signal's frequency and amplitude, it is therefore natural to revert to an analysis in the frequency domain. To this end, both signal and measuring noise are Fourier transformed. When the measuring noise is modelled as white noise (mean zero, constant variance $\sigma_n{}^2$, autocorrelation is zero), the power of the signal is constant over the full frequency domain and is equal to the variance of the signal. This implies that by setting a value for the desired *SNR* and a known value for the power of the signal to be detected, the maximum allowable value for the noise level is obtained, which is translated directly back into a demand to be put on the allowable uncertainty level of the monitoring set-up (largely the choice of the sensor) in conjunction with the sampling frequency. When applying a Fourier transform of a signal there is an upper limit to the frequency that is the well-known Nyqvist frequency. The Nyqvist frequency f_N is defined by the fact that at least two sampling points are needed for each frequency to reconstruct the signal. Therefor f_N is given by Equation (6.8):

$$f_N = \frac{1}{2\Delta t} \tag{6.8}$$

where Δt is the sampling time interval.

Adding frequencies above this limit results in 'aliasing'. In most software development environments, e.g. Matlab®, pre-programmed functions that take this limit into account automatically are available. Parts of the signal having frequencies larger than the Nyqvist frequency cannot be identified from the measurements.

The underlying principle of a Fourier transform is that a signal $f(x)$ can be written as a summation of periodic functions (Equation (6.9)):

$$F(k) = \sum_{j=1}^{j=N_e-1} f(j)e^{\frac{-i2\pi kj}{N_e}} \tag{6.9}$$

where i is the imaginary unit. The real part of $F(k)$ is the amplitude belonging to a frequency of $2\pi k/N_e$, with N_e the number of elements in the time series. The inverse transformation (from frequency domain to time domain) reads as Equation (6.10):

$$f(n) = \frac{1}{N_e} \sum_{k=0}^{k=N_e-1} F(k)e^{\frac{i2\pi k}{N_e}} \tag{6.10}$$

The 'power' of a certain frequency in the signal is defined as Equation (6.11):

$$P(k) = \sqrt{F(k)\overline{F(k)}} \tag{6.11}$$

The overlined $F(k)$ indicates the complex conjugate. For practical applications, built-in functions in packages like Matlab® can be used (for Matlab® that is the Fast Fourier Transform, `F(t)=fft(f)` and its inverse `f=ifft(F)`). For detailed information on Fourier transforms and analysis, the reader is referred to the literature on the subject, e.g. in Grafakos (2014).

6.2.5.2.3 Examples

In order to illustrate the theory discussed in the previous two sections, two examples are given. The first example is on how to use the results of a hydrodynamic model to define lower and upper limits of the measuring frequency related to the process characteristics and to the uncertainty of the measuring device. The second example shows the results when applied to a real-world problem.

Example 1. Figure 6.8 shows a hydrograph (given as water level) for two locations obtained from a hydrodynamic model for a combined sewer system. The process variability is expressed as standard deviation and is calculated to be 0.64 m and 0.16 m for location 1 and location 2, respectively.

Figure 6.9 shows the autocorrelation functions of the hydrographs (the Matlab® commands used for determining sample frequencies are `meas_freq_exp.m` and `min_max_sample.m`, available for download at https://doi.org/10.2166/9781789060102).

In the model a time step of one minute is chosen. For the sake of illustration, let us suppose we want to set a measuring frequency of once per 2 min, so a reading at $t = 0, 2, 4, 6\ldots$ minutes and find out which maximum uncertainty in the measuring device is acceptable to meet the condition $mse_{interp} < 3 \times {\sigma_m}^2$ (Equation (6.6)). For reconstructing by interpolation the value of the water level that occurs at $t = 1, 3, 5\ldots$ minutes, apply Equation (6.12):

$$\frac{1}{2}\sigma_p^2[3 + \rho(0.2) - 4\rho(0.1)] < 3\sigma_m^2 \tag{6.12}$$

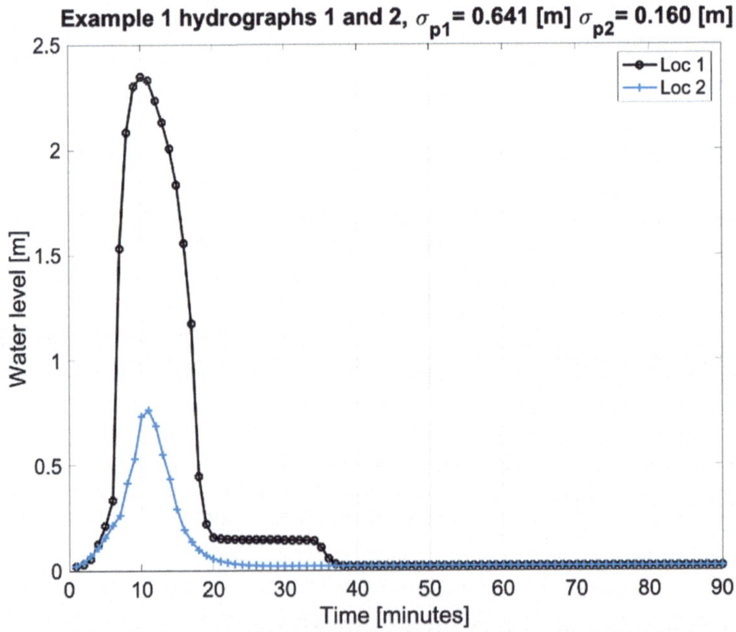

Figure 6.8 Examples of hydrographs. *Source*: Francois Clemens-Meyer (Deltares/TU Delft/NTNU).

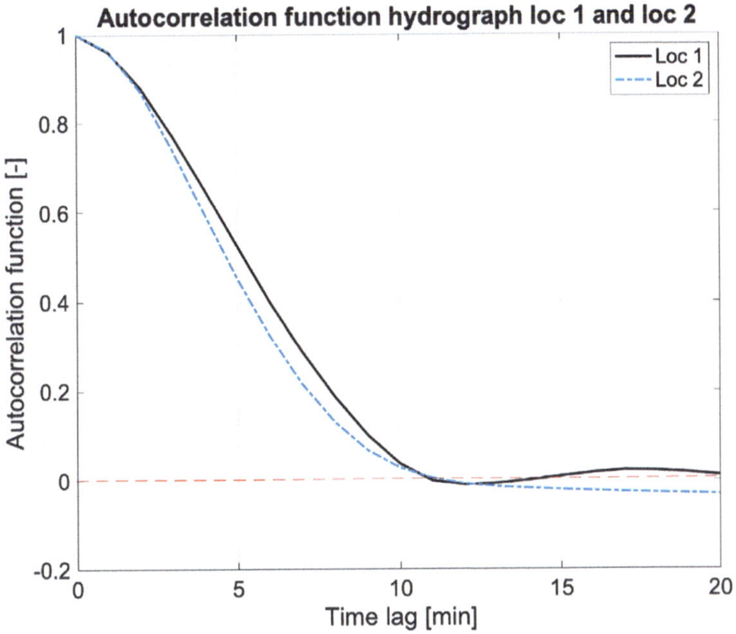

Figure 6.9 Autocorrelation function for the hydrographs given in Figure 6.8. *Source*: Francois Clemens-Meyer (Deltares/TU Delft/NTNU).

Inserting the corresponding values of the autocorrelation functions results in Equations (6.13) and (6.14):

$$\sigma_{m,1} > 0.048\ m \tag{6.13}$$

$$\sigma_{m,2} > 0.036\ m \tag{6.14}$$

So, in this case, for the two locations chosen, no information loss due to interpolation occurs at a sampling rate of once per 2 minutes, provided the uncertainty of the measuring device is less than 0.036 m, which is not a very serious constraint for this case. As the hydrographs, and hence the values of the process variance and the correlation function, are specific for the locations and the load (storm event) imposed on the model, this result can only be regarded as valid for this particular location under the specific load imposed on the model. In order to obtain a more comprehensive indication of the minimum interpolation, an alternative reasoning starts with choosing the measuring uncertainty and works out the maximum allowable time interval between measurements.

In Figure 6.10 the power spectra for both hydrographs are shown. As the time resolution of the hydrographs is one minute, this limits the maximum sampling frequency that can be derived. If one wants to be able to reproduce the signal (the hydrographs), the 'power' of the measuring uncertainty should be less than the smallest value in the spectra of the hydrographs, or more precisely the signal to noise ratio should be such that $SNR > SNR_{min}$. From Figure 6.10, it follows that with $\gamma = 1$ the maximum allowable measuring uncertainty, given the hydrographs and given the frequency of 1/min, is 0.00096 m, which for practical purposes is set at 0.001 m, resulting in a 95% confidence interval of 0.004 m. In an alternative approach, one can chose the measuring uncertainty and determine the 'cut-off'

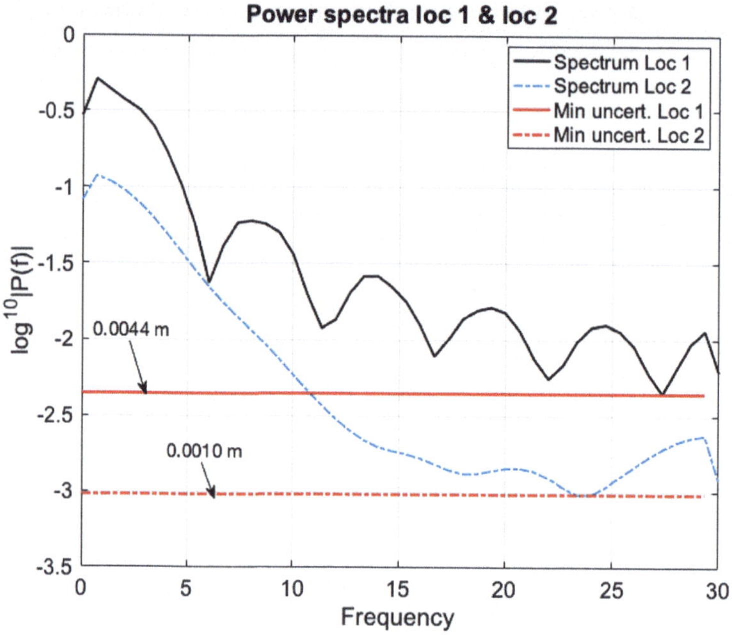

Figure 6.10 Power spectra for the hydrographs given in Figure 6.8. *Source*: Francois Clemens-Meyer (Deltares/TU Delft/NTNU).

frequency (i.e. that frequency of the signal with a power that fulfils the demand $SNR = SNR_{min}$ determining the maximum frequency that corresponds to the chosen SNR_{min}).

Example 2. The example for determining sampling frequencies is based on a study reported by Clemens (2001), for a small catchment with a combined sewer system, known as 'De Hoven' in the centre of the Netherlands. A preliminary rainfall-run-off model and a hydrodynamics model were available. Using the two approaches presented in Section 6.2.5.2 to identify the minimum and maximum sampling frequencies, 10 design storms were used to check for 'load dependencies' of the results. The monitoring set-up was realized and after obtaining data from the monitoring network, the preliminary model was calibrated. A comparison was made between the values obtained in the design phase and what would have been the results when the calibrated model was available during the design (validation using observed storm events). In order to show that the specific process characteristics do not only depend on the storm, but on the location in the system as well, the example is presented for three monitoring locations. The value for γ in Equation (6.6) is set to 1, and the standard deviation of the measuring uncertainty is chosen to be 0.003 mm.

From Tables 6.2 and 6.3, some interesting conclusions are drawn as follows. Per storm, mutually different values for the max and min sampling interval are obtained, but the variation is within a factor of maximum 3. With increasing return period T_r (implying increasing rain intensity, van Luijtelaar & van Rebergen, 1997), the value of the sampling intervals decreases (obviously triggered by an increase in speed of variation of water levels/discharges forced by larger rain intensities). All design storms resulted in a CSO event, while such an event occurred for only two out of the five storms used in the evaluation. It can be seen as well that there is a significant difference in the intervals obtained for real events 1 and 5

Table 6.2 Maximum sampling interval (minimal sampling frequency) for three locations and 15 storm events (10 used for design with T_r the return period, 5 for validation afterwards).

Storm event	Remark	Δt_1 (s)	Δt_2 (s)	Δt_3 (s)
01 $T_r = 0.25$ y	Overflow	767	394	554
02 $T_r = 0.25$ y	Overflow	660	303	391
03 $T_r = 0.5$ y	Overflow	561	472	339
04 $T_r = 0.5$ y	Overflow	547	331	332
05 $T_r = 1$ y	Overflow	419	444	394
06 $T_r = 1$ y	Overflow	548	336	341
07 $T_r = 2$ y	Overflow	365	398	392
08 $T_r = 2$ y	Overflow	446	409	345
09 $T_r = 5$ y	Overflow	327	368	382
10 $T_r = 10$ y	Overflow	324	345	370
Real event 1	Overflow	739	838	849
Real event 2	No overflow	1619	1802	1414
Real event 3	No overflow	2524	2095	2298
Real event 4	No overflow	1490	1422	1628
Real event 5	Overflow	1153	1046	1564

on the one hand and the events 2, 3 and 4 on the other. Clearly a CSO event is a 'fast' process setting the limits for the sampling interval. Using the 10 design storms resulted in choosing a sampling interval of 60 seconds, which turned out to be a 'safe' choice as the values obtained from the recorded events all indicate that a larger value would have been sufficient, it has to be noted here that the output time-resolution of the

Table 6.3 Minimum sampling interval (maximal sampling frequency) for three locations and 15 storm events (10 used for design with T_r the return period, 5 for validation afterwards).

Storm event	Remarks	Δt_1 (s)	Δt_2 (s)	Δt_3 (s)
01 $T_r = 0.25$ y	Overflow	158	135	114
02 $T_r = 0.25$ y	Overflow	171	63	60
03 $T_r = 0.5$ y	Overflow	167	72	75
04 $T_r = 0.5$ y	Overflow	130	60	60
05 $T_r = 1$ y	Overflow	95	64	65
06 $T_r = 1$ y	Overflow	134	60	60
07 $T_r = 2$ y	Overflow	61	89	79
08 $T_r = 2$ y	Overflow	105	74	72
09 $T_r = 5$ y	Overflow	60	60	60
10 $T_r = 10$ y	Overflow	60	60	60
Real event 1	Overflow	199	280	375
Real event 2	No overflow	654	675	686
Real event 3	No overflow	1196	658	1082
Real event 4	No overflow	469	527	505
Real event 5	Overflow	329	300	146

software used (Infoworks™) was set at 60 seconds, this choice automatically determines the lower limit of the sampling interval.

6.2.6 Networks of rain gauges

In virtually every monitoring project related to urban drainage and stormwater management, data on precipitation are indispensable. As discussed in Chapter 2, radar based or microwave attenuation measurements (e.g. Fencl *et al.*, 2015) are employed successfully for covering relatively large areas, also in the urban environment, with up to 5 minutes time resolution and, depending on the technology applied, a spatial resolution of $\sim$1 km^2 or less. There is however still a need for validation with ground-based point measurements. The data for the latter is usually obtained from a network of rain gauges. Installing 'the optimal rain gauge network' in an urban environment is challenging for a number of reasons:

- Demands set on the environmental conditions for a rain gauge set-up as formulated in standards (e.g. WMO, 2018) and Chapter 2 of this book), are hard, if not impossible, to meet.
- Considerations on avoiding vandalism dramatically limits the number of usable locations for installing a ground-based rain gauge.
- Microclimate (especially with respect to wind influences) may cause systematic, difficult to access and/or to compensate for, deviations in the readings.
- Accessibility of locations should be guaranteed, while at the same time discouraging access by trespassers.

As shown by Schellart *et al.* (2012), even on very small time and space scales, a significant variability in observed rain intensities occurs which is due to the fractal characteristics of the rainfall process (Sivakumar, 2010). Variability of rain intensity in time and space reportedly affects the results of urban run-off modelling (Bruni *et al.*, 2015; Ochoa-Rodriguez *et al.*, 2015). The high imperviousness and the small-scale heterogeneity of the urban environment result in very short characteristic response times requiring rain data on a wide range of time and spatial scales. In many cases, though, the demands put on the rain data are not that strict. For example, when discriminating between 'dry' and 'wet' days, e.g. for applying a simple method for determining infiltration rates in drainage networks on a catchment scale (see Section 3.5), one could argue that when an estimate of the daily rain volume is obtained, this would suffice. For a homogeneous catchment (in terms of elevation, land use, etc.) at least one rain gauge in the centre of the catchment would be enough, provided the catchment is small ($\sim$1 km^2). For redundancy reasons, it is suggested to install at least two rain gauges in two close, though separate, locations (e.g. Chartered Institution of Water and Environmental Management [CIWEM], 2017). This allows for cross validation when analysing the data. When a long-lasting systematic deviation between the two observations persists, this may hint at either an issue with the sensors (not uncommon), or indeed a significant difference in the micro-climates between the two locations.

When setting up a monitoring network for rain intensity for a larger urban area (typically larger than 1 km^2), the macro design is more complicated. A first step would be to subdivide the area under study into regions with similar characteristics in terms of land use, presence of vegetation, etc. Further, when the area is hilly, differences in rain microclimate can be expected when there is a predominant wind direction that induces rainfall at the windward side of a hill and a relatively dry microclimate at the leeside. Having done that, a (sub)network per region has to be designed. Basically, this is the same problem as picking locations for water level measurements or discharge as discussed in the previous paragraphs. A reasoning along the same lines is therefore at hand:

- Eliminate all locations that are not eligible for a range of practical reasons.
- Depending on the available budget, decide on the feasible maximum number of gauges.
- Distribute the *N* possible sensors over the *M* potential locations in such a manner that the variability in a spatial sense is accounted for, given the measuring frequency.

It is obvious that the results of the two first actions mentioned depend to a significant extent on the choice of the measuring system and the associated costs for maintenance and operation. The reader is referred to Chapters 2, 7 and 8, respectively, for details on measuring principles, maintenance, and calibration and uncertainty assessment; aspects that affect the choice of the measuring equipment and implicitly their demands in terms of services and environmental requirements.

Grounds for eliminating potential locations for rain gauges are summarized as:

- The presence of leaf abscessing trees, as these will cause significant risk of malfunctioning rain gauges.
- A poor accessibility. The site should preferably allow for a safe, though limited, access, i.e. an employee should be able to safely enter the site. At the same time however, trespassing should be discouraged as much as is feasible to avoid vandalism or theft.
- The non-availability of basic services (i.e. power supply, data communication).
- The presence of tall buildings that significantly influence the local wind field (strength and direction).

Regarding methods for obtaining an 'optimal' network, given the potential locations and a number of locations limited by the available budget, several authors have reported their findings (e.g. Fu *et al.*, 2016; Lei & Schilling, 1993; Rodríguez-Iturbe & Mejía; Shaghaghian & Abedini, 2013; Tiwari *et al.*, 2020). These methods are very diverse in nature and most of them require a substantial background knowledge on mathematical optimization methods, which is beyond the scope of this book to discuss in any detail. The interested reader is referred to the cited references. Ochoa-Rodriguez *et al.* (2019) provide an up-to-date review on both radar rain data and rainfall data obtained from rain gauge networks, and on the influence the configuration/density of the rain gauge network has on the results obtained. Notwithstanding the ongoing developments in the understanding of the properties of rainfall and the manners of recording them, some generic practical insights with respect to network (macro) design are derived from the literature:

- Install at least one, preferably two, gauges per homogeneous (sub)catchment.
- Install at least one gauge per km^2.
- Rain gauges at the periphery of the area under study are a very important source of information on rainfall spatial variability and heterogeneity.
- It is suggested to favour a higher number of measuring locations over a lesser number of locations with high accuracy (and as such more expensive) apparatus when budget issues force a limit on the number of observation stations.

The resolution in time for rain gauges is normally chosen between one day (daily rainfall volume) and one to five minutes, the latter frequency is regarded in practice to be the upper meaningful limit (e.g. for hydrodynamical models: in such a case the level of geometrical detail of the model itself has to be in line with the time-space resolution of the rainfall data), although for research purposes even smaller timescales can be of interest, provided this is combined with a high spatial resolution as well.

When designing a network of rain gauges, a good starting point is to identify properties that provide, or are likely to provide, favourable conditions. Such properties can be:

- Properties owned by the municipality and/or the local water utility, as in most cases these organisations will be involved with the project. Usually, these sites are to a certain extent protected against trespassing and offer a range of basic services for operation.
- Private properties of people involved with the project or other employees from associated parties (consultant, university, municipality, waterboard or involved contractors/construction firms).
- Owned by people having an amateur weather station available. Enquire about their existence and ask them if you can share data. Of course, one cannot expect such people to have top-notch equipment that works with the same protocols as used in the project. Nevertheless, amateur data can be of great value. In some cases, it can even be worthwhile to buy professional equipment and make it available for interested amateurs in exchange for free maintenance and data sharing. Especially in long-term monitoring this can be efficient and effective.

The demands put on the micro design of a rain gauge set-up are discussed in Section 2.2.

6.3 MICRO DESIGN: FROM THE MACRO SAMPLING DESIGN PLAN TO UP AND RUNNING MONITORING STATIONS

The previous section (Section 6.2) provides methods and examples to identify the number of monitoring stations, their locations, and data acquisition frequencies. This section aims to deliver key points to build each monitoring station of the previously designed network, from the sensor selection to the final construction, which must be calibrated and tested to ensure the system is complying with the specifications. After the definition of the goals (Section 6.3.1), Section 6.3.2 suggests a nine-step method for the micro design. Pros and cons of sensor technologies introduced in Chapters 2 to 4 are then detailed with respect to the micro design. Once the monitoring station has been built, Sections 6.3.4 and 6.3.5 introduce few basic tests to ensure the system is running properly prior to its start, and during the first runs, respectively. A few case studies on micro design are briefly presented at the end of this chapter.

6.3.1 Definition of the goals: long-term, mid and short-term installation – 24/7 and event sampling

Once the number of monitoring stations and their locations have been established (Section 6.2), the specifications of each station must be defined prior to the design. To ensure that stations will fulfil the expectations, the goals of each of them should be clearly stated. Several key points deserve some reflection to set reliable specifications.

The first one is the life expectancy of the set-up: a few weeks, months, years up to one or two decade(s). This duration will affect the choice of data communication standards, quality of the equipment and therefore, often, the budget needed, and the time required for the design. For a long-life duration, plausible future needs should be anticipated, data communication standards should still be available until the last refreshment of hardware and software, and a better hardware quality is highly advised to deal with aging. When choosing sensors, one is well advised to balance price, stability and accuracy of the sensor against maintenance, replacements and personnel costs. Ideally one should calculate the cost per data-point of a given minimum quality to make a decision. A temporary monitoring station can be built with lower quality hardware (but not as a necessity) and without consideration of what the future communication standards might be in the coming years. Generally, the design of temporary stations requires less time since reflection on potential refreshment of the station is, in most cases, meaningless.

Table 6.4 Key points for the design of a monitoring station.

		Types of monitoring station	
		24/7	**Event sampling**
Life expectancy	Short term	• Robust components • Stability of the power supply	• Detection of events • Robustness regarding the start/stop procedures
	Long term	• Same as above + quality of the components • Scalability of hardware and software	• Same as above + quality of the components

A second point deserves investigation, regardless of whether the station is designed for the short or long term: 24/7 vs. event monitoring station. A measuring set-up, which does not continuously run, should be designed and built in such a way that the entire system can deal with on and off switches, dry and wet weather modes of use, and automatic and robust procedures to start and stop the entire system. Event monitoring stations are more than likely subject to failures, due to e.g. no suction by a pump when the system starts, pump failures (if the system starts and stops too often due to too sensitive a procedure), biofilms on sensor cells, aging and heat of sensors (Table 6.4). Consequently, such stations should be avoided or, at least, be very carefully designed.

6.3.2 Definition of the needs: hardware, software, maintenance, trained people

Based on the goals defined and the key points to consider, the needs must be clearly identified to ensure the budget and the staff are consistent with the goals. If a main rule for a monitoring station could be stated only once, the authors suggest this one: 'The cheapest investment is not always, in the end, the cheapest solution.'

This remark, which is rather explicit, is to be kept in mind when making decisions regarding hardware, software and maintenance. If, at first sight, cheap components seem to be cost effective solutions to build and run a station given the budget allowance or constraint, it is not always that obvious. Cheap components often suffer from instabilities or frequent failure, either increasing staff costs (to solve the problem) or reducing the data yield of the monitoring set-up. Cheap sensors are often built with cheap components not really able to deal with aging under harsh conditions (e.g. humid atmosphere or temperature variations). This reflection is based on feedback received and personal experience of the authors. The risks mentioned are difficult to assess, as there is a lack of data regarding failure statistics of hardware and the cost of maintenance/reparation is too dependent on the staff/material/travel costs to derive a general rule. Given the development of technologies, complete tests and feedback will most likely be available soon in the literature.

The procedure presented in Figure 6.11 can be recommended for the entire design process. Details for each step are given in the following subsections. The design is organized in nine steps, from the sensors selection to detailed drawing, each step allowing a feedback loop towards the previous ones. For each step, the general idea and a few common sense tips are given, to forecast and avoid potential consequences of incorrect design, poor construction, failures of the components or crisis scenarios.

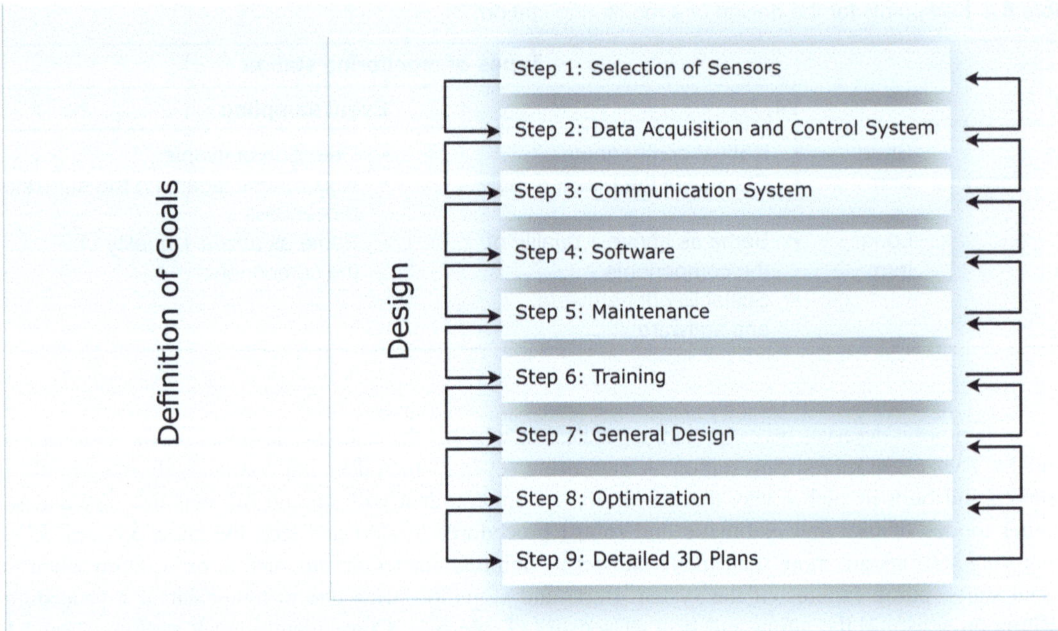

Figure 6.11 Sketch of the recommended nine design steps. *Source*: Mathieu Lepot (TU Delft/Un poids une mesure).

6.3.2.1 Step 1: selection of sensors

For each type of measurement (see Chapters 2, 3 and 4), once the technology(ies) is(are) selected and given the measuring ranges of available sensors, sensors must be selected according to their present and future availability on the market. Given the choice of technologies and manufacturers, the models of each sensor must be selected in accordance with the expected conditions at the measurement location in order to ensure: (i) an overlap between measuring ranges and (ii) a combination of technologies suitable for the expected conditions at the measurement location. If different sensors or technologies are chosen to monitor the same data, but for different parts of the identified range, the total (or even better optimal) measuring ranges should overlap in order to avoid gaps in the complete measurable range by the set-up and increase the data quality. The pros and cons of each type of technology for those sensors are discussed in Chapter 3. The robustness of the probes and the data communication protocols must be considered. The same procedures (except for the measuring range) apply for actuators.

Prior to the choice of the hardware, especially for long-term stations, it is useful to collect information about the system to be monitored: ranges of rain intensity, water level, flow velocity, discharges must be known to obtain information for the sensor selection. Previous or historical data, knowledge, expertise and experience, and modelling must be collected from all reliable sources: system designer, manager, operator, insurance companies, media, citizens, etc. After the verification of those data, the sensors should be chosen according to the expected usual, likely and unlikely conditions that may happen at the measuring location.

Table 6.5 Advantages and disadvantages of different scenarios for sensor redundancy.

Technology	Advantages	Disadvantages
Redundancy with the same hardware	• Comparison between recorded data is easier. • Easier stock management for replacement. • Less skills required to maintain the system.	• The selected sensors should fit the needs, in terms of measuring ranges and conditions. • Sensitivity to unexpected conditions.
Redundancy with the same technology	• Wider measuring range but more sensitive to variations in the measuring conditions.	• Sensitivity to a few special measuring conditions.
Redundancy with different technologies	• Wider measuring range and more robust system regarding the measuring conditions.	• More skills required (for each technology). • More difficulty in data validation.

For data validation purposes and to reduce the likelihood of gaps in data, redundancy (i.e. double or triple sensors for the same data) is one of the key points. Three redundancy solutions can be adopted: (i) by using exactly the same hardware (technology, brand, model), (ii) by using the same technology but different sensors (different brands and/or models), or (iii) by using different technologies. Each of them has advantages and disadvantages, as summarized in Table 6.5.

Ideas for sensor selection

- I 6.1: *Redundancy* – Double or triple the sensors for the same measurement.
- I 6.2: *Variety* – If possible, select different measuring principles for the same measurement, for reducing sensitivity to unusual or unexpected conditions or reasons for failure.
- I 6.3: *Site knowledge* – Grab data and feedback about the site to know the usual, likely and unlikely expected hydraulic and operating conditions.
- I 6.4: *Overlap* – Select your sensors to ensure their measuring ranges are overlapping (e.g. two water levels with the following measuring ranges (0.01 to 0.5 m) and (0.4 to 1 m) – this choice allows measurements between 0.01 and 1 m, with an overlap between 0.4 and 0.5 m).

6.3.2.2 Step 2: data acquisition and control system

Based on the selected sensors (especially their transmitters), the components of the data acquisition system must be chosen. Transmitter outputs (analogue, digital) give strong constraints on the type of data acquisition card(s) or data logger(s) to record the transmitted data, and therefore, on the need for a computer. If required by the design, the same approach remains valid for the choice of control card(s) for actuators. During the selection of the data acquisition system components, an update of the selected sensor(s) might be required to optimize the design (space, cost, availability of hardware): a feedback loop toward step 1 is possible.

Ideas for acquisition and control hardware selection

- I 6.5: *Back to standards* – Be conservative but non-nostalgic and avoid the brand-new promising plug or protocol from a manufacturer not yet tested and evaluated.
- I 6.6: *From now to future* – Try to envisage what standards would be available at the time of the last expected update of the station (this can be done by e.g. keeping track of developments in legislation and/or standards).
- I 6.7: *Scalability* – Select hardware that will allow you to add potential sensors or/and controllers.
- I 6.8: *Modularity* – Select hardware and protocols that will allow you to change individual sensors or/and controllers.

Given the life expectancy, especially for long-term stations, the hardware must be updatable until the last planned refreshment, i.e. the used communication protocols and plugs should be available until this day. For example, lately, the RS232 or RS-485 are almost no longer in use. At the same time, the ethernet ports did not change. The situation is not that clear for USB (Universal Serial Bus): the standards A and B did not change with the generations, the mini-USB (mini A, AB but not B) have been replaced by the micro-USB and, since 2014, the USB-c seems to have become the new standard plug.

Careful attention must be considered for all the components: non-forecasted planned obsolescence might create problems, especially since there is barely any second-hand market for sensors. Technology watchfulness and discussions with experts are the two main means of ensuring the correct decisions regarding this issue. Furthermore, even though it is common sense, proprietary or specific connectors must be avoided.

A communication network is not mandatory if the data can be stored *in situ* and regularly collected. However, the selection of hardware depends on the selected communication network: GSM, internet, radio, etc.

6.3.2.3 Step 3: communication network

The choice of communication network strongly depends on the service availability (Chapter 7) and the selected hardware for the acquisition, and if needed, the control of the station. Maybe the internet (if available) still remains the easiest solution and the most widely applied. The cost and availability of GSM communication is too dependent on the country, the specific location and the supplier to provide a general recommendation. GSM communication might be the best if there is a need for alarm generation and if the staff in charge of the maintenance do not have smartphones. The emerging LoRaWAN (long range wide area network) protocol has been applied successfully in urban hydrology (Blumensaat *et al.*, 2018, 2019; Ebi *et al.*, 2019; Orfeo *et al.*, 2018).

Ideas for communication network

- I 6.9: *Availability* – Which networks are available at the location?
- I 6.10: *Cost* – Actual and forecastable costs.
- I 6.11: *Quantity* – Amount of data to be transferred.
- I 6.12: *Energy* – Energy consumption.
- I 6.13: *Redundancy* – Does the system require redundancy in the communication network?

6.3.2.4 Step 4: software and data storage solutions

The choice of software strongly depends on the selected hardware and the life expectancy. For a short term and simple monitoring station, there is much more flexibility. Usually, such a set-up will not be updated: software updates and availability do not play a relevant role on the choice. Furthermore, if the system will be designed, built, and run by the same people, and if the station is equipped with sensors from the same brand, the choice is only dependent on the skills of the people in charge of the station or the software provided by the sensor manufacturer. However, in the last case, proprietary software often has two serious drawbacks: the lack of information on internal routines and a low flexibility to develop additional routines.

For long term, multi-user and/or multi-sensor manufacturer set-up, the choice of the data acquisition software becomes more critical. The following questions require some deep thought prior to the final choice. Is there a software which can deal with all the sensors and data communication protocols? Is there a software for which the skills and knowledge can be kept over the entire life expectancy, despite turnover of staff? Is there a software which will be updated and available over the same period? If one or more software(s) is(are) an output of the last three questions, the selection is easy. If not, the previous answers should be weighted and analysed to make the final selection. This choice will lead to a widely applied software: there will be people available on the job market with the desired skills; the software will most likely be regularly updated and will follow the change of sensor communication standards.

However, the choice of a new but not widely used software is an option. In such a case, special attention should be devoted to keeping the skills within the organization responsible for the system operation, thus requiring careful management to ensure employees with those skills will stay and/or appropriate training of other staff members is planned.

As for the previous steps, a feedback loop might be activated if the compatibility with the selected hardware is not complete.

Data storage and expected uses of data must be carefully studied during the design step. The following questions should be answered: who will use the data (internal and/or external uses)? What are the skills and knowledge of all the potential data users? Should the data be accessible 24/7 or not? Where must the data be stored and backed up (on site, at the office, in the cloud)? A few common sense tips must be clearly stated.

The first one is associated with the metadata, especially if a network of monitoring stations is set up. Metadata are mandatory to understand the data (importance of metadata is discussed in Chapter 10) and may include at least the following ones: name of the monitoring station, GPS coordinates, type of time used (local or GMT – highly recommended to avoid problems during winter and summer time shifts or with e.g. day-saving time shifts when synching time series from mutual different sources), time step, hardware list (brand, model, serial number), life sheet of all sensors (see Chapter 7) with maintenance data such as: date and type of intervention (including calibration, verification, cleaning, uncertainty data, failures and repairs, after sale service, replacement, etc.), name (action and technician) and comments on each maintenance action, date and calibration/verification data (starting date and validity period). This list can be completed with other topics: their relevance should be discussed with all the people involved in the project.

The second one concerns the data storage location and accessibility (see Chapter 5). On site storage is the easiest method to implement but presents a few drawbacks: such data can be lost and/or stolen on site, are not easily accessible, and require more staff costs for the transfer from the site to the office. A combination of on-site storage and automatic backup at the office is highly recommended. Several options are available to achieve this purpose: continuous or sequential backup (once per hour, per day) while using communication networks (GSM, internet or radio depending on the amount of data to transfer and the corresponding costs).

The third and last one deals with the ease of the data use. Depending on who will use the data and all the plausible mistakes that can be made (changes in the raw data, data deleted, lost, modified, corrupted, etc.), a robust and reliable data management system to ensure the integrity of the data, its backup and a 'user friendly' interface is required. A basic rule: never work with original raw and validated data sets, only on copies.

Software

- CL 6.1: *Cost* – Price of the software and its update?
- CL 6.2: *Open-source* – Or not?
- CL 6.3: *Knowledge* – Available within the team? Widely used?
- CL 6.4: *Compatibility* – With other software or hardware.

Data storage

- CL 6.5: *Separate* – Separate the location of the storage?
- CL 6.6: *Redundancy* – Duplicate the storage hardware.
- CL 6.7: *Read only* – Keep a backup of read-only raw data.
- CL 6.8: *Accessibility* – Easily accessible and efficiently protected.
- CL 6.9: *Experts* – If needed, contact IT experts.

6.3.2.5 Step 5: maintenance

'Plug and play' is most likely the biggest source of frustration in monitoring set-ups. By acting on this commercial promise, a lot of problems can be encountered, especially in urban hydrology, given the constraints related to urban environment and to the (waste)water matrix. In summary, maintenance needs to be considered from the design phase, in order to avoid building an up-to-standard set-up delivering, in the best case, strongly biased data. If monitoring in urban hydrology requires a lot of initial investments, they should not be wasted by insufficient maintenance and data processing costs, which should be evaluated exhaustively. Operation costs to ensure data availability and quality are usually significantly higher than investment costs, especially for long-term monitoring.

The frequency and the duration of maintenance of a monitoring station strongly depend on the type of set-up: number and diversity of sensors, and required maintenance according to the manufacturer specifications (e.g. sensors, pumps, auto-sampler). Combined with the number of monitoring stations of the network, the frequency and duration of maintenance allow estimation of the required maintenance time to be spent *in situ*. A special issue must be pinpointed here: it is not unusual to identify a gap between the required maintenance actions and frequency (claimed or promised by the manufacturer/retailer) and the real ones. It should be no surprise that the claimed required maintenance is only rarely overestimated. On top of this duration, travelling times between the office and the different locations lead to the calculation of the entire required duration to keep the network in good operational condition. In addition to staff costs, trip (fuel, car, insurance, etc.) and consumable (e.g. pump, pipe, calibration standards) costs should be carefully estimated.

Last but not least, the management of the hardware stock needs to be considered. Destruction or malfunction of hardware are unaffordable, and a distinction between critical, sensitive and non-sensitive hardware should be made. On the one hand, pumps and pipes for a by-pass monitoring station can be considered as critical. Data acquisition and control card or hardware (e.g. computer) can be considered as critical or sensitive depending on the goals and the obligations of the set-up. On the other hand, other hardware might be, sometimes, considered as non-sensitive: redundant flowmeters or rain gauges, and calibration standards. The availability and the quality of the data might not be affected if one of these hardware components is temporarily out of order. The key point is data recording and backup. Given the available budget, hardware costs and storage location, a stock of those devices has to be prepared to face upcoming problems and to solve them as fast as the system (and its goals) requires it. A failure risk analysis is highly recommended, e.g. a FMECA – failure mode effects and criticality analysis (Mikulak *et al.*, 2008). Proper management, including an up-to-date list of available hardware and tools, a forecast of the potential needs (e.g. replacement of gaskets every 6 months) and the expected order and delivery times should be planned in advance.

After hardware and software are decided upon, the design of the implementation of all the components must be such that it minimizes the duration of maintenance tasks, i.e. the item to repair or replace should be easily accessible, easy to remove while avoiding (when possible) any disruption in the monitoring. Furthermore, any system that can identify when maintenance operations occur should be designed and set up (e.g. switcher, magnetic contact): such a solution facilitates data validation, while highlighting when the data may be disturbed by the maintenance (e.g. sensors cleaning, calibration).

Additional information on maintenance is provided in Chapter 7.

Objectives to be achieved during the design, for maintenance operations

- CL 6.10: *Operation duration* – Reduce frequency and duration.
- CL 6.11: *Risk* – Minimize the risk for the operators.
- CL 6.12: *Remote* – Allow remote diagnostics to increase maintenance efficiency.

6.3.2.6 Step 6: trained people and training

Since the 'Plug and play' approach is far too optimistic, experienced and trained staff members are essential to ensure the quality of the delivered data. Recruitment and training costs and durations should not be underestimated, especially for complex and long-term set-ups.

After the system is designed, a complete analysis of both existing and required skills and knowledge should be conducted to identify the potential missing ones. Initial training should be planned according to those gaps, while keeping in mind some redundancy between the staff to allow for possible turn overs, like holidays, and ensure there will always be someone able to do the maintenance. For long-term monitoring stations, continuous training of new employees must be set up.

Action to be taken with respect to staff and training

- CL 6.13: *Balance* – Comparison between existing and required skills and knowledge.
- CL 6.14: *Identification* – Identification of suitable training.
- CL 6.15: *Spread* – Spread the skills and knowledge within the team to ensure continuity.
- CL 6.16: *Report* – Report procedures, questions, and comments.

6.3.2.7 Step 7: general design and drawing

In addition to all the foreseeable disfunctions of a system, other unlikely events should be considered to ensure the integrity of the set-up: flooding, thunderstorms and their power supply cuts, vandalism, data communication shutdown, etc. Those are crisis scenarios which require solutions to avoid them or at least minimize risks, damage, and consequences. Table 6.6 summarizes the main solutions for some types of events.

Given the identified possible crises and their solutions, the general set-up and its installation should be designed accordingly. At some locations, regulations and available space might be constraints to this general drawing/layout of the monitoring station: careful attention should be devoted to this step.

Table 6.6 Solutions to avoid crisis.

Type of crises	Solution(s)
Flooding	• Select the right location for the sensitive hardware • Flooding proof set-up
Thunderstorms	• Protection against shutdown • Automatic restart procedures
Vandalism	• Padlock • Alarms • CCTV cameras for dissuasion • Set-up either visible or properly hidden from the surroundings (it is hard to underestimate the creativity and dedication of urban vandals) • Information panel is usually useful to reduce vandalism

Best friends and best enemies

Low- and high-power supplies cabling

A problem that is commonly observed persists: the proximity between low- and high-power cables, i.e. often data transmission and power supply. The alternative high voltage and current of the power supply (typically 220 to 380 V) creates an electromagnetic field. This may affect all the analogue signals delivered by some sensors of the station and, therefore, create some noise in the data. This problem can be encountered with other sources, especially cell phone connections. The proper use of shielded cables (i.e. when the external grid of those cables is connected to the ground) is the most standard and the most efficient solution for this problem.

Electricity and water

Even if it sounds like basic common sense, electricity cables and water pipes should be carefully placed relative to each other. As far as possible, the water should never be in contact with electricity cables: those should be always placed above the water hydraulic head, even in the case of flooding or disruption. The choice of the certified connections (waterproof) helps a lot in reducing plausible consequences of a station malfunctioning.

6.3.2.8 Step 8: optimization

The general drawing is now ready. Can it be optimized? This predesign should be presented to all the people involved in the design, construction, use and maintenance of the monitoring station. Feedback, suggestions, and corrections should be collected, processed, and implemented. In case of hardware failures, especially if the station should run 24/7, the components must be easily replaceable, i.e. the required duration for such an operation must be minimized (see Chapter 7), even if there is(are) (a) redundant component(s). If the acquired data are crucial (for any kind of reason), special attention must be devoted during the design to lead to a fast-to-repair set-up.

Given the general drawing established at step 7, the accessibility of the different components must be carefully studied to minimize the maintenance duration (including safety protocol, e.g. the method to replace components must minimize the need for access within the sewer – always more time consuming and expensive than maintenance on the ground).

Ideas on optimization

- I 6.14: *Collective* – The optimization should be a collective work.
- I 6.15: *User-friendly* – Adopt user-friendly approaches, for any potential type of users.
- I 6.16: *Cost* – Reduce the cost.
- I 6.17: *Question* – Question all the decisions made since the beginning, were they worth it.

6.3.2.9 Step 9: detailed 3D drawing of the monitoring station

After step 8, everything should be clear on paper. The best way to ensure the design reliability is probably to draw in 3D the entire set-up in its environment. This is especially advised when a complex system of cables and pipes is present. A 3D drawing is a powerful means to reveal construction conflicts. This will help the validation of the entire design, can help the authorization that may be required prior to the construction, and may be used for communication purposes with the different actors involved in the project.

3D drawing

- I 6.18: *Dynamic* – Implement and study the dynamics of every component.
- I 6.19: *Sub-contracting* – Those sketches can be used by sub-contractors.
- I 6.20: *Cost* – Reduce the cost.

There are plenty of 3D drawing software packages available, often for free and with comprehensive libraries of objects. The kinematics and the available space for maintenance can be accurately verified at this step (Figure 6.12).

6.3.3 First tests

6.3.3.1 Calibrate

Prior to the data collection, all the sensors and the data acquisition system should be properly checked and calibrated. Chapter 7 is, amongst other issues, devoted to calibration methods and, therefore, this section is

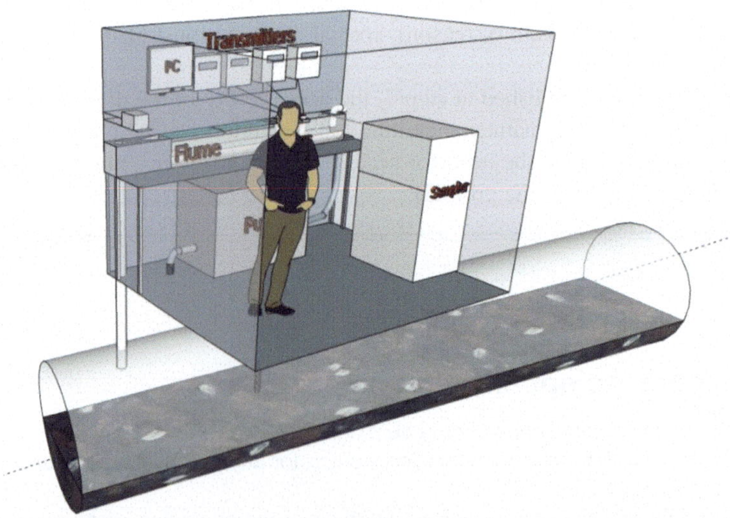

Figure 6.12 3D sketch of the OTHU monitoring station. *Source*: Nicolas Walcker (INSA Lyon).

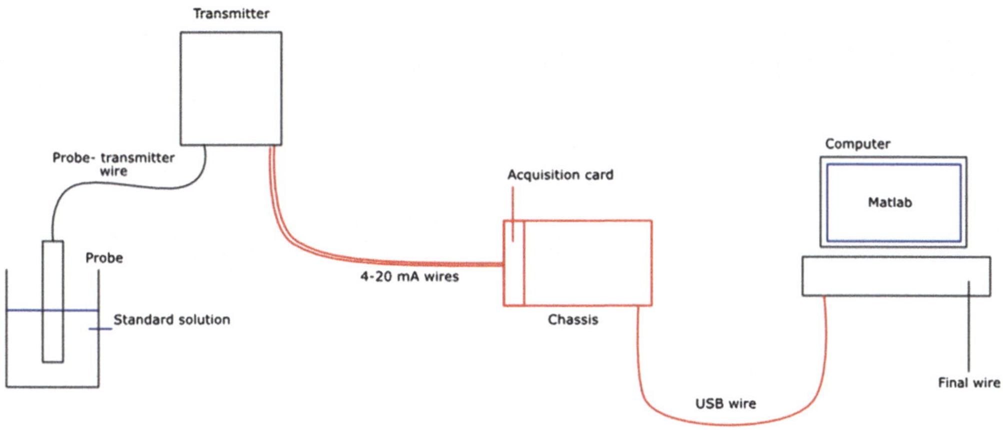

Figure 6.13 Sketch of a measuring chain during an *in situ* calibration process. *Source*: adapted from Lepot (2012).

just a reminder. The calibration of sensors is highly recommended, as a standard best practice, like the periodic checks and re-calibration when necessary (after repair, drifts, etc.).

However, even if sensors are calibrated, the entire data acquisition system should be calibrated as well. In fact, sensors are often calibrated between the measured sample, the sensor and the transmitter (value plotted or recorded within the sensors). Other bias may occur along the measuring chain shown in Figure 6.13.

A typical measuring chain (Figure 6.13) includes different components: sensor, transmitter, data acquisition and control hardware, computer with software and cables. The most frequent calibration procedures are done between the sensors and the transmitter, while using relatively blind internal procedures given by the sensor manufacturer. Bias may also occur between the output of the transmitter and the computer. Therefore, calibration of the full measuring chain is required, from the reference value until the final file (see Chapter 7), not only between the reference value and the screen of the transmitter. In the USA, the standard way to collect data with analogue outputs is the voltage. Between the output of the transmitter and the analogue to digital converter, there are cables. Due to the linear electrical resistance of the cable, the voltage might decrease along the length of the cable.

6.3.3.2 *Run, test, verify and correct*

After calibration, the system has to be stressed to verify that the given values are consistent with the current conditions. Several methods (detailed hereafter) are available for such purposes.

6.3.3.2.1 Getting the data as expected?

The corrected values (i.e. after the calibration correction, see Chapter 7) should be consistent with the real conditions, while considering the uncertainty. A few questions can help the reader to realize those tests. Does the system provide all the expected data? Are the zeros (water level at zero, velocity at zero, no rain) properly measured? Is it possible to create artificial events (rain, discharge in a pipe) to verify the values given by the system?

6.3.3.2.2 Redundancy, tracing experiment

If it is rather easy to verify the zeros for each sensor, the creation of artificial values (rain, discharge, water level) is slightly more complicated but achievable: injection of known discharge, simulation of different

Table 6.7 Recommended methods to check the data recorded by the measuring system (except for zeros).

Measurements	Technology	Recommended methods
Rain	• Rain gauges • Radar • Others	• Mariotte bottle or calibrated pump • Comparison with rain gauges (if available) • Comparison with rain gauges (if available)
Water level	• US or optical measurements • Pressure	• Horizontal plate under the sensors • Automatic restart procedures
Velocity	• All technologies	• Velocity at the free surface
Discharges	• All technologies	• Cumulated volumes • Mass balance • Tracing experiment

water levels with pumps or fire hydrants, known rain event with a Mariotte bottle, etc. The value recorded in the final file by each sensor should be consistent with the generated artificial value. Table 6.7 suggests some methods to verify recorded values.

Once each sensor has been independently verified, the consistency between each other (if available) must be checked, i.e. the test of the comparison between pairs or sets of values must be applied to all redundant (e.g. two water levels sensors) or correlated (e.g. water level and flow velocity for flow without backwater effects) sensors. The potential difference(s) between the values given by different sensors or calculated by a combination of them must be investigated: installation errors, defaults during the calibration procedures, site or hydraulic condition effects, errors in settings of data loggers, unit conversion, etc. Hypotheses should be tested and verified while avoiding tuning some magic parameters to get consistent values. It is important to provide evidence of proper functioning based on a series of tests rather than simply hoping that the system will work as expected or designed. Proper qualification of the monitoring system is thus fundamental.

6.3.3.2.3 Verify with data

If the system to be verified has several monitoring stations or locations (like inlet, outlet and volume of storage tank or flowmeter in pipe downstream of a pumping station), or several sensors to measure the same value at the same place, other tests can be performed: mass balance at the inlet/outlet of a tank, comparison of cumulated volume (to identify potential drifts), discharges between two points without connection, etc. The design of such tests is necessarily site dependent. There are numerous tests to achieve such comparisons (see Chapter 9 on data validation). A single piece of advice: question, be critical and challenging, and be impartial regarding the design that has just been built.

6.3.3.3 Conclusion after the first tests

After all qualification tests are run (calibration and verification) and assuming the station successfully passed them, the station is ready to be used as designed. Otherwise, understanding the reasons for errors, failures and disagreements with initial design is an important and necessary phase until all corrections are made.

6.3.4 Once the monitoring station is operational

The previously proposed tests can be applied periodically to ensure the system is working fine, even independently from the highly recommended data validation procedures. Two other recommendations should be followed to ensure the right use of the facilities: a site book, to keep track of all the events that occur (expected or unexpected) and continuous training of the employees operating the station.

6.3.4.1 Site book

Derived from a rather old-fashion laboratory book, a physical or digital site book is useful to keep track of observations made by the user. Both versions have pros and cons: a physical book is easier to set up, and efficient for writing down the observations. However, this type of book can be lost, destroyed (e.g. by flooding), and does not allow recording of pictures/photos. The two versions used at the same site can combine the advantages of both, while leading to the need of having to look at both versions. All kind of events should be noted in such a tool: defect of components, misfunctioning, abnormal discharges/events, strange smells, etc.: the list is unlimited. Those data may seem useless at first glance, but they can be helpful for later understanding of misfunctioning, improvements of future designs, and data validation. Some key points need to be written down: names of the people who recorded the observation, date, hour, duration, observations (direct or indirect), reasons or hypotheses to explain the observations (see Table 6.8). The notion of direct and indirect observations needs to be explicit. As an example, an abnormally high discharge can be directly (e.g. the observers can see the flow) or indirectly (e.g. noise in the pipe, new floating materials dumped at high elevation) observed. Indirect observations require more details in the section 'reasons or hypotheses'.

There are numerous solutions available for a digital site book: e.g. SharePoint, file exchange systems, an e-mail address for each station or an electronic calendar. Each solution presents pros and cons, and the selected solution should fit everyone: from the technician in charge of the daily maintenance to the manager.

6.3.4.2 Continuous training of involved people

The importance of continuous training is often underestimated by managers. This is the only way to keep every employee involved in measurements updated: technology progresses fast, new measuring methods or data transmission protocols are available, and standards change. This technological watchfulness requires few actions: exchange between practitioners, visits of/by manufacturers, workshops, readings, internal and external trainings, etc. and permanent questioning.

Table 6.8 Example of content in a physical site book.

Date	29/02/2016
Hour	19:07 UT
Observer	Iris Pear & JP Manoeuvre
Observations	Inconsistencies between water level on the transmitter screens
Action taken	Verification of each sensor with horizontal plate
Identified reason	Obstacle under the US sensor
Problem solved?	Yes
When did the problem start?	I don't know (to be checked with the data)
Additional info	Picture sent by e-mail to the manager (29/02/2016, 19h33 UT)
Suggestion to avoid this problem occurring again	Installation of a grid with large mesh upstream of the measuring location

Figure 6.14 Photo of the leaping weir in Chassieu, France. *Source*: Nicolas Walcker (INSA Lyon).

6.3.5 Example of micro design

Measuring discharges in large pipes is rather challenging: the wide range of hydraulic conditions often requires different technologies to ensure good measurements. The Chassieu OTHU site in Lyon, France has been monitored since the early 2000s. The inlet of the stormwater settling tank in Chassieu is a concrete pipe (circular, diameter of 1600 mm, slope of 1%). Some velocity and water level sensors have been placed in the pipe to measure discharges.

After 10 years of use, feedbacks highlight rather doubtful values when the water level is below 7 cm, due to the disturbances created by the Doppler probes on the measured velocities in their close surroundings. In order to improve the knowledge on the dynamics of small rain events, a leaping-weir has been installed at the pipe outlet, downstream of the previously installed sensors: when the discharge is lower than 4 L/s in the pipe, all the flow passes through the leaping weir (Figure 6.14) and is measured by an accurate electromagnetic flowmeter in a pressurized pipe. For higher discharges, the existing set-up (ultra sound and pressure sensors coupled to flow velocity Doppler sensors) is used, since a part of the flow goes straight, without passing through the electromagnetic flowmeter.

6.4 ADVANCED AND EMERGING MONITORING TECHNOLOGIES

This section briefly introduces some emerging techniques, mainly used in research and sometimes only in the laboratory: the authors lack comprehensive feedback on these techniques. However, they seem promising and once more and robust experience is obtained they will be available in future literature reviews.

6.4.1 Event detection

A recent study introduced a rather robust and simple method to detect overflow events and record their duration (Hofer *et al.*, 2018). This method requires two thermometers: one in the main pipe and one at the invert level of the overflow pipe: once the temperatures are equal at both sites, there is an overflow. As soon as temperatures diverge again, while taking into account the thermal inertia of both sensors, the overflow is finished.

6.4.2 DTS for infiltration

Over the past decade the application of distributed temperature sensing (DTS) using glass fibre cables to detect infiltration in sewer systems or the presence of misconnections (stormwater discharges into wastewater systems or vice-versa) has gained popularity (Hoes *et al.*, 2009; Nienhuis *et al.*, 2013). The measuring principle is described in Section 3.5.3.1.8. In practice there are several issues to reckon with when planning to perform a DTS measurement. A first and very important issue is the attenuation of the optical signal when using long cables, especially the signal loss in connectors between cables is a point for attention (Tyler *et al.*, 2009). In most practical cases one needs to use a number of cables that have to be connected (Figure 6.15) to cover the whole length of the sewer reach one wishes to monitor, typically some kilometres. The reason for this is that inserting huge lengths (in the order of 1–2 kilometres) of cables asks for the deployment of a relatively large team of workers and the availability of heavy equipment. In addition, the risk of a cable getting stuck or broken increases with the length of the cable (Schilperoort *et al.*, 2016). A trade-off has to be made between reducing the number of connectors and the length of the cables used. In this sense the quality of the connectors improved over time allowing for use of shorter cables. At present the application of 200 m long cables is possible given the strongly increased quality of optical connectors applicable under field conditions. These cables are relatively easy to handle (reduced weight) and allow for relatively simple adaptations in the sewer reach covered during the measuring campaign, as only one or a few cable parts have to be replaced.

Further, one is advised to plan the sewer reach well in advance; correcting the route of an optical cable once installed is a tedious and costly task. In most cases the duration of the monitoring campaign is in the order of weeks to months. Therefore, in planning the route, make sure there is a location where a container holding the measuring computer can be stationed, while accounting for traffic and operator safety, and the availability of power supply and preferably data-communication facilities. When choosing the route, make sure there are enough locations at which the cable can be accessed for temperature calibration.

Figure 6.15 Connecting cable segments using an optical connector for DTS measurements. *Source*: courtesy Remy Schilperoort (Partners4UrbanWater).

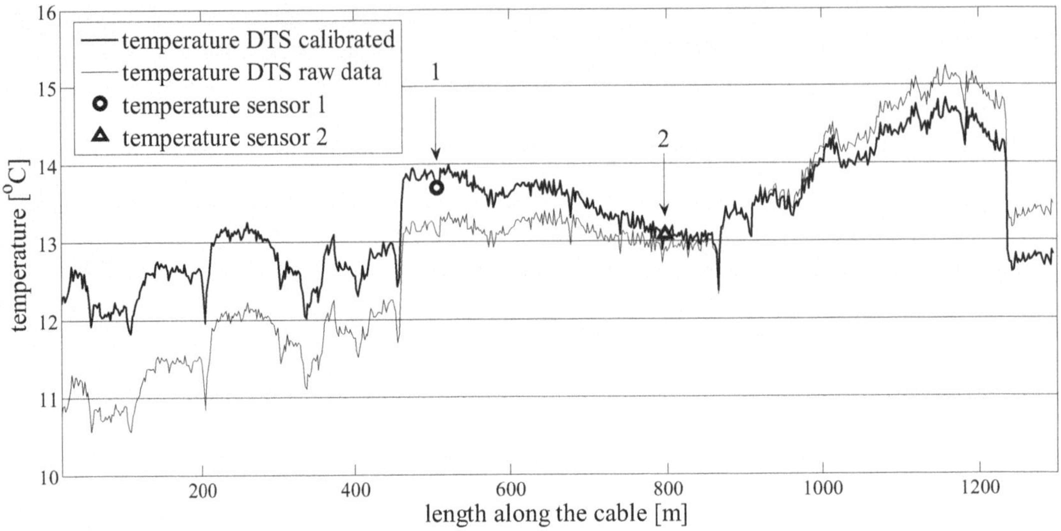

Figure 6.16 Example of a calibration read-out of a DTS cable, along with the corrected result. *Source*: courtesy Remy Schilperoort (Partners4UrbanWater).

For calibration, the temperature is measured at some points along the cable using an accurate thermometer (Pt 100) or, alternatively one can insert parts of the cable (at known distances from the measuring computer) in a bucket of melting ice. Figure 6.16 shows an example of a calibration result.

In most cases one is not really interested in the absolute temperature but in the change over time and its location, making an accurate calibration for temperature values unnecessary. However, one still needs to verify the indicated distances. This can be done by exposing the cable at known locations to a low (or) high temperature, e.g. by simply pouring water with a temperature that is likely to differ from the water present in the system onto the cable (Figure 6.17).

6.4.3 Optical methods for determining flow velocity fields

The application of optical based methods for determination of flow velocity fields in urban drainage has been applied for about the last decade (e.g. Jeanbourquin *et al.*, 2011). LS-PIV (large scale particle image velocimetry) and LS-PTV (large scale particle tracking velocimetry) are non-invasive technologies for determining flow velocity fields. This type of method is becoming more and more popular and emerged from a range of optical measuring methods developed and applied in the experimental study of fluid flow fields since the 1980–1990s. Without going into details of the technologies, a brief description is given hereafter.

Using video footage of floating objects on the surface of a river, channel or conduit in which small light-reflecting or light-emitting objects are present (either introduced as a seeding with known, tailored properties or naturally present), the displacement of these objects between two successive frames is determined and forms the basis for determining a local velocity vector. The difference between PIV and PTV is that the former is based on correlation between successive images, while the latter is based on the movement of individual particles. For a measuring set-up in the field, these differences are unimportant as the differences are mainly found in the post-processing of the raw data (images/video-footage). In

Figure 6.17 Pouring cold water onto the cable, in the underground sewer, using a hose and a funnel for validation of the measured distances by the DTS system. *Source*: courtesy Remy Schilperoort (Partners4UrbanWater).

Figure 6.18 an example of a (field) result from LS-PTV measurement in a wastewater pumping station is shown.

An important issue is ensuring the right light conditions. There is no general rule for this, the settings of light intensity, aperture and exposure time need to be determined by a trial-and-error process. There are however some practical limitations when doing so:

- The exposure time should be relatively short to avoid motion blur as this reduces the useability of the footage for post-processing.
- Depending on the position and orientation of the camera, the aperture has to be chosen in such a manner that the whole region of interest is within focus, this may limit the exposure time.
- The time interval between two frames has to be known and must be significantly larger than the exposure time and in such a magnitude that between frames a noticeable change in position of the markers/seedings has occurred. When observing a pumping station under dry weather conditions, a time interval of 60 seconds between frames is sufficient, while when observing flow in a channel during a storm event, a framerate of 30 frames per second or more may be needed. An optimal, or at least a robust, combination of exposure time, framerate, light conditions and image post-processing settings has to be determined by trial and error.

For an in-depth treatise on PIV, LS-PIV and (LS-)PTV the reader is referred to literature (see e.g. Adrian & Westerweel, 2010; Le Coz *et al.*, 2010).

For practical applications, the following issues have to be considered:

- Camera lenses suffer from optical distortion. To avoid systematic deviations in the result this should be compensated for. Heikkila & Silvén (1997) describe a method to do so in some detail, which is implemented in e.g. a Matlab® function.

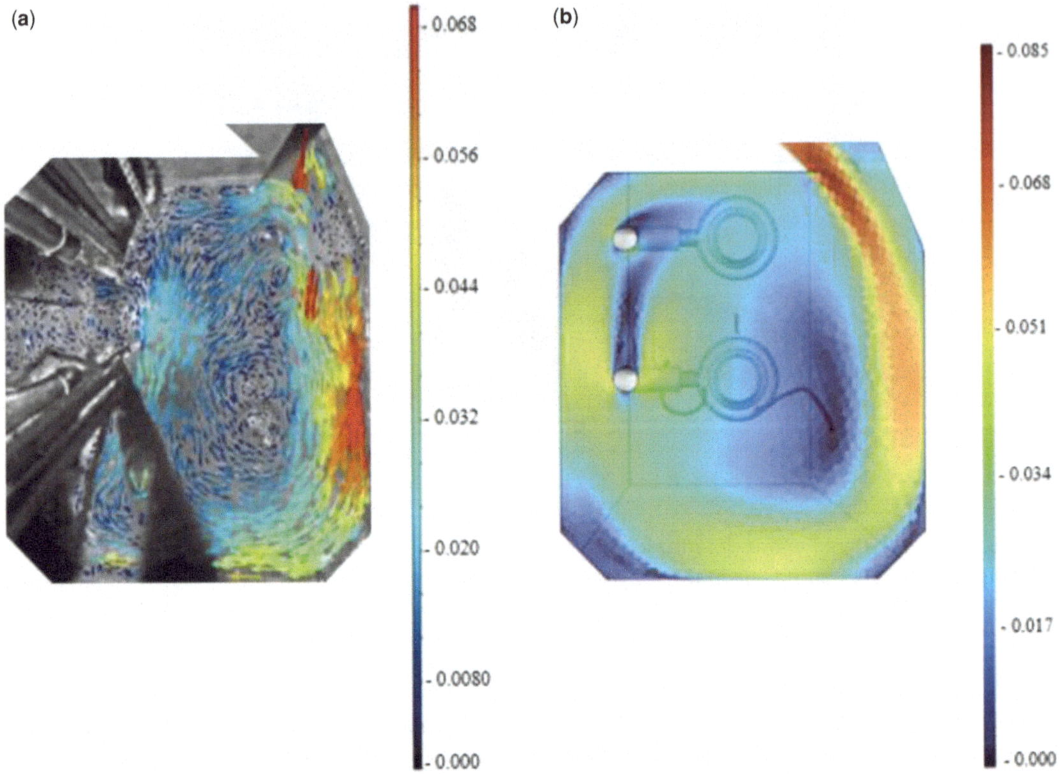

Figure 6.18 Measured and CFD (computational fluid dynamics) modelled surface flow velocities in a pump sump: (a) measured velocities; (b) CFD computed velocities. *Source*: courtesy Alex Duinmeijer (Municipality Rotterdam/TU Delft), based on Duinmeijer (2020).

- There must be a quantified link between pixel positions in the camera's sensor and the real world in 3D. This implies that the exact locations of a number (at least four, but preferably more) of reference points in the field of view (FOV) of the camera have to be known. These reference points should preferably be in, or in the direct vicinity of, the plane on which the camera focusses.
- Make sure the field of interest is covered by the camera's FOV after mounting it.
- In underground spaces, an external light source is needed. Safety issues, especially ATEX requirements (see Chapter 7) are a matter of concern.
- The camera should be mounted in such a manner that its position is and remains fixed throughout the whole length of the monitoring campaign. One is well advised to recheck the reproduction of the position of known elements (reference points) in the image directly after a site visit for maintenance purposes, or to do so on a regular basis e.g. at least prior to each post-processing session.
- Especially in wastewater systems, corrosion is often an issue. Therefore, make sure the equipment is mounted in a protective casing (Figure 6.19).
- When applying LS-PIV or LS-PTV in the field, make sure the effect of wind on the floating markers is minimal, as this can result in significant systematic errors in the end-result.
- Open-source software for post-processing and additional information on LS-PIV can be found in INRAE (2020).

Figure 6.19 Example of cameras installed in a wastewater pumping station: (a) Camera including electronics; (b) Camera housing, (c) Two cameras installed at a grate in a wastewater pumping station. *Source*: courtesy Antonio Moreno-Rodenas (Deltares).

6.5 SUMMARY AND TRANSITION

In this chapter, macro and micro design of measuring networks and measuring set-ups are discussed, as they are a mixture of theoretical considerations and a compilation of practical experience. This topic is prone to rapid development. It is and remains a challenge to communicate developments and share practical experience. In this respect, readers, scientists and practitioners alike, are invited to share their finding through (inter)national platforms, e.g. through working groups (e.g. the IWA/IAHR WG on metrology in Urban Drainage). Key to making a sound macro design of a monitoring network is that all knowledge available about the network should be used, be it model results, historic observations on e.g. flooding events and/or citizens' complaints. The latter sources provide important clues on 'weak spots' in the system that are locations where additional information on the processes taking place are of importance for the systems manager. In developing the macro design some aspects of the micro design are being decided upon, e.g. when the availability of infrastructure like power supply is not taken in account, this automatically implies that equipment needs to have its own power supply unit. These choices in turn have consequences on the operational phase which is discussed in Chapter 7 on operation and maintenance.

REFERENCES

Adrian R. J. & Westerweel J. (2010). *Particle Image Velocimetry*. Cambridge University Press, New York (USA), 586 p. ISBN 978-0521440080.

Blumensaat F., Dicht S. & Ebi C. (2019). Niedrigenergiefunk im Untergrund - Möglichkeiten und Grenzen einer neuen Datenfernübertragung in der Siedlungsentwässerung [Underground low energy radio communication - Possibilities and limitations for data acquisition in wastewater systems]. *Aqua & Gas*, **99**(3), 52–60. (in German).

Blumensaat F., Scheidegger A., Ebi C., Dicht S., Schaltegger F., Rüst A., Schmitt U. & Maurer M. (2018). Digitalization meets reality – Concept and experiences from long-term wireless data collection with 50+ sewer monitors. *Proceedings of the 11th International Conference on Urban Drainage Modelling*, 23–26 Sept., Palermo, Italy, 360–363.

Boomgaard M. E., Langeveld J. G. & Clemens F. H. L. R. (2001). Wastewater system optimisation using genetic algorithms. *Proceedings of the Specialty Symposium on Urban Drainage Modeling, at the World Water and Environmental Resources Congress*, May 20–24, Orlando, Florida, United States, 788–796. doi: 10.1061/40583(275)74.

Bruni G., Reinoso R., van de Giessen N. C., Clemens F. H. L. R. & ten Veldhuis J. A. E. (2015). On the sensitivity of urban hydrodynamic modelling to rainfall spatial and temporal resolution. *Hydrology and Earth Systems Sciences*, **19**(2), 691–709. doi: 10.5194/hess-19-691-2015.

Chacon-Hurtado J., Alfonso L. & Solomatine D. (2017). Rainfall and streamflow sensor network design: a review of applications, classification, and a proposed framework. *Hydrology and Earth System Sciences*, **21**(6), 3071–3091. doi: 10.5194/hess-21-3071-2017.

Cherqui F., James R., Poelsma P., Burns M. J., Szota C., Fletcher T. & Bertrand-Krajewski J.-L. (2020). A platform and protocol to standardise the test and selection of low-cos sensors for water level monitoring. *H2Open Journal*, **3**(1), 437–456. doi: 10.2166/h2oj.2020.050.

CIWEM (2017). *Urban Drainage Group - Code of Practice for the Hydraulic Modelling of Urban Drainage*. CIWEM - Chartered Institution of Water and Environmental Management, London (UK), 169 p. Available at https://www.ciwem.org/assets/pdf/Special%20Interest%20Groups/Urban%20Drainage%20Group/Code%20of%20Practice%20for%20the%20Hydraulic%20Modelling%20of%20Ur.pdf (accessed 05 Dec. 2020).

Clemens F. H. L. R. (2001). *Hydrodynamic Model in Urban Drainage: Calibration and Application*. PhD thesis, Delft University of Technology, Delft, The Netherlands, 388 p. ISBN 90-407-2163-7.

Desimoni E. & Brunetti B. (2015). About estimating the limit of detection by the signal to noise approach. *Pharmaceutica Analytica Acta*, **6**(3), 4 p. doi: 10.4172/2153-2435.1000355.

Diaz-Delgadillo F. J., Montiel O. & Sepulveda R. (2016). Reducing the size of traveling salesman problems using vaccination by fuzzy selector. *Expert Systems with Applications*, **49**, 20–30. doi: 10.1016/j.eswa.2015.11.026.

Duinmeijer S. P. A. (2020). *On the Free-Surface Vortex Driven Motion of Buoyant Particles*. PhD thesis, Delft University of Technology, Delft, The Netherlands, 176 p. ISBN 978-94-6366-271-0.

Ebi C., Schaltegger F., Rüst A. & Blumensaat F. (2019). Synchronous LoRa mesh network to monitor processes in underground infrastructure. *IEEE Access*, **7**, 57663–57677. doi: 10.1109/ACCESS.2019.2913985.

Fencl M., Rieckermann J., Sykora P., Stransky D. & Bares V. (2015). Commercial microwave links instead of rain gauges: fiction or reality? *Water Science and Technology*, **71**(1), 31–37. doi: 10.2166/wst.2014.466.

Fu Y., Jing C. & Du M. (2016). Urban rain gauge siting selection based on GIS-multicriteria analysis. *The International archives of the Photogrammetry, Remote Sensing and Spatial Sciences, Volume XLI-B2, XXIII ISPRS Congress*, 12–19 July, Prague, Czech Republic, 185–190. doi: 10.5194/isprsarchives-XLI-B2-185-2016.

Grafakos L. (2014). *Classical Fourier Analysis, 3rd edn*. Springer, New York (USA), 655 p. ISBN 978-1-4939-1193-6. doi: 10.1007/978-1-4939-1194-3.

Heikkila J. & Silvén O. (1997). A four-step camera calibration procedure with implicit image correction. *Proceedings of the IEEE Computer Society Conference on Computer Vision and Pattern Recognition*, 17–19 June, San Juan, Porto Rico, USA, 1106–1112. doi: 10.1109/CVPR.1997.609468.

Henckens G. J. R. & Clemens F. H. L. R. (2004). Design and optimisation of monitoring networks in urban drainage. *Proceedings of the 19th European Junior Scientists Workshop on "Process Data and Integrated Urban Water Modelling"*, 11–14 March, Meaux-la-Montagne, France, 8 p.

Hoes O. A. C., Schilperoort R. P. S., Luxemburg W. M. J., Clemens F. H. L. R. & van de Giessen N. C. (2009). Locating illicit connections in storm water sewers using fiber optic distributed temperature sensing. *Water Research*, **43**(20), 5187–5197. doi: 10.1016/j.watres.2009.08.020.

Hofer T., Montserrat A., Gruber G., Gamerith V., Corominas L. & Muschalla D. (2018). A robust and accurate surrogate method for monitoring the frequency and duration of combined sewer overflows. *Environmental Monitoring and Assessment*, **190**, 209, 18 p. doi: 10.1007/s10661-018-6589-3.

INRAE (2020). Fudaa-LSPIV Information Webpage. Available at https://riverhydraulics.inrae.fr/en/tools/measurement-software/fudaa-lspiv-2/ (accessed 04 Dec. 2020).

Jeanbourquin D., Sage D., Nguyen L., Schaeli B., Kayal S., Barry D. A. & Rossi L. (2011). Flow measurements in sewer systems based on image analysis: automatic flow velocity algorithm. *Water Science and Technology*, **64**(5), 1108–1114. doi: 10.2166/wst.2011.176.

Kapelan Z., Savic D. A. & Walters G. A. (2003). Multi-objective sampling design for water distribution model calibration. *Journal of Water Resources Planning and Management*, **129**(6), 466–479. doi: 10.1061/(ASCE)0733-9496(2003)129:6(466).

Kapelan Z., Savic D. A. & Walters G. A. (2005). Optimal sampling design methodologies for water distribution model calibration. *Journal of Hydraulic Engineering*, **131**(3), 190–200. doi: 10.1061/(ASCE)0733-9429(2005)131:3(190).

Le Coz J., Hauet A., Pierrefeu G., Dramais G. & Camenen B. (2010). Performance of image-based velocimetry (LSPIV) applied to flash-flood discharge measurements in Mediterranean rivers. *Journal of Hydrology*, **394**(1–2), 42–52. doi: 10.1016/j.jhydrol.2010.05.049.

Lei J. & Schilling W. (1993). Requirements of spatial rain data resolution in urban runoff simulation. *Proceedings of the 6th ICUD – International Conference on Urban Drainage*, 13–17 Sept., Niagara Falls, Canada, Vol. 1, 447–452.

Lepot M. (2012). *Mesurage en continu des flux polluants en MES et DCO en réseau d'assainissement [Continuous monitoring of TSS and COD fluxes in sewers]*. PhD thesis, INSA Lyon, Villeurbanne, France, 257 p. (in French).

Lepot M., Aubin J.-B. & Clemens F. (2017). Interpolation in time series: an introductive overview of existing methods, their performance criteria and uncertainty assessment. *Water*, **9**(10), 796, 20 p. doi: 10.3390/w9100796.

Mikulak R. J., McDermott R. E. & Beauregard M. R. (2008). *The Basics of FMEA, 2nd edn*. Productivity Press, New York (USA), 90 p. ISBN 978-1563273773.

Nienhuis J., de Haan C., Langeveld J., Klootwijk M. & Clemens F. (2013). Assessment of detection limits of fiber-optic distributed temperature sensing for detection of illicit connections. *Water Science and Technology*, **67**(12), 2712–2718. doi: 10.2166/wst.2013.176.

Ochoa-Rodriguez S., Wang L.-P., Gires A., Pina R. D., Reinoso-Rondinel R., Bruni G., Ichiba A., Gaitan S., Cristiano E., van Assel J., Kroll S., Murlà-Tuyls D., Tisserand B., Schertzer D., Tchiguirinskaia I., Ono C., Willems P. & ten Veldhuis M.-C. (2015). Impact of spatial and temporal resolution of rainfall inputs on urban hydrodynamic modelling outputs: A multi-catchment investigation. *Journal of Hydrology*, **531**, 389–407. doi: 10.1016/j.jhydrol.2015.05.035.

Ochoa-Rodriguez S., Wang L.-P., Willems P. & Onof C. (2019). A review of radar-rain gauge data merging methods and their potential for urban hydrological applications. *Water Resources Research*, **55**(8), 6356–6391. doi: 10.1029/2018WR023332.

Orfeo D., Burns D., Farrell R., Qin M., Mitchell H., Ou C., Xia T. & Huston D. R. (2018). Mechano-magnetic telemetry for underground water infrastructure monitoring. *Frontiers in Built Environment*, **4**, 29, 14 p. doi: 10.3389/fbuil.2018.00029.

Raspe R. E. (1895). *The Surprising Adventures of Baron Munchausen*. Available at https://www.gutenberg.org/files/3154/3154-h/3154-h.htm (accessed 05 Dec. 2020).

Rodríguez-Iturbe I. & Mejía J. M. (1974). The design of rainfall networks in time and space. *Water Resources Research*, **10**(4), 713–728. doi: 10.1029/WR010i004p00713.

Ruiz-Cardenas R., Ferreira M. A. R. & Schmidt A. M. (2010). Stochastic search algorithms for optimal design of monitoring networks. *Environmetrics*, **21**(1), 102–112. doi: 10.1002/env.989.

Schellart A. N. A., Wang L. & Onof C. (2012). High resolution measurement and analysis in a small urban catchment. *Proceedings of the 9th International Workshop on Precipitation in Urban Areas: Urban Challenges in Rainfall Analysis*, 6–9 Dec., St. Moritz, Switzerland, 115–120. ISBN 9783906031217.

Schilperoort R. P. S., de Haan C., Liefting E., Nijhof H., Hof A., Meijer C. & Langeveld J. (2016). Advances in practical applicability of DTS in sewer systems. *Proceedings of the 8th International Conference on Sewer Processes and Networks*, 31 Aug.–2 Sept., Rotterdam, The Netherlands, 33–37.

Schilperoort R. P. S., Dirksen J. & Clemens F. H. L.R. (2008). Practical aspects for long-term monitoring campaigns: pitfalls to avoid to maximize data yield. *Proceedings of the 11th International Conference on Urban Drainage*, 31 Aug.–5 Sept., Edinburgh, Scotland, UK, 10 p.

Schilperoort R. P. S., Dirksen J., Langeveld J. G. & Clemens F. H. L. R. (2012). Assessing characteristic time and space scales of in-sewer processes by analysis of one year of continuous in-sewer monitoring data. *Water Science and Technology*, **66**(8), 1640–1620. doi: 10.2166/wst.2012.115.

Shaghaghian M. R. & Abedini M. J. (2013). Rain gauge network design using coupled geostatistical and multivariate techniques. *Scientia Iranica*, **20**(2), 259–269. doi: 10.1016/j.scient.2012.11.014.

Sivakumar B. (2010). Fractal analysis of rainfall observed in two different climatic regions. *Hydrological Sciences Journal*, **45**(5), 727–738. doi: 10.1080/02626660009492373.

Thompson K. E., Vamvakeridou-Lyroudia L., Kapelan Z. & Savic D. (2011). *Optimal macro-location methods for sensor placement in urban water systems – Literature review*. EU FP7 project PREPARED, report 2011.004, 75 p. Available at https://emps.exeter.ac.uk/media/universityofexeter/emps/research/cws/downloads/PREPARED_Deliverable_3_5_1_final.pdf (accessed 05 Dec. 2020).

Tiwari S., Jha S. K. & Singh A. (2020). Quantification of node importance in rain gauge network: influence of temporal resolution and rain gauge density. *Scientific Reports*, **10**(1), 9761, 17 p. doi: 10.1038/s41598-020-66363-5.

Tscheikner-Gratl F., Bellos V., Schellart A., Moreno-Rodenas A., Muthusamy M., Langeveld J., Clemens F., Benedetti L., Rico-Ranires M. A., Fernandez de Carvalho R., Breuer L., Shucksmith J., Heuvelink G. B. M. & Tait S. (2019). Recent insights on uncertainties present in integrated catchment water quality modelling. *Water Research*, **150**, 368–379. doi: 10.1016/j.watres.2018.11.079.

Tyler S. W., Selker J. S., Hausner M. B., Hatch C. E., Torgersen T., Thodal C. E. & Schladow S. G. (2009). Environmental temperature sensing using Raman spectra DTS fiber-optic methods. *Water Resources Research*, **45**, W00D23, 11 p. doi: 10.1029/2008WR007052.

van Luijtelaar H. & van Rebergen E. W. (1997). Guidelines for hydrodynamic calculations on urban drainage in The Netherlands: backgrounds and examples. *Water Science and Technology*, **36**(8–9), 253–258. doi: 10.1016/S0273(97)00590-8.

Walcker N., Bertrand-Krajewski J.-L., Vacherie S., Lepot M., Castebrunet H., Barraud S. & Lipeme Kouyi G. (2018). Une nouvelle station de mesure pour l'acquisition de séries chronologiques en hydrologie urbaine [A new monitoring station for the acquisition of time series in urban hydrology]. *Techniques Sciences Méthodes*, **3**, 55–64. doi: 10.1051/tsm/201803055. (in French).

WMO (2018). *Guide to instruments and methods of observation - Volume I: Measurement of meteorological variables*. World Meteorological Organization, Geneva (Switzerland), 573 p. ISBN 978-92-63-10008-5. Available at https://library.wmo.int/doc_num.php?explnum_id=10179 (accessed 05 Dec. 2020).

Chapter 7

Operation and maintenance

Jakob Benisch[1], Björn Helm[1], Jean-Luc Bertrand-Krajewski[2], Simon Bloem[3], Frédéric Cherqui[2], Uwe Eichelmann[1], Stefan Kroll[4] and Peter Poelsma[5]

[1]*TU Dresden, Institut für Siedlungs- und Industriewasserwirtschaft, Dresden, Germany*
[2]*University of Lyon, INSA Lyon, Laboratory DEEP, Villeurbanne, France*
[3]*Eawag, Swiss Federal Institute of Aquatic Science and Technology, Dübendorf, Switzerland*
[4]*Aquafin NV, Aartselaar, Belgium*
[5]*University of Melbourne, Waterway Ecosystem Research Group, School of Ecosystem and Forest Sciences, Melbourne, Australia*

ABSTRACT

This chapter first provides information on general health and safety rules to be applied by operators in monitoring urban drainage and stormwater management (UDSM) systems, especially in the harsh confined environment of underground sewer systems. Second, it presents experience-based key recommendations for best practice and quality in operation, management and maintenance of sensors and installations for rainfall and discharge measurements. In the last part, three numerical methods (ordinary least squares, Williamson least squares, non-linear regression) are presented, with detailed examples of application, to establish calibration functions which are necessary for all sensors used in UDSM monitoring.

Keywords: Discharge measurement, health and safety rules, operation and maintenance, rain gauge, sensor calibration and verification.

SYMBOLS

aw_d	mean values of b_{jd} in the Williamson least squares (WLS) regression
b_j	coefficients of calibration functions, to be fitted by regression

doi: 10.2166/9781789060119_0203

b_{jd}	b_j coefficients for order d polynomial function in the WLS regression
b_{j0d}	preliminary estimates of b_{jd} in the WLS regression for order d
B	$d \times 1$ vector of the b_j coefficients
$control_d$	intermediate calculation control matrix in the WLS regression
$cov(a,b)$	covariance of two quantities a and b
d	order of a polynomial function
$diag(M)$	main diagonal of any matrix M
d_{opt}	optimal order of a polynomial calibration function
e	$N \times 1$ vector of the residuals
f	calibration function $y = f(x)$
f_{vald}	final S_d value in the iterative WLS regression
F	$N \times (d+1)$ intermediate matrix in the ordinary least squares (OLS) regression
F_{calc}	experimental variance used in the Snedecor test to estimate d_{opt}
F_{theo}	theoretical variance used in the Snedecor test to estimate d_{opt}
h_m	water level measured by a sensor during regular monitoring (m)
H_0	null hypothesis to be tested
i	index of the standard or certified values in a calibration experiment
I_m	measured rainfall intensity in a rain gauge calibration experiment (mm/h)
I_{mt}	maximum acceptable error in a verification experiment
I_t	true (standard) value of rainfall intensity in a rain gauge calibration experiment (mm/h)
j	index from 0 to d of the calibration function coefficients
k	index of repeated sensor measurement values in a calibration experiment
l	second index of the calibration function coefficients
m	as index: indicates a value measured by a sensor during regular monitoring
M	intermediate matrix in the OLS regression
M_{bjd}	matrix of b_{jd} vectors in the WLS regression
N	total number of experimental measurements in a calibration experiment
N_{MC}	number of Monte Carlo simulations in the WLS regression
N_r	pre-set number of tips in a rain gauge calibration experiment
N_x	number of standard or certified values used in a calibration experiment
N_y	number of repeated sensor measurement values in a calibration experiment
$PreRel$	relative accuracy for estimating the b_j and X_i values in the WLS regression
Q	intermediate matrix in the OLS regression
r	index from 1 to N_{MC} of Monte Carlo simulations in the WLS regression
R	intermediate matrix in the OLS regression
s_i	standard deviation of y_{ik} repeated measurements with $k = 1{:}N_y$ for a given x_i
$s_{i\ lim}$	minimum significant value of s_i
$s_{i\ not\ nul}$	intermediate quantity in calculation of final s_i values in the WLS regression
S	upper bound value of s_i^2 for y_m used in estimating $u(\hat{x})$
S_d	sum of Williamson least squares to be minimized for an order d polynomial function
S_{1d}, S_{2d}	components of S_d
S_{OLS}	sum of ordinary least squares to be minimized in the OLS regression
S_r	residual variance
SSR	sum of the squares of the residuals
t	as quantity: Student t value

t	as index: true (standard) value in a rain gauge calibration experiment
T	as exponent: indicates the transpose of a vector or matrix
$u(x)$	standard uncertainty of any quantity x
V_R	reference value for H_0 testing
$W(x)$	inverse of $u(x)^2$
$Willorder_d$	intermediate quantity in the WLS regression
x	standard or certified values in calibration/verification experiments
$\hat{x}$	best estimate of any quantity x
x_i	i-th standard or certified value in calibration/verification experiments
x_{ir}	x_i values created in Monte Carlo samples in the WLS regression
X_i	predicted x_i value in the WLS regression
Y	sensor measured values in calibration/verification experiments
y_{ik}	k-th repeated measurement for the i-th standard value in a calibration experiment
y_m	value measured by a sensor during regular monitoring
y_{irk}	y_{ik} values created in Monte Carlo samples in the WLS regression
Y	$N \times 1$ vector of the sorted y_{ik} values
Y_{ik}	predicted y_{ik} value in the WLS regression
$z(p)$	p-th element of a vertical vector z with $(d + 1 + N_x)$ elements in the WLS regression
$Z_{0\,d}$	initial intermediate vector in the iterative WLS regression
$Z_{0\,bisd}$	another initial intermediate vector in the iterative WLS regression
Z_{nd}	n-th intermediate vector in the iterative WLS regression
$Z_{n\,bisd}$	another n-th intermediate vector in the iterative WLS regression
α	level of coverage between 0 and 1
∂	symbol of partial derivative
ν	degrees of freedom
$\bar{}$	above any quantity: indicates the mean value of this quantity
$\hat{}$	above any quantity: indicates the best estimate of this quantity

7.1 INTRODUCTION

This chapter provides useful and practice-proven approaches for the operation and maintenance (O&M) of typical measurement set-ups in urban drainage and stormwater management (UDSM) systems. It is not designed to provide step-by-step procedures for performing O&M on any specific piece of monitoring equipment. Rather, the manufacturer's specifications should be the primary source of information and our recommendations intend to supplement general approaches and considerations.

Operation and maintenance combines decisions and actions regarding the upkeep and quality assurance of monitoring equipment in a structured and organized way based on standards like those of the ISO 9000 family (ISO, 2015*a*, *b*, 2018). Actions in equipment *operation* include, but are not limited to, scheduling, procedures and standard protocols, control, and set-up optimization. Actions in equipment *maintenance* aim at preventing equipment failure or decline. Properly managed O&M actions increase the monitoring efficiency and reliability; they are a key factor for maximizing the service life of equipment and safety of monitoring operators.

Figure 7.1 Some of the safety equipment required to enter UDSM systems. *Source*: courtesy of Robert Lohse Photography (2014).

Some UDSM systems, especially underground infrastructure such as sewers or underground basins, are a harsh environment: equipment and personnel may be exposed to chemical and hydraulic risks (Figure 7.1). Other risks not related to water should also be considered in many UDSM systems such as fall or collision hazard. However, to obtain information on flow conditions and process dynamics from the system, working within such an environment is unavoidable. The hazards and risks related to this endeavour as well as corresponding mitigation measures are shown in the first part of this chapter.

Reliable rainfall and flow data are a basis for the operation, planning and optimization of urban drainage systems. For example, the German Association for Water, Wastewater and Waste (DWA) suggests a symbolic price of 0.5 Eurocent per data point to value the benefits provided by such data, but it also reflects the effort related to data acquisition (DWA, 2014). Appropriate O&M of measurement devices as described in Sections 7.4 and 7.6 ensures accurate results. Various operation conditions and their specific implications are illustrated. Finally, routines and best practices for regular maintenance provide guidance, to ensure proper operation and reliable results.

This chapter will thus introduce you to:

- Health and safety in UDSM.
- Operation, maintenance, and management of:
  - o Rain gauges.
  - o Discharge measurement devices.
- Methods for calibration and verification of sensors.

7.2 HEALTH AND SAFETY

The operation and maintenance of monitoring equipment in UDSM infrastructures implies dangers to workers' well-being and equipment. Knowledge of possible risks is as important as measures to mitigate them. Before any action is undertaken, make sure that you are aware of your sewer and sustainable urban drainage systems (SUDS) related health and safety (H&S) national guidelines as well as the regulations from the local sewer operator.

Typical regulations include, but are not restricted to the following topics:

- Work safety in UDSM systems.
- Occupational health care in UDSM systems.
- Road works and traffic regulation.
- Working in confined spaces.
- Working with hazardous substances or biological agents.
- Working in potentially explosive environments.

The following guidance does not replace in-depth examination of local or national norms and regulations. They are adapted to the specific technical and environmental boundary conditions and may well exceed the approaches presented here. The authors hope that this introduction to the topic proves to be useful, but disclaim responsibility arising from unlawful or negligent actions in a concrete monitoring situation. McManus (2019) presents potentially the most comprehensive resource on health and safety in confined spaces and deserves further reading.

7.2.1 Health and safety management

Health and safety management generally follows a progressive cycle of four steps: Plan – Do – Check – Act:

- 'Planning' comprises of organizational (responsibilities for health and safety related tasks and equipment), staff related (provision of preventive medical check-ups, instructions and training) and activity related (plan safety measures, organize teams, develop check lists for safety equipment) tasks, which promote health and safety.
- 'Doing' relates to the actual work activities: provision of safety resources and equipment, follow through safety measures and instruction or maintenance schedules.
- 'Checking' safeguards at organizational level, that staff and activity related deadlines are met. A reporting system, e.g. on the state of equipment, incidents or near misses, might be helpful to keep record.
- 'Acting' transmits lessons learned from incidents to the next level of planning and doing. It makes sure new safety approaches are incorporated in the safety planning.

The sequence of those steps should be continuously perused and enhanced. Documentation and communication between staff groups fosters the application of site-specific best practices. If an office or department that is responsible for occupational health and safety is available within your organization, it is highly recommended to integrate this into your H&S management.

7.2.1.1 Risk mitigation planning

Risk mitigation measures should be selected based on a site-specific risk assessment during the planning phase. This consists of:

- Definition of working locations and activities, including: public and/or traffic area location; below ground or confined space location; in or above water; prone to strong flow; prone to sediment or mud accumulation; prone to hazardous gas development; activities involving heavy lifting;

activities involving handling or contact with hazardous materials; and activities involving handling electrical equipment.

- Assessment of hazards: hazards at the workplaces in UDSM facilities include physical, chemical, biological, and ergonomic hazards.
- Deviation of risks: risks result from the probability of exposure and the extent of damage caused by a specific hazard. Risks are typically grouped into low, intermediate, and high risks. Risk mitigation priorities are defined by the risk level.
- Selection of risk mitigation measures: the hierarchy of measures results from their preventive power. Structural (sometimes referred to as technical) measures prevent risks from becoming effective, organizational measures separate the threat from the worker while personal protective measures reduce the probability of exposure or the extent of damage.

Measures against falling at an in-sewer site shall exemplify the selection process of work-related risk mitigation: structural measures, like railings or platforms are a preferable option but not always feasible in constraint spaces. Installing parts of the equipment, such as loggers or energy supply units, outside the sewer represents an organizational measure and reduces the work time within the sewer. Tripod hoists and falling protection gear are personal protective measures. They mitigate the damage of an occurred incident, for example of a fall due to wet conditions (slippery hazard) or assist in rescuing unconscious people from the sewer. When working close to a detention or infiltration basin, favour locations with limited slope or equipped with railings or other safeguards.

7.2.1.2 Temporal dimension of safety management

The second step of the safety management cycle includes all actions taken by field staff and throughout the fieldwork to mitigate risks. Long-term preparation, short-term preparation, on-site measures, and in-action measures temporally structure the risk mitigation approach.

- Long-term preparation includes activities that increase the general risk awareness, safety skills, preparedness and protection of staff: regular instructions or training internalize the threats of working in sewer networks and provide practice for emergencies. Medical checks and immunization against waterborne diseases increase the protection level. Cooperation with the local operator is essential. It is likely that the operator manages and regulates the access rights and provides allowance for measurement campaigns. Agreements about location, set-up, duration, and safety requirements promote planning security. They also protect your measurement devices and monitoring scope because it prevents the loss of sensors during high-pressure pipe cleaning or other building operations.
- Short-term preparation involves activities before the field trip. Staff planning ensures sufficient labour force. Never undertake works in UDSM structures unaccompanied. Additional staff may be required for road blockage, operation of safety equipment and emergency aid, checking and packing of site-specific safety equipment and communication devices. Ensure that weather conditions remain dry and imply safe working conditions. Recapitulate safety briefing, work distribution, and emergency plans.
- On-site measures provide a safe environment for the actual O&M work. The area for work site access may be blocked to the public, depending on the associated levels of risks (underground sewer systems and infrastructures are blocked, SUDS may be completely or partly open to the public). Verify signal connection of the communication device. Test gas levels and use ventilation equipment if indicated. Check access structure and implement additional falling

protection if indicated. Check water levels in the different sections of the system (if applicable) and potential sudden changes.

- In-action measures safeguard during the O&M work. They may include (i) use of falling protection equipment, (ii) use of personal protection equipment, including skin protection, respiratory protection and protective helmet, (iii) use of electrical equipment save in a potentially explosive atmosphere, (iv) use of life vests in and above water, and (v) rescue equipment in place for trained staff.

7.2.2 Situation specific risk mitigation

The most relevant hazards to entering and working in storm- and wastewater structures are assembled in the following. As introduced in the risk assessment approach above, they will be linked to monitoring locations or situations and mitigation measures. The compilation is not meant to cover every potential threat but reflects common dangers based on the risk assessment guideline for sewer-related work sites of the German Social Accident Insurance (DGUV) and reflects years of experience with measurements in sewer networks (DGUV, 2007). Additional information is provided when applicable, to consider the specific risks related to SUDS.

7.2.2.1 Working in traffic-affected areas

7.2.2.1.1 Location and activities

A large portion of UDSM systems, and in particular sewer pipes, is located on public ground and often below traffic areas. Selecting a monitoring site without pedestrian or traffic influence greatly simplifies site access, but constraints in the site configuration may not always permit this. Even if the works take place below the surface, equipment is transported to the site and supporting staff remain above ground. SUDS are rarely below traffic areas but can be close to traffic areas: for example, monitoring a swale located in the median strip of an arterial road should be avoided.

7.2.2.1.2 Hazards and risks

Traffic causes the risk of intrusion of the measuring site or its access point by vehicles or persons. Moving objects, like vehicles, cause threats while entering or exiting the measurement site. Passers-by might push objects towards people on site. Supportive staff can be hit by vehicles. The risks strongly relate to the volume, type and speed of traffic and visibility of the site.

7.2.2.1.3 Measures

When working in traffic-affected areas, traffic should be blocked or deviated around the access site. Usually, a previous announcement of the traffic interference is placed at the traffic authority in charge for the road. National guidelines for road works safety provide guidance on blocking and marking zones. Figure 7.2 gives a short-term construction blockage as an example. At sites with heavy traffic, additional signalling staff may be necessary. All staff working above ground in the traffic area should wear high-visibility clothing (e.g. according to ISO, 2013b). It is favourable to park the equipment transport vehicle in front of the access spot, it shields the working area from the traffic and provides an additional barrier.

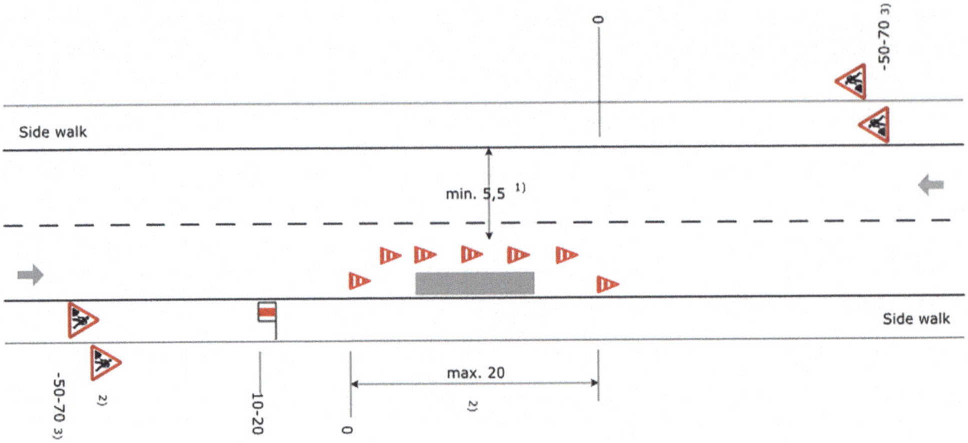

Figure 7.2 An example of regular safety zones and signs for short-duration road blockage. *Source*: adapted from Gehlen (2018).

7.2.2.2 Accessing underground structures and confined spaces

7.2.2.2.1 Location and activities

Sewers, underground detention basins, retention tanks, storage chambers and overflow structures are predominantly underground structures. Vertical manholes and shafts usually provide access to these structures. Commonly, manhole covers are made of solid cast iron and are very heavy.

7.2.2.2.2 Hazards and risks

Stumbling and falling into those narrow openings might cause serious injury. Especially, combined systems tend to have a corrosive environment; corroded steel rungs and ladders might break or exhibit sharp edges. Risks depend on the age and structural condition of the shaft, and depth, size, and geometry of the structure. Lifting the manhole cover poses an ergonomic and injury threat if done inappropriately. Working in the confined environment may cause additional pressures due to poor visibility, noise, and claustrophobia.

7.2.2.2.3 Measures

Open the shaft with two persons or apply lifting support tools (see Figure 7.3 for an example). Hooks or wheel-supported levers reduce ergonomic burdens.

Open shafts are never to be left unattended and security measures/barriers must be implemented. A safeguarding person above ground needs to stay in permanent contact with the person working in the confined space. The safeguard must be able to call emergency rescue without leaving his/her post. Well-fitting and robust gear prevents entanglement or injury on access. Mobile ladders with stay bar and handle may substitute for corroded or missing rungs. Helmets and personal falling protection equipment mitigate the effect of falling incidents while lifelines and lifting equipment support rescue after an incident. Light and hearing protection should be always at hand. Safety tripods as shown in Figure 7.4 provide a combination of falling protection and lifting equipment.

Figure 7.3 Ergonomic lifter for manhole covers developed by Aquafin NV. *Source*: courtesy of Jan Swankaert (Aquafin NV).

Figure 7.4 Entering the sewer network with a tripod. *Source*: Jakob Benisch (TU Dresden).

7.2.2.3 Working in and above water

7.2.2.3.1 Location and activities

Channels, basins, and wetlands are commonly filled with water during work activities. Activities may require standing in still or flowing water; floating on platforms or boats; and handling equipment above or below water level.

7.2.2.3.2 Hazards and risks

Waters rich in fat and grease, organic matter and sediments favour biofilm and mud accumulation on the ground of wastewater structures and create slippery conditions. Strong currents and/or buoyancy at high water levels reduce standing stability. Structures with pronounced bottom slope and bottom materials with a low surface roughness, like plastics or fibreglass, increase slipping risk. Furthermore, higher water levels increase the risk of drowning. Retention basins, which are emptied after heavy rain events, can unexpectedly increase discharges by a great extent. Water in urban drainage systems is often turbid. Obstacles and ledges may not be seen and trigger stumbling and falling. During summer period and in exposed locations, sun protection gear is recommended as water reflects ultra-violet rays.

7.2.2.3.3 Measures

Waterproof gear prevents direct contact with hazardous water constituents (Figure 7.5). Safety harnesses, swimming or floating devices or rescue ropes should be deployed according to the work situation and task. Harnesses accomplish falling protection and enable pull back in case of a fall. In locations where water depth is or may instantaneously rise higher than knee-deep, lifejackets should be worn. A waistband on waders prevents quick filling with water. Rubber boots should allow slipping out in case they get trapped. Rescue ropes and lifting equipment are required in restricted access sites.

7.2.2.4 Working in potentially hazardous atmospheres

7.2.2.4.1 Location and activities

Underground sites with limited air exchange or ventilation favour the development and accumulation of potentially hazardous gases. Typically, the atmosphere is not the object of work

Figure 7.5 Use of waterproof equipment to work in a wetland in Melbourne, Australia. *Source*: Frédéric Cherqui (INSA Lyon).

activities but a condition to be considered. It affects all activities both below and above the water surface.

7.2.2.4.2 Hazards and risks

Sewer gases may contain hydrogen sulphide (H_2S), ammonia (NH_3), methane (CH_4), volatile esters, carbon oxides (CO/CO_2), sulphur dioxide (SO_2), nitrogen oxides (NO_x) and other substances. McManus (2019) gives a detailed account of the constituents and concludes on their properties and adverse effects. Hazards arise from toxic, asphyxiating and flammable/explosive properties of these gases. Gases may accumulate at the bottom or in high points of tanks and sewers, depending on their density. Gases like carbon oxides are odourless and may cause faint without premonitory symptoms. Hydrogen sulphide (H_2S) causes most intoxications in sewer workspaces. Concentrations of about 0.1% (1000 ppm) are nearly instantly lethal, but at prolonged exposure concentrations as low as 20 ppm can affect health. The above-mentioned gases arise from the decomposition of organic matter, a specific smell of rotten eggs is an indicator for their presence. However, this can be fallacious since human odour receptors are stunned at higher concentrations (about 200 ppm, Brandes & Möller, 2003). High concentrations or accumulation of these substances promote the development of hazardous atmospheres. Catchment areas with intense emission or specific industrial effluent, flat sewer slopes and backwater influenced structures (like culverts) increase the risk of the presence of hazardous gases. High temperatures accelerate organic decomposition and sewer gas development. Sewers with more than 15% wetted perimeter host more biofilm and tend to exhibit elevated sewer gas concentrations. Pumping stations and transition zones from pressurized sewer pipes to gravity flow sections are especially prone to sewer gas release. Accidental or illicit disposal of flammable hydrocarbons (e.g. gasoline, solvents but H_2S as well) into sewers can create acutely explosive conditions, such incidents should be communicated as an alert immediately at notice.

7.2.2.4.3 Measures

At work sites with a known or suspected risk of hazardous atmospheres, ventilation is mandatory. Opening neighbouring manholes increases natural ventilation. Technical ventilation with blowers or suction ventilators may further improve conditions at locations prone to ventilation problems. Due to the increased technical and surveillance effort, three persons should be on site, including the *in situ* worker, a safeguard and an aide. Gas detectors measure concentrations of oxygen and several hazardous gases (CO_2, CO, CH_4, H_2S) simultaneously in real-time and give warnings. Before entering the sewer network, the gas detector should be lowered to the level where works will be carried out, and to low and high points in the system. The sensor should remain with the worker throughout his activities. CO_2 in the exhaled air should be prevented from triggering the sensor. Self-rescue respirators provide breathable air in hazardous atmospheres. These devices are typically enclosed in a rugged case and attached to the workers belt or harness throughout the work. If the person *in situ* reports or is suspected of health detriment by gas, emergency rescue should be called immediately. The safeguarding person is strictly not allowed to enter the hazardous atmosphere for any rescue attempts. The rescuer (e.g. firefighters) must be equipped with a separate set of breathing protection and respirator upon access. The falling protection and lifting equipment, as introduced in the confined spaces section, enables rescue of unconscious affected persons from hazardous atmospheres with the help of a rope winch. The ATEX product directive 2014/34/EU (ATEX, 2016) regulates equipment and protective systems intended for use in potentially explosive atmospheres. If measurement equipment or tools are procured, ATEX-compliant versions should be preferred where available or might even be required, depending on national regulations.

Fatal accidents in sewers

The most dramatic accidents happen when people are not aware or underestimate the dangers of toxic gases and too little oxygen in sewer systems, especially in manholes and pumping stations. The screen copy in Figure 7.6 shows the introduction of a report about two sewer workers who lost their lives in a sewer manhole, one after another. The second one tried to rescue the first person down in the manhole. In addition to detection and prevention measures, the best thing you can do under such circumstances is to call for help, making sure backup support is on site and protecting yourself, before trying to save the life of your colleagues.

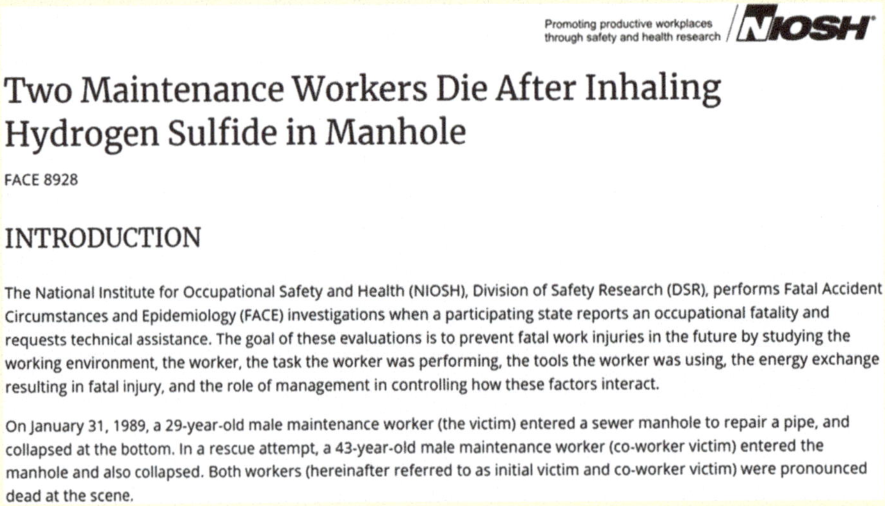

Promoting productive workplaces through safety and health research / NIOSH

Two Maintenance Workers Die After Inhaling Hydrogen Sulfide in Manhole

FACE 8928

INTRODUCTION

The National Institute for Occupational Safety and Health (NIOSH), Division of Safety Research (DSR), performs Fatal Accident Circumstances and Epidemiology (FACE) investigations when a participating state reports an occupational fatality and requests technical assistance. The goal of these evaluations is to prevent fatal work injuries in the future by studying the working environment, the worker, the task the worker was performing, the tools the worker was using, the energy exchange resulting in fatal injury, and the role of management in controlling how these factors interact.

On January 31, 1989, a 29-year-old male maintenance worker (the victim) entered a sewer manhole to repair a pipe, and collapsed at the bottom. In a rescue attempt, a 43-year-old male maintenance worker (co-worker victim) entered the manhole and also collapsed. Both workers (hereinafter referred to as initial victim and co-worker victim) were pronounced dead at the scene.

Figure 7.6 Introduction of a US National Institute for Occupational Safety and Health investigation report on two maintenance workers who died from H_2S inhalation. *Source*: https://www.cdc.gov/niosh/face/in-house/full8928.html (accessed 25 Aug. 2020).

7.2.2.5 Working with hazardous substances and biologic agents

7.2.2.5.1 Location and activities

Wastewater and to a lesser extent stormwater as well as their respective debris and sediments contain hazardous substances. All monitoring-related activities in UDSM systems involve direct contact with or exposure to waterborne substances most often through skin contact, inhalation or ingestion. Cleaning of clogged, rusted or scaled sensors may require the use of hazardous cleaning agents. Aerobic bacteria metabolize hydrogen sulphide into sulfuric acid, which condenses and accumulates at walls and pipes.

7.2.2.5.2 Hazards and risks

Hazardous chemicals in the workplace cause physiological and physicochemical threats. Physiological hazards include toxic substances, carcinogens, skin irritants or respiratory sensitizers. Physicochemical hazards generally result from flammable, corrosive, oxidizing or explosive properties of a substance. Biological hazards are biogenic substances or organisms, which threaten the health of humans and other living organisms. Biological hazards include pathogenic microorganisms, viruses, toxins (from biological sources), spores, fungi, and bio-active substances. Carr (2001) gives an overview on sewage-related infections transmitted through wastewater. Sulfuric acid can cause corrosive injuries. Inhalation of substances is more likely at locations with spray or dust formation, e.g. at invert drops or in temporarily flooded structures.

7.2.2.5.3 Measures

Respiratory protection and protective clothing prevent exposure to waterborne or work-related hazardous materials. Before and after the work tasks, items in direct contact with wastewater and stormwater should be stored and transported apart from other gear, preferably in separate containers. Personal protective clothing and equipment should be selected according to expected work situations and tasks. Typically, liquid tight or spray tight gear that covers the body partly or fully is chosen. Regular cleaning and disinfection of hands, exposed body parts and equipment prevents spread of contaminants. Cleaning and disinfection agents should be available on site. Frequent use of skin care prevents contamination through lesions and fissures. Skin protection plans provide a framework for selection and scheduling of measures. Long-term work-related health management requires regular preventive medical check-ups according to the level of exposure. Vaccinations against poliomyelitis, typhoid fever, Hepatitis A and Hepatitis B are recommended or compulsory. In consultation with local health authorities, additional immunizations may be required according to regional disease prevalence. Illness after sewer work activities might relate to wastewater specific pathogens, for example leptospira (or Weil's disease) is such a germ and it is likely to be overlooked in common anamnesis – therefore the working area should be reported specifically during the doctor's visit.

On-site health and safety equipment

- CL 7.1: *Communication device* – mobile phone, radio transceiver
- CL 7.2: *Traffic warning/safety equipment* – signs, posts, cones, high visibility clothing
- CL 7.3: Manhole lid lifters
- CL 7.4: *Personal protective equipment* – rubber boots, waders, hazardous material suit, protective helmet, safety harness
- CL 7.5: Mobile shaft access ladder
- CL 7.6: Portable multi gas detector
- CL 7.7: Falling protection and lifting device
- CL 7.8: Respiratory protection mask
- CL 7.9: Self-rescue respirators/ventilation device
- CL 7.10: Explosion protected (head)light
- CL 7.11: First aid kit
- CL 7.11: Fire extinguisher
- CL 7.12: Washing, cleaning and disinfection agents.

7.3 OPERATION

7.3.1 General ideas on operation

The operation of measurement strongly depends on local circumstances. Also, the choice of the right sensor set-up to solve the task at hand is of paramount importance (see Section 6.3). This final decision for (or against) a specific monitoring set-up is heavily influenced by existing experience with available equipment or, when new devices are to be purchased, on the sensor specifications. Besides publicly available information on suitable devices, personal conversation with manufacturers and colleagues from research groups and government departments who have experience using these devices can be of great help not only during the procurement process but also for problems occurring later. Sensor manuals tend to frequently overstrain the operator but are a must-read before application. Exchange of experiences with other professionals in this area is always mutually beneficial!

If the installation and maintenance are carried out by different persons/organizations, it is important to involve the maintenance technician in the installation process. This ensures efficient transition from installation to operation and that knowledge on the general functioning, operation and handling of the devices is maintained.

The sensors down in the sewer network are usually well protected by the heavy manhole covers but devices on the ground, like loggers and data transmission equipment, rain gauges or water level sensors in SUDS and urban streams can be prone to criminal acts. It is not uncommon that such equipment is stolen, even though its reusability as well as its (black-)market value is low to negligible. To avoid losing valuable equipment, a protected and inaccessible measurement site is of paramount importance. Fenced infrastructures (like treatment plants, pumping stations or storage basins) or any other enclosed area can serve well as such safe sites.

However, compromises frequently have to be made between a specific, interesting location, availability of electricity and security, in order to fulfil the monitoring aim. If the risk of vandalism and theft has to be taken, it is advisable to choose spots that are close to buildings with a high pedestrian traffic and are lit during night hours, making them less attractive for thieves. Furthermore, placing devices at heights, which only can be accessed by ladders reduces casual theft. Chains can secure movable objects, like auto samplers or rain

Figure 7.7 Metal cabinet housing monitoring devices. *Source*: Peter Poelsma (University of Melbourne).

Figure 7.8 Ultrasonic (left) and capacitive (right) sensors for water level monitoring installed in PVC pipes and above the water to avoid any vandalism – wetland in Melbourne, Australia. *Source*: Frédéric Cherqui (INSA Lyon).

gauges. Metal cabinets can also be installed that can house all equipment safely. They can be bolted to existing concrete or to concrete slabs installed for the cabinet. Cables and sample hoses can be secured in conduit connecting the base of the cabinet to the sewer (Figure 7.7). Small systems can be hidden in non-attractive enclosure and installed above the water (Figure 7.8).

A short explanation of the monitoring activity should be given to curious people, passing by during maintenance, to raise awareness and maybe prevent them from exploring the measurement site themselves. Big explanation signs at the measurement site might unnecessarily attract attention, however small stickers on the devices with a contact number can be reasonable (Figure 7.9). Another idea is the

Figure 7.9 Rain gauge with protection and explanation sign. *Source*: Simon Bloem (Eawag).

installation of fake cameras – if legally possible – that can be another cheap and effective means of discouraging theft and vandalism.

7.3.2 Operation of rain measurement equipment

7.3.2.1 Site specification

The World Metrological Organization (WMO) has released the CIMO-Guide to standardize the measurement of metrological parameters including precipitation (WMO, 2018). In an urban area it is especially difficult to find a representative location, as a certain amount of open space is required, which is harder to find in densely populated areas. Flat rooftops within fenced areas are suggested by experience.

7.3.2.2 Wind shields and bird spikes

Wind speed is an important environmental factor in precipitation measurement. This is especially true if the gauge is installed in a flat area with short vegetation, like meadows or fields. A common method to dampen this effect is the use of wind shields e.g. metal plates arranged around the rain gauge to create a field of smooth airflow (Figure 7.10).

Erroneous measurement results caused by blocked outlets from bird droppings are very common. Bird spikes on the upper edges of the collection cylinder can be an upgrade to overcome this issue.

7.3.2.3 Monitoring duration

Rain monitoring over a time span of years or decades requires further planning efforts. All high growing vegetation and changes to the surroundings have to be anticipated and taken into consideration. It can become reasonable to change the location of the rain gauge, so alternative locations should be also planned in advance.

Meadows and fields areas are causing problems not only by being prone to wind errors but also due to high grown vegetation during the summer months. In this case, any vegetation in the area around the gauge should be kept short. In locations where a huge accumulation of snow is expected, shovelling snow or raising the gauge are possible countermeasures.

Figure 7.10 Rain gauge with wind shield. *Source*: Uwe Eichelmann (TU Dresden).

Frozen rain and hail, and especially snow, are problematic for rain gauges. If a considerable amount of snow is expected (and this is of importance for the planned monitoring), gauges with a heating option need to be used. However, heated rain gauges do not only require a grid connection, but they also evaporate water. Savina *et al.* (2011) compared a heated tipping-bucket gauge with heated weighing precipitation gauges for the measurement of snow and found that the tipping-bucket gauge measured less precipitation due to the bigger heated surface and the required time for filling up of the tipping bucket in comparison with the immediate measurement of the weighing system. Overall, the heating-related losses of the tipping bucket amounted to 23.7% and it showed a mean delay of about 30 minutes in recording the beginning of the events.

7.3.3 Operation of discharge measurement

7.3.3.1 Preparation

Before the installation of a sensor, a functionality check of the measurement system (sensor and logger) is suggested. A visual inspection of the mechanical components should be carried out, and the general functionality and the manufacturer's specification should be crosschecked. This system test saves time during the field deployment and avoids possible errors that lead to data gaps. A close look at cables, connectors, antennas, sealings, drying capsules and debris on the sensor is recommended. If the system does not have time synchronization, it is advisable to check on offsets and the time drift of the internal real-time clock. The simplest case for the functionality checks of *in situ* water level measurement is a sink, where different levels can be adjusted easily. Level measurement for *ex situ* sensors (such as radar or ultrasonic sensors) can be carried out by measuring to the ground from an elevated position. Note that radar level sensors may not work by placing a plastic or cardboard surface in front of them (they require a permittivity higher than 1.2 F/m for reflection). Ultrasonic and radar sensors both have a blocking distance which should be known and taken into account before the installation.

After the installation, another check should follow a few days later, to ensure that the installation was carried out correctly, and settings like storing frequency and offset are correctly set and that the sensor performs well under the given conditions (e.g. water level minima during the night or peak flows during an storm event). The accuracy can be verified by a second, preferably more accurate measurement device (see Sections 3.2, 3.3 and 3.4 for available options), which can also give insights into the on-site measurement performance of the used sensor. This is of particular importance for the following subsection on continuous monitoring.

7.3.3.2 Continuous monitoring

After the initial decisions on site are made, long term, continuous measurement is usually preferred over monitoring campaigns. As the focus of this book is on measurements within the urban area, off-grid locations for continuous monitoring sites, without an option for remote data transfer, are not to be expected. This would significantly increase the initial planning efforts as well as the maintenance. The use of large battery packs, solar panels, and wireless data transfer via GPRS, or LPWAN (low-power wide-area network) may help expand the range of the monitoring project if available for the type of device to be put in operation. During decade-long monitoring campaigns the deteriorating effect of sunlight must be taken into consideration. High quality cables need to be used or the cables/wires need to be checked from time to time, as the cable insulation may start to crack because of UV-light influence, motion at cable holders or animal bites.

Ideas on possibilities for on-site discharge test

In cases where there is little or no flow, a test can be carried out by releasing water via a hydrant or large tank. Electromagnetic flow meters can be used to accurately measure the flow from a hydrant at various flow rates maintained for several minutes to test the flow meter *in situ* (Figure 7.11).

Figure 7.11 Experimental set-up to check discharge measurement devices. *Source*: Peter Poelsma (University of Melbourne).

From an operator's or practitioner's point of view, both remote access to the station and availability of real time visualization of the recorded time series are highly desirable conditions as they can serve as the basis for planning site visits. Under these conditions, a daily checking of the recordings is strongly recommended as a first step in the data validation process (see Chapter 9). However, if the measurement network contains many sensors, an automated reporting or alarming (e.g. by email or SMS) is recommended. Setting threshold values or an increase in standard deviation of the signal (if the signal gets noisy) can serve as such warning parameters (see Chapter 9). An alert if no data is received after a specific amount of time helps enable a rapid reaction to avoid important loss of data. Under certain conditions, such as for demonstration purposes or for better understanding of complex hydraulic conditions, the installation of a camera can be a useful amendment.

7.3.3.3 Monitoring campaigns

In contrast to continuous monitoring, monitoring campaigns are shorter in duration. Usually, such campaigns determine the amount of extraneous water or record a time series for model calibration with a certain number of rain events. Well-chosen spots for installation of sensors should be taken into account (see Sections 3.3, 3.4 and 6.2) as well as required safety and precaution measures (see Section 7.2). Remote data access and visualization may not always be possible, although recent developments in telemetry and signal coverage further enhance wireless data transmission capabilities (see Section 5.2).

Ideas on installing *ex situ* sensors

Especially for *ex situ* sensors measuring water level or velocity by radar and/or ultrasonics, it is important that the sensor is installed perpendicularly above the water surface. To make the installation and following control in narrow manholes easier and more precise, it is recommended to extend the sensor with a self-adhesive bubble level (blue circles on Figure 7.12).

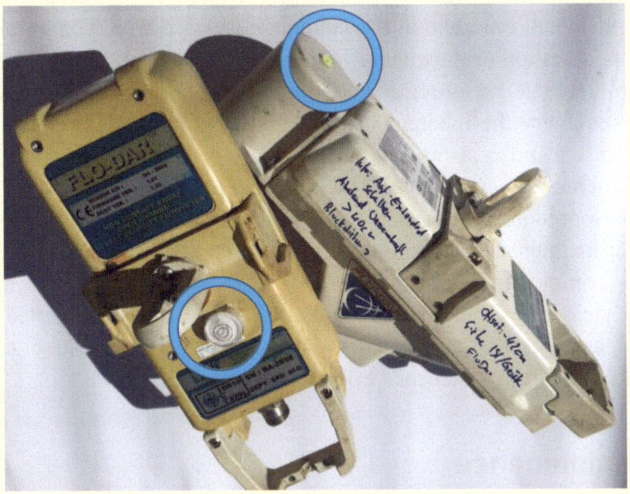

Figure 7.12 *Ex situ* flow meters (FloDar, Marsh-McBirney/Hach) with a bubble level. *Source*: Jakob Benisch (TU Dresden).

7.4 MAINTENANCE

7.4.1 General ideas on maintenance

Actual or impending failure, loss in accuracy or increase of uncertainty predicate the need for maintenance of monitoring systems. Ideally, maintenance is performed to keep equipment running efficiently for either the operation period of the monitoring activity or the design life of the component. The maintenance effort of a monitoring system follows a bathtub curve. A high failure rate characterizes the initial monitoring period because site conditions and equipment configuration may require adjustments of design, installation, and sensor selection. For example, minimum flow during night-time might be lower than expected and inferior to the lower measurement range. With debugging the measurement set-up, failure rate decreases and a productive period with lower maintenance requirements sets in. However, maintenance during this phase is still required to ensure the reliability of the results, e.g. through appropriate scheduling of sensor cleaning. During the wear out period failure rate increases with time and cannot be compensated by maintenance measures and is mostly related to mechanical or electrical failure of equipment components. There are different approaches distinguished by Sullivan *et al.* (2010) for reactive, preventive, and predictive maintenance. Reactive maintenance is considered as the 'run-to-failure' mode, which is obviously undemanding in labour and costs but will not yield reliable results. In contrast to that preventive and predictive maintenance are either based on a fixed schedule or on expected or predicted maintenance intervals (e.g. estimation of lifetime of batteries, online sensor results or taking boundary conditions like heavy rainfalls into account). Reliability centred maintenance prioritizes the importance of certain measurement locations over others allowing for the optimized use of financial and human resources. Finally, the aim of well-organized sensor maintenance is to reduce the on-site time as much as possible, without a loss of data or a deterioration in data quality. The amount of maintenance strongly depends on site conditions and design, sensor type, required data quality and reliability.

Experience shows that a state of perfect working online sensors will not be achieved, and well recorded events often require a considerable amount of preparation and experience. To optimize this process and the required maintenance, the question of the quality of the desired data is of major importance and should be asked before setting up elaborate measurement equipment and choosing sensor types. Furthermore, the results from online sensors carry the inherent risk of overwhelming their end-user by the sheer amounts of (erroneous) data. Also 'backup' sensors can create such problems if two sensors disagree; there is the difficulty of deciding which one is correct and which one should be ignored.

On the other hand, sensor redundancy is useful in detecting failures and errors (see Chapter 9). A last, but very important, point for maintenance of a monitoring station is the exchange and upgrade of sensors and loggers. Manufacturers commonly release firmware updates for their devices as well as new sensor generations, that could yield better recorded data. Walcker *et al.* (2018) reported how an improvement with new devices and arrangement at measurement stations had a positive effect on reliability of collected data and could decrease maintenance efforts after rehabilitation.

7.4.2 Planning maintenance

The reliable and continuous operation of measurement stations is crucial to obtain reasonable data for whatever purpose. If one's job is to investigate suddenly occurring, local rain events it becomes obvious that the system must always be in a steady state, which requires maintenance of measurement devices on a permanent basis. In contrast to that, short measurement campaigns with less complex sensors are also not that demanding to operate. In organizing maintenance efforts, several factors have an influence on the frequency of site visits and should be taken into consideration before and during the operation of the measurement devices. The most important factors are going to be explained hereafter, however the range of available products in this field and their specific maintenance demands still require further training.

7.4.3 Maintenance of rain measurement equipment

The maintenance of rain gauges usually depends on the location. In open areas like fields problems with spiders, bird droppings and nestings as well as debris can disturb the measurement. Pollen and leaves also cause problems and need to be removed on a seasonal (pollen, leaves) and regular basis. Information on maintenance for a rain gauge can found in the WMO guide as well (WMO, 2018).

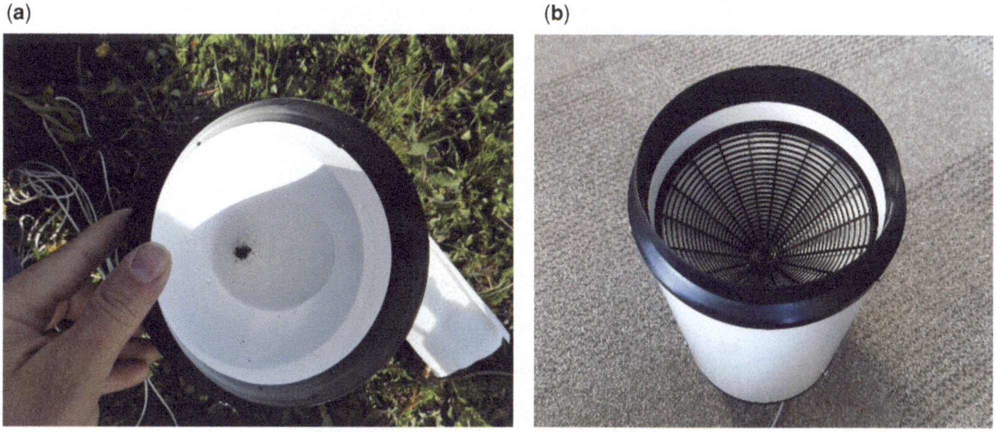

Figure 7.13 (a) blocked R.M. Young tipping bucket rain gauge; (b) removable grid for rain gauge. *Source*: (a) Simon Bloem (Eawag), (b) Stefan Kroll (Aquafin NV).

In an urban area, e.g. on a flat roof without surrounding vegetation, less problems are to be expected. As for discharge measurement devices, if the gauges are operated autonomously from the grid and without data transfer, battery changes and re-setting of internal clocks are of high importance.

7.4.3.1 Tipping bucket rain gauges

Tipping bucket rain gauges need individual care depending on the location. A common problem is clogging of the inlets by bird droppings, dead animals, and debris (Figure 7.13(a)). Removable grids as shown in Figure 7.13(b) can help with this. The outlets of rain gauges may also become blocked, especially during long-term rain monitoring at one site, and they should not be forgotten in the cleaning procedure. Most of the time, a cross-comparison with another rain gauge nearby is useful to determine if it is clogged or not, if online data are available.

The tipping bucket mechanism and the funnel as well as the outlet must be cleaned on a regular basis. Also, the mechanical levelling of the rain gauge should be checked at the same time. The calibration process has to be conducted with an appropriate volume calibration bottle (Figure 7.14). It is also recommended to apply a dynamic intensity calibration to evaluate the systematic under-estimation of intensities by traditional tipping bucket rain gauges (see Section 7.6 on sensor calibration).

Regular calibration can also be carried out using a bulk collector (Figure 7.15). All the rainfall measured by the gauge is collected and weighed at the end of a set period, e.g. monthly. Evaporation is minimized by using a sealed container with the inlet tube reaching the bottom of the bulk container and a breather tube several metres long as per rain collectors used for isotopic analysis described in Gröning *et al.* (2012).

7.4.3.2 Weighing rain gauges

Weighing rain gauges demand less frequent maintenance and are easier to test or calibrate as compared to tipping bucket rain gauges. To check the functionality, the manufacturer provides manuals with an additional weight to test the scale (e.g. Ott Hydrometrie). During wintertime, internal antifreeze agent dosing or heating should be used to prevent bursting of the collecting vessel. In the maintenance process it is recommended to check pollution (nesting) and mechanical levelling as well as the functionality of the scale with an accuracy test.

Figure 7.14 R.M. Young tipping bucket rain gauge with calibration bottle. *Source*: Simon Bloem (Eawag).

Figure 7.15 Rain gauge with bulk collector used for regular calibration. *Source*: Peter Poelsma (University of Melbourne).

Figure 7.16 Self-powered water level monitoring system using three AA rechargeable batteries and a 0.5 W solar panel (on top). The data are sent every 15 minutes using the SigFox network – wetland in Melbourne, Australia. *Source*: Frédéric Cherqui (INSA Lyon).

7.4.4 Maintenance for discharge measurement

7.4.4.1 Battery exchange/energy management

A power connection provides the most reliable and least maintenance intense source of energy. However, it may not be available at all monitoring sites and the usage of batteries becomes inevitable. As the lifetime of the battery highly depends on the measurement interval, the desired temporal resolution is important and addressed in the macro design chapter (Section 6.2.5). Wireless data communication usually consumes a considerable share of available energy. If real time data collection is not essential, the use of large communication intervals can help mitigate this problem. Also, the use of communication protocols optimized for low energy consumption can prove very beneficial. Low-power wide-area networks (LPWANs) such as LoRaWAN, SigFox or NB-IoT requires less energy to connect to a network and send the data, however they offer a limited bandwidth. The surrounding (air) temperature also affects the lifetime of batteries; especially during the winter months a shorter life span can be expected. To overcome those limitations, some sensors offer the possibility to connect external batteries to expand the measurement duration. After several years of use, batteries have to be replaced. Solar panels and wind turbines may provide additional energy sources and must be adapted to the batteries and the system consumption (Figure 7.16). Their contribution to the total energy supply varies seasonally and event wise and should be included with high safety margins.

7.4.4.2 Readout of data

Like batteries, the available memory on the loggers is a further limitation; especially older devices working at high temporal monitoring resolution quickly fill up their available memory. Some devices allow the possibility to switch to a higher temporal resolution during events by setting a specific threshold (discharge, water level). This option prolongs the service life of the memory and battery to some extent but can also create problems in data evaluation (see Section 9.3.5). If a stable remote data transfer is possible, this is also a very reasonable solution to this issue (providing real-time data visualization with all its advantages for maintenance).

7.4.4.3 Reparation or improvement of the systems

If the monitoring is installed for a long period, it could be useful to facilitate the replacement of any part of the system. Potential replacement concerns the sensor, the logger, the communication module, the batteries, or any other component. To ease the replacement, parts must be easy to access and the sensor holder should guarantee that the replacement sensor will be at the exact same location when measurements are position dependent. Considering reparation or improvement is even more important when the monitoring system has been developed in-house. DIY (Do It Yourself) monitoring systems (Figure 7.17) offer very high modularity but are also more prone to failure during their development, or to frequent improvements (there is always something to improve!). Transparent lids help to easily diagnose the system with LEDs or a screen. The enclosure should be easy to open and close without compromising the watertightness. The DIY system can also be built in a way that allows easy change of defective electronical components. A simple system is always easier to maintain than a system with many components.

7.4.4.4 Calibration

Discharge measurement calibration, which is based on the measurement of velocity and water level, is a complex procedure. Very often, the user of a device is limited to make changes only on the water level measurement and apply a factor for the correction of the velocity measurement, which is not fully satisfactory to ensure reliable flow data. Water level sensors can be calibrated with usual calibration protocols (see Section 7.6), but most flow velocity sensors require sophisticated facilities like a test

Figure 7.17 DIY system used to monitor water level, trigger sampling, save the data on SD card and send it using the GSM network. The OLED (organic light-emitting diode) screen displays the last actions. *Source*: Frédéric Cherqui (INSA Lyon).

Figure 7.18 Measuring water level with a laser meter and milk powder. *Source*: courtesy of Francois Clemens-Meyer (Deltares/TU Delft/NTNU).

channel with an accurate reference discharge (most often a magnetic flow meter in a completely filled pipe) or other reference installation where certificated flow velocity values can be provided. Such calibrations can be carried out by manufacturers or at specialized test institutes but usually not on site by users. Methods for on-site estimation of discharge, besides the usage of secondary, accurate instruments mentioned above (see Sections 3.4.3 and 7.3.3), can be tracer experiments or combining portable velocity meters with electromagnetic sensors and water level. For locations without backwater effects, a stage discharge relation can be derived from periods with reliable measurements and used for cross-checking instant readouts or uncertain data. The recordings of water level sensors can be calibrated with the help of a ruler, or temporary or permanent staff gauges. Water may not be sufficiently reflective for the application of a laser distance meter. Results can be improved by using milk powder (Figure 7.18).

Ideas on calibrating water level sensors

Using a laser distance meter (mounted on a beam with a gimbal) for the calibration of a water level sensor is a very accurate reference, with an uncertainty in the millimetre range. To get correct measurements however, the laser beam needs to be reflected on the water surface. That is not always the case in sewers or pumping stations. To overcome this problem, milk powder can be spread on the water to obtain a reflective surface, which allows for an accurate determination of the height reference of a water level sensor after (re)installing.

7.4.4.5 Cleaning

Submerged sensors, in particular, should be cleaned on a regular basis; ultrasonic profilers will not work properly when they are covered with sediments or debris. Build-up of sediment is strongly dependent on characteristics of the sewer (e.g. site geometry, slope) and the catchment (slope, surface use, soil type). This has to be taken into account when cleaning intervals are going to be planned. A strategy for identifying those cleaning intervals can be that very frequent cleaning is a starting point, followed by an increase of these intervals, together with a documentation on the build-up process at the site visit. Based on that, an optimum can be found – remote access can optimize this process further. Very often, sediment movement occurs during heavy rain events, a reasonable occasion for (additional) maintenance.

Ideas on sediment affected sites

Thick layers of sewer sediments at discharge measurement sites are problematic, especially if the sediment is rather loose and unstable. The offset or sludge level-function of most *ex situ* measurement devices allows these constant, dense, and steady layers of sediment to be taken into account. *In situ* sensors can be mounted on an elevated base if sediment levels appear on a constant level and the manufacturer settings allow such an option (Figure 7.19).

However, such elevated bases, and also lateral or on side walls installation of sensors, may lead to significant errors in discharge measurements, as the velocity field measured by the sensor is modified and cannot be interpreted to estimate the discharge as it would be with a typical central bottom location. Some experiments have shown e.g. that lateral installation of Doppler sensors to avoid sediments may lead to errors of 30% in the discharge (Lepot *et al.*, 2014). When such non-typical sensor locations cannot be avoided (e.g. by changing the measurement site, the measurement technology, etc.), it is of crucial importance to carry out a local specific investigation (e.g. with tracing experiments or velocity field exploration when and where possible, or with 3D flow modelling) to check the discharge values and derive correction factors if necessary.

Discharge peaks from heavy storm events are however also able to remove such 'stable' layers. Another possibility is to narrow the cross section, this increases the flow velocity and thereby reduces the amount of settleable particles. In this case, care must be taken not to influence the overall flow characteristics of the monitored conduit section. There are rare cases in bigger sewers, where it can be an idea to mount a wedge sensor on the bottom of a pontoon floating on water and to measure the velocity profile from there. The installation of pontoons, however, bears the disadvantage that floating debris will accumulate on the nose if the ship's hull is not sufficiently streamlined, thus potentially increasing the cleaning frequency.

Figure 7.19 Elevated base for *in situ* wedge sensors. *Source*: Jakob Benisch (TU Dresden).

7.4.4.6 *Set internal logger clocks*

Synchronicity of data is crucial, as shown by Schilling (1991): a modelled peak run-off deviated by 16% with only a 2-minute synchronization error in precipitation measurement. Logger clocks are more or less prone to differ to the real time, depending on quality of the crystal oscillators used. The use of data transmission may provide synchronization with a time server or network time but not all controllers and loggers allow for this. According to experience, time drifts are in the range of less than 1 minute over one week. They should be noted and corrected immediately within this initial step or later on, mostly linear time drifts are assumed. It is also possible to read the time from a GPS module: this solution provides both an accurate time and the location of the sensor. Sensor clocks should be working according to the Coordinated Universal Time (UTC), and take care of daylight savings between summertime and wintertime (especially when the logger clocks are updated by the time of the laptop used for reading out!).

7.4.4.7 *Exchange desiccant of ventilated water level sensors*

Water level measurement working with pressure cells needs to compensate for the changes in atmospheric pressure on the recorded levels. To do so, pressure compensation is usually carried out either by a second sensor, recording the atmospheric pressure (unventilated systems) or via an opening by which the atmospheric pressure is compensated immediately (ventilated system). These systems are usually more accurate and require less pre-processing, but they are more vulnerable to humidity entering in the form of condensed water or buckling of the tube. To protect those openings, silica gel is used to keep them dry. Changing of these capsules is necessary and often indicated by a change of colour of the silica (Figure 7.20). If the opening is installed in a shaft, which tends to be filled to high level or is severely humid, unventilated systems should be the preferred solution.

Important factors for the operation of discharge/level devices

- CL 7.13: *Change of batteries* – Depending on capacity of batteries, measurement interval and sensor type.
- CL 7.14: *Readout data* – Depending on measurement interval and capacity of available memory.
- CL 7.15: *Calibration* – Age of sensor, cleaning, site characteristics.
- CL 7.16: *Cleaning* – Site characteristics.
- CL 7.17: *Set logger clock* – Time stability of internal clock.
- CL 7.18: *Exchange of desiccant for water level sensors* – Manufacturer recommendation and site characteristics.

7.4.4.8 *Standard procedure*

To organize sensor maintenance in a comparable and uniform way, a generalized procedure is suggested for single steps and documentation. An example protocol is also given hereafter.

Figure 7.20 Half-used silica capsule of a water level sensor. *Source*: Jakob Benisch (TU Dresden).

- Site appearance before cleaning: the first impression of the site should be documented, this can be done by notes and also taking pictures in extraordinary cases. If the logger has a display the recordings should be noted from there before starting the cleaning process. Also, the actual weather conditions or the high water marks in the shaft from a previous rain event can give a good impression of maximal water levels during the previous rain event. In a similar way, the amount and height of sediment in the channel give an idea of remobilization forces during rain events.
- Cleaning: besides health and safety equipment, a toilet brush usually suffices for the cleaning itself. Taking hot water with you for washing afterwards is also reasonable (during wintertime it can also be used for thawing solid frozen manhole covers). Effects of the cleaning can be easily checked by changes in the displayed measurement results or by the readout data which should be checked as soon as possible, best done on site or in the following days (not weeks!).
- Data validity check (see Chapter 9): after the cleaning is done, the validity of the sensor results should be checked. Sensor drifts and offsets are often occurring phenomena. Correcting those errors later on can be very tedious. Often, such a correction is so uncertain, that it becomes hard to decide whether it is more reasonable to correct or delete those data.

Example (to be adapted) for a maintenance protocol: such protocols can be on paper or preferably digitalized and recorded in a database allowing the history of all actions to be maintained and some posterior analysis to be conducted

- Site:
- Name of maintenance technician:
- Date and time of
  - Arrival:
  - Departure:

(1) *Appearance at arrival*

Exceptional things/comments/pictures:
Weather condition:
Time and measurement results before cleaning:

Time	Water level	Velocity	Discharge

Check of system time (UTC-time zone)

- Correction: yes/no
- Time-offset:

(2) *Cleaning*

Time of cleaning:
Comments (sediment height/changes of measurements after cleaning):

(3) *Control and data verification/calibration*

Time and measurement results after cleaning:

Time	Water level	Velocity	Discharge

Reference measurement taken:

Time	Water level	Velocity	Discharge

Calibration: yes/no, if yes:

Time	Water level	Velocity	Discharge

Site specifications (should be done initially or when sensors are exchanged etc.):

- Coordinates/shaft number:
- Profile:
- Contact details of responsible site owner/local authority:
- Device passwords:

Used equipment:

- Sensor:
- Serial-Number:
- Firmware:
- Logging interval:
- Critical minimum voltage:

7.4.4.9 Flow control structure maintenance

Measuring water level upstream of a control structure which is rated (i.e. has a well-established relationship between flow and level) or has been built according to standards which allow the use of a formula to convert level to discharge, is a common way to determine flows in open channels. In these cases, it is very important to maintain the flow control structure so that the rating or formula used to convert level to discharge is applicable. Usually, the control structure is a cross section that is very stable and unlikely to change. Any changes need to be noted and rectified if possible, e.g. removal of sediment or debris. Where weirs are used, the appropriate approach and pool dimensions also need to be maintained which may require periodic desilting of the weir pool. More info is given in ISO (2013a).

Other flow structures such as sluice gates or orifices can also be used to estimate flow based on water level measurements (see Chapter 3). It is essential that these structures are maintained and remain clean and unobstructed for the estimation of flows using standard equations, and also for the proper functioning of the system upstream (e.g. storage basin, wetland, etc.).

7.5 SITE VISITS

Site visits for maintenance processes are crucial, not only for cleaning and calibration procedures but also for learning and understanding phenomena and processes which happen on site and would never be revealed.

Typically, weather forecasts should be checked before leaving. Even though the duration and intensity of rain events is very hard to predict, a general idea on whether rain is to be expected can be quickly drawn. In some countries online rain radar maps are available, which can be very accurate for a short forecasting period. As mentioned earlier, entering UDSM facilities (sewers, stormwater detention tanks, infiltration basins, etc.) shortly before or during a rain event may be extremely dangerous, so make sure that you have read and informed yourself in the previous sections about your health and safety as well as your local regulations.

Besides the maintenance protocol, a packing list is highly recommended too. It can be used as a document for other practitioners as well as a summary of your experiences so far. Especially for far away sites, a lap of honour because of a forgotten tool is an unnecessary waste of time.

Written documentation is very important during maintenance processes – a diary in a digital version (preferably) or on paper should be always available. The most relevant steps for typical maintenance of discharge measurement devices are given in Section 7.4.4. These steps should be also written down in the diary, along with the timestamp, when the action has been taken. Visual documentation using a digital camera can be a valuable extension of a diary.

Ideas on documentation of site visits

Similar to the light-switch in a fridge, a switch at the door of the monitoring station, which is connected to a logger, can give a great record about when the station has been opened and closed. It can be a bit elaborate to install but on a site which is supposed to measure over a long period it is worth doing!

7.6 SENSOR CALIBRATION AND VERIFICATION

7.6.1 Introduction

It is of crucial importance in general metrology, and thus in the UDSM field, to check that sensors function according to their specifications and that measuring chains deliver reliable results with known uncertainties. Consequently, all sensors used in UDSM metrology, either during short-term measurement campaigns or for long-term continuous monitoring stations, must be periodically calibrated and verified.

Some preliminary definitions are necessary. According to JCGM (Joint Committee for Guides in Metrology, 2012), calibration is an 'operation that, under specified conditions, in a first step, establishes a relation between the quantity values with measurement uncertainties provided by measurement standards and corresponding indications with associated measurement uncertainties and, in a second step, uses this information to establish a relation for obtaining a measurement result from an indication.' As a complement to this definition, JCGM (2012) adds two notes: '(1) A calibration may be expressed by a statement, calibration function, calibration diagram, calibration curve, or calibration table. In some cases, it may consist of an additive or multiplicative correction of the indication with associated measurement uncertainty', and '(2) Calibration should not be confused with adjustment of a measuring system, often mistakenly called 'self-calibration', nor with verification of calibration.'

JCGM (2012) defines verification as the 'provision of objective evidence that a given item fulfils specified requirements.' In the case of sensor calibration, it may be interpreted as the fact that, after a new calibration experiment, a given sensor provides measurement values which satisfy a maximum tolerated deviation compared to the previous calibration.

Lastly, JCGM (2012) defines a measuring chain as a 'series of elements of a measuring system constituting a single path of the signal from a sensor to an output element.'

Even if some manufacturers may claim their sensors are automatically calibrated ('self-calibration') or even do not require calibration at all, calibration and verification by the user/operator remain absolutely necessary for various reasons:

- Calibration and verification are basic requirements for usual best practices and quality assurance in metrology. It is thus the responsibility of the person in charge of UDSM monitoring to work according to these high-quality standards and to deliver reliable measured values. He/she cannot rely on external and non-locally representative information and data.
- We strongly recommend calibration of the entire measuring chain, under its real operation conditions, from the sensor transducer to the final measured values stored in data loggers or databases, which we are interested in and which will be used for analysis, interpretation, modelling, etc., and not only the sensor itself. This is very important as sources of errors, bias and uncertainties are not only attached to the sensors themselves, but also to signal amplification, conditioning, conversion, re-scaling, transfers, etc.
- Measurement values and data must be given with their uncertainties, as described in Section 8.1. Calibration experiments are a primary source of information to estimate the uncertainties of values delivered by sensors.

In practice, there are no ideal or perfect sensors, but only real ones affected by imperfections. In this case, calibration allows:

- Estimation, quantification, and correction of these errors, and estimation of the resulting uncertainties in measurements.
- Determination of a calibration function to correct the residual errors of the sensor and to estimate uncertainties.

BOX 0: EXAMPLES WITH MATLAB®

Detailed examples of calculations with Matlab® are shown in dedicated boxes further on in this chapter: the instructions and codes are written with the Matlab® syntax and `courier new` font to distinguish them from the rest of the text. Instructions and code lines can be copied-pasted directly by the reader who would like to replicate them for training or to adapt them to his/her own needs.

Numerical results in boxes are usually given with four digits (`format short`). In the main text, numerical values are rounded to the number of significant digits. It is also important to note, for readers who would like to reproduce them, that all calculations have been run without rounding in the successive steps.

Matlab® codes and associated data csv files are available for download at https://doi.org/10.2166/9781789060102.

In the following sections, numerical methods for calibration are introduced, and examples of application are presented.

7.6.2 Principle of calibration

For a given sensor, the principle of calibration consists of observing the sensor outputs y (i.e. measured values, preferably as outputs at the end of the measuring chain) delivered when the sensor is submitted to known input values x, and establishing the corresponding calibration function $y = f(x)$. The calibration shall be carried out under such specified conditions that the effects of external influential quantities (like temperature, pressure, pH, radiation, etc.) are negligible, or quantifiable, or set to standard values (e.g. calibration should be carried out at 20°C).

For simple sensors, x and y are scalars (e.g. rain gauge, water level sensors). For more complex sensors, x and y may be vectors or matrices (e.g. UV-visible spectrophotometers). The principle does not change, but the determination of the calibration function f may be more complicated and requires more sophisticated mathematical tools.

The known values x shall, as far as possible, be certified standards, provided with valid certificates indicating their attachment to higher level national or international standards. When standards exist, they shall be used (e.g. for mass measurements, distance measurements). In the case where standards do not exist, surrogate certified materials or solutions shall be used, e.g. primary standard formazin solutions for turbidimeters, certified solutions for pH and conductimeters, certified solutions for UV-visible spectrophotometers, etc. In the case where no certified materials or solutions exist, it is the responsibility of the user to define alternative tools with known uncertainties and to prepare them by means of traceable and reproducible protocols, e.g. calibrated pumps or Mariotte bottle for rain gauges, tracing experiments for discharge and/or flow velocity measurements, etc.

In theory, a perfect sensor has a very simple calibration function: it should be $y = x$, i.e. the measured values are equal to the standard values. In practice, calibration functions are different, to account for unavoidable errors:

offset (or zero) error $$y = b_0 + x \quad \text{with } b_0 \neq 0 \tag{7.1}$$

sensitivity (or slope) error $$y = b_1 x \quad \text{with } b_1 \neq 1 \tag{7.2}$$

offset and sensitivity error $$y = b_0 + b_1 x \quad \text{with } b_0 \neq 0,\ b_1 \neq 1 \tag{7.3}$$

linearity error $y = b_0 + b_1x + b_2x^2$ with $b_0 \neq 0,\ b_1 \neq 1,\ b_2 \neq 0$ (7.4)

or $y = b_0x^{b_1}$ with $b_0 \neq 1,\ b_1 \neq 1$ (7.5)

Polynomial functions (Equations (7.1) to (7.4)) are used for various types of sensors, while a power function (Equation (7.5)) is used typically for tipping bucket rain gauges.

Numerical and statistical methods (regression, variance analysis, variance tests, Monte Carlo simulations) are used to:

- Estimate the values of the parameters b_j.
- Estimate their standard uncertainties (see Section 8.2.1 for the definition of standard uncertainty).
- Estimate if they are significantly different from 0 or 1 depending on the equation used.
- Estimate the optimal order of the polynomial function which is necessary and sufficient to establish a meaningful calibration function.

Once the calibration function f is established, it is then used to estimate the most likely true value $\hat{x}$ from a measured value y_m, by using the reciprocal function f^{-1}:

$$\hat{x} = f^{-1}(y_m) \tag{7.6}$$

The calibration function is established as $y = f(x)$ and not as $x = f(y)$. This is due to theoretical conditions which are required for applying the ordinary least squares regression method: x values should have no uncertainty or uncertainties which are negligible compared to uncertainties in y values. In other words, uncertainty $u(x)$ in standard values should be lower than uncertainties $u(y)$ in values given by the sensor. In theory, it is recommended to use standards such that $u(x)$ is less than one tenth of $u(y)$. In practice, this is no always possible: one fifth is acceptable. If $u(x)$ is greater than this threshold, the ordinary least squares regression is in theory no longer applicable and alternative methods like e.g. the Williamson least squares regression should be used (see Section 7.6.4.4).

In the case where the calibration function is a first order polynomial function (Equation (7.3)):

$$y = b_0 + b_1x \tag{7.7}$$

one calculates

$$\hat{x} = \frac{y_m - b_0}{b_1} \tag{7.8}$$

As the coefficients b_j are affected by uncertainties which are evaluated by the regression, it is then possible, by using either the law of propagation of uncertainties or Monte Carlo simulations, according to international standards (see Section 8.2), to estimate the standard uncertainty in $\hat{x}$.

7.6.3 Calibration and verification protocols

In practice, independently from the numerical and statistical methods to be applied, a calibration is basically carried out as follows:

- Choose N_x standards (or surrogate certified materials and solutions) to evenly cover the sensor range of measurement (or the sub-range of interest for the user).
- For each x_i standard value, take repeated readings of output values y_{ik}, with $k = 1$ to N_y. The repeated measurements y_{ik} are necessary to properly estimate uncertainties in sensor outputs from the observed standard deviation of the y_{ik} values (see Figure 7.21). As an indicative value, N_y could be between

10 and 25. If only one single value y_i is measured for each x_i standard value, uncertainties in sensor measurements cannot be evaluated.

- Determine the calibration function (Equations (7.1) to (7.5)) and related quantities (uncertainties, etc.).

For verification, the protocol is similar except for the 3rd step (Figure 7.22). If all measured values y_{ik} are within a specified interval defined by the maximum acceptable error I_{mt} on each side of the previous calibration function, then the verification is accepted, and the sensor is used with the previous calibration function. If some measured values y_{ik} are outside the specified interval, then the verification is rejected. A new calibration function shall be established (or the sensor shall be re-adjusted, repaired or replaced if necessary). The maximum acceptable error I_{mt} and the corresponding specified interval are defined by the user, according to his/her criteria which must be clearly indicated in the verification protocol. Usual values for I_{mt} are between 2 and 3 times the standard deviation of the measured values y_{ik} of the previous calibration, i.e. the square root of the variance shown in Figure 7.27 in Box 2 (see Section 7.6.4.3).

Calibration and verification

Experiments are usually sensitive to several external influential quantities, the most important one being temperature, along with pressure, humidity, electromagnetic interferences, etc. as indicated in manufacturer's manuals or information documents.

Certified standards should always be used under the required conditions (e.g. at 20°C). Sensors should be at the same temperature. In all cases, temperature should be measured, and values corrected accordingly. Experience has shown that using standards at different temperatures or at changing temperature during the calibration experiment may lead to additional errors and uncertainties and to biased calibration functions.

Protocols must be available in a written format, either on paper or in a digital document. It is very important to ensure that successive calibrations or verifications are always carried out under the same conditions. Revisions of written protocols must be made with the usual traceability required by quality assurance.

7.6.4 Regression methods for calibration functions

7.6.4.1 Introduction

In order to establish calibration functions (Equations (7.1) to (7.5)), regression methods are used. Three methods are presented, accompanied with detailed examples of application:

- The ordinary least squares (OLS) regression for polynomial functions (Equations (7.1) to (7.4)), which is the most frequently used method (Section 7.6.4.2).
- The extended Williamson least squares (WLS) regression for polynomial functions (Equations (7.1) to (7.4)) that needs to be applied in cases where uncertainties in both x and y values have similar orders of magnitude or cannot be neglected and have to be accounted for (Section 7.6.4.4).
- The non-linear regression method used for power functions (Equation (7.5)) (Section 7.6.4.6).

Other methods exist for multivariate or non-linear calibration functions, based on more sophisticated methods, e.g. the Marquart-Levenberg algorithm or the partial least squares (PLS) regression, to be used

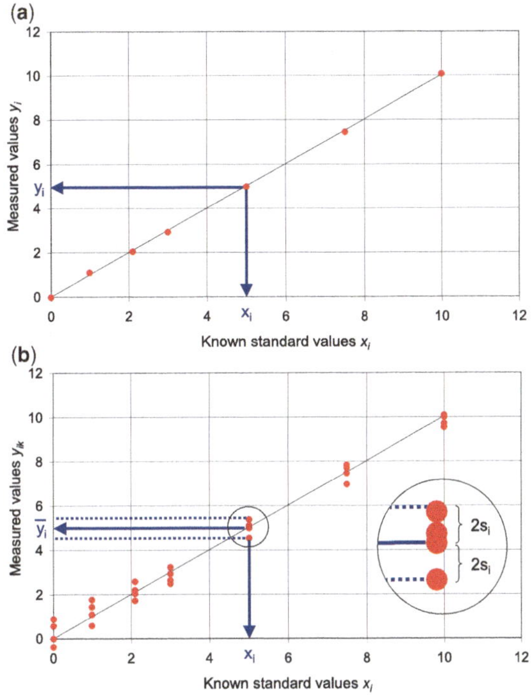

Figure 7.21 Sensor calibration with single y_i measurements (a) and repeated measurements y_{ik} (b). *Source*: Jean-Luc Bertrand-Krajewski (INSA Lyon).

for specific sensors (e.g. UV-visible spectrophotometers): they are beyond the scope of this chapter. More information can be found in e.g. Xin *et al.* (2011).

7.6.4.2 Ordinary least squares regression (OLS)

7.6.4.2.1 Introduction

The calibration data set is composed of $k = 1{:}N_y$ repeated measurements y_{ik} for $i = 1{:}N_x$ standard values x_i. For convenience in calculation, N_y should be identical for all N_x standard values. Three polynomial functions with order d ranging from 1 to 3 are systematically evaluated by the ordinary least squares (OLS) regression:

$$y = \sum_{j=0}^{d} b_j x^j \tag{7.9}$$

The values of the coefficients b_j are calculated, with their respective standard uncertainties $u(b_j)$ and covariances $cov(b_j, b_l)$ $\forall j,\ l = 0:d$. The most appropriate polynomial order is determined by means of the Snedecor variance test.

In addition, the graph showing the variance s_i^2 of the y_{ik} values vs. the x_i values is important (see Figure 7.27 in Box 2): if the variance s_i^2 is not constant (or approximately constant) over the calibration range, the OLS regression is theoretically not applicable and shall be replaced e.g. by a Williamson regression (Section 7.6.4.4).

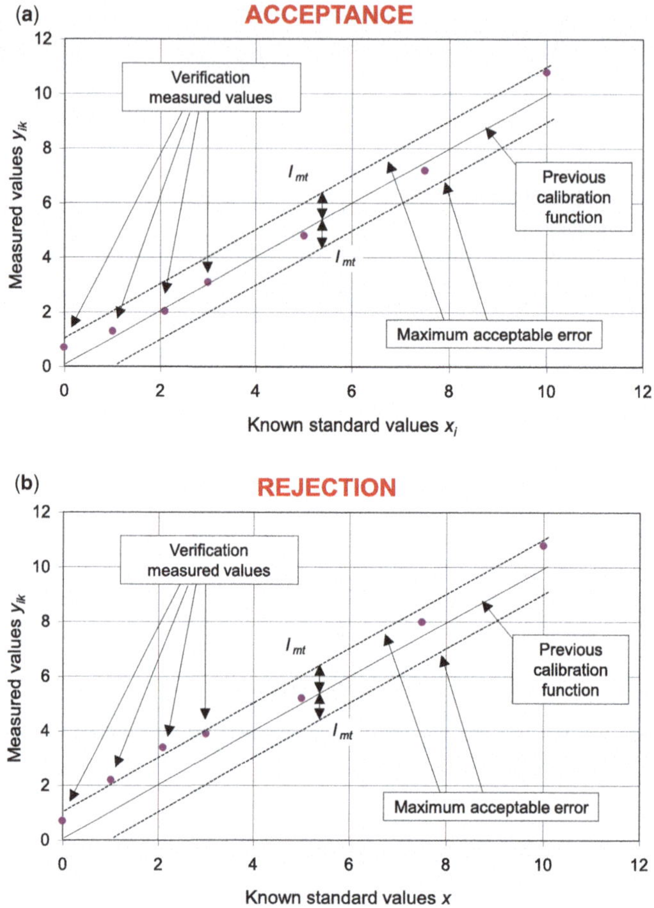

Figure 7.22 Verification accepted (a) or rejected (b). *Source*: Jean-Luc Bertrand-Krajewski (INSA Lyon).

7.6.4.2.2 Calculation of coefficients, standard uncertainties and covariances

With d the polynomial order, the determination of the coefficients b_j requires minimization of the sum S_{OLS}

$$S_{OLS} = \sum_{i=1}^{N_x} \sum_{k=1}^{N_y} \left(\sum_{j=0}^{d} b_j x_i^j - y_{ik} \right)^2 \tag{7.10}$$

Minimizing S_{OLS} is equivalent to Equation (7.11)

$$\frac{\partial S_{OLS}}{\partial b_l} = 0 \qquad \forall l = 0 : d$$
$$\Leftrightarrow \sum_i \sum_k \left(\sum_j b_j x_i^j \right) x_i^l = \sum_i \sum_k y_{ik} x_i^l \tag{7.11}$$

With the following matrix notations

$$\boldsymbol{B} = \begin{bmatrix} b_0 \\ b_1 \\ \vdots \\ b_d \end{bmatrix}, \boldsymbol{Y} = \begin{bmatrix} y_{11} \\ y_{12} \\ \vdots \\ y_{N_xN_y} \end{bmatrix}$$

and with $\boldsymbol{F}$ a matrix built with concatenated single column vectors α_i, β_i, … each one including N_y repeated values

$$\boldsymbol{F} = \begin{bmatrix} \alpha_1 & \beta_1 & \cdots & \delta_1 \\ \alpha_2 & \beta_2 & \cdots & \delta_2 \\ \vdots & \vdots & & \vdots \\ \alpha_{N_x} & \beta_{N_x} & \cdots & \delta_{N_x} \end{bmatrix} \quad \text{with } \alpha_i = \begin{bmatrix} 1 \\ 1 \\ \vdots \\ 1 \end{bmatrix}, \beta_i = \begin{bmatrix} x_i \\ x_i \\ \vdots \\ x_i \end{bmatrix}, \ldots \delta_i = \begin{bmatrix} x_i^d \\ x_i^d \\ \vdots \\ x_i^d \end{bmatrix}$$

Equation (7.11) can be simply written as Equation (7.12):

$$(\boldsymbol{F}^T\boldsymbol{F})\boldsymbol{B} = \boldsymbol{F}^T\boldsymbol{Y} \tag{7.12}$$

which can be solved directly as Equation (7.13):

$$\boldsymbol{B} = (\boldsymbol{F}^T\boldsymbol{F})^{-1}\boldsymbol{F}^T\boldsymbol{Y} \tag{7.13}$$

However, the matrix $\boldsymbol{F}$ is badly scaled: there are $N = N_x \times N_y$ rows and only $d + 1$ columns (usually with $N \gg d$). This may generate numerical problems in calculating precise values of variances and covariances. Thus Equation (7.12) is solved by means of the QR orthogonal-triangular decomposition (see e.g. Nougier, 2001).

Equation (7.12) can be simplified as Equation (7.14):

$$\boldsymbol{FB} = \boldsymbol{Y} \tag{7.14}$$

With the QR decomposition, $\boldsymbol{F}$ can be replaced by Equation (7.15):

$$\boldsymbol{QRB} = \boldsymbol{Y} \tag{7.15}$$

where

$$\boldsymbol{M} = \boldsymbol{R}^{-1}\boldsymbol{Q}^T \tag{7.16}$$

and one gets directly

$$\boldsymbol{B} = \boldsymbol{MY} \tag{7.17}$$

The residuals e and SSR (sum of squared residuals) are given, respectively, by Equations (7.18) and (7.19):

$$\boldsymbol{e} = \boldsymbol{FB} - \boldsymbol{Y} \tag{7.18}$$

$$SSR = \boldsymbol{e}^T\boldsymbol{e} = \sum_i \sum_k e_{ik}^2 = \sum_i \sum_k (y_{ik} - \hat{y}_{ik})^2 \tag{7.19}$$

The final mean residual variance S_r is given by Equation (7.20):

$$S_r = \frac{SSR}{N-d-1} \tag{7.20}$$

The covariance matrix of $\boldsymbol{B}$ is calculated by Equation (7.21):

$$cov(\boldsymbol{B}) = \frac{SSR}{N-d-1}\boldsymbol{M}\boldsymbol{M}^T \tag{7.21}$$

Consequently, the standard uncertainty in $\boldsymbol{B}$ is equal to Equation (7.22):

$$u(\boldsymbol{B}) = \sqrt{diag(cov(\boldsymbol{B}))} \tag{7.22}$$

The 95% coverage interval for $\boldsymbol{B}$ is given by Equation (7.23):

$$\left[\boldsymbol{B} - t_{\frac{1+\alpha}{2}}(\nu) \times u(\boldsymbol{B}),\ \boldsymbol{B} + t_{\frac{1+\alpha}{2}}(\nu) \times u(\boldsymbol{B})\right] \tag{7.23}$$

where $t_{\frac{1+\alpha}{2}}(\nu)$ is the Student t value with $\alpha = 0.95$ and $\nu = N - d - 1$ the number of degrees of freedom. The Student t value is found in statistics tables and can be obtained from software tools like Excel®, Matlab® or Octave® (see Equation 8.27 in Table 8.2).

Equation (7.23) can be rewritten differently and used to test the null hypothesis H_0 that any b_j is equal to a reference value V_R (usually zero or one). For example, with Equation (7.3), in order to test the null hypothesis that the coefficient b_0 is equal to $V_R = 0$, i.e. $H_0(b_0 = V_R)$, one checks if Equation (7.24) is verified:

$$\frac{|b_0 - V_R|}{u(b_0)} \leq t_{\frac{1+\alpha}{2}}(\nu) \tag{7.24}$$

If Equation (7.24) is true, the hypothesis H_0 is accepted. Otherwise, the hypothesis H_0 is rejected with a risk $(1-\alpha)$ of being wrong. This test is useful as it allows confirmation that the zero offset of the calibration curve is significant.

Similarly, in order to test the null hypothesis that the coefficient b_1 is equal to $V_R = 1$, i.e. $H_0(b_1 = V_R)$, one checks if Equation (7.25) is verified

$$\frac{|b_1 - V_R|}{u(b_1)} \leq t_{\frac{1+\alpha}{2}}(\nu) \tag{7.25}$$

7.6.4.2.3 Determination of the optimum polynomial order

In order to choose the most appropriate polynomial order d, a Snedecor variance test is carried out (Bertrand-Krajewski *et al.*, 2000; Neuilly & Cetama, 1998). One calculates, with $N = N_x N_y$,

$$F_{1calc} = (N-3)\frac{SSR_{(d=1)} - SSR_{(d=2)}}{SSR_{(d=2)}} \quad \text{to compare orders 1 and 2} \tag{7.26}$$

$$F_{2calc} = (N-4)\frac{SSR_{(d=2)} - SSR_{(d=3)}}{SSR_{(d=3)}} \quad \text{to compare orders 2 and 3.} \tag{7.27}$$

The above values are compared, respectively, with the following Snedecor theoretical values:

$$F_{1theo} = F_{0.95}(1, N-3) \tag{7.28}$$

$$F_{2theo} = F_{0.95}(1, N-4) \tag{7.29}$$

If $F_{1calc} < F_{1theo}$, then the optimal order is $d = 1$. Else, if $F_{2calc} < F_{2theo}$, the optimal order is $d = 2$. Else the optimal order is $d = 3$. In practice, orders d higher than 3 are not used as it would imply that the sensor is strongly non-linear, which is not expected for usual sensors.

7.6.4.2.4 Implementation of OLS

The above OLS regression calculations can be run step by step with various tools allowing matrix algebra, including e.g. Excel®, Matlab®, Octave®, etc. In the next section, a detailed example of application is given.

7.6.4.3 OLS regression example with Matlab®

Note: Examples and codes written for Matlab® (https://fr.mathworks.com) can also be used without any modification with the free software tool Octave® (https://www.gnu.org/software/octave/). The compatibility has been checked by the authors with Matlab® 2017b and Octave 5.1.0.

Let us consider a water level pressure sensor with a measuring range of 0–2 m. After initial adjustment according to specifications given in the manufacturer's user manual, the sensor has been calibrated in the laboratory. It has been set at the bottom of 3.5 m high Perspex column (Figure 7.23). The column is equipped with a 4 m long Class II certified metallic meter and filled with water at different levels x_i

Figure 7.23 Calibration of a water level pressure sensor in a Perspex water column. *Source*: Jean-Luc Bertrand-Krajewski (INSA Lyon).

Table 7.1 Measured values y_{ik} of the pressure sensor calibration experiment for each water level x_i. All values are in mm.

x_i	$u(x_i)$	y_{i1}	y_{i2}	y_{i3}	y_{i4}	y_{i5}	y_{i6}	y_{i7}	y_{i8}	y_{i9}	y_{i10}	y_{i11}	y_{i12}
399	0.25	399	400	400	400	400	400	400	399	399	399	399	399
799	0.25	800	800	800	800	800	800	800	800	800	800	800	800
1201	0.25	1201	1201	1202	1202	1202	1202	1201	1201	1201	1201	1201	1201
1600	0.25	1601	1601	1601	1601	1601	1602	1600	1600	1600	1600	1600	1600
2000	0.25	2002	2002	2002	2002	2002	2002	2001	2001	2001	2001	2001	2001

measured with an uncertainty of 0.5 mm (i.e. standard uncertainty $u(x_i) = 0.25$ mm). This uncertainty is lower than the expected uncertainty of the sensor that is estimated around a few millimetres.

The sensor is located carefully at the bottom of the column to ensure its exact position at the zero reference level. All the equipment is installed a few hours in advance to ensure that all elements are at the same temperature of approx. 19°C. The 4–20 mA values delivered by the sensor are recorded by a laptop to simulate the *in situ* data logger installation used in the monitoring station; the values displayed directly on the sensor transmitter are visually observed during the experiment only to check that the sensor functions well, but they are not used for the calibration.

The sensor is calibrated for $N_x = 5$ water levels ranging from 399 to 2000 mm, corresponding respectively to 20, 40, 60, 80 and 100% of the measurement range. For each water level, $N_y = 12$ measured values are recorded every 30 s. The experimental values are given in Table 7.1. They are also available in the file `piezo1.csv`.

Application of the OLS repression gives the results shown in Table 7.2 (detailed calculations for order $d = 2$ are given in Box 1, and for all calculations in Box 2). The Snedecor test indicates that the optimal

Table 7.2 Results of the OLS regression.

Order $d = 1$		Order $d = 2$		Order $d = 3$	
b_0	0.5088	b_0	0.3841	b_0	−2.1707
b_1	1.0004	b_1	1.0007	b_1	1.0096
$u(b_0)$	0.1775	b_2	-1.1154×10^{-7}	b_2	-8.6821×10^{-6}
$u(b_1)$	1.3383×10^{-4}	$u(b_0)$	0.3652	b_3	2.3815×10^{-9}
$cov(b_0, b_1)$	-2.1485×10^{-5}	$u(b_1)$	6.9629×10^{-4}	$u(b_0)$	0.7536
S_r	0.3444	$u(b_2)$	2.8476×10^{-7}	$u(b_1)$	2.4651×10^{-3}
		$cov(b_0, b_1)$	-2.3932×10^{-4}	$u(b_2)$	2.2886×10^{-6}
		$cov(b_0, b_2)$	9.0675×10^{-8}	$u(b_3)$	6.3195×10^{-10}
		$cov(b_1, b_2)$	-1.9453×10^{-10}	$cov(b_0, b_1)$	-1.8105×10^{-3}
		S_r	0.3495	$cov(b_0, b_2)$	1.6154×10^{-6}
				$cov(b_0, b_3)$	-4.2842×10^{-10}
				$cov(b_1, b_2)$	-5.5795×10^{-9}
				$cov(b_1, b_3)$	1.5065×10^{-12}
				$cov(b_2, b_3)$	-1.4372×10^{-15}
				S_r	0.2838

BOX 1: STEP BY STEP APPLICATION OF OLS REGRESSION WITH MATLAB®

With the data given in Table 7.1, the Matlab® instructions are as follows for the regression with order $d = 2$.

Create the vertical vector x_i with the five standard values:

```
xi=[399 799 1200 1600 2000]'
```

Create the matrix y_{ik} with the $N = 5 \times 12 = 60$ measured values and the same structure as in Table 7.1:

```
yik=[399,400,400,400,400,400,400,399,399,399,399,399;
800,800,800,800,800,800,800,800,800,800,800,800;
1201,1201,1202,1202,1202,1202,1201,1201,1201,1201,1201,1201;
1601,1601,1601,1601,1601,1602,1600,1600,1600,1600,1600,1600;
2002,2002,2002,2002,2002,2002,2001,2001,2001,2001,2001,2001]
```

Set

```
N=60
```

Create the single column vector $\boldsymbol{Y}$ containing all y_{ik} values:

```
Y=(reshape(yik',[ ],1))
```

Create the matrix $\boldsymbol{F}$ with its successive columns:

```
F(:,1)=ones(N,1)
F(:,2)=reshape(repmat(xi,1,size(yik,2))',[ ],1)
F(:,3)=F(:,2).^2;
```

Apply the QR decomposition and calculate $\boldsymbol{M}$:

```
[Q,R] =qr(F,0)
M=R\Q'
```

Calculate the coefficients b_j:

```
B=M*Y
```

One gets (with six digits):

$b_0 = 0.384124$, $b_1 = 1.000663$ and $b_2 = -1.115445 \times 10^{-7}$.

Calculate the residual variance S_{r2}:

```
Sr2=((F*B-Y)'*(F*B-Y))/(N-3)
```

One gets $S_{r2} = 0.3495\ \text{mm}^2$.

Calculate the covariance and the standard uncertainties of the coefficients b_j:

```
covB = Sr2* (M*M')
```

and

```
uB=sqrt(diag(covB))
```

One gets respectively

$cov(b_0, b_1) = -2.3932 \times 10^{-4}$, $cov(b_0, b_2) = 9.0675 \times 10^{-8}$, $cov(b_1, b_2) = -1.9453 \times 10^{-10}$

and

$u(b_0) = 0.3652$, $u(b_1) = 6.9629 \times 10^{-4}$, $u(b_2) = 2.8476 \times 10^{-7}$.

Determine the optimal order d_{opt}.

Calculate SSR with $d = 2$:

```
SSR2 = (F*B-Y)'*(F*B-Y)
```

One gets $SSR_2 = 19.9214$.

Similar calculations for order $d = 1$ (not shown here – see Box 2) gives $SSR_1 = 19.9750$.

Calculate F_{1calc} and F_{1theo}:

```
F1calc=(N-3)*(SSR1-SSR2)/SSR2 = 0.1534
F1theo=finv(0.95,1,N-3) = 4.0098
```

As $F_{1calc} < F_{1theo}$, $d_{opt} = 1$.

BOX 1: (Continued)

Check the null hypotheses $b_0 = 0$ and $b_1 = 1$ for the straight line ($d_{opt} = 1$).

```
b0=0.5088
b1=1.0004
ub0=0.1775
ub1=1.3383e-4
N=60
dopt=1
alpha=0.95
```

Calculate t:

```
tinv((1+alpha)/2,N-dopt-1)
```

One gets $t = 2.0017$.

Then

`abs(b0-0)/ub0` $= 2.8665 > t$

`abs(b1-1)/ub1` $= 2.9555 > t$

Both null hypotheses are rejected.

BOX 2: APPLICATION OF THE OLS REGRESSION WITH THE MATLAB® CODE `OLS123`

(Matlab® codes and csv files available for download at https://doi.org/10.2166/9781789060102).

All calculations presented in Box 1 are automated in the Matlab® code `OLS123`, which (i) runs the OLS regression for all orders $d = 1$ to 3, (ii) determines the optimal order d_{opt}, and (iii) provides four graphs showing the three calibration functions and the variance of the repeated experimental measurements y_{ik} for each standard value x_i.

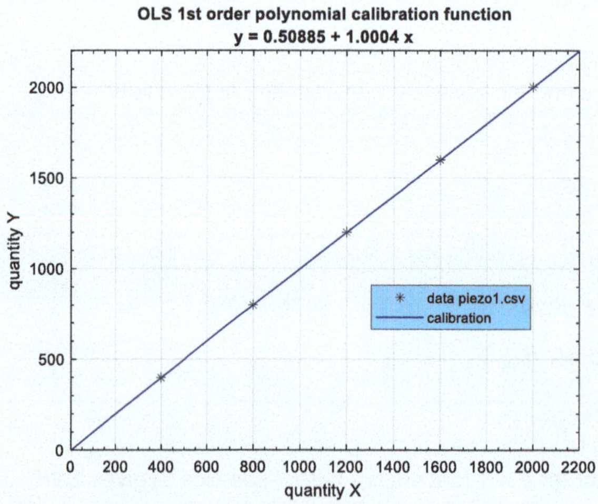

Figure 7.24 Graph of the OLS 1st order calibration function. *Source*: Jean-Luc Bertrand-Krajewski (INSA Lyon).

For the calibration data set in the csv file `piezo1.csv`, type

BOX 2: (Continued)

```
B=OLS123('piezo1')
```

The results shown in Table 7.2 are displayed in the Matlab® command window and saved as the Matlab® variable named `OLS123.mat`, with d_{opt} in the first row.

Four graphs are displayed (Figures 7.24 to 7.27):

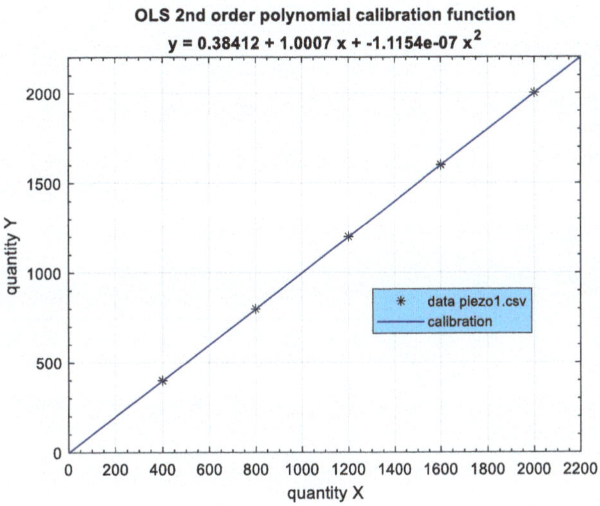

Figure 7.25 Graph of the OLS 2nd order calibration function. *Source*: Jean-Luc Bertrand-Krajewski (INSA Lyon).

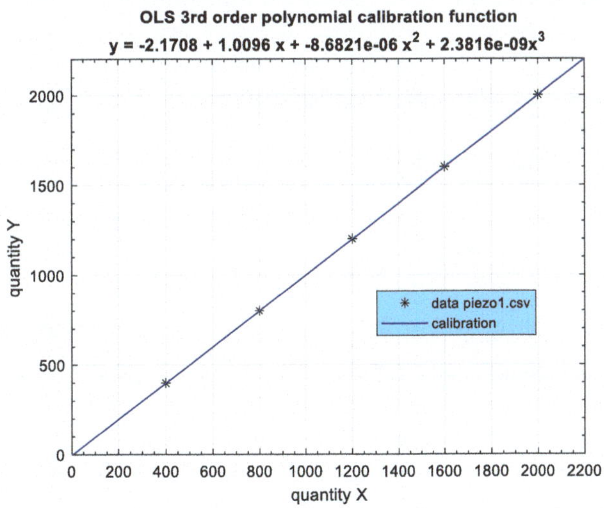

Figure 7.26 Graph of the OLS 3rd order calibration function. *Source*: Jean-Luc Bertrand-Krajewski (INSA Lyon).

BOX 2: (Continued)

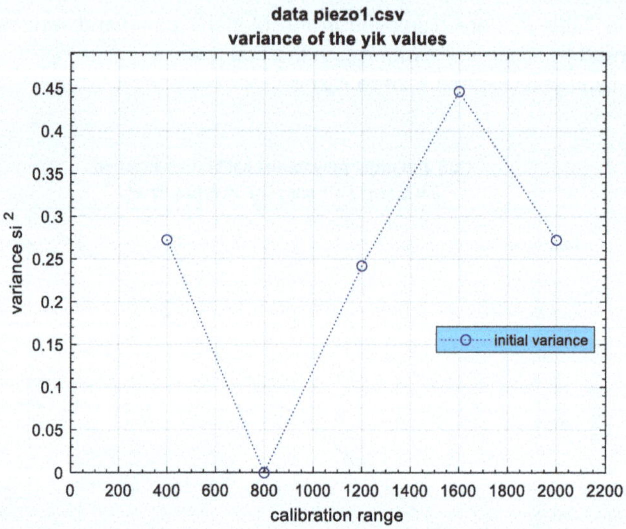

Figure 7.27 Graph of the variance s_i2 of the y_{ik} experimental measurements. *Source*: Jean-Luc Bertrand-Krajewski (INSA Lyon).

order is $d_{opt} = 1$, i.e. the straight line function. Consequently, for this sensor, the calibration function to be used is:

$$y = b_0 + b_1 x = 0.5088 + 1.0004\,x \tag{7.30}$$

One can check if $b_0 = 0.5088$ and $b_1 = 1.004$ are significantly different from 0 and 1, respectively. Applying Equations (7.24) and (7.25), one gets, respectively

$$\frac{|b_0 - 0|}{u(b_0)} = \frac{|0.5088|}{0.1775} = 2.8664$$

$$\frac{|b_1 - 1|}{u(b_1)} = \frac{|1.0004 - 1|}{1.3383 \times 10^{-4}} = 2.9555$$

Both values are larger than $t_{\frac{1+\alpha}{2}}(\nu) = 2.0017$. Consequently, one rejects the null hypotheses and concludes that b_0 and b_1 are significantly different from 0 and 1 respectively, with a risk of 5% of being wrong in this conclusion.

Figure 7.27 is of particular importance, as it allows estimation of the uncertainty in sensor output values. In this example, the variance s_i^2 ranges between approximately 0.2 and 0.5 mm^2, with the exception of the second standard value $x_2 = 799$ mm where the variance is zero as the 12 repeated measurements y_{2k} are all equal to 800 mm (see Table 7.1). In this case, one can reasonably assume that the value $S = 0.5$ mm^2 is an upper bound of s_i^2 over the entire measuring range, which corresponds to a maximum standard uncertainty $u(y)$ in sensor output values given by Equation (7.31):

$$u(y) = \sqrt{S} \tag{7.31}$$

i.e. $u(y) = 0.7071$ mm.

A sensor output value y is thus assumed to lie, with a 95% probability, in the interval

$$[y - 1.96 \times u(y),\ y + 1.96 \times u(y)] \tag{7.32}$$

or to be equal to $y \pm 1.96 \times u(y)$, i.e. $y \pm 1.3859$ mm in this example, to be rounded to 1.4 mm in practice.

In addition, the standard deviation $u(y) = \sqrt{S}$ can be used later on as the maximum acceptable error I_{mt} in the verification process (see Section 7.6.3 and Figure 7.22).

According to Figure 7.22, using a constant value S over the entire measuring range as a global upper bound of s_i^2 is acceptable. For some sensors, s_i^2 may significantly increase, decrease, or vary over the measuring range. In such cases, applying a constant value S is not appropriate and it can be replaced by upper bound sub-ranges S_i which vary with x_i. But, preferably, the OLS method should be replaced by e.g. the Williamson regression (Section 7.6.4.4).

7.6.4.3.1 Application of the calibration function

Let us assume the data logger of a monitoring station gives a measured water level $h_m = 482$ mm. According to the variance analysis (Figure 7.27), the standard uncertainty $u(h_m)$ is assumed to be equal to $S = 0.7071$ mm. By applying the reciprocal function of Equation (7.30), the estimate of the true water level $\hat{h}$ is given by:

$$\hat{h} = \frac{h_m - b_0}{b_1} = \frac{482 - 0.5088}{1.0004} \tag{7.33}$$

One gets $\hat{h} = 481.3$ mm.

The standard uncertainty $u(\hat{h})$ can be estimated by the law of propagation of uncertainties (also named the Type B method): see Chapter 8 on Uncertainty Assessment, Section 8.2.3 and also Box 3. By accounting for the covariance between the coefficients b_0 and b_1, $u(\hat{h})$ is calculated by:

$$u(\hat{h})^2 = u(h_m)^2\left(\frac{\partial \hat{h}}{\partial h_m}\right)^2 + u(b_0)^2\left(\frac{\partial \hat{h}}{\partial b_0}\right)^2 + u(b_1)^2\left(\frac{\partial \hat{h}}{\partial b_1}\right)^2 + 2cov(b_0, b_1)\left(\frac{\partial \hat{h}}{\partial b_0}\right)\left(\frac{\partial \hat{h}}{\partial b_1}\right) \tag{7.34}$$

Replacing the partial derivatives by their literal expressions, Equation (7.34) becomes

$$u(\hat{h})^2 = u(h_m)^2\left(\frac{1}{b_1}\right)^2 + u(b_0)^2\left(-\frac{1}{b_1}\right)^2 + u(b_1)^2\left(-\frac{h_m - b_0}{b_1^2}\right)^2 + 2cov(b_0, b_1)\left(-\frac{1}{b_1}\right)\left(-\frac{h_m - b_0}{b_1^2}\right) \tag{7.35}$$

One gets $u(\hat{h}) = 0.7173$ mm. The major contribution to this value is the uncertainty $u(h_m)$.

One concludes that, from the measured value $h_m = 482$ mm, the calibration function gives the best estimate $\hat{h} = 481.3$ mm and that, accounting for the uncertainties in both the measured value h_m and the calibration function coefficients b_j, the 95% coverage interval for $\hat{h}$ is equal to [479.9, 482.7] mm.

7.6.4.4 Williamson least squares regression (WLS)

7.6.4.4.1 Introduction

The OLS regression (Section 7.6.4.2) is applicable under some conditions, one of them being that uncertainties in x_i values (i.e. in standards or certified materials used for calibration) are either negligible

BOX 3: CALCULATION OF $\hat{h}$ AND $u(\hat{h})$ WITH THE MATLAB® CODE uTypeB

Read first Chapter 8 Section 8.2.3 for details on the Type B method and the `uTypeB` code. (Matlab® codes and csv files available for download at https://doi.org/10.2166/9781789060102).

Note that all calculations in this Box have been carried out with non-rounded numerical values even if only four digits are shown here.

Type

```
hm=782
uhm=0.7071
b0=0.5088
ub0=0.1775
b1=1.0004
ub1=1.3383e-4
covb=-2.1485e-5
```

Then type

```
Z=[hm uhm]
A=[b0 ub0 b1 ub1]
chain='(Z(:,1)-A(:,1))./A(:,2)'
alpha=0.95
MatCor=[1 0 0; 0 1 covb/ub0/ub1; 0 covb/ub0/ub1 1]
```

Lastly type

```
hest=uTypeB(Z,A,chain,alpha,MatCor)
```

One gets $\hat{h} = 481.3$ mm and $u(\hat{h}) = 0.7173$ mm, which gives [479.9, 482.7] mm as the 95% coverage interval for $\hat{h}$.

or, at least, very low compared to uncertainties in y_{ik} values measured by the sensor which is calibrated. This is usually the case when high quality standards are available. Another condition, more rarely checked, is that the variance of the y_{ik} over the measuring range of the sensor (see Figure 7.27) is constant or approximately constant. If these required conditions are not respected, the OLS regression should be replaced by more elaborate methods, for example those derived from the WLS regression method (Williamson, 1968).

The following paragraphs describe how to calculate first to third order polynomial Williamson regressions for two quantities x and y, both being affected with significant or equivalent uncertainties, in the case of sensor calibration data sets where the number N_x of x_i values is lower than the total number $N = N_x N_y$ of points.

7.6.4.4.2 Standard uncertainties $u(y_i)$ and weights W

The calibration data set is composed of $k = 1{:}N_y$ repeated measurements y_{ik} for $i = 1{:}N_x$ standard values x_i. For convenience of calculations, N_y should be identical for all the N_x standard values. Each value x_i has a standard uncertainty $u(x_i)$ certified by the standard manufacturer or estimated by the user. Standard uncertainties $u(y_i)$ are estimated from standard deviations s_i (see Section 8.2.2 for the Type A method for uncertainty assessment):

$$u(y_i) = s_i = \left(\frac{\sum_{k=1}^{N_y}(y_{ik} - \overline{y_i})^2}{N_y - 1}\right)^{\frac{1}{2}} \quad \forall\, i = 1{:}N_x \tag{7.36}$$

where

$$\overline{y_i} = \frac{\sum_{k=1}^{N_y} y_{ik}}{N_y} \tag{7.37}$$

It may happen that all y_{ik} values are identical for a given standard x_i (see e.g. Table 7.1 for $x_i =$ 799 mm).

$$y_{ik} = \overline{y_i} \qquad \forall\, k = 1{:}N_y \tag{7.38}$$

In this case, $s_i = 0$, which is inappropriate (division by zero) for further calculations where the inverse of $u(y_i)^2$ is used (see Equation (7.45)). In order to solve this potential problem, the principle consists of replacing zero s_i values by another low value from the same data set, equal to the minimum of non-zero s_i values. First create an intermediate quantity $s_{i\ not\ nul}$ equal to the initial quantity s_i. Then detect all s_i values lower than a critical threshold value close to zero $s_{i\ lim}$, for example 10^{-6} (the value should be adapted for each case). Replace the corresponding $s_{i\ not\ nul}$ values by an arbitrarily extremely high value like, e.g. 10^{+10}. Calculate the minimum value $min(s_{i\ not\ nul})$. Lastly, replace the s_i values such that $s_i < s_{i\ lim}$ by $min(s_{i\ not\ nul})$:

$$s_i = \left| \begin{array}{ll} \min(s_{i\ not\ nul}) & \text{if } s_i < s_{i\ lim} \\ s_i & \text{if } s_i \geq s_{i\ lim} \end{array} \right. \qquad \forall\, i = 1{:}N_x \tag{7.39}$$

Lastly, $\forall\, i = 1{:}N_x$, one defines the weights to be used later in the WLS regression:

$$W(x_i) = \frac{1}{u(x_i)^2} \tag{7.40}$$

$$u(y_i) = s_i \tag{7.41}$$

$$W(y_i) = \frac{1}{u(y_i)^2} \tag{7.42}$$

7.6.4.4.3 Calculation of the coefficients b_j

The polynomial functions are written

$$y = \sum_{j=0}^{d} b_j x^j \tag{7.43}$$

where $d = 1{:}3$.

The OLS method (see Section 7.4.6.2) gives the values of the coefficients b_j by minimizing the sum S_{OLS}:

$$S_{OLS} = \sum_{i=1}^{N_x}\sum_{k=1}^{N_y}(Y_{ik} - y_{ik})^2 = \sum_{i=1}^{N_x}\sum_{k=1}^{N_y}\left(\sum_{j=0}^{d} b_j x_i^j - y_{ik}\right)^2 \tag{7.44}$$

where all uncertainties are ignored and where Y_{ik} are the predicted values and x_i and y_{ik} are the measured values.

The WLS regression consists of determining the values of both the coefficients b_j ($j = 0{:}d$) and the predicted values X_i ($i = 1{:}N_x$) which minimize the sum S_d:

$$
\begin{aligned}
S_d &= \sum_{i=1}^{N_x}\sum_{k=1}^{N_y}\left(\frac{1}{u(x_i)^2}(X_i - x_i)^2 + \frac{1}{u(y_i)^2}(Y_{ik} - y_{ik})^2\right) \\
&= \sum_{i=1}^{N_x}\sum_{k=1}^{N_y}\left(W(x_i)(X_i - x_i)^2 + W(y_i)\left(\sum_{j=0}^{d} b_j X_i^j - y_{ik}\right)^2\right) \\
&= N_y \sum_{i=1}^{N_x} W(x_i)(X_i - x_i)^2 + \sum_{i=1}^{N_x}\sum_{k=1}^{N_y}\left(W(y_i)\left(\sum_{j=0}^{d} b_j X_i^j - y_{ik}\right)^2\right)
\end{aligned}
\tag{7.45}
$$

where (x_i, y_{ik}) are the measured values, (X_i, Y_{ik}) are the predicted values, $u(x_i)$ and $u(y_i)$ are the standard uncertainties in x_i and y_i, and N_x and N_y are, respectively, the number of standards and the number of repeated measurements for each standard. In the above equation, it is assumed that the predicted value X_i should be unique for a given standard.

Minimizing S_d corresponds to solving the $(d + 1 + N_x)$ equations

$$
\frac{\partial S_d}{\partial b_j} = 0 \qquad j = 0{:}d \tag{7.46}
$$

$$
\frac{\partial S_d}{\partial X_i} = 0 \qquad i = 1{:}Nx \tag{7.47}
$$

In practice, S_d is minimized numerically. In the Matlab® code `WLS123cal` (see Box 4), this is achieved by means of two successive steps:

- Step 1: determination of the initial coefficients b_{j0} by the OLS regression.
- Step 2: determination of the final coefficients b_j and of the predicted values X_i by non-linear least squares minimization.

Step 1: OLS regression for initial values of b_j

In the first step, all uncertainties are ignored and S_{OLS} (Equation 7.44) is minimized to find a vector of initial values b_{j0d} by using the `regress` Matlab® function:

$$
b_{j0d} = \begin{vmatrix} b_{00} \\ b_{10} \\ \vdots \\ b_{d0} \end{vmatrix} \tag{7.48}
$$

Step 2: WLS regression for final values of b_j and X_i

In the second step, the final optimum values of b_j and X_i are calculated by minimizing S_d: the third line of Equation (7.45) is written in a Matlab® function named `Willorderd`:

$$
\begin{aligned}
Willorder_d(z) &= S_{1d} + S_{2d} \\
&= N_y \sum_{i=1}^{N_x} W(x_i)(z(d+1+i) - x_i)^2 \\
&+ \sum_{i=1}^{N_x} \sum_{k=1}^{N_y} \left(W(y_i) \left(\sum_{j=0}^{d} z(j) z(d+1+i)^j - y_{ik} \right)^2 \right)
\end{aligned} \tag{7.49}
$$

where $z(p)$ corresponds to the p-th element of a vertical vector z with $(d+1+N_x)$ elements.

The minimization is obtained by the unconstrained Nelder-Mead simplex algorithm. Initial values are given in the concatenated vector Z_{0d} with $(d+1+N_x)$ elements:

$$
Z_{0d} = \begin{vmatrix} b_{00} \\ b_{10} \\ \vdots \\ b_{d0} \\ X_{10} \\ X_{20} \\ \vdots \\ X_{N_x 0} \end{vmatrix} \quad \text{where} \quad \begin{vmatrix} X_{10} \\ X_{20} \\ \vdots \\ X_{N_x 0} \end{vmatrix} = \begin{vmatrix} x_1 \\ x_2 \\ \vdots \\ x_{N_x} \end{vmatrix} \tag{7.50}
$$

The minimization output vector is Z_{1d}:

$$
Z_{1d} = \begin{vmatrix} b_0 \\ b_1 \\ \vdots \\ b_d \\ X_1 \\ X_2 \\ \vdots \\ X_{N_x} \end{vmatrix} \tag{7.51}
$$

One continues with a second iteration to get Z_{2d}, etc. Iterations are terminated when relative variations in all b_j and X_i values are below a threshold value *PreRel* set by the user. Typically, one can suggest $PreRel = 10^{-2}$, i.e. iterations are terminated when all b_j and X_i values show relative variations of less than 1% between two successive iterations. The vector $control_d$ is used for the termination test between iterations m and $(m+1)$:

$$control_d = \begin{vmatrix} (b_{0(m+1)} - b_{0(m)})/b_{0(m)} \\ (b_{1(m+1)} - b_{1(m)})/b_{1(m)} \\ \vdots \\ (b_{d(m+1)} - b_{d(m)})/b_{d(m)} \\ (X_{1(m+1)} - X_{1(m)})/X_{1(m)} \\ (X_{2(m+1)} - X_{2(m)})/X_{2(m)} \\ \vdots \\ (X_{N_x(m+1)} - X_{N_x(m)})/X_{N_x(m)} \end{vmatrix} \tag{7.52}$$

If, for example, $|(X_{2(m+1)} - X_{2(m)})/X_{2m}| \leq PreRel$, then $control_d(X_2) = 1$, else $control_d(X_2) = 0$. Iterations continue while the sum of all $control_d$ Boolean values remains less than $(d + N_x)$.

7.6.4.4.4 Numerical calculation of variances and standard uncertainties

Variances and standard uncertainties are estimated by means of Monte Carlo simulations.

7.6.4.4.5 Generation of samples

N_{MC} is the number of Monte Carlo simulations to be run. The first step consists of generating N_{MC} data sets of points (x_i, y_{ik}) for Monte Carlo simulations. For each standard value x_i, a sample of N_{MC} values x_{ir} with $r = 1{:}N_{MC}$ is generated in such a way that the x_{ir} values are normally distributed, with the mean value $\overline{x_{ir}} = x_i$ and the standard deviation $s(x_{ir}) = u(x_i)$. Samples of y_{irk} values are generated in the same way.

7.6.4.4.6 Calculation of b_j and X_i for the N_{MC} data sets

For the r-th data set of points (x_{ir}, y_{irk}), the coefficients b_j and the predicted values X_i are calculated as described above by Equations (7.45) and (7.49). The process is repeated for all $r = 1{:}N_{MC}$. The matrix M_{bjd} contains all transposed sub-vectors with the b_j values:

$$Mb_{jd} = \begin{vmatrix} b_{01} & b_{11} & \cdots & b_{d1} \\ b_{02} & b_{12} & \cdots & b_{d2} \\ \vdots & \vdots & \vdots & \vdots \\ b_{0r} & b_{1r} & \cdots & b_{dr} \\ \vdots & \vdots & \cdots & \vdots \\ b_{0N_{MC}} & b_{1N_{MC}} & \cdots & b_{dN_{MC}} \end{vmatrix} \tag{7.53}$$

Then mean values aw_d of b_{jd} are calculated:

$$aw_d = \frac{1}{N_{MC}} \left| \sum_{r=1}^{N_{MC}} b_{0r} \sum_{r=1}^{N_{MC}} b_{1r} \cdots \sum_{r=1}^{N_{MC}} b_{dr} \right| \tag{7.54}$$

Finally, the X_i values which minimize S_d (Equation (7.45)) are estimated again by setting the vector

$$b_{jd} = aw_d^T = \begin{vmatrix} \overline{b_0} \\ \overline{b_1} \\ \vdots \\ \overline{b_d} \end{vmatrix} \tag{7.55}$$

and by defining a new initial vector $Z_{0\,bisd}$ with N_x elements

$$Z_{0\,bisd} = \begin{vmatrix} X_{10} \\ X_{20} \\ \vdots \\ X_{N_x0} \end{vmatrix} = \begin{vmatrix} x_1 \\ x_2 \\ \vdots \\ x_{N_x} \end{vmatrix} \tag{7.56}$$

for the minimization of the new quantity $Willorder_{d\,final}$

$$\begin{aligned} Willorder_{d\,final}\,(z) &= S_{1d} + S_{2d} \\ &= N_y \sum_{i=1}^{N_x} W(x_i)(z_{bis}(i) - x_i)^2 + \sum_{i=1}^{N_x}\sum_{k=1}^{N_y}\left(W(y_i)\left(\sum_{j=0}^{d} b_{jd} z_{bis}(i)^j - y_{ik}\right)^2\right) \end{aligned} \tag{7.57}$$

where $z_{bis}(p)$ corresponds to the p-th element of a vertical vector z_{bis} with N_x elements.

The minimization procedure is applied once again to determine the final X_i values. One gets the final output vector $Z_{1\,bisd}$ with N_x elements:

$$Z_{1\,bisd} = \begin{vmatrix} X_1 \\ X_2 \\ \vdots \\ X_{N_x} \end{vmatrix} \tag{7.58}$$

Additionally, the sum S_d (Equation (7.45)) is calculated for the final b_j and X_i values: this final S_d value is named f_{vald}.

7.6.4.4.7 Calculation of variances and standard uncertainties

The variance matrix is calculated and multiplied by the correction factor f_{vald} /(N-d-1):

$$\begin{vmatrix} var(b_0) & cov(b_0, b_1) & \cdots & cov(b_0, b_d) \\ cov(b_1, b0) & var(b_1) & \cdots & cov(b_1, b_d) \\ \vdots & \vdots & \cdots & \vdots \\ cov(b_d, b_0) & cov(b_d, b_1) & \cdots & var(b_d) \end{vmatrix} = \frac{f_{vald}}{N-d-1} cov(Mb_{jd}) \tag{7.59}$$

Finally, standard uncertainties $u(b_j)$ are calculated:

$$u(b_j) = \sqrt{var(b_j)} \qquad j = 0{:}d \tag{7.60}$$

7.6.4.5 WLS regression example with Matlab®

BOX 4: APPLICATION OF THE WLS REGRESSION WITH THE MATLAB® CODE `WLS123cal`

(Matlab® codes and csv files available for download at https://doi.org/10.2166/9781789060102).

All calculations presented in Section 7.6.4.4 are automated in the Matlab® code `WLS123cal`, which (i) runs the WLS regression for all orders $d = 1$ to 3, (ii) determines the optimal order d_{opt}, and (iii) provides four graphs showing the three calibration functions and the variance of the repeated experimental measurements y_{ik} for each standard value x_i.

Table 7.3 Measured values y_{ik} of the turbidity sensor calibration experiment for each standard solution x_i. All values are in NTU.

x_i	$u(x_i)$	y_{i1}	y_{i2}	y_{i3}	y_{i4}	y_{i5}	y_{i6}	y_{i7}
0	0.1	7.81	7.81	7.81	8.78	7.81	7.81	7.81
50	0.3	55.66	55.66	55.66	55.66	55.66	55.66	55.66
100	0.5	105.47	104.49	104.49	104.49	104.49	105.47	105.47
300	1.5	304.69	304.69	304.69	304.69	304.69	304.69	304.69
500	2.5	502.93	502.93	502.93	502.93	502.93	502.93	502.93
1000	5.0	1005.87	1007.82	1005.87	1005.87	1007.82	1007.82	1005.87
2000	10.0	2158.23	2157.25	2156.28	2157.27	2157.25	2157.25	2158.23
3000	15.0	3938.53	3935.6	3935.6	3937.56	3935.6	3938.53	3939.51

y_{i8}	y_{i9}	y_{i10}	y_{i11}	y_{i12}	y_{i13}	y_{i14}	y_{i15}	y_{i16}
7.81	7.81	7.81	7.81	7.81	7.81	7.81	7.81	7.81
55.66	55.66	55.66	55.66	55.66	55.66	55.66	55.66	55.66
105.47	104.49	105.47	105.47	105.47	105.47	104.49	105.47	105.47
304.69	304.69	304.69	304.69	304.69	304.69	304.69	304.69	304.69
502.93	502.93	501.96	501.96	501.96	502.93	502.93	502.93	502.93
1005.87	1007.82	1007.82	1005.87	1005.87	1004.89	1005.89	1004.89	1005.87
2150.42	2147.49	2150.42	2151.39	2151.39	2151.39	2151.42	2151.39	2152.39
3939.51	3938.53	3937.56	3937.56	3935.6	3938.53	3939.53	3938.53	3937.56

y_{i17}	y_{i18}	y_{i19}	y_{i20}	y_{i21}	y_{i22}	y_{i23}	y_{i24}	y_{i25}
7.81	7.81	7.81	7.81	7.81	7.81	8.78	7.81	8.78
55.66	55.66	55.66	55.66	55.66	55.66	53.71	53.71	55.66
105.47	105.47	104.49	104.49	105.47	105.47	105.47	104.49	105.47
304.69	304.69	304.69	304.69	304.69	304.69	304.69	304.69	304.69
502.93	502.93	502.93	502.93	502.93	502.93	502.93	502.93	502.93
1005.87	1007.82	1004.89	1005.87	1007.82	1005.87	1005.87	1004.89	1005.87
2150.42	2148.47	2151.39	2150.42	2150.42	2151.39	2151.42	2148.47	2148.47
3935.6	3935.6	3934.63	3933.65	3933.65	3933.65	3932.67	3930.72	3929.74

BOX 4: (Continued)

This example deals with the calibration of a turbidity sensor in the range 0–3000 NTU (nephelometric turbidity unit) used for continuous monitoring in a sewer system to estimate TSS (total suspended solids) concentrations (Métadier & Bertrand-Krajewski, 2012). Eight class 1 NIST certified standard solutions have been used to calibrate the sensor, with $N_y = 25$ repeated measurements for each standard solution. Experimental data are given in Table 7.3. They are also available in the file `turbi251.csv`.

Table 7.4 Results of the WLS regression for the turbi251 data set.

d_{opt}	3
d	1
b_0	3.9508296
b_1	1.08102263
$u(b_0)$	2.1714809
$u(b_1)$	0.03924718
$cov(b_0,b_1)$	−0.03078075
f_{val1}	61068.4646
d	2
b_0	8.60338819
b_1	0.93360608
b_2	0.00010355
$u(b_0)$	0.59234984
$u(b_1)$	0.0166773
$u(b_2)$	1.32E-05
$cov(b_0,b_1)$	−0.0048215
$cov(b_0,b_2)$	2.04E-06
$cov(b_1,b_2)$	−1.44E-07
f_{val2}	7425.66587
d	3
b_0	7.30150367
b_1	0.99625498
b_2	−5.25E-05
b_3	5.06E-08
$u(b_0)$	0.22578187
$u(b_1)$	0.00871392
$u(b_2)$	1.82E-05
$u(b_3)$	6.22E-09
$cov(b_0,b_1)$	−0.00114698
$cov(b_0,b_2)$	1.86E-06
$cov(b_0,b_3)$	−5.29E-10
$cov(b_1,b_2)$	−1.32E-07
$cov(b_1,b_3)$	3.78E-11
$cov(b_2,b_3)$	−1.07E-13
f_{val3}	504.912917
PreRel	0.01
NMC	3000

BOX 4: (Continued)

For the calibration data set in the csv file `turbi251.csv`, type

`NMC = 3000` (3000 runs of Monte Carlo simulation, the minimum value is 500)

`PreRel = 1e-2` (relative accuracy of final b_j values $PreRel = 10^{-2}$)

`silim = 1e-6` (threshold minimum variance in case $s_i = 0$: $s_{i\,lim} = 10^{-6}$)

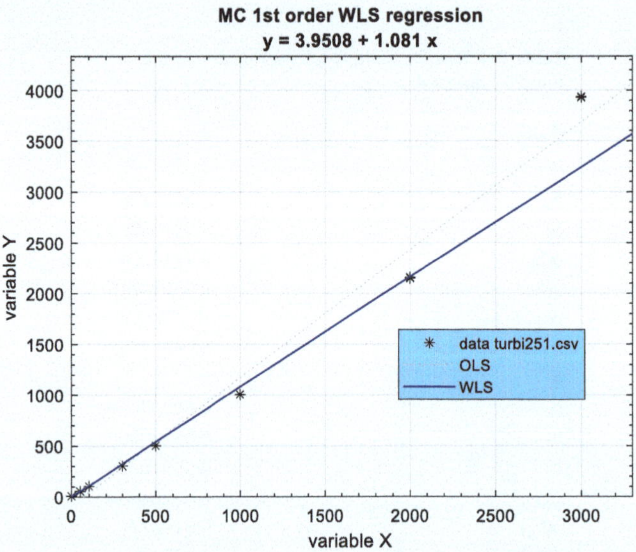

Figure 7.28 Graph of the WLS 1st order calibration function. *Source*: Jean-Luc Bertrand-Krajewski (INSA Lyon).

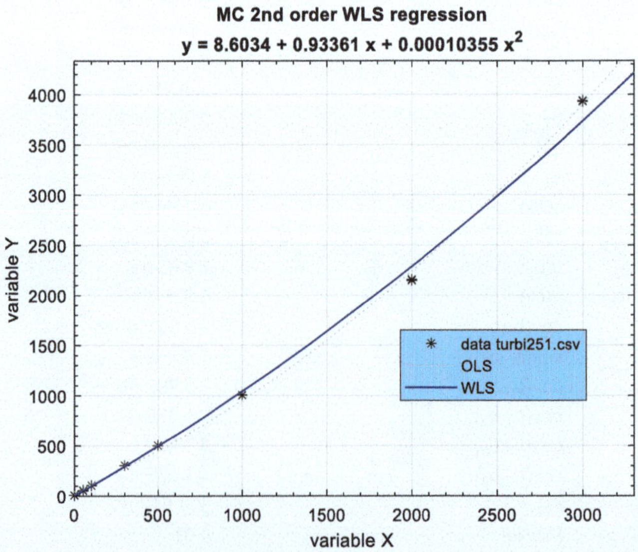

Figure 7.29 Graph of the WLS 2nd order calibration function. *Source*: Jean-Luc Bertrand-Krajewski (INSA Lyon).

BOX 4: (Continued)

Then type

```
B=WLS123cal('turbi251',NMC,PreRel,silim)
```

The results shown in Table 7.4 are displayed in the Matlab® command window and saved as the Matlab® variable named `WLS123cal.mat`, with d_{opt} in the first row. In this example, the optimal order is $d_{opt} = 3$ as the sensor response is strongly non-linear (Figure 7.30).

Four graphs are displayed (Figures 7.28 to 7.31), allowing comparison of the OLS and WLS regressions. In this case (Figure 7.31), the variance s_i increases significantly over the measuring range.

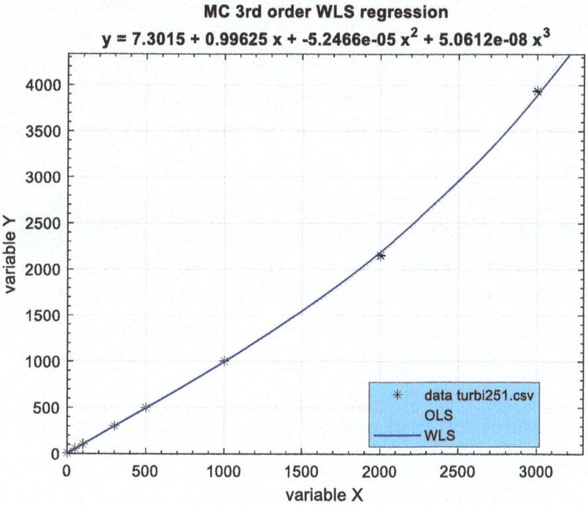

Figure 7.30 Graph of the WLS 3rd order calibration function. *Source*: Jean-Luc Bertrand-Krajewski (INSA Lyon).

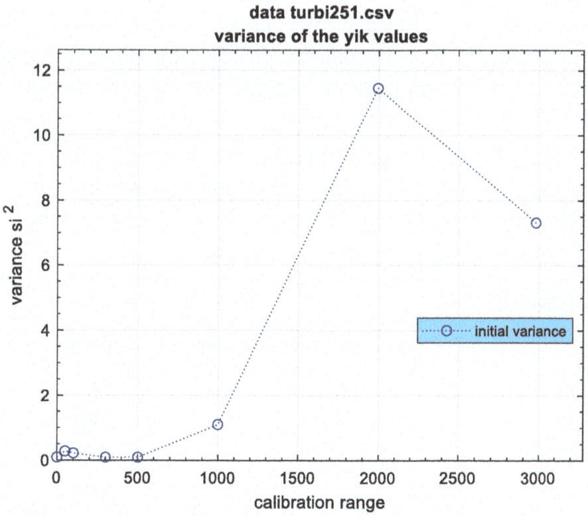

Figure 7.31 Graph of the variance $s_i 2$ of the y_{ik} experimental measurements. *Source*: Jean-Luc Bertrand-Krajewski (INSA Lyon).

7.6.4.6 Non-linear regression of power function for tipping bucket rain gauge calibration

7.6.4.6.1 Calibration protocol and results

Tipping bucket rain gauges (see Section 2.2) are usually affected by overfilling errors when rainfall intensity increases, which leads to a systematic underestimation of the true intensity, especially beyond 50–60 mm/h (Bertrand-Krajewski *et al.*, 2000). To estimate and correct this underestimation, a dynamic calibration of each rain gauge is necessary.

The dynamic (or intensity) calibration of a rain gauge requires some specific equipment. The protocol is based on counting the number of tips of the rain gauge bucket when it is supplied by a perfectly known constant intensity, for example the supply at constant level from a Mariotte bottle or the discharge from a previously calibrated peristaltic pump connected to a stabilized power supply to guarantee discharge stability.

A number N_r of tips is planned, for example 20 or 30 tips. The first tip indicates the start of a watch which will be stopped at the time of the N_r-th tip. The time elapsed to reach the N_r-th tips allows calculation of the intensity I_m measured by the rain gauge. I_m is then compared to the true intensity I_t determined by weighing (with an enlarged uncertainty of a tenth of a mg) or by rigorous volumetric measurement (with certified laboratory flasks or graduated cylinders) at controlled temperature. In the usual approach, the operation is repeated several N_y times for each pre-determined intensity and the average values of I_t and I_m are used to establish the calibration function by regression. But it is also possible to keep all individual measured pairs (I_{tk}, I_{mk}) in the regression.

An example of tipping bucket rain gauge calibration data is given in Table 7.5. They are also available in the file `ciaponi1.csv`. Figure 7.32 displays the data and shows clearly how the rain gauge underestimates intensities beyond 50–70 mm/h.

7.6.4.6.2 Example of determination and use of the calibration function

The usual choice for tipping bucket rain gauge calibration function is the power function (Equation 7.5) which can be rewritten with the notation of the above example:

$$I_m = b_0 I_t^{b_1} \tag{7.61}$$

As Equation (7.61) is not linear for b_1, linear regressions with OLS and WLS methods are not applicable and non-linear regression is necessary. It is beyond the scope of this section to introduce in detail non-linear regression. The reader can find further information in textbooks, e.g. Seber and Wild (2003).

Table 7.5 Experimental data set of a tipping bucket rain gauge calibration. The rain gauge interception area is 1000 cm^2 and each tip corresponds to 0.2 mm of rain (source of data: Ciaponi *et al.*, 1993).

I_t (mm/h)	I_m (mm/h)
43.8	43.8
110.4	108.0
177.6	165.7
255.0	226.8
322.3	280.8
410.0	346.0

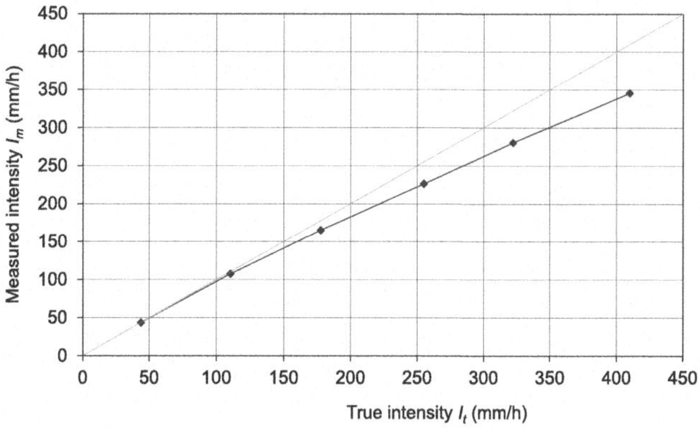

Figure 7.32 Example of dynamic calibration of a tipping bucket rain gauge. *Source*: Jean-Luc Bertrand-Krajewski (INSA Lyon).

Results obtained with the Matlab® function `nlinfit` are given hereafter and detailed in Box 5. The non-linear regression gives $b_0 = 1.5988$, $b_1 = 0.8943$, $u(b_0) = 0.0870$, $u(b_1) = 0.0095$ and $cov(b_0, b_1) = -0.0008$.

Let us consider that the rain gauge indicates a measured intensity $I_m = 120$ mm/h, with a standard uncertainty $u(I_m) = 5$ mm/h. The best estimate of the true intensity $\widehat{I_t}$ is given by

$$\widehat{I_t} = \left(\frac{I_m}{b_0}\right)^{\frac{1}{b_1}} \tag{7.62}$$

One gets $\widehat{I_t} = 125.0$ mm/h and $u(\widehat{I_t}) = 5.9$ mm/h.

BOX 5: CALCULATION OF $\widehat{I_t}$ AND $u(\widehat{I_t})$ WITH THE MATLAB® FUNCTION `nlinfit`

Read first Chapter 8 Section 8.2.3 for details on the Type B method and the `uTypeB` code.

(Matlab® codes and csv files available for download at https://doi.org/10.2166/9781789060102).

Note that all calculations in this Box have been carried out with non-rounded numerical values even if only four digits are shown here.

Type

```
D=dlmread([ 'ciaponi1','.csv'] ,';',1,0)
It=D(:,1)
Im=D(:,2)
Bjinit=[1;1]
```

Then type

```
[Bj,R,J,covB] =nlinfit(It,Im,@(bj,x)(bj(1).*x.^bj(2)),Bjinit)
```

One gets

$$Bj = \begin{bmatrix} b_0 \\ b_1 \end{bmatrix} = \begin{bmatrix} 1.5988 \\ 0.8943 \end{bmatrix}$$

BOX 5: (Continued)

$$covB = \begin{bmatrix} 0.0076 & -0.0008 \\ -0.0008 & 0.0001 \end{bmatrix}$$

Type

```
uBj=sqrt(diag(covB))
```

One gets

$$uBj = \begin{bmatrix} u(b_0) \\ u(b_1) \end{bmatrix} = \begin{bmatrix} 0.0870 \\ 0.0095 \end{bmatrix} \quad \text{and} \quad cov(b_0, b_1) = covB(1,2) = -0.0008.$$

Calculate $\widehat{I_t}$ and $u(\widehat{I_t})$ for $I = 120$ mm/h and $u(I) = 5$ mm/h with the Matlab® code `uTypeB`

Type

```
I=120
uI=5
```

Then type

```
Z=[I, uI]
A=[Bj(1) uBj(1) Bj(2) uBj(2)]
chain='(Z(:,1)./A(:,1)).^(1/A(:,2))'
alpha=0.95
MatCor=[ 1 0 0; 0 1 covB(1,2)/uBj(1)/uBj(2); 0 covB(1,2)/uBj(1)/uBj(2) 1]
```

Lastly type

```
Itest=uTypeB(Z,A,chain,alpha,MatCor)
```

One gets $\widehat{I_t} = 125.0$ mm/h and $u(\widehat{I_t}) = 5.9$ mm/h which gives [113.3, 136.7] mm/h as the 95% coverage interval for $\widehat{I_t}$.

7.7 SUMMARY AND TRANSITION

This chapter has given information about three essential aspects of monitoring UDSM systems: (i) health and safety rules for access to monitoring sites and stations, (ii) experience-based recommendations for operation and maintenance, and (iii) methods for sensor calibration. Maintenance and calibration of sensors are required steps in the following necessary estimation of measurement uncertainties which is detailed in Chapter 8, and pre-requisites for data validation presented in Chapter 9.

REFERENCES

ATEX (2016). *ATEX 2014/34/EU Guidelines – Guide to the application of the Directive 2014/34/EU of the European Parliament and of the council of 26 February 2014 on the harmonization of the laws of the Member States relating to equipment and protective systems intended for use in potentially explosive atmospheres.* European Commission, Brussels (Belgium). Available at http://www.industry-finder.com/download/2014-34-UE-ATEX-directive/ATEX-2014-34-EU_Guidelines-1st_Edition_April_2016.pdf (accessed 13 Aug. 2020).

Bertrand-Krajewski J.-L., Laplace D., Joannis C. & Chebbo G. (2000). *Mesures en hydrologie urbaine et assainissement [Measurements in urban hydrology and sewerage].* Technique et Documentation, Paris (France), juin 2000, 794 p. ISBN 2-7430-0380-4. (in French).

Brandes E. & Möller W. (2003). *Brennbare Flüssigkeiten und Gase [Flammable liquids and gases].* Wirtschaftsverlag NW – Wirtschaftsverlag NW, Verlag für neue Wissenschaft, Sicherheitstechnische Kenngrößen Band 1, Bremerhaven (Deutschland), 603 p. ISBN 3-89701-745-8. (in German).

Carr R. (2001). Chapter 5 – Excreta-related infections and the role of sanitation in the control of transmission. In *Water Quality: Guidelines, Standards and Health*, L. Fewtrell & J. Bartram (eds.), IWA Publishing, on behalf of WHO, London (UK), pp. 89–113. ISBN 1900222280. Available at https://www.who.int/water_sanitation_health/dwq/iwachap5.pdf (accessed 13 Aug. 2020).

Ciaponi C., Moisello U. & Papiri S. (1993). Rainfall measurements and spatial variability in a small urban catchment. *Proceedings of the 6th International Conference on Urban Storm Drainage*, 13–17 September 1993, Niagara Falls, Canada, **1**, 158–163.

DGUV (2007). *GUV-i 8755/DGUV-Information 203-063 – Beurteilung von Gefährdungen und Belastungen am Arbeitsplatz in der Abwasserentsorgung – Gefährdungs- und Belastungs-Katalog [Assessment of hazards and pollution at the workplace in sewage disposal – hazard and pollution catalog]*. DGUV – Deutsche Gesetzliche Unfallversicherung (German Statutory Accident Insurance), Sankt Augustin (Deutschland), 20 p. (in German).

DWA (2014). *Merkblatt DWA-M 151 Messdatenmanagementsysteme (MDMS) in Entwässerungssystemen [Measurement data management system (MDMS) in drainage systems]*. DWA – Deutsche Vereinigung für Wasserwirtschaft, Abwasser und Abfall e.V., Hennef (Deutschland), 51 p. ISBN 978-3-944328-67-6. (in German).

Gehlen R. (2018). *RSA 95 – Richtlinien für die Sicherung von Arbeitsstellen an Straßen [Guidelines for securing workplaces on roads]*. Moravia Verlag, Wiesbaden (Deutschland). Available at http://www.moravia-verlag.de/seminare/rsa/rsa_inhalt.htm (accessed 25 Aug. 2020). (in German).

Gröning M., Lutz H. O., Roller-Lutz Z., Kralik M., Gourcy L. & Pöltenstein L. (2012). A simple rain collector preventing water re-evaporation dedicated for $\delta^{18}O$ and $\delta^{2}H$ analysis of cumulative precipitation samples. *Journal of Hydrology*, **448–449**, 195–200. doi: 10.1016/j.jhydrol.2012.04.041.

ISO (2013a). *ISO 18365:2013 – Hydrometry — Selection, establishment and operation of a gauging station*. ISO, Geneva (Switzerland), 17 p.

ISO (2013b). *ISO 20471:2013 – High visibility clothing — Test methods and requirements*. ISO, Geneva (Switzerland), 22 p.

ISO (2015a). *ISO 9000:2015 – Quality management systems — Fundamentals and vocabulary*. ISO, Geneva (Switzerland), 51 p.

ISO (2015b). *ISO 9001:2015 – Quality management systems — Requirements*. ISO, Geneva (Switzerland), 29 p.

ISO (2018). *ISO 9004:2018 – Quality management — Quality of an organization — Guidance to achieve sustained success*. ISO, Geneva (Switzerland), 59 p.

JCGM (2012). *JCGM 200:2012 – International vocabulary of metrology — Basic and general concepts and associated terms (VIM) (Vocabulaire international de métrologie — Concepts fondamentaux et généraux et termes associés (VIM)*. Joint Committee for Guides in Metrology, Geneva (Switzerland), 108 p. Available at https://www.bipm.org/documents/20126/2071204/JCGM_200_2012.pdf (accessed 26 April 2021).

Lepot M., Momplot A., Lipeme Kouyi G. & Bertrand-Krajewski J.-L. (2014). Rhodamine WT tracer experiments to check flow measurements in sewers. *Flow Measurement and Instrumentation*, **40**, 28–38. doi: 10.1016/j.flowmeasinst.2014.08.010.

McManus N. (2019). *Safety and Health in Confined Spaces*. CRC Press, Boca Raton, FL (USA), 928 p. ISBN 978-0367400248.

Métadier M. & Bertrand-Krajewski J.-L. (2012). The use of long-term on-line turbidity measurements for the calculation of urban stormwater pollutant concentrations, loads, pollutographs and intra-event fluxes. *Water Research*, **46**(20), 6836–6856. doi: 10.1016/j.watres.2011.12.030.

Neuilly M. & Cetama (1998). *Modélisation et estimation des erreurs de mesure [Modelling and estimating measurement errors]*. Technique et Documentation, Paris (France), 704 p. ISBN 2-7430-0272-7. (in French).

Nougier J.-P. (2001). *Méthodes de calcul numérique – Volume 2: fonctions, équations aux dérivées [Numerical methods – Volume 2 : functions, derivatives equations]*. Hermès Science Publications, Paris (France), 406 p. ISBN 2-7462-0279-4. (in French).

Savina M., Schäppia B., Molnara P., Burlandoa P. & Sevruk B. (2011). Comparison of a tipping-bucket and electronic weighing precipitation gage for snowfall. *Atmospheric Research*, **103**, 45–51. doi: 10.1016/j.atmosres.2011.06.010.

Schilling W. (1991). Rainfall data for urban hydrology: what do we need? *Atmospheric Research*, **27** (1–3), 5–21. doi: 10.1016/0169-8095(91)90003-F.

Seber G. A. F. & Wild C. J. (2003). *Nonlinear Regression*. Wiley Interscience, Hoboken, NJ (USA), 768 p. ISBN 978-0471471356.

Sullivan G., Pugh R., Melendez A. P. & Hunt W. D. (2010). *Operations & Maintenance Best Practices – A Guide to Achieving Operational Efficiency (Release 3)*. Pacific Northwest National Lab. (PNNL), Richland, WA (USA), report No. PNNL-19634. 321 p. Available at https://www.energy.gov/sites/prod/files/2013/10/f3/omguide_complete.pdf (accessed 14 Aug. 2020).

Walcker N., Bertrand-Krajewski J.-L., Vacherie S., Lepot M., Castebrunet H., Barraud S. & Lipeme Kouyi G. (2018). Une nouvelle station de mesure pour l'acquisition de séries chronologiques en hydrologie urbaine [A new monitoring station for time series data acquisition in urban hydrology]. *TSM*, **3**, 55–64. doi: 10.1051/tsm/201803055. (in French).

Williamson J. H. (1968). Least-squares fitting of a straight line. *Canadian Journal of Physics*, **46**, 1845–1847. doi: 10.1139/p68-523.

WMO (2018). *Guide to Meteorological Instruments and Methods of Observation, 2018 Edition*. WMO – World Meteorological Organization, Geneva (Switzerland), ISBN 978-92-63-10008-5. Available at https://library.wmo.int/?lvl=notice_display&id=12407 (accessed 13 Aug. 2020).

Xin G., Bertrand-Krajewski J.-L. & Juutilainen M. (2011). *EU FP-7 Project PREPARED Deliverable 3.1.5: A unified Protocol for Sensor Calibration and Verification*. Aquateam, Oslo (Norway), PREPARED report 2011.002, March 2011, 46 p. Available at http://jlbkpro.free.fr/publications.html#rd8vcUjo (accessed 16 Aug. 2020).

Chapter 8

Uncertainty assessment

Jean-Luc Bertrand-Krajewski[1], Mathias Uhl[2] and Francois H. L. R. Clemens-Meyer[3,4,5]

[1]University of Lyon, INSA Lyon, Laboratory DEEP, Villeurbanne, France
[2]Muenster University of Applied Sciences, Institute for Infrastucture·Water·Resources·Environment (IWARU), Münster, Germany
[3]Delft University of Technology, Water Management Department, Delft, The Netherlands
[4]Deltares, Unit Hydraulic Engineering, Dept. Experimental Facility Support, Delft, The Netherlands
[5]Norwegian University of Science & Technology, Faculty of Engineering, Dept. Civil & Environmental Engineering, Trondheim, Norway

ABSTRACT

Assessing uncertainties in measurements must become a standard practice in the field of urban drainage and stormwater management. This chapter presents three standard methods to estimate uncertainties: the Type A method (repeated measurements), the Type B method (law of propagation of uncertainties) and the MC method (Monte Carlo method). Each method is described with its fundamental principles and equations, various examples are presented in detail and Matlab® codes are given to facilitate the calculations for routine applications. An advanced method to account for partial autocorrelation in time series is presented. Lastly, typical orders of magnitude of standard uncertainties for usual sensors used in urban drainage and stormwater management are given.

Keywords: Coverage interval, error, guide for uncertainty in measurements, law of propagation of uncertainties, Monte Carlo method, standard uncertainty.

doi: 10.2166/9781789060119_0263

SYMBOLS

a low boundary of an interval or index of the iterative calculation of δ or numerical coefficient
A matrix containing data related to constant quantities
b high boundary of an interval or vector of coefficients b_i or numerical coefficient
b_i regression coefficients
B channel width (m)
B_c notch width (m)
B_e effective width (m)
c integer value used to calculate δ
c_i sensitivity coefficient related to the quantity x_i in the measurement function f
C_d discharge coefficient (-)
COV covariance matrix
d_{low} distance between low boundaries of Type B and MCM coverage intervals
d_{high} distance between high boundaries of Type B and MCM coverage intervals
D pipe diameter (m)
e_{fs} numerical factor to calculate V from V_{fs} (-)
e_{max} numerical factor to calculate V from V_{max} (-)
f function of quantities x_i representing the measurement process
fc as index: full autocorrelation
F_y numerical coefficient for velocity-area methods
F_z numerical coefficient for velocity-area methods
g gravity (m/s^2)
h water level (m)
h_e effective head (m)
h_p crest height (m)
i index
I slope of a channel or a pipe (m/m)
I_{est} estimated rainfall intensity (mm/h)
I_m measured rainfall intensity (mm/h)
I_r reference rainfall intensity (mm/h)
IC_{95min} shortest 95% coverage interval calculated with the Monte Carlo method
j index
J smallest integer greater than or equal to $100/\alpha$
k coverage factor
K Manning-Strickler coefficient (m$^{1/3}$/s)
K_b correction factor in the calculation of Q_{RW} (-)
K_h correction factor in the calculation of Q_{RW} (-)
l integer value used to calculate δ
L distance upstream a weir where the water level is measured (m)
m mean value of a normal distribution or number of quantities in the matrix Z
M number of Monte Carlo simulations
MC as index: refers to the Monte Carlo method
n number of repeated measurements in the Type A method
nc as index: no autocorrelation

n_{dig}	number of significant digits used to calculate δ
N	number of quantities x_i used in the function f
$N(m,s)$	normal (Gaussian) probability distribution with mean value m and standard deviation s
p	number of quantities in the matrix A
pc	as index: partial autocorrelation
q	integer used for estimating the narrowest coverage interval in Monte Carlo simulations
Q	discharge (m^3/s)
Q_c	discharge in a circular pipe (m^3/s)
Q_{MS}	discharge calculated with the Manning-Strickler formula (m^3/s)
Q_p	perimeter flow for velocity-area methods (m)
Q_{RW}	discharge over a rectangular weir (m^3/s)
r	coefficient of correlation
r	as index: index of Monte Carlo simulations
$r(x_i, x_j)$	coefficient of correlation of x_i and x_j
r_{ij}	coefficient of correlation of x_i and x_j
R_c	circular pipe radius (m)
R_h	hydraulic radius (m)
$s(y)$	standard deviation of y
S	wet cross section (m^2)
t	Student t value
TB	as index: refers to the Type B method
$Trap(a,b,\beta)$	trapezoidal probability distribution in the interval $[a,b]$ with the coefficient β
$Tri(a,b)$	triangular probability distribution in the interval $[a,b]$
$u(x_i, x_j)$	covariance of x_i and x_j
$u(Y)$	standard uncertainty of Y
$u^*(Y)$	relative standard uncertainty of Y
$U(a,b)$	uniform probability distribution in the interval $[a,b]$
$U(Y)$	enlarged uncertainty of Y
v	flow velocity at a given position within a wet cross section (m/s)
V	cross section mean flow velocity (m/s)
V_d	daily volume (m^3)
V_{fs}	free surface flow velocity (m/s)
V_{max}	maximum flow velocity (m/s)
x_i	quantities used in the measurement function f to calculate y
X_i	random variable corresponding to the quantity x_i
$\bar{y}$	mean value of y
Y	measured or calculated quantity
$Y_{\alpha,low}$	low boundary of a coverage interval for Y calculated by the Monte Carlo method for a level of confidence α
$Y_{\alpha,high}$	high boundary of a coverage interval for Y calculated by the Monte Carlo method for a level of confidence α
z	matrix containing data related to time varying quantities (time series)
α	level of confidence
β	numerical coefficient of a trapezoidal probability distribution
δ	tolerance

Δt	time step (s)
ε_i	finite difference used in the 2nd order approximation of c_i
ν	degree of freedom
ν_{eff}	effective degree of freedom

8.1 INTRODUCTION

Why is uncertainty assessment important and should be systematically done? In urban drainage and stormwater management (UDSM), like in numerous other professional fields and disciplines, information, knowledge, performance analysis, modelling, scenario analysis, planning and decision making are based on or use measurement results. However, measurements are never perfect and cannot be carried out without uncertainties. Consequently, 'when reporting the result of a measurement of a physical quantity, it is obligatory that some quantitative indication of the quality of the result be given so that those who use it can assess its reliability. Without such an indication, measurement results cannot be compared, either among themselves or with reference values given in a specification or standard. [...] When all of the known or suspected components of error have been evaluated and the appropriate corrections have been applied, there still remains an uncertainty about the correctness of the stated result, that is, a doubt about how well the result of the measurement represents the value of the quantity being measured' (ISO, 2008b, p. vii).

Uncertainty assessment (UA) should thus become a standard professional practice in UDSM, aiming to comply with laws and regulations, quality control requirements, expected professional skills, basic and applied research needs, etc. This chapter aims to provide information, concepts, methods, tools, and detailed examples facilitating knowledge transfer and implementation of uncertainty assessment. However, as UA is not always obvious and requires some training, 'critical thinking, intellectual honesty and professional skills' (ISO, 2008b, p. 8) remain fundamental.

This chapter is organized in three main sections:

- Section 8.2 presents the methods and international standards for UA, with their principles, conditions of application, step by step explanations and basic examples of application.
- Section 8.3 provides some additional examples for various aspects of UDSM.
- Section 8.4 gives complements including *in situ* uncertainties and some reference values for typical sensors and measurement methods used in UDSM.

BOX 0: EXAMPLES WITH MATLAB®

Detailed examples of calculations with Matlab® are shown in dedicated boxes throughout this chapter: the instructions and codes are written with the Matlab® syntax and `courier new` font to distinguish them from the rest of the text. The instructions and code lines can be copied-pasted directly by the reader who would like to replicate them for training or to adapt them to his/her own needs.

Numerical results in boxes are usually given with 4 digits (`format short`). In the main text, numerical values are rounded to the number of significant digits. It is also important to note, for readers who would like to reproduce them, that all calculations have been run without rounding in the successive steps.

Matlab® codes and associated data csv files are available for download at https://doi.org/10.2166/9781789060102.

8.2 INTERNATIONAL STANDARDS AND METHODS FOR UNCERTAINTY ASSESSMENT

8.2.1 Introduction and common rules of application

The first internationally unified frame for UA in measurements was published in 1993 as an ISO (International Organization for Standardization) guide entitled GUM – Guide for Uncertainty in Measurements (ISO, 1993), re-published with revisions in 1995 and also as a European standard in 1999 (CEN, 1999). It was later on revised, adapted and completed as parts of a new Guide for Uncertainty in Measurement, abbreviated hereafter as the ISO Guide 98, elaborated at international level by the JCGM – Joint Committee for Guides in Metrology – convened by the Bureau International des Poids et Mesures (BIPM), the International Electrotechnical Commission (IEC), the International Organization for Standardization (ISO), and the International Organization of Legal Metrology (OIML). The Supplement 1 published in 2008 introduces the Monte Carlo method for uncertainty assessment.

The ISO Guide 98 is based on a statistical approach to estimate uncertainties in measurements, in agreement with definitions given in Chapter 12.

In this chapter, we refer to the following parts of the ISO Guide 98:

- As general introduction for all concepts and methods:
  ISO (2009a). *ISO/IEC Guide 98-1:2009(E) Uncertainty of measurement – Part 1: Introduction to the expression of the uncertainty in measurement.* Geneva (Switzerland): ISO, September 2009, 32 p.
- As Guide for uncertainty in measurements method (abbreviated as GUM):
  ISO (2008a). *ISO/IEC Guide 98-3:2008(E) Uncertainty of measurement – Part 3: Guide to the expression of uncertainty in measurement (GUM: 1995).* Geneva (Switzerland): ISO, December 2008, 130 p.
- As Monte Carlo method (abbreviated as MCM):
  ISO (2008b). *ISO/IEC Guide 98-3/Suppl.1:2008(E) Uncertainty of measurement – Part 3: Guide to the expression of uncertainty in measurement (GUM: 1995) Supplement 1: Propagation of distributions using a Monte Carlo method.* Geneva (Switzerland): ISO, December 2008, 98 p. and
  ISO (2009b). *ISO/IEC Guide 98-3/S1/AC1:2009(E) Uncertainty of measurement – Part 3: Guide to the expression of uncertainty in measurement (GUM: 1995), Supplement 1: Propagation of distributions using a Monte Carlo method, Technical corrigendum 1.* Geneva (Switzerland): ISO, May 2009, 2 p.

GUM and MCM may also be referred to as the 'propagation of uncertainties' method and the 'propagation of distributions' method, respectively.

This chapter does not reproduce the full content of the above detailed standards. A brief introduction is presented below in Section 8.2, and additional examples are given in Section 8.3.

For any measured or calculated quantity Y, there are three steps in UA:

(1) Estimation of the true value* of Y. **(Note: the symbol * indicates that the definition of the word or the expression is given in Chapter 12).**

(2) Estimation of the standard uncertainty* of Y noted $u(Y)$.

(3) Estimation of the coverage interval* of Y for a given level of probability α (typically 95%):

$[Y - ku(Y), Y + ku(Y)]$ in the case of methods A and B (described respectively in Sections 8.2.2 and 8.2.3), where k is the coverage factor*, or

$[Y_{\alpha,low}, Y_{\alpha,high}]$ in the case of MCM (described in Section 8.2.4).

The third step is optional but is almost systematically applied in practice.

Reporting UA should be done systematically, and the following information should be provided (ISO, 2008b, p. 25):

- A detailed and clear description of (i) the measurement process, and (ii) the methods used.
- A list of all uncertainty components that are accounted for and how they are evaluated.
- All values, constants, corrections used in the UA analysis process, so that it could independently repeated if necessary.

A test of the foregoing list is to ask oneself 'Have I provided enough information in a sufficiently clear manner that my result can be updated in the future if new information or data become available?' (ISO, 2008b, p. 25).

An important precondition for UA is the absence of coarse errors and systematic deviations in measurements. This is ensured by the rigorous application of metrological best practices, including sensor calibration, and periodic maintenance and checking (see Chapter 7, especially Section 7.6).

8.2.2 Type A method for uncertainty assessment of repeated measurements

8.2.2.1 Principle

The Type A method assumes that the quantity of interest Y can be measured directly and repeatedly, according to repeatability conditions*. It is applicable to stationary quantities that do not change with time (at least at the timescale of measurements), and to dynamic processes provided they are repeatable. Examples are the diameter of a pipe, the width of a channel, the angle of a weir, the hydraulic conductivity of a soil, etc.

One assumes that the measurement of the quantity Y is a random process, due to all possible sources of variabilities attached to the instruments used, the measurement conditions, the operator and to the quantity itself. Each measurement y_i is assumed to be an independent observation of Y. The best estimate of the true value of Y is given by the mean $\bar{y}$ of the $i = 1{:}n$ repeated measurements y_i:

$$\bar{y} = \frac{1}{n}\sum_{i=1}^{n} y_i \tag{8.1}$$

The unbiased standard deviation $s(y)$ of the measured values y_i is calculated as follows:

$$s(y) = \sqrt{\frac{1}{n-1}\sum_{i=1}^{n}(y_i - \bar{y})^2} \tag{8.2}$$

The unbiased standard deviation $s(\bar{y})$ of the mean value $\bar{y}$ is given by:

$$s(\bar{y}) = \sqrt{\frac{1}{n(n-1)}\sum_{i=1}^{n}(y_i - \bar{y})^2} = \frac{s(y)}{\sqrt{n}} \tag{8.3}$$

The standard uncertainty* $u(\bar{y})$ is then assumed to be equal to the standard deviation $s(\bar{y})$.

The expanded uncertainty* $U(\bar{y})$ with a given level of probability α (typically 95%) is calculated with the coverage factor* k. The value of k depends on α. If n is lower than 30, the level of information about Y and the distribution of its measurements y_i is limited. One assumes that the distribution of the y_i values is a Student t distribution. In this case, which is frequent in practice as making more than 30 repeated measurements may be too long or too expensive, the value of k is given by:

$$k = t_{\frac{1+\alpha}{2}}(\nu) \tag{8.4}$$

where $t_{\frac{1+\alpha}{2}}(\nu)$ is the Student t value with $\alpha = 0.95$ for a symmetric probability level of 95% and $\nu = n - 1$ degrees of freedom. The Student t value is found in statistics tables and can be obtained from software tools like Excel®, Matlab® or Octave® (see Table 8.2).

The expanded uncertainty is:

$$U(\bar{y}) = ku(\bar{y}) \tag{8.5}$$

and the coverage interval with the probability level α is then calculated as follows:

$$[\bar{y} - U(\bar{y}), \bar{y} + U(\bar{y})] = [\bar{y} - ku(\bar{y}),\ \bar{y} + ku(\bar{y})] \tag{8.6}$$

If n is above 30, the distribution of the measurements y_i is usually assumed to be normal (i.e. Gaussian) and, in this case, $k = 1.96$ for $\alpha = 0.95$ (Table 8.3).

In practice, due to (i) the unavoidable approximations in the measurement process, (ii) the fact that the measured values are not necessarily exactly normally distributed and (iii) 'the impracticality of trying to distinguish between intervals having levels of confidence that differ by one or two percent', the ISO Guide 98 (ISO, 2008a, appendix G) indicates that it is also acceptable to approximate $k = 1.96$ by $k = 2$ (which corresponds to the exact value $\alpha = 0.9545$ in case of the normal distribution). The 95% coverage interval of $\bar{y}$ is then approximated by:

$$[\bar{y} - 2u(\bar{y}),\ \bar{y} + 2u(\bar{y})] \tag{8.7}$$

In this chapter, we use $k = 1.96$ to approximate 95% coverage intervals with the hypothesis of the normal distribution. It is recommended to systematically apply Equations (8.4) and (8.6) for any number of measurements n.

The coverage interval (Equation (8.6)) is usually interpreted, in a simplified way, as 'the true value of the mean $\bar{y}$ of the quantity Y has an approximately 95% probability to lie between $\bar{y} - ku(\bar{y})$ and $\bar{y} + ku(\bar{y})$'. This can be acceptable only if (i) there is no bias (systematic error) in the measurements, which is ensured only by proper calibration of the sensor used for measurements and careful checking of the complete measurement process, and (ii) the number of measurements is high enough to ensure that the mean of the measured values is reasonably close to the true value of Y.

As indicated by Equation (8.3), increasing n allows decreasing the uncertainty in $\bar{y}$ proportionally to the square root of n. Multiplying n by 4 and 10 leads to dividing the uncertainty respectively by 2 (i.e. $\sqrt{4}$) and 3.16 (i.e. $\sqrt{10}$).

8.2.2.2 Basic example with Matlab®

The diameter D of a 1 m circular sewer pipe has been measured four times with a 2 m long class II meter (i.e. true length of this meter is between 1.9993 and 2.0007 m, according to the class definition given in OJEU, 2014). The four measured values D_i with $i = 1{:}4$ are given in Table 8.1.

Table 8.1 Four measurements of the pipe diameter *D*.

D_i (mm)
1002
1000
997
1002

Note: Examples and codes written for Matlab® (https://fr.mathworks.com) can also be used without any modification with the free software tool Octave® (https://www.gnu.org/software/octave/). The compatibility has been checked by the authors with Matlab® 2017b and Octave 5.1.0.

As shown in Box 1 below, the best estimate of the pipe diameter is $\bar{D} = 1000.2$ mm and its 95% coverage interval is [996.5, 1004.0] mm, with only one meaningful digit. Box 2 shows how to apply the Type A method with the Matlab® code `uTypeA`.

BOX 1: STEP BY STEP APPLICATION OF THE TYPE A METHOD WITH MATLAB®

For the data given in Table 8.1, the Matlab® instructions are as follows.

Create the vertical vector D_i with the four measured values:

```
Di=[1002 1000 997 1002]'
```

Calculate the mean value $\bar{D}$:

```
Dbar=mean(Di)
```

One gets $\bar{D} = 1000.2500$ mm.

Calculate the standard uncertainty $u(\bar{D})$:

```
uDbar=std(Di)/sqrt(length(Di))
```

One gets $u(\bar{D}) = 1.1814$ mm.

Calculate k with $\alpha = 95\%$ and $\nu = n\text{-}1$ degrees of freedom:

```
alpha=0.95
k=tinv((1+alpha)/2, length(Di)-1)
```

One gets $k = 3.1824$. The expanded uncertainty $k \times u(\bar{D}) = 3.7599$ mm.

Calculate the coverage interval with the probability level α:

```
Dbar-k*uDbar
Dbar+k*uDbar
```

One gets respectively 996.4900 mm and 1004.0099 mm.

BOX 2: APPLICATION OF THE TYPE A METHOD WITH THE MATLAB® CODE uTypeA

(Matlab® codes and csv files available for download at https://doi.org/10.2166/9781789060102).
All calculations presented in Box 1 are automated in the Matlab® code `uTypeA`.
With the previous notations defined in Box 1, enter `uTypeA(yi,alpha)` with `yi` the vertical vector of n measured values y_i and `alpha` the level of probability. The `uTypeA` function provides respectively the mean value $\bar{y}$, the standard uncertainty $u(\bar{y})$, and the boundaries of the coverage interval $[\bar{y} - ku(\bar{y}),\ \bar{y} + ku(\bar{y})]$ with the level of probability α and the coverage factor k calculated with the Student t value.
For the Box 1 example, type

```
Dbar=uTypeA(Di,alpha)
```

One gets $\bar{D} = 1000.2500$ mm, $u(\bar{D}) = 1.1814$ mm, $a = 996.4900$ mm and $b = 1004.0099$ mm.

8.2.3 Type B method for uncertainty assessment by the law of propagation of uncertainties

8.2.3.1 Principle

The Type B method is applied in cases where the quantity of interest Y can be measured neither directly nor repeatedly, i.e. when the Type A method cannot be applied. Some examples: rainfall intensities calculated from measured tips of a rain gauge bucket, discharge calculated from the water level measured over a weir, discharge calculated from both measured water level and mean flow velocity, infiltration flow in a stormwater infiltration tank calculated from water level and mass balance, pollutant load calculated from measured water quality and discharge, etc. In many cases in urban hydrology, the quantities of interest vary with time: repeated measurements are not possible. Measured process data are usually recorded as time series.

It is assumed that Y is determined from N other quantities X_i by means of a function f representing the measurement process. All quantities are assumed to be random quantities. All measured, estimated or known values x_i of the quantities X_i and their standard uncertainties $u(x_i)$ shall be known from Type A repeated measurements, previous applications of the Type B method, sensor calibration, experiments, expertise, standards, scientific literature, textbooks, etc. Previous knowledge on uncertainties $u(x_i)$ and their distribution is a pre-requisite for the Type B method and is discussed in Section 8.2.3.3.

The estimate y of the quantity Y is given by:

$$y = f(x_1, x_2, \ldots x_i, \ldots x_N) \tag{8.8}$$

The combined standard uncertainty $u(y)$ is obtained using the following equation, also referred to as the Law of Propagation of Uncertainties (LPU):

$$u(y)^2 = \sum_{i=1}^{N} u(x_i)^2 \left(\frac{\partial f}{\partial x_i}\right)^2 + 2\sum_{i=1}^{N-1} \sum_{j=i+1}^{N} u(x_i, x_j)\left(\frac{\partial f}{\partial x_i}\right)\left(\frac{\partial f}{\partial x_j}\right) \tag{8.9}$$

where $u(x_i, x_j)$ is the covariance of x_i and x_j:

$$u(x_i, x_j) = r(x_i, x_j)u(x_i)u(x_j) \tag{8.10}$$

where $r(x_i, x_j)$ is the correlation coefficient of x_i and x_j.

The partial derivatives, also called sensitivity coefficients c_i, are evaluated at $X_i = x_i$ using:

$$c_i = \left(\frac{\partial f}{\partial X_i}\right) \tag{8.11}$$

In the case where f has a complicated expression, its derivatives may be difficult to establish analytically. They can be replaced by numerical second order approximations:

$$c_i = \left(\frac{\partial f}{\partial X_i}\right) \approx \frac{f(x_i + \varepsilon_i) - f(x_i - \varepsilon_i)}{2\varepsilon_i} \tag{8.12}$$

where ε_i is very small compared to x_i. Typically, one can use $\varepsilon_i = u(x_i)/1000$.

The expanded uncertainty $U(\bar{y})$ with a given level of probability α (typically 95%) is calculated with the coverage factor k:

$$U(y) = ku(y) \tag{8.13}$$

The value of k depends on α. Ideally, uncertainty estimates $u(x_i)$ are based upon reliable Type A and Type B evaluations with a sufficient number n of observations such that using the coverage factor of $k = 1.96$ will ensure a confidence level close to 95%.

If the above assumption is not valid, the effective degree of freedom ν_{eff} needs to be estimated using the Welch-Satterthwaite formula:

$$\nu_{eff} = u(y)^4 \left(\sum_{i=1}^{N} \frac{[c_i u(x_i)]^4}{\nu_i}\right)^{-1} \tag{8.14}$$

If $u(x_i)$ is determined from a Type A estimation based on n repeated measurements, then $\nu_i = n-1$. If $u(x_i)$ is determined from a previous Type B estimation, and if the distribution of x_i is exactly known (i.e. the type and the boundaries of the distribution are known), which is frequent in practice, then $\nu_i \rightarrow \infty$. Otherwise, ν_i is estimated from the following equation:

$$\nu_i = \frac{1}{2}\frac{u(x_i)^2}{\sigma[u(x_i)]^2} \approx \frac{1}{2}\left[\frac{\Delta u(x_i)}{u(x_i)}\right]^{-2} \tag{8.15}$$

where $\sigma[u(x_i)]$ is the standard deviation of the standard uncertainty $u(x_i)$.

The quantity between large brackets in the last part of Equation (8.15) corresponds to the relative uncertainty of the standard uncertainty $u(x_i)$, i.e. how exactly the standard uncertainty $u(x_i)$ itself is known. This is usually based on scientific judgement and expertise.

In the case where the value of ν_{eff} obtained from Equation (8.14) is not an integer, then round ν_{eff} to the nearest lower integer.

The value of k is then calculated from the Student t distribution:

$$k = t_{\frac{1+\alpha}{2}}(\nu_{eff}) \tag{8.16}$$

and the coverage interval with the probability level α is calculated as follows:

$$[y - U(y), y + U(y)] = [y - ku(y),\ y + ku(y)] \tag{8.17}$$

Note that the coverage interval is, by definition, symmetric around the estimate y of the measurand Y.

8.2.3.2 Covariances in the type B method

The main equation of the LPU is Equation (8.9). The right-hand part contains two sums. The first one is always applied and corresponds to cases where all measured values x_i and their uncertainties $u(x_i)$ are fully independent of each other and are not correlated: all covariances $u(x_i, x_j) = 0$ or all coefficients of correlation $r(x_i, x_j) = 0$. This is the case when all x_i values are estimated with separate sensors or independent measurement processes and sources of information.

The second double sum has to be taken into account when measured values x_i or their uncertainties $u(x_i)$ are not independent of each other and are correlated: covariances $u(x_i, x_j) \neq 0$ or coefficients of correlation $r(x_i, x_j) \neq 0$ shall be (i) detected by means of a detailed analysis of the measurement process and (ii) quantified to be included in the calculation. Detection and quantification of covariances are not always obvious and should receive special attention. As covariances and coefficients of correlation may be positive or negative, they may contribute, sometimes in a very high proportion, to respectively increasing or decreasing the standard uncertainty $u(y)$. More details are given in Section 8.2.6.

It is frequent in practice that estimating the value y of the measurand Y includes intermediate quantities X_i in the function f which are themselves based on common measured quantities X_j. This shall be avoided as it generates covariance and complicates the estimation of the standard uncertainty $u(y)$. It is thus very important to avoid intermediate quantities as much as possible, and to write the function f in a way which may be less usual but reflects closely the measurement process with independent quantities.

Example 1: Calculation of flow by the Manning-Strickler equation

The discharge Q (m^3/s) in a rectangular channel is commonly calculated using the Manning-Strickler equation as:

$$Q = f(K, I, S, R_h) = KI^{\frac{1}{2}}SR_h^{\frac{2}{3}} \tag{8.18}$$

where K ($m^{1/3}/s$) is the Manning-Strickler coefficient, I (m/m) is the pipe invert slope, S (m^2) is the flow cross section and R_h (m) is the hydraulic radius.

The coefficient K is estimated either from field experiments or, more frequently, from tables or textbooks. The slope I is estimated from field measurements, maps or GIS data. Both the cross section S and the hydraulic radius R_h are calculated from the channel width B (m) and the measured water level h (m):

$$S = Bh \tag{8.19}$$

$$R_h = \frac{Bh}{B + 2\,h} \tag{8.20}$$

K, I, B and h are the truly independent quantities estimated by means of different and independent instruments and information. But S and R_h are obviously not independent quantities and are highly correlated as both depend on B and h. Clearly Equation (8.18) is not appropriate to be used in Equation (8.9) as covariance between S and R_h has been introduced. It is thus recommended to rewrite Equation (8.18) without correlated intermediate quantities and only with the truly independent quantities in a way which reflects closely the real measurement process, as:

$$Q = f(K, I, B, h) = KI^{\frac{1}{2}}(Bh)\left(\frac{Bh}{B+2\,h}\right)^{\frac{2}{3}} = KI^{\frac{1}{2}}(Bh)^{\frac{5}{3}}(B+2\,h)^{-\frac{2}{3}} \tag{8.21}$$

8.2.3.3 Estimation of standard uncertainties $u(x_i)$ from prior information on distributions

As the Type B method requires all values $u(x_i)$ as inputs in Equation (8.9), and as these values are not necessarily obtained from Type A estimations (i.e. series of repeated observations), it is necessary to provide additional information on the probability distributions of the quantities X_i.

If the value x_i of the quantity X_i is known from repeated measurements and if the Type A method is applied to estimate $u(x_i)$, the degrees of freedom $\nu(x_i)$ are known and $u(x_i)$ can be used directly in Equation (8.9). In the case where x_i is not given with its standard uncertainty $u(x_i)$ but with (i) a coverage interval $[a, b]$ and (ii) a coverage factor k or a level of probability α, then $u(x_i)$ is calculated from Equation (8.6), by assuming the values of the quantity X_i are distributed according to a normal (i.e. Gaussian) distribution:

$$u(x_i) = \frac{(b-a)}{2k} \tag{8.22}$$

or

$$u(x_i) = \frac{(b-a)}{2k(\alpha)} \tag{8.23}$$

where $k(\alpha)$ is calculated either from the normal distribution or from the Student t distribution for an infinite number of degrees of freedom $\nu = +\infty$. The value $k(\alpha)$ is found in statistics tables and can be obtained from software tools like Excel®, Matlab® or Octave® (Table 8.2). Most typical values are given in Table 8.3.

Example 2: If $x_i = 100$ and $[a, b] = [99, 101]$ with $k = 2$ (i.e. $\alpha \approx 0.95$), then Equation (8.22) gives $u(x_i) = 0.5$. If $x_i = 100$ and $[a, b] = [99, 101]$ with $\alpha = 0.99$, then $k(\alpha) = 2.58$ and Equation (8.23) gives $u(x_i) = 0.39$.

Table 8.2 Excel® and Matlab®/Octave® functions to calculate the coverage factor k from the probability level α and reciprocally.

	Normal distribution		**Student distribution**	
From α to k				
Excel®*	$k(\alpha)$ = NORMINV((1 + α)/2, 0, 1)	(8.24)	$k(\alpha)$ = TINV((1 − α), 1e6)	(8.25)
Matlab®	$k(\alpha)$ = norminv((1 + α)/2)	(8.26)	$k(\alpha)$ = tinv((1 + α)/2, inf)	(8.27)
From k to α				
Excel®*	$\alpha(k)$ = (NORM.DIST(k, 0, 1, 1) ∗ 2) − 1	(8.28)	$\alpha(k)$ = (T.DIST(k, 1e6) ∗ 2) − 1	(8.29)
Matlab®	$\alpha(k)$ = (normcdf(k) ∗ 2) − 1	(8.30)	$\alpha(k)$ = (tcdf(k, inf) ∗ 2) − 1	(8.31)

* With Excel®, the infinite value is replaced by one million (1e6).

Table 8.3 Most typical values of probability level α and corresponding coverage factor k for the normal distribution. Note: as indicated in ISO (2008a) appendix G, it is acceptable to replace $k = 1.96$ by $k = 2$ to approximate 95% coverage intervals.

Probability level α	Coverage factor k
0.68	1.00
0.90	1.64
0.95	1.96
0.99	2.58

The probability distribution of the quantities X_i can take a variety of other forms. In addition to the normal distribution, three typical examples are the uniform (or rectangular), the triangular and the trapezoidal distributions (Figure 8.1).

The symmetric uniform distribution corresponds to cases with limited information about the value x_i of the quantity X_i. One knows for example that the value x_i lies in the interval $[a, b]$ with a probability close to 1 but without any additional information about the shape of the distribution. It is thus assumed that (i) any value within the interval $[a, b]$ has the same probability to be the most likely value of x_i, and (ii) any value outside this interval is almost unlikely. In this case, x_i and $u(x_i)$ are given respectively by:

$$x_i = \frac{a+b}{2} \tag{8.32}$$

and

$$u(x_i) = \frac{b-a}{2\sqrt{3}} \tag{8.33}$$

Example 3: If $[a, b] = [99, 101]$, then Equations (8.32) and (8.33) give respectively $x_i = 100$ and $u(x_i) = 0.58$.

The symmetric triangular distribution corresponds to cases with more information. One knows for example that the most likely value x_i is the central value of the interval $[a, b]$ and that the probability declines regularly towards the lower and upper bounds a and b, with the assumption that any value outside the interval is unlikely. In this case, x_i is given by Equation (8.32) and $u(x_i)$ is given by:

$$u(x_i) = \frac{b-a}{2\sqrt{6}} \tag{8.34}$$

Example 4: If $[a, b] = [99, 101]$, then Equations (8.32) and (8.34) give respectively $x_i = 100$ and $u(x_i) = 0.41$.

The symmetric trapezoidal distribution is used to account for the fact that in the uniform distribution, the abrupt probability step below a and above b is likely unphysical. Slopes on each side of the distribution are thus included to get a more realistic distribution (Figure 8.1). The trapezoidal distribution is characterized by both the bottom interval $[a, b]$ and the coefficient β which represents the width of the

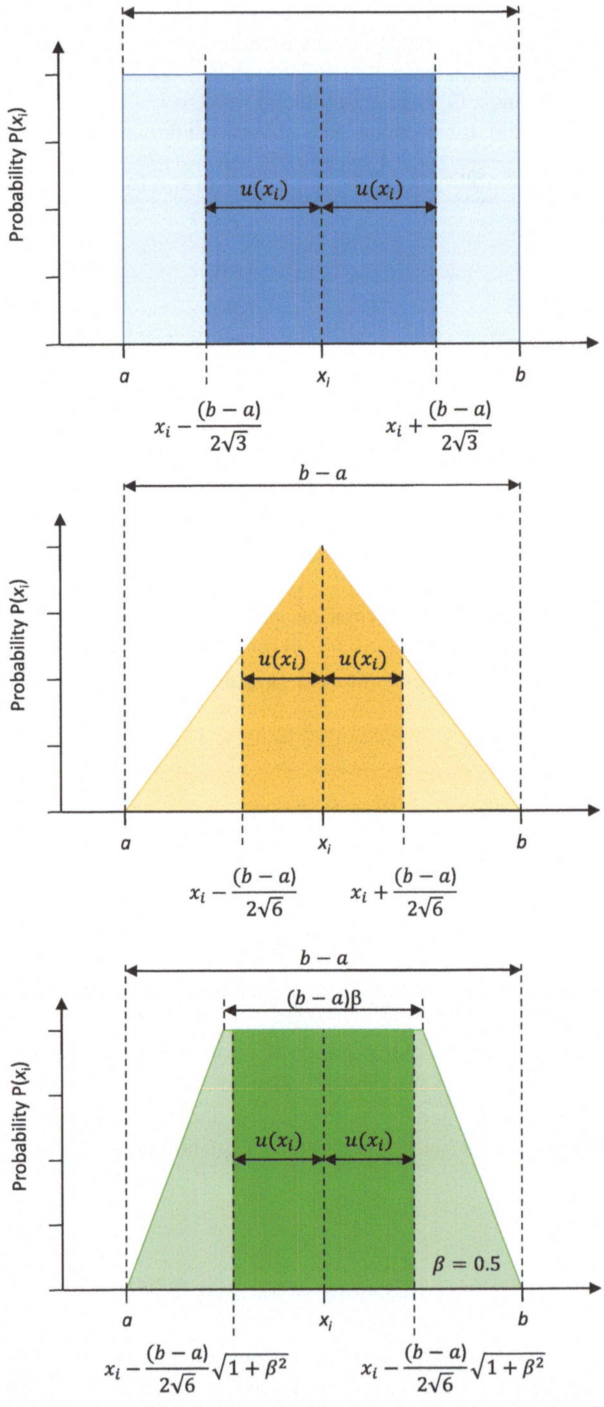

Figure 8.1 Uniform, triangular and trapezoidal distributions for the Type B method. *Source*: adapted from ISO (2008a) by Jean-Luc Bertrand-Krajewski (INSA Lyon).

Table 8.4 Choice of typical distributions for the Type B method.

Distribution	Examples/available information
Normal	• Calibration certificates, handbooks, material or sensor specifications, knowledge quoting either (i) a probability level α or a coverage factor k with the expanded uncertainty, (ii) a number of standard deviations, or (iii) a given probability level interval α. • Information from Type A estimations based on repeated measurements.
Uniform	• Maximum bounds within which all values of the quantity are assumed to lie with equal probability. • Maximum instrument drift between calibrations. • Error due to limited resolution of an instrument's display or digitizer. • Manufacturers' tolerance limits.
Triangular	• Maximum bounds within which all values of the quantity are assumed to lie with higher probability for the central value and decreasing probabilities towards the interval bounds.
Trapezoidal	• Maximum bounds within which all values of the quantity are assumed to lie with equal probability in the central part of the interval.

top of the distribution as a fraction of the interval $[a, b]$. In this case, x_i is given by Equation (8.32) and $u(x_i)$ is given by:

$$u(x_i) = \frac{b-a}{2\sqrt{6}}\sqrt{1+\beta^2} \tag{8.35}$$

Example 5: If $[a, b] = [99, 101]$ and $\beta = 0.5$, then Equations (8.32) and (8.35) give respectively $x_i = 100$ and $u(x_i) = 0.46$.

It is worth noting that Equation (8.35) is similar to Equation (8.33) when $\beta = 1$ (uniform distribution) and to Equation (8.34) when $\beta = 0$ (triangular distribution).

Other distributions, including non-symmetrical ones, are described in ISO (2008a) and also in textbooks (e.g. Gentle, 2003; Thomopoulos, 2018). The choice of an appropriate distribution for each quantity X_i in running the Type B method is based on knowledge and experience. Some examples are given in Table 8.4. In practice, normal and uniform distributions are among the most frequently used ones.

8.2.3.4 Basic example with Matlab®

The discharge Q (m^3/s) in an open rectangular channel is calculated by means of the Manning-Strickler formula, written as discussed in Section 8.2.3.2:

$$Q = f(K, I, B, h) = KI^{\frac{1}{2}}(Bh)\left(\frac{Bh}{B+2\,h}\right)^{\frac{2}{3}} = KI^{\frac{1}{2}}(Bh)^{\frac{5}{3}}(B+2\,h)^{-\frac{2}{3}} \tag{8.36}$$

where K ($m^{1/3}/s$) is the Manning-Strickler coefficient, I (m/m) is the channel invert slope, B (m) is the channel width and h (m) is the water level in the channel.

The channel is made of smooth concrete, with no deposits, no biofilm, and no surface degradation. Textbooks (e.g. Lencastre, 1999) indicate that the value of K is usually between 70 and 80 $m^{1/3}/s$. Consequently, in the absence of *in situ* measurements, it is reasonable to assume that the value of K

lies with a symmetric uniform probability in the interval $[a, b] = [70, 80]$ $m^{1/3}/s$ (uniform distribution $U(70, 80)$).

The mean slope I of the open channel has been measured by a land surveyor along a reach of 50 m with the direct levelling method. The result is $I = 0.0032$ m/m, with a standard uncertainty $u(I) = 6 \times 10^{-6}$ m/m (normal distribution $N(0.0032, 6\times10^{-6})$).

The channel width B has been measured four times with a class II meter, similarly as the pipe diameter in Section 8.2.2.2. The results are $B = 0.805$ m, $u(B) = 0.002$ m and $\nu_B = 3$ degrees of freedom.

The water level $h = 0.32$ m is measured by means of a calibrated ultrasound sensor. *In situ* sensor calibration accounting for uncertainties related to both the sensor itself and local measurement conditions results in a whole standard uncertainty $u(h) = 1.5$ mm (normal distribution $N(0.32, 0.0015)$).

With the above values, the discharge is $Q = 0.346$ m^3/s (detailed calculations are given in Box 3).

The next step consists of calculating the standard uncertainty $u(Q)$ by applying Equation (8.9). As the quantities K, I, B and h are measured with independent sensors, there is no covariance between them and only the first part of Equation (8.9) is applied. This leads to:

$$u(Q)^2 = \sum_{i=1}^{4}(u(x_i))^2\left(\frac{\partial Q}{\partial x_i}\right)^2 = u(K)^2\left(\frac{\partial Q}{\partial K}\right)^2 + u(I)^2\left(\frac{\partial Q}{\partial I}\right)^2 + u(B)^2\left(\frac{\partial Q}{\partial B}\right)^2 + u(h)^2\left(\frac{\partial Q}{\partial h}\right)^2 \tag{8.37}$$

The partial derivatives of Q need to be estimated first. There are two possibilities: algebra derivation or second order numerical approximation. In this example, both approaches are used and compared.

The algebra derivation gives (with 6 digits for illustrative purpose):

$$\frac{\partial Q}{\partial K} = I^{\frac{1}{2}}(Bh)^{\frac{5}{3}}(B+2\,h)^{-\frac{2}{3}} = \frac{Q}{K} = 0.004615 \tag{8.38}$$

$$\frac{\partial Q}{\partial I} = \frac{1}{2}KI^{-\frac{1}{2}}(Bh)^{\frac{5}{3}}(B+2\,h)^{-\frac{2}{3}} = \frac{Q}{2I} = 54.090477 \tag{8.39}$$

$$\frac{\partial Q}{\partial B} = \frac{5}{3}hKI^{\frac{1}{2}}(Bh)^{\frac{2}{3}}(B+2\,h)^{-\frac{2}{3}} - \frac{2}{3}KI^{\frac{1}{2}}(Bh)^{\frac{5}{3}}(B+2\,h)^{-\frac{5}{3}} = \frac{Q}{3}\left(\frac{5}{B} - \frac{2}{B+2\,h}\right) = 0.557013 \tag{8.40}$$

$$\frac{\partial Q}{\partial h} = \frac{5}{3}BKI^{\frac{1}{2}}(Bh)^{\frac{2}{3}}(B+2\,h)^{-\frac{2}{3}} - \frac{4}{3}KI^{\frac{1}{2}}(Bh)^{\frac{5}{3}}(B+2\,h)^{-\frac{5}{3}} = \frac{Q}{3}\left(\frac{5}{h} - \frac{4}{B+2\,h}\right) = 1.483588 \tag{8.41}$$

The second order numerical approximation, calculated according to Equation (8.12), gives:

$$\frac{\partial Q}{\partial K} = \frac{Q(K+\varepsilon_K, I, B, h) - Q(K-\varepsilon_K, I, B, h)}{2\varepsilon_K} = 0.004615 \tag{8.42}$$

$$\frac{\partial Q}{\partial I} \approx \frac{Q(K, I+\varepsilon_I, B, h) - Q(K, I-\varepsilon_I, B, h)}{2\varepsilon_I} = 54.090477 \tag{8.43}$$

$$\frac{\partial Q}{\partial B} \approx \frac{Q(K, I, B+\varepsilon_B, h) - Q(K, I, B-\varepsilon_B, h)}{2\varepsilon_B} = 0.557013 \tag{8.44}$$

$$\frac{\partial Q}{\partial h} \approx \frac{Q(K, I, B, h+\varepsilon_h) - Q(K, I, B, h-\varepsilon_h)}{2\varepsilon_h} = 1.483588 \tag{8.45}$$

Both approaches provide results which, in this example, are identical to the 6th digit at least. The advantage of the numerical approximation is the possibility to run calculations automatically without algebra.

The resulting standard uncertainty is $u(Q) = 0.013$ m^3/s and the relative standard uncertainty is $u^*(Q) = u(Q)/Q = 0.039$, i.e. 3.9%. All calculations with Matlab® are given in Box 3.

BOX 3: STEP BY STEP CALCULATIONS OF *Q* AND *u*(*Q*) IN A RECTANGULAR CHANNEL WITH MATLAB®

Let us first set the values and the standard uncertainties for *K*, *I*, *B*, and *h*.
Define the interval for *K*:

```
intK=[70 80]
```

Calculate *K* and its standard uncertainty *u*(*K*):

```
K=mean(intK)
uK=diff(intK)/2/sqrt(3)
```

One gets $K = 75$ m$^{1/3}$/s and $u(K) = 2.8867$ m$^{1/3}$/s.
Then type

```
I=3.2e-3
uI=6e-6
B=0.805
uB=2e-3
h=0.32
uh=1.5e-3
```

Calculate the discharge *Q*(*K*,*I*,*B*,*h*) by Equation (8.36):

```
Q=K.*power(I,1/2).*power(B.*h,5/3).*power(B+2.*h,-2/3)
```

One gets $Q = 0.3462$ m^3/s.
Calculate the standard uncertainty *u*(Q) by Equation (8.37).
Let us first define the quantities `cK`, `cI`, `cB` and `ch` respectively equal to the numerical values of the partial derivatives of Q (Equations 8.42 to 8.45):

```
epsK=uK/1000
epsI=uI/1000
epsB=uB/1000
epsh=uh/1000
cK=((K+epsK).*power(I,1/2).*power(B.*h,5/3).*power(B+2*h,-2/3)-
(K-epsK).*power(I,1/2).*power(B.*h,5/3).*power(B+2*h,-2/3))/epsK/2
cI=(K.*power(I+epsI,1/2).*power(B.*h,5/3).*power(B+2*h,-2/3)-
K.*power(I-epsI,1/2).*power(B.*h,5/3).*power(B+2*h,-2/3))/epsI/2
cB=(K.*power(I,1/2).*power((B+epsB).*h,5/3).*power(B+epsB+2*h,-2/3)-
K.*power(I,1/2).*power((B-epsB).*h,5/3).*power(B-epsB+2*h,-2/3))/epsB/2
ch=(K.*power(I,1/2).*power(B.*(h+epsh),5/3).*power(B+2*(h+epsh),-2/3)-
K.*power(I,1/2).*power(B.*(h-epsh),5/3).*power(B+2*(h-epsh),-2/3))/
epsh/2
```

Then create two intermediate vectors `Vu` and `Vc` and apply Equation (8.37):

```
Vu=[uK uI uB uh]'
Vc=[cK cI cB ch]'
uQ=sqrt(sum(power(Vu.*Vc, 2)))
```

One gets
$u(Q) = 0.0136$ m^3/s
The relative standard uncertainty is calculated by

```
uQ/Q
```

One gets $u^*(Q) = 0.0392$, i.e. 3.9%.
The 95% coverage interval is given by

```
[ Q-1.96*uQ, Q+1.96*uQ]
```

One gets [0.3196, 0.3728] m^3/s.

Let us now calculate the 95% coverage interval. In the first and most simple way, one may assume that the coverage factor $k = 1.96$ for the probability level $\alpha = 0.95$. In this case, the 95% coverage interval for Q is determined from Equation (8.17):

$$[Q - 1.96 \times u(Q),\ Q + 1.96 \times u(Q)] = [0.320,\ 0.373]\ \mathrm{m^3/s} \tag{8.46}$$

A more detailed approach is based on the estimation of the effective degree of freedom from the Welch-Satterthwaite formula (Equations 8.14 to 8.16). Detailed Matlab® calculations are given in Box 4. The effective degrees of freedom ν_i are estimated as follows.

As $u(K)$ can be considered to be known itself with a relative uncertainty of 20% according e.g. to hydraulic textbooks, then, according to Equation (8.15):

$$\nu(K) = \frac{1}{2}[0.20]^{-2} = 12.49 \tag{8.47}$$

to be approximated to the nearest lower integer, i.e. $\nu(K) = 12$.

$u(I)$ is given by the land surveyor, based on repeated calibrations of the sensors. One assumes here that $\nu(I) = \infty$.

$u(B)$ is calculated from a Type A estimation based on $n = 4$ repeated measurements: $\nu(B) = n - 1 = 3$.

$u(h)$ is estimated from a sensor calibration based on $n = 60$ measurements (12 repeated measurements for 5 values along the sensor measurement range). Thus $\nu(h) = 60 - 1 = 59$. For comparison purposes, one may also assume that $\nu(h) = \infty$.

Applying Equation (8.14) gives $\nu_{eff} = 12$, with both $\nu(h) = 59$ or $\nu(h) = \infty$.

BOX 4: STEP BY STEP CALCULATIONS OF THE COVERAGE INTERVAL OF Q WITH THE EFFECTIVE DEGREE OF FREEDOM

Set the respective values of the degrees of freedom for K, I, B, and h:

```
nuK=floor(0.5*power(0.20, -2))
```

One gets $\nu(K) = 12$.

Type

```
nuI=Inf
nuB=3
nuh=59 (or nuh=Inf)
```

Define `Vnu` an intermediate vertical vector:

```
Vnu=[nuK nuI nuB nuh]'
```

The effective degree of freedom ν_{eff} is calculated by Equation (8.14), with `Vu` and `Vc` defined in Box 3:

```
nueff=floor(power(uQ,4)/sum(power(Vu.*Vc,4)./Vnu))
```

One gets $\nu_{eff} = 12$, with both $\nu(h) = 59$ or $\nu(h) = \infty$.

Then type

```
alpha=0.95
k=tinv((1+alpha)/2, nueff)
```

One gets $k = 2.1788$.

The coverage interval is then calculated by:

```
[Q-k*uQ,Q+k*uQ]
```

One gets [0.3166, 0.3757] m^3/s.

The coverage factor k corresponding to $\nu_{eff} = 12$ is equal to 2.18 (Equation (8.16)), which is a little bit higher than the default value $k = 1.96$ used previously.

Lastly, applying Equation (8.17), the coverage interval for the discharge Q is:

$$[Q - k \times u(Q), Q + k \times u(Q)] = [0.317, \quad 0.376] \text{ m}^3/\text{s}.$$

Looking at the details of the calculations reveals that the strongest contribution to the combined uncertainty is due to the uncertainty in the Manning-Strickler coefficient K. If one assumes that the relative uncertainty of $u(K)$ is 10% instead of 20%, then $\nu(K)$ increases from 12 to 49. Consequently, ν_{eff} increases from 12 to 52, and $k = 2.006$, which is now equivalent to the default value. This emphasizes the importance of a reliable assessment of all components contributing to the estimation of the coverage interval, including the degrees of freedom.

All calculations presented in this example can be run automatically with the Matlab® code `uTypeB` presented in Box 5.

BOX 5: APPLICATION OF THE TYPE B METHOD WITH THE MATLAB® CODE uTypeB

(Matlab® codes and csv files available for download at https://doi.org/10.2166/9781789060102).

The calculations of the discharge Q and its standard uncertainty $u(Q)$ shown in Box 3 and Box 4 are automated in the Matlab® code `Y = uTypeB(Z,A,chaine,alpha,MatCor,NuZ,NuA)`, where the quantities X_i of the measurement process are divided into two groups Z and A, where:

`Z` is the matrix containing data related to time varying quantities (time series). Its structure is as follows: Z has as many lines as time steps in the time series. Each line contains, in successive columns from left to right, m pairs of data for each time varying quantity Z: Z_1, $u(Z_1)$, Z_2, $u(Z_2)$,...Z_i, $u(Z_i)$, ...Z_m, $u(Z_m)$.

`A` is the matrix containing data related to constant quantities. The single line of A contains, in the successive columns from left to right, p pairs of data for each constant quantity A: A_1, $u(A_1)$, A_2, $u(A_2)$,...A_j, $u(A_j)$,...A_p, $u(A_p)$.

The sum $m + p$ is equal to N, the number of quantities X_i in Equation (8.9).

`chaine` is the equation string describing the measurement process of Y using Z_i and A_j quantities.

`alpha` is the level of probability.

`MatCor` is the matrix of correlation between Z_i and A_j quantities.

`NuZ` is the vector of the degrees of freedom $\nu(Z_i)$ of each quantity Z_i, in the same order as in the matrix Z.

`NuA` is the vector of the degrees of freedom $\nu(A_j)$ of each quantity A_j, in the same order as in the matrix A.

`NuZ` and `NuA` are optional inputs in the function `uTypeB`: if they are not used, the coverage interval is estimated solely with the default values of k given in Tables 8.2 and 8.3 for an infinite degree of freedom.

The `uTypeB` function provides respectively the following results in successive columns of the output matrix Y, with one line per time step: the value y of Y, its standard uncertainty $u(Y)$, the boundaries of the coverage interval calculated with an infinite degree of freedom and the boundaries with the effective degree of freedom ν_{eff}.

For the above example, there is one time-varying quantity: the water level h, and three constant quantities: respectively the Manning-Strickler coefficient K, the channel slope I and the channel width B. Consequently, type

BOX 5: (Continued)

`Z=[h uh]`
with only one line as there is only one single value of h.
`A=[K uK I uI B uB]`
The previous measurement process equation (see Box 3):
```
Q=K.*power(I,1/2).*power(B.*h,5/3).*power(B+2.*h,-2/3)
```
is then re-written with notations indicating the quantities with their rank in matrices Z and A.
h is the first quantity in the matrix Z: `h` is replaced by `Z(:,1)` in the above expression of Q. Similarly, `K`, `I` and `B` are replaced, respectively, by `A(:,1)`, `A(:,2)` and `A(:,3)` as they are, respectively, the first, second and third quantities in the matrix A. It is important to note that the indices refer to the rank of the quantities in matrices Z and A, and not to the rank of the columns.
Consequently, type
```
chaine='A(:,1).*power(A(:,2),0.5).*power(A(:,3).*Z(:,1),5/3).*power
        (A(:,3)+2.*Z(:,1),-2/3)'
alpha=0.95
```
In this example, all four quantities in Z and A are independent as there is no correlation or covariance between them. Consequently, type
`MatCor=eye(4)`
which gives

$$MatCor = \begin{bmatrix} 1 & 0 & 0 & 0 \\ 0 & 1 & 0 & 0 \\ 0 & 0 & 1 & 0 \\ 0 & 0 & 0 & 1 \end{bmatrix}$$

where the columns from left to right and the lines from top to bottom refer successively to the Z_i and A_j quantities.
In addition, type
```
NuZ=[59]
NuA=[12 inf 3]
```
Lastly type
```
Q=uTypeB(Z,A,chaine,alpha,MatCor,NuZ,NuA)
```
One gets $Q = 0.3462$ m^3/s, $u(Q) = 0.0136$ m^3/s, and
coverage interval with an infinite degree of freedom: [0.3196, 0.3728]
coverage interval with the effective degree of freedom: [0.3166, 0.3757].

8.2.4 Monte Carlo method for uncertainty assessment

8.2.4.1 Principle

The Monte Carlo method (MCM) is a generic simulation method, which can be applied to estimate uncertainties under various conditions and in particular when the conditions of the Type B method are not satisfied (non-linearity, non-symmetric distributions, significance of second order terms in the derivation of the LPU) or are very difficult to apply. It is considered as the reference method.

The MCM consists basically of repeatedly simulating the measurement process f calculating the measurand Y from the quantities X_i as described by Equation (8.8) in Section 8.2.3.1:

$$y = f(x_1, x_2, \ldots x_i, \ldots x_N) \tag{8.48}$$

Samples of size M are built for all quantities X_i involved in the measurement process, according to their distributions and with appropriate correlation coefficients. Then the samples are used to calculate M times the value y of the measurand Y and the M values of y allows calculation of the mean value $\bar{y}$ of the measurand Y, the standard uncertainty $u(y)$ and the coverage interval corresponding to the defined probability level α.

As for the Type B method, it is of particular importance:

- To write the measurement process $Y = f(X_1, X_2, \ldots X_N)$ in a way which closely reflects the measurement process with independent quantities X_i.
- To define the probability distributions for all quantities X_i.
- To carefully analyse the possible correlations between quantities and to quantify them when they exist.

The principle of the MCM is illustrated in Figure 8.2. The sample of each quantity X_i for $i = 1{:}N$ in the measurement process f is represented by a vector of M values with $r = 1{:}M$: $x_{i,1}, x_{i,2}, \ldots x_{i,r}, \ldots x_{i,M}$. Within each sample, the successive values $x_{i,r}$ are sorted neither in ascending nor in descending order but are randomly listed.

The samples are built according to two requirements: (i) each sample distribution should represent the information and knowledge about each quantity X_i, as in the Type B method (see Section 8.2.4.2), and (ii) the correlation between samples should represent the correlation as described in Equation (8.10) between the quantities X_i in the measurement process f (see Section 8.2.4.3). Creating large samples cannot be done manually and software tools are required.

Once all samples of the quantities X_i are built appropriately, M values of y are calculated as follows:

$$y_r = f(x_{1,r}, x_{2,r}, \ldots x_{i,r}, \ldots x_{N,r}) \tag{8.49}$$

with $r = 1{:}M$.

Sample for X_1	Sample for X_2	⋯	Sample for X_i	⋯	Sample for X_N	$f(X_1, X_2, \ldots X_i, \ldots X_N)$	Sample for Y
$x_{1,1}$	$x_{2,1}$		$x_{i,1}$		$x_{N,1}$	→	y_1
$x_{1,2}$	$x_{2,2}$		$x_{i,2}$		$x_{N,2}$	→	y_2
$x_{1,3}$	$x_{2,3}$		$x_{i,3}$		$x_{N,3}$	→	y_3
⋮	⋮		⋮		⋮		⋮
$x_{1,r}$	$x_{2,r}$		$x_{i,r}$		$x_{N,r}$	→	y_r
⋮	⋮		⋮		⋮		⋮
$x_{1,M}$	$x_{2,M}$		$x_{i,M}$		$x_{N,M}$	→	y_M

Figure 8.2 Principle of the Monte Carlo method. *Source*: Jean-Luc Bertrand-Krajewski (INSA Lyon).

The estimate of the measurand Y is given by the mean $\bar{y}$ of the y_r values:

$$\bar{y} = \frac{1}{M}\sum_{r=1}^{M} y_r \tag{8.50}$$

The estimate of the standard uncertainty $u(y)$ is given by the standard deviation $s(y)$ of the values y_r:

$$u(y) = \sqrt{\frac{1}{M-1}\sum_{r=1}^{M}(y_r - \bar{y})^2} \tag{8.51}$$

One of the advantages of the MCM, contrarily to the Type B method, is its ability to deal with non-symmetric distributions. As a consequence, the distribution of the y_r values may be also non-symmetric, and thus the coverage interval is not necessarily symmetric around the mean value $\bar{y}$, as this is the case in Equation (8.17) with the Type B method.

With the MCM, the coverage interval $[y_{\alpha,low}, y_{\alpha,high}]$ for the given level of probability α corresponds to the narrowest interval containing the fraction α of the values y_r. Its estimation requires some preliminary steps for processing of the values y_r:

(1) Sort all values y_r in ascending and non-decreasing order (in case of possible equalities among values y_r). The sorted values are then noted $y_{(r)}$ with $r = 1{:}M$.
(2) Define an integer $q = \alpha M$ if αM is an integer. Otherwise, define q as the integer part of $(\alpha M + 1/2)$.
(3) Determine the confidence interval $[y_{\alpha,low\ r}, y_{\alpha,high\ r}]$ for Y where, for any $r = 1{:}M\text{-}q$, $y_{\alpha,low\ r} = y_{(r)}$ and $y_{\alpha,high\ r} = y_{(r+q)}$.
(4) The shortest coverage interval with probability level α is obtained with r^* such that, for $r = 1{:}M\text{-}q$,

$$y_{(r*+q)} - y_{(r*)} \leq y_{(r+q)} - y_{(r)} \tag{8.52}$$

8.2.4.2 Creating non-correlated samples

Creating samples of the quantities X_i according to a given probability density function (pdf) requires a random number generator. Basic Matlab® functions to create uniform, normal, triangular and trapezoidal samples are given in Box 6. Similar functions exist with other software tools. They are parameterized and used to create samples following predefined pdfs.

Other pdfs can be used with the MCM: Student t pdf, exponential pdf for values which cannot be higher or lower than a threshold (typically non-negative values), gamma pdf, empirical pdfs based on experiments, truncated normal pdf, lognormal pdf, etc.

Additional information to generate samples according to these pdfs can be found in ISO (2008b) and in textbooks (e.g. Gentle 2003; Press *et al.*, 2007; Robert & Casella, 2005).

Non-correlated samples can be created individually one after another or simultaneously by using appropriate software functions. It is however important to ensure that they are really not correlated, usually by controlling the parameters of the random number generator.

8.2.4.3 Creating correlated samples

In the case where two quantities X_i and X_j with $i \neq j$ are correlated in the measurement process f with a correlation coefficient $r(x_i, x_j)$ as given in Equation (8.10), their corresponding samples shall be built in such a way that (i) each quantity is distributed according to its own pdf and (ii) the correlation of the two samples is equal to $r(x_i, x_j)$. Generating correlated samples may be complex and presenting the details of

BOX 6: MATLAB® FUNCTIONS AND CODES TO CREATE NON-CORRELATED SAMPLES WITH PRE-DEFINED PDFS

Two basic Matlab® functions are available: `rand` and `randn`.

```
rand(M,1)
```

generates a vector of random numbers *x* uniformly distributed between 0 and 1, noted *U*(0, 1) with *M* lines and 1 column. An example of results is given in Figure 8.3 with $M = 10{,}000$.

```
randn(M,1)
```

generates a vector of random numbers *x* normally distributed with mean value $m = 0$ and standard deviation $s = 1$, noted *N*(0,1). An example of results is given in Figure 8.4 with $M = 10{,}000$.

These basic functions can be parameterized and used to create samples following predefined pdfs.

Uniform (rectangular) pdf

To create a sample of *M* values *x* uniformly distributed in the interval [*a*, *b*] (see Figure 8.1), noted *U*(*a*, *b*):

```
x=a+(b-a).*rand(M,1)
or
x=unifrnd(a,b,M,1)
```

Normal (Gaussian) pdf

To create *M* values *x* normally distributed with mean value *m* and standard deviation *s*, noted *N*(*m*, *s*):

```
x=m+s.*randn(M,1)
```

Triangular pdf

To create a sample of *M* values *x* distributed according to a triangular pdf in the interval [*a*, *b*] (see Figure 8.1), noted *Tri*(*a*, *b*):

0.223770404697041
0.373563807642645
0.087500349576586
0.640116548246715
0.180616887753108
0.045051107473574
0.723173479183095
0.347437645581790
0.660616824502904
0.383868601071971
0.627346502443467
0.021649814630306
0.910569988523029
0.800558656278811
0.745847484342721
0.813112813610761
0.383306318625529
0.617279232316449
0.575494859702814
0.530051704765016
0.275069755821935
0.248628959661970
0.451638770451973
0.227712826026548
0.804449583613070

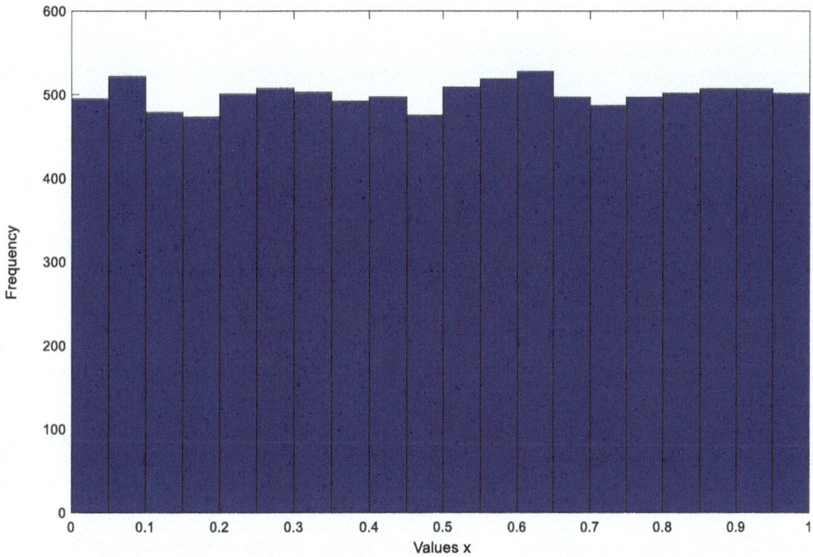

Figure 8.3 First 25 values (left) and histogram (right) of the 10,000 values *x* created by `rand(10000,1)`. *Source*: Jean-Luc Bertrand-Krajewski (INSA Lyon).

BOX 6: (Continued)

-0.110223485241791
0.414258701179269
0.230095272512933
0.857030996309090
0.048208732255379
1.023348450918008
-0.181992873626858
-0.290128384322202
0.815203911371096
0.322332331608368
0.138374675429518
-0.451012495444746
-0.153040599206172
-0.879002127680258
-0.192706939633842
0.194885569504571
1.594878084883789
-0.441943288721280
-0.421970829486422
0.225809147824111
0.146405492458328
0.009690168833214
0.107598715326320
-2.049531488596142
1.314029894987952

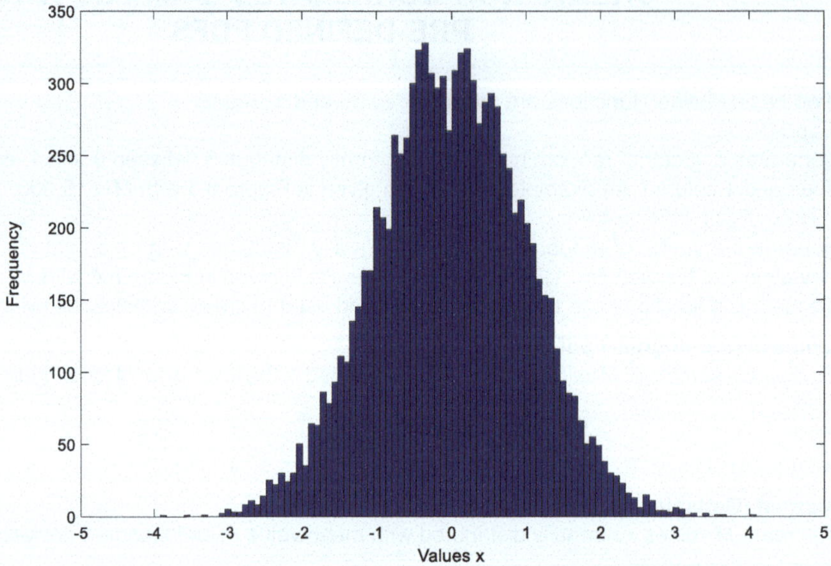

Figure 8.4 First 25 values (left) and histogram (right) of the 10,000 values *x* created by `randn (10000,1)`. *Source*: Jean-Luc Bertrand-Krajewski (INSA Lyon).

generate two independent samples

```
s1=rand(M,1)
s2=rand(M,1)
```

and then

```
x=a+(b-a)/2.*(s1+s2)
```

Trapezoidal pdf

To create a sample of *M* values *x* distributed according to a trapezoidal pdf in the interval [*a*, *b*] with the top coefficient β (see Figure 8.1), noted *Trap*(*a*, *b*, β):

generate two independent samples

```
s1=rand(M,1)
s2=rand(M,1)
```

and then

```
x=a+(b-a)/2.*((1+β)*s1+(1-β)*s2)
```

The function `rng` allows controlling of the independence (no correlation) of samples.

Parametric functions `makedist` and `random` allow creation of samples with other pre-defined pdfs.

available methods (e.g. copulas) is beyond the scope of this chapter (Gentle, 2003; Nelsen, 2010). Creating samples with various pdfs and predefined correlation coefficients requires specific methods. Among them, the copula functions are very convenient (Nelsen, 2010) and can generate bi- and multivariate samples with correlated marginal distributions. Generation of correlated uniform and normal samples with Matlab® is described in Box 7. Other examples are given below in Section 8.3.

BOX 7: MATLAB® FUNCTIONS AND CODES TO CREATE UNIFORM AND NORMAL CORRELATED SAMPLES

Bivariate normal (Gaussian) pdf

To create two correlated samples of size M for two quantities X_i and X_j normally distributed, with their own respective means m_i and m_j, and standard deviations s_i and s_j, noted $N(m_i, s_i)$ and $N(m_j, s_j)$, with a defined coefficient of correlation $r_{ij} = r(x_i, x_j) = r_{ji} = r(x_j, x_i)$ (knowing that $r_{ii} = r_{jj} = 1$):
create the vector of means $m = [m_i, m_j]$
create the covariance matrix:

$$COV = \begin{bmatrix} cov(x_i, x_i) & cov(x_i, x_j) \\ cov(x_j, x_i) & cov(x_j, x_j) \end{bmatrix} = \begin{bmatrix} r_{ii}s_is_i & r_{ij}s_is_j \\ r_{ji}s_js_i & r_{jj}s_js_j \end{bmatrix} = \begin{bmatrix} s_i^2 & r_{ij}s_is_j \\ r_{ji}s_js_i & s_j^2 \end{bmatrix}$$

and then type

```
X=mvnrnd(m,COV,M)
```

The output matrix X contains 2 columns, respectively with the values of X_i and X_j.

Example 1: two samples X_1 and X_2 normally distributed with $M = 10{,}000$ values, with $m_1 = 10$, $s_1 = 25$, $m_2 = 25$, $s_2 = 1.2$ and $r_{12} = 0.45$.

First define

```
m=[10 25]
s1=0.5
s2=1.2
r12=0.45
COV=[s1^2 r12* s1* s2; r12* s2* s1 s2 ^2]
M=10000
```

and then

```
X=mvnrnd(m,COV,M)
```

The results are illustrated in Figure 8.5.

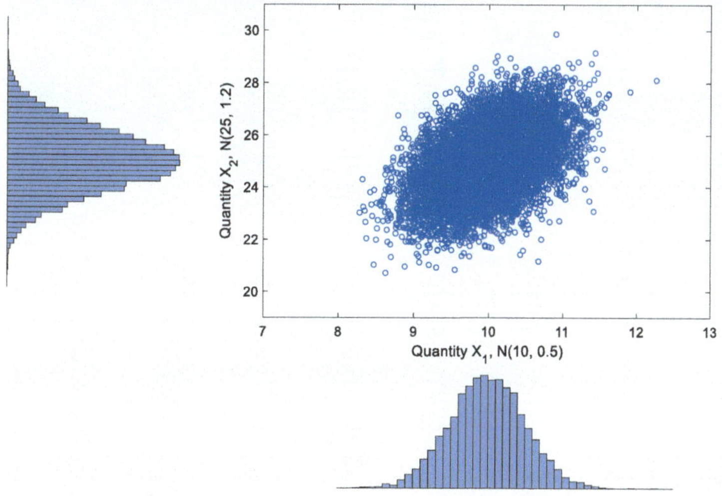

Figure 8.5 Scatter plot and marginal histograms of two correlated normally distributed samples with 10,000 values x_1 ($m_1 = 10$, $s_1 = 0.5$) and x_2 ($m_2 = 25$, $s_2 = 1.2$) and $r_{12} = 0.45$ created by `mvnrnd`. *Source*: Jean-Luc Bertrand-Krajewski (INSA Lyon).

BOX 7: (Continued)

Bivariate uniform (rectangular) pdf

The easiest way to generate two correlated uniformly distributed samples X_i and X_j is to start with two correlated standard normally distributed samples $N(0, 1)$ and then to use the normal cumulative probability function (cdf) usually noted Φ. The two new standard uniform samples $U(0, 1)$ are then re-scaled as $U(a_i, b_i)$ and $U(a_j, b_j)$ on their respective intervals.

Example 2: two samples X_1 and X_2 uniformly distributed with $M = 10{,}000$ values, with $[a_1, b_1] = [5, 15]$, $[a_2, b_2] = [18, 32]$ and $r_{12} = 0.45$.

First create two normally distributed samples $N(0, 1)$ in the matrix Z_1 :

```
m=[0 0]
COV=[1 0.45; 0.45 1]
M=10000
Z1=mvnrnd(m,COV,M)
```

Then generate the two uniformly distributed samples $U(0, 1)$ in the matrix Z_2 by means of the function Φ, noted `normcdf` in Matlab®, which transforms a standard normal random sample $N(0, 1)$ into a random sample that is uniform $U(0, 1)$:

```
Z2=normcdf(Z1)
```

Lastly, create the two samples $U(5, 15)$ and $U(18, 32)$ from the two samples in Z_2 :

```
X1=5+(15-5).*Z2(:,1)
X2=18+(32-18).*Z2(:,2)
```

The results are illustrated in Figure 8.6.

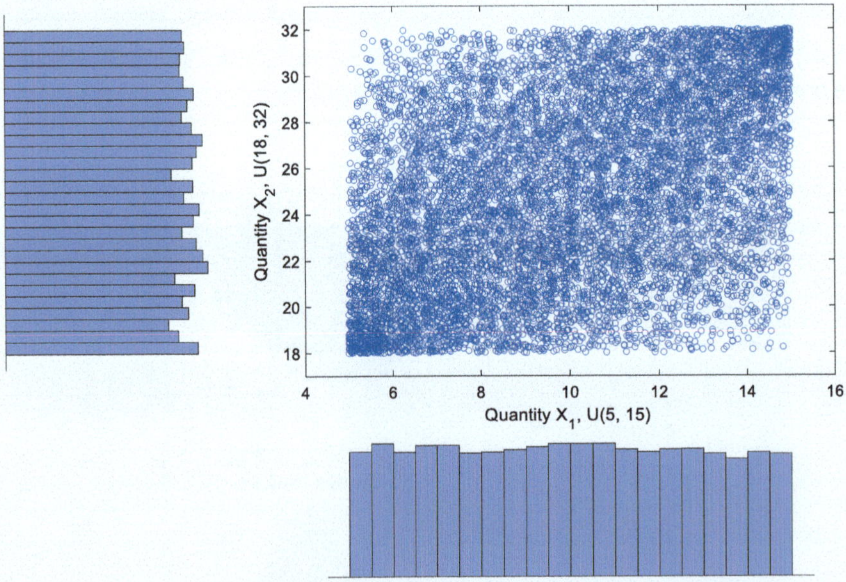

Figure 8.6 Scatter plot and marginal histograms of two correlated uniformly distributed samples with 10,000 values x_1 ($a_1 = 5$, $b_1 = 15$) and x_2 ($a_2 = 18$, $b_2 = 32$) and $r_{12} = 0.45$ created by `mvnrnd` and `normcdf`. *Source*: Jean-Luc Bertrand-Krajewski (INSA Lyon).

BOX 7: (Continued)

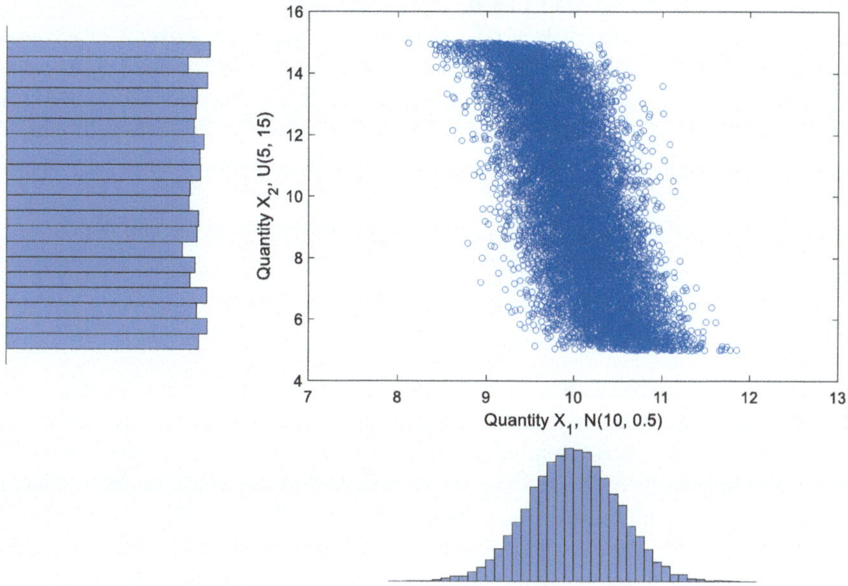

Figure 8.7 Scatter plot and marginal histograms of two correlated samples with $M = 10{,}000$ values and $r_{12} = -0.7$: x_1 as $N(10, 0.5)$ and x_2 as $U(5, 15)$ created by `copularnd`. *Source*: Jean-Luc Bertrand-Krajewski (INSA Lyon).

Other bi- and multivariate pdfs

Example 3: create two samples X_1 normally distributed with mean $m = 10$ and standard deviation $s = 0.5$, and X_2 uniformly distributed in the interval $[a, b] = [5, 15]$, with $M = 10{,}000$ values and a correlation coefficient $r_{12} = -0.7$.

First create the copula Z:

```
M=10000
r12=-0.7
Z=copularnd('gaussian',r12,M)
```

Then create the two samples X_1 and X_2 in the matrix X:

```
X=[norminv(Z(:,1),10,0.5) 5+(15-5).*Z(:,2)]
```

The results are illustrated in Figure 8.7.

The correlation coefficient $r(X_1, X_2) = -0.69$, which is close to the target value $r_{12} = -0.7$.

8.2.4.4 Size of samples

As the MCM is stochastic in nature, each run of the MCM giving a sample of M estimates y_r of Y from the random samples of the quantities X_i will generate slightly different outputs for $\bar{y}$, $u(y)$ and $[y_{\alpha,low}, y_{\alpha,high}]$. If M is too small (some tens or hundreds), the results may be significantly different for each run, especially for the interval $[y_{\alpha,low}, y_{\alpha,high}]$ which strongly depends on the representativeness and exhaustiveness of the possible combinations of the values $x_{i,r}$ involved in the calculation of the values y_r. Large values of M ensure that successive runs of the MCM deliver successive output estimates which are stable and do not differ from each other more than a required numerical tolerance to be defined by the operator.

ISO (2008b) indicates in its Section 7.2.2 that 'the choice of a value of M that is large compared with $1/(1-\alpha)$, e.g. M at least 10^4 times greater than $1/(1-\alpha)$, should be made'. For example, with $\alpha = 0.95$, a minimum value of $M = 2 \times 10^5$ should be used.

In practice, in many cases, 'a value of $M = 10^6$ can often be expected to deliver a 95% coverage interval for the output quantity such that its length is correct to one or two significant decimal digits' (note in Section 7.2.1 of ISO, 2008b). However, it is always recommended to check that this prior choice of M is appropriate.

A procedure that selects M adaptively by iterations can be used (Section 7.9 of ISO, 2008b). It is based on the number of significant digits n_{dig} considered as meaningful in a numerical value z by the operator. The numerical tolerance δ is then defined as follows:

Express z in the form $c \times 10^{\,l}$ where c is a n_{dig} decimal digit integer and l is an integer.

$$\text{Set } \delta = \frac{1}{2} 10^l \tag{8.53}$$

Example: tolerance

If a discharge $Q = 0.346\ \text{m}^3/\text{s}$ with a standard uncertainty $u(Q) = 0.013\ \text{m}^3/\text{s}$ (taken from Section 8.2.3.4), and if the last two digits are significant, then $n_{dig} = 2$ and $u(Q)$ can be written as 13×10^{-3}, therefore $c = 13$ and $l = -3$. Consequently, the tolerance $\delta = ½ \times 10^{-3} = 0.0005 = 5 \times 10^{-4}\ \text{m}^3/\text{s}$. If only one digit is significant, then $Q = 0.35\ \text{m}^3/\text{s}$, $u(Q) = 0.01\ \text{m}^3/\text{s}$, $u(Q)$ can be written as 1×10^{-2}, therefore $c = 1$ and $l = -2$. Consequently, the tolerance $\delta = ½ \times 10^{-2} = 0.005 = 5 \times 10^{-3}\ \text{m}^3/\text{s}$.

The objective of the MCM adaptive procedure is to estimate $\bar{y}$, $u(y)$ and $[y_{\alpha,low}, y_{\alpha,high}]$ in a way that ensures they meet the numerical tolerance required. The adaptive procedure includes the following steps (ISO, 2008b):

(1) Set n_{dig}, usually 1 or 2.
(2) Set $M = \max(J, 10^4)$ where J is the smallest integer greater than or equal to $100/\alpha$.
(3) Set $a = 1$, first application of the MCM in the sequence.
(4) Carry out the M Monte Carlo simulations.
(5) From the M outputs, calculate $\bar{y}^{(a)}$, $u(y)^{(a)}$, $y_{\alpha,low}^{(a)}$ and $y_{\alpha,high}^{(a)}$ respectively, for the ath member of the sequence.
(6) If $a = 1$, increase a by one and return to step 4.
(7) Calculate the standard deviation s_y associated with the average estimates $y^{(1)}$, $y^{(2)}$, …$y^{(a)}$ by:

$$s_y^2 = \frac{1}{a(a-1)} \sum_{r=1}^{a} (y^{(r)} - y_m)^2 \tag{8.54}$$

where

$$y_m = \frac{1}{a} \sum_{r=1}^{a} y^{(r)} \tag{8.55}$$

(8) Calculate the counterpart of this statistic also for $u(y)^{(a)}$, $y_{\alpha,low}^{(a)}$ and $y_{\alpha,high}^{(a)}$ respectively.
(9) Use all $a \times M$ values available so far to calculate $\bar{y}$, $u(y)$, $y_{\alpha,low}$ and $y_{\alpha,high}$.
(10) Calculate the numerical tolerance δ according to Equation (8.53) respectively for $\bar{y}$, $u(y)$, $y_{\alpha,low}$ and $y_{\alpha,high}$.
(11) If any of $2s_y$, $2s_{u(y)}$, $2s_{y\alpha,low}$ and $2s_{y\alpha,high}$ exceeds its respective tolerance δ, increase a by one and return to step 4. Otherwise, go to step 12.
(12) As all outputs of the MCM are now stabilized, use all $a \times M$ values available to calculate $\bar{y}$, $u(y)$, and $[y_{\alpha,low}, y_{\alpha,high}]$.

The above adaptive method is appropriate to minimize the total number $a \times M$ of simulations to be run to achieve a given tolerance. However, if the choice of M in step 2 is too low, it will be necessary to increase the number a of applications, and the global time required for simulations will increase. So, in practice, it is suggested to use preferably $M = 10^5$ rather than 10^4 in step 2. In addition, for repetitive calculations (e.g. calculations of uncertainties in discharge in a sewer with measurements every 2 or 5 minutes), the adaptive procedure does not need to be carried out for each time step, which is counterproductive. A first estimation can be done to estimate the minimum necessary total number $a \times M$ of simulations to reach the required tolerance, and then run this total number of Monte Carlo simulations, possibly multiplied by a safety factor, to save computation times. For example, if preliminary assessments indicate that a total number $a \times M = 4 \times 10^5$ simulations is necessary, the routine number can be increased to 10^6 for repetitive simulations. This is only a case-by-case decision.

8.2.4.5 Basic example with Matlab®

For this example, let us take the same case as for the Type B example in Section 8.2.3.4. The four measured quantities K, I, B and h involved in the measurement process given by Equation (8.21) are independent and not correlated, so independent samples will be created. Detailed Matlab® calculations are given in Box 8.

BOX 8: STEP BY STEP APPLICATION OF THE MCM WITH MATLAB®

Define $M = 10^6$:

```
M=1e6
```

For K, create a sample U(70, 80):

```
K=unifrnd(70,80,M,1)
```

One gets

```
mean(K)=74.99
std(K)=2.8854
```

For I, create a sample $N(0.0032, 6 \times 10^{-6})$:

```
I=normrnd(0.0032, 6e-6,M,1)
```

One gets

```
mean(I)=0.003200
std(I)=6.003e-06
```

For B, create a sample N(0.8005, 0.002):

```
B=normrnd(0.805, 0.002,M,1)
```

One gets

```
mean(B)=0.80500
std(B)=0.00200
```

For h, create a sample N(0.32, 0.0015) (as with a degree of freedom $\nu = 59$, the Student t distribution is equivalent to a normal distribution):

```
h=normrnd(0.32,0.0015,M,1)
```

One gets

```
mean(h)=0.3199
std(h)=0.001499
```

The histograms of the four samples are shown in Figure 8.8.

BOX 8: (Continued)

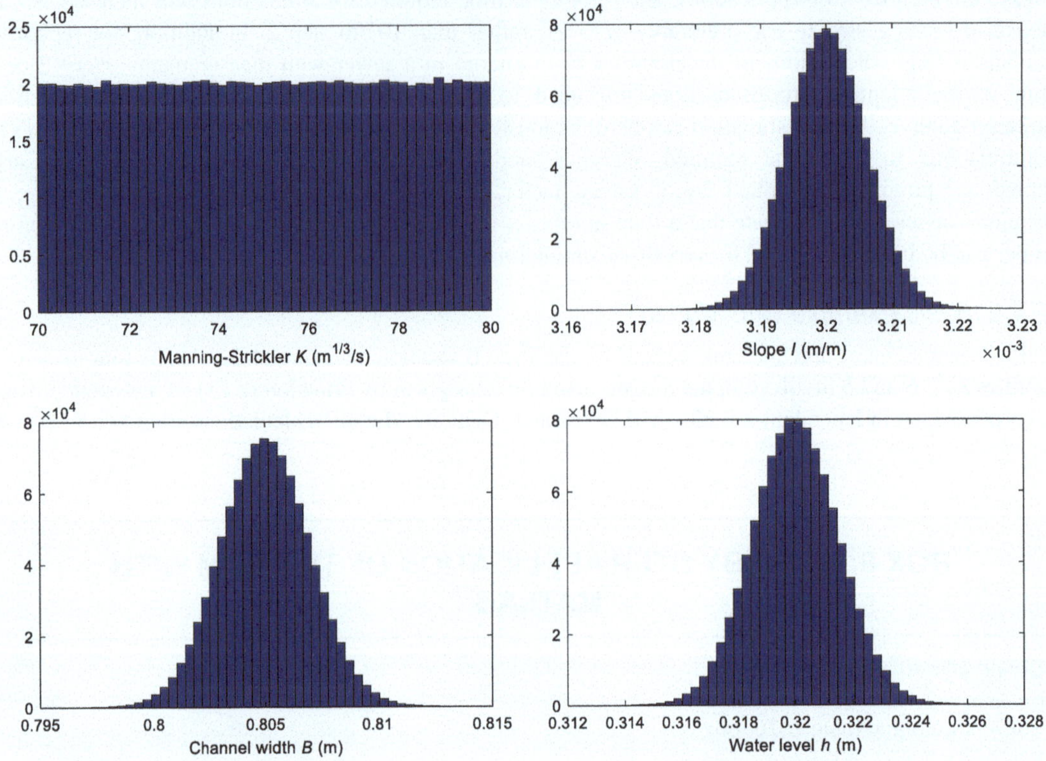

Figure 8.8 Histograms of K, I, B and h used in the MCM with $M = 1{,}000{,}000$ values. *Source*: Jean-Luc Bertrand-Krajewski (INSA Lyon).

Then calculate the vector Q with M values:

```
Q=K.*power(I,1/2).*power(B.*h,5/3).*power(B+2.*h,-2/3)
```

(Note the importance of the dot in front of the * operators in the above expression to ensure sample vectors are multiplied term by term according to the Matlab® syntax).

The histogram of the M values of Q is shown in Figure 8.9.

Calculate the mean value:

```
mean(Q)
```

One gets Q = 0.3462 m^3/s.

Calculate the standard deviation:

```
std(Q)
```

One gets $u(Q) = 0.0136$ m^3/s.

Calculate the shortest 95% coverage interval:

```
alpha=0.95
Qsort=sort(Q)
q=round(alpha*M)
```

which is an integer.

One gets $q = 950{,}000$.

BOX 8: (Continued)

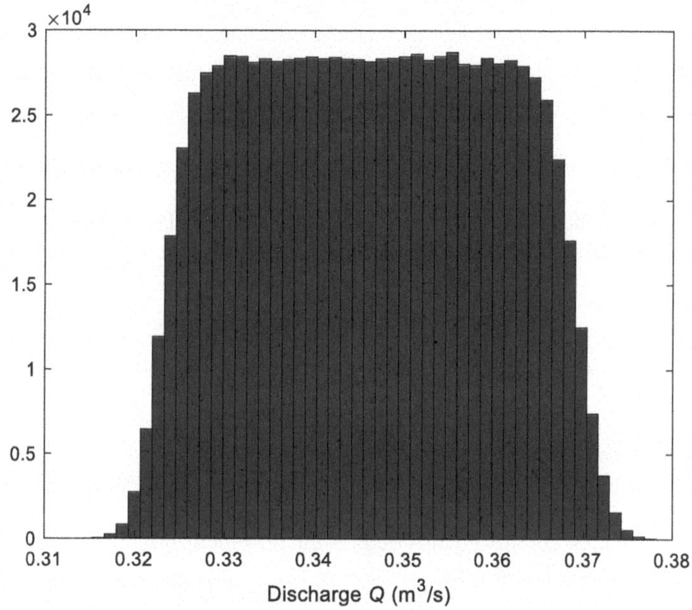

Figure 8.9 Histogram of the discharge Q calculated with the MCM with $M = 1{,}000{,}000$ values. *Source*: Jean-Luc Bertrand-Krajewski (INSA Lyon).

Create the matrix *IC95* with three columns (respectively low boundary, high boundary and width of the interval) as follows:

```
for r=1:M-q
IC95(r,1:3)=[Qsort(r) Qsort(r+q) Qsort(r+q)-Qsort(r)];
end
```

Find the boundaries (in columns 1 and 2) of the shortest interval (in column 3) in *IC95*:

```
IC95(find(IC95(:,3)==min(IC95(:,3))),1:2)
```

The result is $[Q_{95,\ low}, Q_{95,\ high}] = [0.3233, 0.3689]$ m^3/s.

Let us create non-correlated samples with size $M = 10^6$ for the four quantities:

- K with a uniform distribution between 70 and 80 m$^{1/3}$/s.
- I, B and h with normal distributions given, respectively, by $N_I(0.0032, 6\times10^{-6})$, $N_B(0.8005, 0.002)$ and $N_h(0.32, 0.0015)$.

The histograms of the four samples are shown in Figure 8.8. The M values of the discharge Q are then calculated by Equation (8.21). The resulting histogram of the Q values is shown in Figure 8.9.

The mean value is $\bar{Q} = 0.346$ m^3/s, the standard uncertainty is $u(Q) = 0.014$ m^3/s and the shortest 95% coverage interval is $[Q_{95,low}, Q_{95,high}] = [0.323, 0.369]$ m^3/s. This interval is shorter than the interval [0.320, 0.373] obtained with the Type B method in Section 8.2.3.4. This difference is discussed in Section 8.2.5.

Box 9 shows how to apply the Monte Carlo method with the Matlab® code `uMCM`.

BOX 9: APPLICATION OF THE MONTE CARLO METHOD WITH THE MATLAB® CODE uMCM

(Matlab® codes and csv files available for download at https://doi.org/10.2166/9781789060102). The calculations of the discharge Q, its standard uncertainty $u(Q)$ and its shortest coverage interval shown in Box 8 are automated in the Matlab® code `Y = uMCM(Z,A,chaine,alpha,MatCor,Mmc,distrib)`, where the quantities X_i of the measurement process are divided into two groups Z and A, where:

`Z` is the matrix containing data related to time varying quantities (time series). Its structure is as follows: Z has as many lines as time steps in the time series. Each line contains, in successive columns from left to right, m pairs of data for each time varying quantity Z: Z_1, $u(Z_1)$, Z_2, $u(Z_2)$,…Z_i, $u(Z_i)$, …Z_m, $u(Z_m)$.

`A` is the matrix containing data related to constant quantities. The single line of A contains, in the successive columns from left to right, p pairs of data for each constant quantity A: A_1, $u(A_1)$, A_2, $u(A_2)$,…A_j, $u(A_j)$,…A_p, $u(A_p)$.

`chaine` is the equation string describing the measurement process of Y using Z_i and A_j quantities.

`alpha` is the level of probability.

`MatCor` is the $N \times N$ matrix of correlation between Z_i and A_j quantities, with $N = m + p$.

`Mmc` is the length of the samples (number of Monte Carlo simulations).

`distrib` is the vertical vector containing the codes of the pdfs for all quantities Z_i and A_j: 0 for no pdf, 1 for normal pdf, 2 for uniform pdf, 3 for triangular pdf.

The `uMCM` function provides, respectively, the following results in successive columns of the output matrix Y, with one line per time step: the value y of Y, its standard uncertainty $u(Y)$, and the boundaries of the shortest coverage interval.

For the above example, there is one time-varying quantity: the water level h, and three constant quantities: respectively, the Manning-Strickler coefficient K, the channel slope I and the channel width B. Consequently, type

`Z=[h uh]` with only one line as there is only one single value of h.

```
A=[K uK I uI B uB]
```

The measurement process equation:

```
Q=K.*power(I,1/2).*power(B.*h,5/3).*power(B+2.*h,-2/3)
```

is then re-written with notations indicating the quantities with their rank in matrices Z and A.

h is the first quantity in the matrix Z: `h` is replaced by `Z(:,1)` in the above expression of Q. Similarly, `K`, `I` and `B` are replaced, respectively, by `A(:,1)`, `A(:,2)` and `A(:,3)` as they are, respectively, the first, second and third quantities in the matrix A. It is important to note that the indices refer to the rank of the quantities in matrices Z and A, and not to the rank of the columns.

Consequently, type

```
chaine='A(:,1).*power(A(:,2),0.5).*power(A(:,3).*Z(:,1),5/3).*power(A
(:,3)+2.*Z(:,1),-2/3)'
```

and

```
alpha=0.95
```

In this example, all four quantities in Z and A are independent as there is no correlation or covariance between them. Consequently, type

```
MatCor=eye(4)
```

BOX 9: (Continued)

which gives

$$MatCor = \begin{bmatrix} 1 & 0 & 0 & 0 \\ 0 & 1 & 0 & 0 \\ 0 & 0 & 1 & 0 \\ 0 & 0 & 0 & 1 \end{bmatrix}$$

Type

```
Mmc=1e6
```

Distribution of quantities in *Z* and *A* are, respectively, normal (for *h*), uniform (for *K*), normal (for *I* and *B*).
Consequently, type

```
distrib=[ 1 2 1 1]'
```

Lastly type

```
Q=uMCM(Z,A,chaine,alpha,MatCor,Mmc,distrib)
```

One gets $Q = 0.346\ m^3/s$, $u(Q) = 0.0136\ m^3/s$, and the 95% coverage interval is [0.3233, 0.3688] m^3/s.
The values are slightly different from those given in Box 8 as each MC run generates new samples.

8.2.5 Comparison of uncertainties estimated with Type B and Monte Carlo methods

The Type B method is less demanding than the MCM in term of calculations: once the literal expression of the Law of Propagation of Uncertainties (Equation (8.9)) for the quantity *Y* of interest is established, it can be applied instantaneously to any value of *Y*. On the contrary, the MCM imposes that the simulations are run systematically for each value of *Y*, which requires more computation time. However, the MCM is much more flexible and does not require the conditions of application of the Type B method. This is why MCM is the reference method.

In practice, it is frequent that the operator compares the Type B method and the MCM: if both methods give equivalent results, then the Type B method can be applied, which may be more convenient, especially with repetitive calculations. If the results are not equivalent, then the MCM should be used.

The comparison of both methods, based on their coverage intervals, is made with the following steps:

(1) Apply the Type B method to calculate y and the interval $[y - U(y), y + U(y)]$ (Equation (8.13)) for the given level of probability α.
(2) Apply the MCM to calculate $u(y)$ and the interval $[y_{\alpha,low}, y_{\alpha,high}]$ for the same level of probability α.
(3) Set the required tolerance δ for the coverage interval according to Equation (8.53).
(4) Compare the above coverage intervals by determining:

$$d_{low} = |y - U(y) - y_{\alpha,low}| \tag{8.56}$$

$$d_{high} = |y + U(y) - y_{\alpha,high}| \tag{8.57}$$

(5) If both d_{low} and d_{high} are not larger than the tolerance δ for $u(y)$, then the comparison is positive, and the Type B method can be applied. Otherwise, the MCM should be applied. It is important to note that the comparison applies only for the specified probability level α.

Example 1:

The comparison is made with the results obtained in Sections 8.2.3.4 and 8.2.4.5. The probability level α is 0.95.

Coverage interval obtained with the Type B method: $[0.320, 0.373]\ m^3/s$
Coverage interval obtained with the MCM: $[0.323, 0.369]\ m^3/s$
Standard uncertainty $u(y) = 0.013\ m^3/s$.
If $n_{dig} = 2$, then $u(y) = 13 \times 10^{-3}$, $l = -3$ and $\delta = 0.5 \times 10^{-3}$.
$d_{low} = |0.320–0.323| = 3 \times 10^{-3} > \delta$
$d_{high} = |0.373–0.369| = 4 \times 10^{-3} > \delta$

Consequently, the Type B method and the MCM give non-equivalent results and the MCM shall be applied.

The same approach can be used to compare the coverage intervals obtained in Section 8.2.3.4 with both an infinite degree of freedom and the effective degree of freedom. With $n_{dig} = 2$ and $\delta = 0.5 \times 10^{-3}$, the conclusion is that the results are different.

Example 2:

Let us modify the above example by assuming now that all quantities follow normal distributions, for both Type B and MCM calculations, with

$K = 75\ m^{1/3}/s$, $u(K) = 2.88\ m^{1/3}/s$
$I = 3.2 \times 10^{-3}\ m/m$, $u(I) = 6 \times 10^{-6}\ m/m$
$B = 0.805\ m$, $u(B) = 2 \times 10^{-3}\ m$
$h = 0.32\ m$, $u(h) = 1.5 \times 10^{-3}\ m$

One gets (with 4 digits for illustrative purpose):

Coverage interval with the Type B method: $Q_{TB} = [0.3196, 0.3728]$
Standard uncertainty with the Type B method: $u(y)_{TB} = 0.0136\ m^3/s$
Coverage interval with the MCM: $Q_{MC} = [0.3195, 0.3726]\ m^3/s$
Standard uncertainty with the MCM: $u(y)_{MC} = 0.0136\ m^3/s$
If $n_{dig} = 2$, then $u(y) = 13 \times 10^{-3}$, $l = -3$ and $\delta = 0.5 \times 10^{-3}$.
$d_{low} = abs(Q_{TB,low} - Q_{MC,low}) = 1.5287 \times 10^{-4} < \delta$
$d_{high} = abs(Q_{TB,high} - Q_{MC,high}) = 1.0596 \times 10^{-4} < \delta$

In this case, the Type B method and the MCM give equivalent results and can be both used indifferently.

8.2.6 Correlation between quantities

Correlation (or covariance) between quantities involved in a measurement process may have a significant effect on the estimation of the resulting uncertainty. Therefore, special attention should be devoted to checking and estimating correlations when they exist. The best practice in uncertainty assessment thus requires systematic checking and accounting for the possible correlation and estimating it. Only if its influence is proved to be negligible over the entire range of expected values of all quantities, may it be neglected in further calculations. It is worth mentioning that detecting and quantifying correlation is not always obvious and may require expertise.

Correlation can occur:

- Between different input quantities X_i used to calculate an output variable Y.
- As autocorrelation in time series used to calculate aggregated values.

8.2.6.1 Correlation between input quantities

According to ISO (2008a, Sections 5.2.4 and 5.2.5, p. 22–23), 'there may be significant correlation between two input quantities if the same measuring instrument, physical measurement standard, or reference datum having a significant standard uncertainty is used in their determination. [...] Correlations between input quantities cannot be ignored if present and significant. The associated covariances should be evaluated experimentally if feasible by varying the correlated input quantities, or by using the pool of available information on the correlated variability of the quantities in question.'

As explained in Section 8.2.3.2, it is recommended to write the function f

$$Y = f(X_1, X_2, \ldots X_i, \ldots X_N) \tag{8.58}$$

with the quantities X_i corresponding to separate and thus uncorrelated measurement instruments.

Other frequent cases where covariance should be accounted for occur when quantities X_i used in Equation (8.58) are obtained from previous regression functions like e.g. calibration functions, rating curves, etc.

If covariance remains and cannot be avoided between some quantities X_i, it should then be quantified. As covariance may be positive or negative, it may increase or decrease the resulting uncertainty in Y. Some examples are given in Section 8.2.6.3.

8.2.6.2 Autocorrelation in time series

Time series data of flow, water level or water quality are not randomly distributed as they result from dynamic processes where each value can be considered as equal to the previous one with a small increase or decrease depending on the evolution of the quantity. In such cases, the successive values of a given quantity may be considered as partly autocorrelated. Analogous to geostatistics, the chronostatistics approach (Gy, 1988, 2012) may be applied in such cases, using the variograph of the data to estimate the time horizon over which the autocorrelation between successive values is meaningful. Details of the variograph approach are given in Bertrand-Krajewski & Bardin (2001).

The uncertainty in the sum of a time series can be calculated with the Type B method under three different assumptions:

(1) No autocorrelation ($r = 0$) between successive values of the time series.
(2) Full autocorrelation ($r = 1$) between successive values of the time series.
(3) Partial autocorrelation estimated between successive values of the time series calculated from the variograph.

The third assumption is the recommended approach. The two first ones are given for comparison only. The Matlab® code `uTypeBsum` given in Box 10 allows calculation of the uncertainty in the sum of a time series with the three above assumptions.

8.2.6.3 Examples of correlation

8.2.6.3.1 Covariance in measurements

The previous example of a discharge measurement in a rectangular open channel (Section 8.2.3.4) is revisited, with the following equations:

$$Q = f(K, I, S, R_h) = KI^{\frac{1}{2}}SR_h^{\frac{2}{3}} \tag{8.59}$$

$$S = Bh \tag{8.60}$$

BOX 10: CALCULATING THE UNCERTAINTY IN THE SUM OF A TIME SERIES WITH THE MATLAB® CODE uTypeBsum

(Matlab® codes and csv files available for download at https://doi.org/10.2166/9781789060102).
Let us consider a times series of volume (m^3) measured during 24 hours with a two-minute time step. The three columns of the `vol1.csv` file (separator ;) contain successively the date and time, the volume *Vol* (m^3) and the standard uncertainty *u*(*Vol*) (m^3).
One should first import the data of the `vol1.csv` file in a matrix, without the column of date and time:

```
data=dlmread('vol1.csv',';',1,1)
```

Calculate the daily volume V_d and its standard uncertainty $u(V_d)$ with the three assumptions about correlation (i.e. no autocorrelation – nc, full autocorrelation – fc, and partial autocorrelation – pc):

```
Vd=uTypeBsum(data)
```

One gets, respectively, $V_d = 328.0588$ m^3, $u(V_d)_{nc} = 0.4495$ m^3, $u(V_d)_{fc} = 12.0623$ m^3 and $u(V_d)_{pc} = 4.6842$ m^3.

$$R_h = \frac{Bh}{B + 2\,h} \tag{8.61}$$

$$Q = f(K, I, B, h) = KI^{\frac{1}{2}}(Bh)\left(\frac{Bh}{B + 2\,h}\right)^{\frac{2}{3}} = KI^{\frac{1}{2}}(Bh)^{\frac{5}{3}}(B + 2\,h)^{-\frac{2}{3}} \tag{8.62}$$

To ensure no covariance between the quantities *K*, *I*, *B* and *h*, Equation (8.62) was used for the previous uncertainty calculations with the Type B method and the MCM (see Sections 8.2.3.4 and 8.2.4.5). The minimum 95% coverage interval of *Q* given by the MCM was [0.323, 0.369] m^3/s. It can be considered as the reference value.

The most usual writing of Equation (8.59) uses the wet section *S* and the hydraulic radius R_h. Both values are calculated by Equations (8.60) and (8.61), respectively, from the same uncorrelated measured quantities *B* and *h*, and are thus correlated.

With the previous values

$B = 0.805$ m, $u(B) = 2\times10^{-3}$ m
$h = 0.32$ m, $u(h) = 1.5\times10^{-3}$ m

one gets with MCM:

$S = 0.2576$ m^2, $u(S) = 0.0014$ m^2, $R_h = 0.1783$ m and $u(R_{h\cdot}) = 5\times10^{-4}$ m.

One can assume that *S* and R_h are positively fully correlated, with $r(S, R_h) = +1$. This assumption can be tested by means of a basic Monte Carlo simulation with Matlab® (see Box 11).

If this correlation is not accounted for, applying the MCM gives $Q_1 = 0.3462$ m^3/s and $u(Q_1) = 0.0134$ m^3/s. If this correlation is accounted for, the MCM gives $Q_2 = 0.3462$ m^3/s and $u(Q_2) = 0.0135$ m^3/s (see detailed calculations in Box 12).

The covariance very slightly increases $u(Q)$ (from 0.0134 to 0.0135 m^3/s) and the width of the 95% coverage interval (from [0.3199, 0.3726] to [0.3198, 0.3728] m^3/s). In this example, the increase may be considered as negligible for operational use but reflects the influence of the writing of the equation *f* used in uncertainty assessment for representing the measurement process.

BOX 11: ESTIMATING THE CORRELATION BETWEEN S AND R_h WITH THE MATLAB® CODE uMCM

(Matlab® codes and csv files available for download at https://doi.org/10.2166/9781789060102).

Type

```
B=0.805
uB=2e-3
h=0.32
uh=1.5e-3
M=1e6
Bmc=normrnd(B, uB, M, 1)
hmc=normrnd(h, uh, M, 1)
Smc=Bmc.*hmc
Rhmc=Bmc.*hmc./(Bmc+2*hmc)
```

Calculate the coefficient of correlation between S and R_h:

```
corr(Smc, Rhmc)
```

One gets $r(S, R_h) = 0.9961$, which is equivalent to one in practice.

BOX 12: IGNORING OR ACCOUNTING FOR THE CORRELATION BETWEEN S AND R_h WITH THE MATLAB® CODE uMCM

(Matlab® codes and csv files available for download at https://doi.org/10.2166/9781789060102).

In addition to the variables created in Box 11, type

```
S=mean(Smc)
uS=std(Smc)
Rh=mean(Rhmc)
uRh=std(Rhmc)
K=75
uK=2.88
I=3.2e-3
uI=6e-6
Z=[S uS Rh uRh]
A=[K uK I uI]
distrib=[1 2 1 1]'
```

If the correlation $r(S, R_h)$ between S and R_h is ignored, the correlation matrix is

```
MatCor1=eye(4)
```

$$MatCor1 = \begin{bmatrix} 1 & 0 & 0 & 0 \\ 0 & 1 & 0 & 0 \\ 0 & 0 & 1 & 0 \\ 0 & 0 & 0 & 1 \end{bmatrix}$$

Type

```
chaine='A(:,1).*power(A(:,2),0.5).*Z(:,1).*power(Z(:,2),2/3)'
```

BOX 12: (Continued)

```
Mmc=1e6
alpha=0.95
```

and lastly

```
Q1=uMCM(Z,A,chaine,alpha,MatCor1,Mmc,distrib)
```

One gets $Q_1 = 0.3462$ m^3/s, $u(Q_1) = 0.0134$ m^3/s and $IC95\ min_1 = [0.3199, 0.3726]$ m^3/s.

If the correlation $r(S, R_h) = +1$ between S and R_h is accounted for, applying the MCM requires modification of the correlation matrix *MatCor*:

```
MatCor2=MatCor1
MatCor2(1,2)=1
MatCor2(2,1)=1
```

$$MatCor2 = \begin{bmatrix} 1 & 1 & 0 & 0 \\ 1 & 1 & 0 & 0 \\ 0 & 0 & 1 & 0 \\ 0 & 0 & 0 & 1 \end{bmatrix}$$

Then type

```
Q2=uMCM(Z,A,chaine,alpha,MatCor2,Mmc,distrib)
```

One gets $Q_2 = 0.3462$ m^3/s, $u(Q_2) = 0.0135$ m^3/s and $IC95\ min_2 = [0.3198, 0.3728]$ m^3/s.

The initial minimum 95% coverage interval of Q obtained with the MCM and the four separate and independent quantities (Section 8.2.4.5) was [0.323, 0.369] m^3/s. It can be considered as the reference value as it does not include any correlation between the input quantities.

8.2.6.3.2 Covariance resulting from a regression function

Let us consider the discharge in a 1.8 m high A180 egg-shape sewer measured by means of both a water level sensor and a flow velocity sensor. *In situ* measurements by a land surveyor have been carried out to collect a series of 21 pairs of points (h_i, S_i) with h_i (m) the water level and S_i (m^2) the corresponding wet section (see Table 8.5, Figure 8.10, and the file `eggshape1.csv`). They are used to establish a function $S(h)$ defined as a third order polynomial function:

$$S = b_1\, h + b_2\, h^2 + b_3\, h^3 \tag{8.63}$$

The regression (detailed calculations are given in Box 13) gives the following results:

- Values of the parameters b_i: $[b_1, b_2, b_3] = [0.7825, 0.3601, -0.1473]$.
- Standard uncertainties $u(b_i)$: $[u(b_1), u(b_2), u(b_3)] = [0.0173, 0.0287, 0.0114]$.
- Correlation matrix of the parameters b: $\mathrm{MatCor}(b) = \begin{bmatrix} 1 & -0.9656 & 0.9098 \\ -0.9656 & 1 & -0.9848 \\ 0.9098 & -0.9848 & 1 \end{bmatrix}$.

The values of the coefficients of correlation in $MatCor(b)$ show that the parameters b_i are strongly correlated, either positively (b_1 with b_3) or negatively (b_1 with b_2, b_2 with b_3).

Let us now calculate the discharge Q and $u(Q)$ when the measured water level is $h = 0.45$ m and the mean flow velocity $V = 0.42$ m/s (detailed calculations are given in Box 14). Their respective standard

Table 8.5 Couples of points (h_i, S_i) for the A180 egg-shape sewer (file `eggshape1.csv`).

h_i (m)	S_i (m^2)
0.00	0.00
0.10	0.07
0.11	0.08
0.12	0.09
0.20	0.17
0.30	0.27
0.40	0.37
0.50	0.47
0.60	0.57
0.70	0.78
0.80	0.68
0.90	0.88
1.00	0.99
1.10	1.09
1.20	1.20
1.30	1.30
1.40	1.40
1.50	1.49
1.60	1.58
1.70	1.66
1.83	1.72

uncertainties are $u(h) = 3 \times 10^{-3}$ m (from sensor calibration and *in situ* conditions) and $u(V) = 0.05$ m/s (from expertise).

Q is given by the following equation:

$$Q = S(h)V = (b_1 h + b_2 h^2 + b_3 h^3)V \tag{8.64}$$

One gets $Q = 0.1729$ m^3/s.

The quantities h and V are measured independently and are not correlated. They are also not correlated with the parameters b_i which have been established from the land surveyor's data. However, *MatCor*(b) shows that the parameters b_i are correlated with each other.

$u(Q)$ and the 95% coverage intervals are calculated with the MCM for two cases: case 1 accounting for the correlation of the parameters b_i, and case 2 neglecting the correlation. The results given in Table 8.6

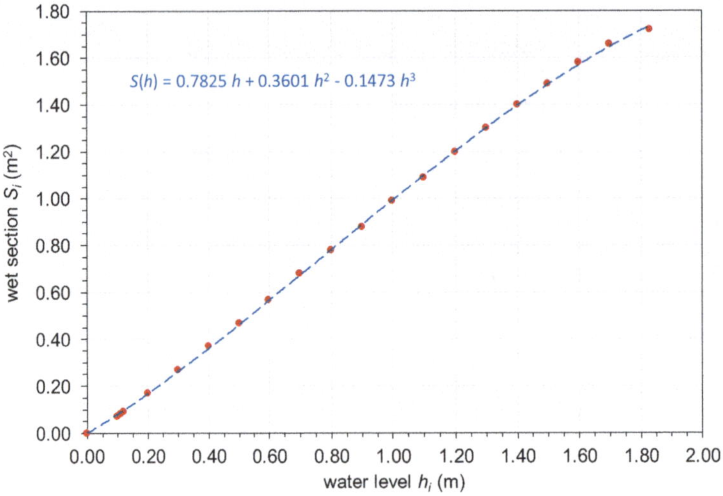

Figure 8.10 Plot of measured points (h_i, S_i) for the A180 egg-shape sewer (red dots) and fitted 3rd order polynomial regression function $S(h)$ (dashed blue line). *Source*: Jean-Luc Bertrand-Krajewski (INSA Lyon).

BOX 13: REGRESSION OF THE *S*(*h*) FUNCTION FOR AN A180 EGG-SHAPE SEWER WITH MATLAB®

With `hi` and `Si` the vertical vectors containing respectively the 21 values of h_i and S_i in Table 8.5, one can estimate the values of the parameters b, their standard uncertainties $u(b)$, their covariance matrix $COV(b)$ and the mean squared error of the regression mse.

Type

```
[b ub mse COVb] = lscov ([hi hi.^2 hi.^3] , Si)
```

One gets

$[b_1, b_2, b_3] = [0.7825, 0.3601, -0.1473]$

$[u(b_1), u(b_2), u(b_3)] = [0.0173, 0.0287, 0.0114]$

$mse = 7.04 \times 10^{-5}$

and

$$COV(b) = 10^{-3} \times \begin{bmatrix} 0.2982 & -0.4789 & 0.1795 \\ -0.4789 & 0.8251 & -0.3232 \\ 0.1795 & -0.3232 & 0.1306 \end{bmatrix}$$

By applying Equation (8.10), the correlation matrix of b, noted $MatCor(b)$, is calculated by

```
MatCorb=COVb./(ub* ub')
```

which gives

$$MatCor(b) = \begin{bmatrix} 1 & -0.9656 & 0.9098 \\ -0.9656 & 1 & -0.9848 \\ 0.9098 & -0.9848 & 1 \end{bmatrix}$$

Table 8.6 Comparison of results of uncertainty in discharge Q in the A180 egg-shape sewer, by accounting for the correlation between the parameters b (case 1) or by neglecting it (case 2).

	Case 1 With correlation	**Case 2 Without correlation**
Q (m^3/s)	0.1729	0.1729
$u(Q)$ (m^3/s)	0.0207	0.0211
IC95 min (m^3/s)	[0.1322, 0.2131]	[0.1316, 0.2141]
Relative standard uncertainty (%)	11.9%	12.2%

BOX 14: CALCULATION OF *u*(*Q*) AND 95% COVERAGE INTERVALS WITH THE MATLAB® CODE uMCM

(Matlab® codes and csv files available for download at https://doi.org/10.2166/9781789060102).
In addition to the variables defined in Box 13, type

```
h=0.45
V=0.42
Q=sum(b.*[h h.^2 h.^3]')*V
```

One gets $Q = 0.1729$ m^3/s.
To calculate the standard uncertainty $u(Q)$ and the 95% coverage interval with the MCM, type

```
uh=3e-3
uV=0.05
Z=[h uh V uV]
A=[b(1) ub(1) b(2) ub(2) b(3) ub(3)]
chaineQ='(A(:,1).*Z(:,1)+A(:,2).*Z(:,1).^2+A(:,3).*Z(:,1).^3).*Z(:,2)'
alpha=0.95
Mmc=1e6
distrib=[1 1 1 1 1]'
```

For case 1, type:

$$MatCor1 = \begin{bmatrix} 1 & 0 & 0 & 0 & 0 \\ 0 & 1 & 0 & 0 & 0 \\ 0 & 0 & 1 & -0.9656 & 0.9098 \\ 0 & 0 & -0.9656 & 1 & -0.9848 \\ 0 & 0 & 0.9098 & -0.9848 & 1 \end{bmatrix}$$

For case 2, type:

```
MatCor2=eye(5)
```

$$MatCor2 = \begin{bmatrix} 1 & 0 & 0 & 0 & 0 \\ 0 & 1 & 0 & 0 & 0 \\ 0 & 0 & 1 & 0 & 0 \\ 0 & 0 & 0 & 1 & 0 \\ 0 & 0 & 0 & 0 & 1 \end{bmatrix}$$

BOX 14: (Continued)

Lastly type
```
Q1=uMCM(Z,A,chaineQ,alpha,MatCor1,Mmc,distrib)
```
and
```
Q2=uMCM(Z,A,chaineQ,alpha,MatCor2,Mmc,distrib)
```
Results are given in Table 8.6.

indicate that, in this case, accounting for the correlation of the parameters b_i slightly reduces the resulting uncertainty in the discharge Q. Similar results are obtained with the Type B method.

8.2.6.3.3 Covariance in rain gauge calibration function

Tipping bucket rain gauges are affected by systematic underestimation when rainfall intensities increase (see e.g. Bertrand-Krajewski *et al.*, 2000). Therefore, in addition to the bucket static calibration, it is recommended to establish a dynamic calibration function to account for and then correct the underestimation error (see Chapter 7).

An experiment is carried out with known and controlled rainfall intensities I_r (mm/h) over a range of measurement. The response of the rain gauge, i.e. the measured rainfall intensity I_m (mm/h) for each applied intensity I_r, is recorded (the protocol is described in Bertrand-Krajewski *et al.*, 2000 and in Chapter 2). An example of data for a rain gauge in the Mediterranean region where high intensities are expected is given in the file `raingauge1.csv` and shown in Table 8.7.

The typical calibration function is a power function

$$I_r = b_1 I_m^{b_2} \tag{8.65}$$

Let us estimate the parameters b_1 and b_2, their standard uncertainties, and their co-variance and correlation (detailed calculations are given in Box 15). One gets:

$$[b_1, b_2] = [2.0826, 0.8401]$$
$$[u(b_1), u(b_2)] = [0.4440, 0.0398]$$
$$\text{and } MatCor = \begin{bmatrix} 1.0000 & -0.9969 \\ -0.9969 & 1.0000 \end{bmatrix}$$

Table 8.7 Rain gauge dynamic calibration data set (file `raingauge1.csv`).

I_r (mm/h)	I_m (mm/h)
0	0
30	30
60	60
110	109
160	155
220	199
290	237

BOX 15: REGRESSION OF A RAIN GAUGE CALIBRATION FUNCTION WITH MATLAB®

This non-linear regression can be done by means of the Matlab® function `nlinfit`.
Type

```
Ir=[0 30 60 110 160 220 290]'
Im=[0 30 60 109 155 199 237]'
fun=@(b,Ir) b(1).*power(Ir,(b(2)))
b0=ones(1,2)
```

and lastly

```
[b,R,J,CovB]=nlinfit(Ir,Im,fun,b0)
```

One gets
$b = [b_1, b_2] = [2.0826, 0.8401]$
Then

```
ub=sqrt(diag(CovB))
```

gives
$u(b) = [u(b_1), u(b_2)] = [0.4440, 0.0398]$
The covariance matrix is

$$CovB = \begin{bmatrix} 0.1971 & -0.0176 \\ -0.0176 & 0.0016 \end{bmatrix}$$

The correlation matrix is given by

```
MatCor=CovB./(ub*ub')
```

$$MatCor = \begin{bmatrix} 1.0000 & -0.9969 \\ -0.9969 & 1.0000 \end{bmatrix}$$

The parameters b_i are strongly negatively correlated.

From these results, one can now estimate the most likely true rainfall intensity I_{est} from a measured value I_{mes} and its standard uncertainty $u(I_{est})$ by using the Type B method and the inverse function

$$I_{est} = \left(\frac{I_{mes}}{b_1}\right)^{\frac{1}{b_2}} \tag{8.66}$$

Let us consider $I_{mes} = 152$ mm/h with $u(I_{mes}) = 5\%$.

The standard uncertainty $u(I_{est})$ is calculated with the Type B method for two cases: case 1 accounting for the correlation of the parameters b_i, and case 2 neglecting the correlation (detailed calculations are given in Box 16). One gets, respectively:

$I_{est1} = 165$ mm/h and $u(I_{est1}) = 10$ mm/h.
$I_{est2} = 165$ mm/h and $u(I_{est2}) = 59$ mm/h.

In this example, not accounting for the correlation between quantities leads to a standard uncertainty multiplied by a factor close to 6.

BOX 16: CALCULATION OF $u(I_{est})$ WITH THE MATLAB® CODE uTypeB

(Matlab® codes and csv files available for download at https://doi.org/10.2166/9781789060102). In addition to the variables defined in Box 15, type

```
Imes=152
uImes=0.05* Imes
b1=b(1)
ub1=ub(1)
b2=b(2)
ub2=ub(2)
Z=[Imes uImes]
A=[b1 ub1 b2 ub2]
alpha=0.95
```

Case 1: correlation is accounted for.

Type

```
MatCor1=eye(3)
MatCor1(2:3,2:3)=MatCor
```

$$MatCor1 = \begin{bmatrix} 1 & 0 & 0 \\ 0 & 1 & -0.9969 \\ 0 & -0.9969 & 1 \end{bmatrix}$$

```
chaineIest='(Z(:,1)./A(:,1))^(1/A(:,2))'
```

and then

```
Iest1=uTypeB(Z,A,chaineIest,alpha,MatCor1)
```

One gets $I_{est1} = 165$ mm/h and $u(I_{est1}) = 10$ mm/h.

Case 2: correlation is not accounted for.

Type

```
MatCor2=eye(3)
```

$$MatCor2 = \begin{bmatrix} 1 & 0 & 0 \\ 0 & 1 & 0 \\ 0 & 0 & 1 \end{bmatrix}$$

```
Iest2=uTypeB(Z,A,chaineIest,alpha,MatCor2)
```

One gets $I_{est2} = 165$ mm/h and $u(I_{est2}) = 59$ mm/h.

8.3 EXAMPLES OF APPLICATIONS

All data and the following examples, along with the Matlab® codes, are available at https://doi.org/10.2166/9781789060102, so that the readers can re-do the examples themselves and also use the software codes for their own needs and applications.

8.3.1 Uncertainty in discharge calculation with a thin plate rectangular weir formula

Let us consider a thin plate rectangular weir with lateral contraction and the following characteristics (Figure 8.11):

- Channel width $B = 0.8$ m.
- Notch width $B_c = 0.48$ m.
- Crest height $h_p = 0.25$ m.
- Water head $h = 0.15$ m, measured at the distance L upstream of the weir.

The Kindsvater-Carter formula and the other parameters to estimate the discharge Q_{RW} over the rectangular weir are given in hydraulics textbooks and in the international standard ISO (2017):

$$Q_{RW} = \frac{2}{3} C_d \sqrt{2\,g} B_e h_e^{\frac{3}{2}} \tag{8.67}$$

where C_d (-) is the discharge coefficient, B_e (m) is the effective width and h_e (m) is the effective head, which themselves depend on other parameters related to the geometry of the weir:

$$C_d = f\left(\frac{B_c}{B}, \frac{h}{h_p}\right) \tag{8.68}$$

$$B_e = B + K_b \tag{8.69}$$

$$h_e = h + K_h \tag{8.70}$$

where K_b and K_h are quantities compensating for combined effects of viscosity and surface tension of water.

***Step 1*: Check of the applicability of the formula**

Before applying the Kindsvater-Carter formula (Equation (8.67)), the operator shall check that the numerical values of the variables respect the five conditions given in ISO (2017), as listed and checked in Table 8.8. In the present case, all conditions are satisfied, and the formula can be applied.

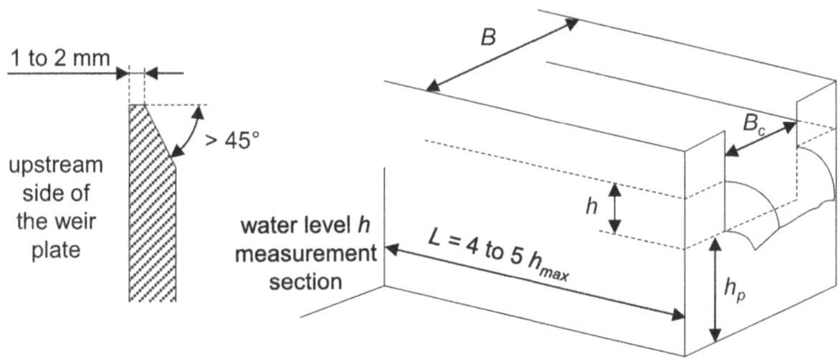

Figure 8.11 Rectangular weir with lateral contraction. *Source*: Jean-Luc Bertrand-Krajewski (INSA Lyon).

Table 8.8 Checking of the conditions of applicability of the Kindsvater-Carter formula.

Condition required by ISO 1438 (2017)	Example data	Check
h/h_p shall be less than 2.5	$h/h_p = 0.6 < 2.5$	☑
h shall be higher than 0.03 m	$h = 0.15 > 0.03$ m	☑
B_c shall be larger than 0.15 m	$B_c = 0.48 > 0.15$ m	☑
h_p shall be higher than 0.10 m	$h_p = 0.25 > 0.10$ m	☑
(B-Bc)/2 shall be larger than 0.10	(B-Bc)/2 = 0.16 > 0.10	☑

***Step 2*: Calculation of the discharge Q_{RW}**

For $B_c/B = 0.6$, and $h/h_p = 0.6$, the ISO standard gives (ISO 2017, pp. 11–13):

$$C_d = 0.593 + 0.018\,\frac{h}{h_p} \tag{8.71}$$

$K_b = 3.6 \times 10^{-3}$ m
$K_h = 10^{-3}$ m

Accordingly, Equation (8.67) can be rewritten as

$$Q_{RW} = \frac{2}{3}\left(a + b\frac{h}{h_p}\right)\sqrt{2\,g}(B_c + K_b)(h + K_h)^{\frac{3}{2}} \tag{8.72}$$

where $a = 0.593$ and $b = 0.018$ (see Equation (8.71)).

The discharge over the rectangular weir is $Q_{RW} = 0.506$ m^3/s (detailed calculations are given in Box 17).

***Step 3*: Calculation of the 95% coverage interval of the discharge Q_{RW}**

This step includes the estimation of the standard uncertainties of all quantities in Equation (8.72). Water level sensor calibration, *in situ* repeated measurements of weir geometry and the ISO standard give, respectively:

$u(h) = 1$ mm
$u(h_p) = 1$ mm
$u(B_c) = 0.5$ mm
$u(a) = 0.75\%$
$u(b) = 0.75\%$
$u(K_b) = 0.15$ mm
$u(K_h) = 0.15$ mm.

One assumes that the uncertainty in the gravity $g = 9.81$ m/s is negligible.

Applying the MCM (detailed calculations are given in Box 17) gives the following results:

$u(Q_{RW}) = 0.006$ m^3/s
$IC95\ min(Q_{RW}) = [0.049, 0.052]$ m^3/s.

BOX 17: CALCULATION OF Q_{RW} AND OF ITS STANDARD UNCERTAINTY WITH THE MATLAB® CODE uMCM

(Matlab® codes and csv files available for download at https://doi.org/10.2166/9781789060102).

Type

```
h=0.15
uh=1e-3
hp=0.25
uhp=1e-3
Bc=0.48
uBc=0.5e-3
a=0.593
ua=0.75e-2*a
b=0.018
ub=0.75e-2*b
Kb=3.6e-3
uKb=0.15e-3
Kh=0.001
uKh=0.15e-3
g=9.81
Z=[h uh]
A=[hp uhp Bc uBc a ua b ub Kb uKb Kh uKh]
distrib=ones(7,1)
Nmc=1e6
alpha=0.95
MatCor=eye(7)
chaineRW='2/3*sqrt(2*9.81).*(A(:,3)+A(:,4).*Z(:,1)./A(:,1)).*
(A(:,2)+A(:,5)).*(Z(:,1)+A(:,6)).^(3/2)'
```

and lastly

```
QRW=uMCM(Z,A,chaineRW,alpha,MatCor,Nmc,distrib)
```

One gets $Q_{RW} = 0.5059$ m^3/s, $u(Q_{RW}) = 6.3699 \times 10^{-4}$ m^3/s and *IC95 min*(Q_{RW}) = [0.0493, 0.0518] m^3/s.

8.3.2 Uncertainty in discharge calculation with both water level and flow velocity measurements

Let us consider a circular sewer with radius R_c (Figure 8.12) equipped with a sensor measuring the water depth h (m) and another sensor measuring the mean flow velocity V (m/s).

The discharge Q_c (m^3/s) is calculated by the following equation:

$$Q_c = S(h)V = R_c^2\left[\arccos\left(1-\frac{h}{R_c}\right) - \left(1-\frac{h}{R_c}\right)\sin\left(\arccos\left(1-\frac{h}{R_c}\right)\right)\right]V \tag{8.73}$$

According to four repeated measurements (Type A method), the radius R_c is equal to 0.6 m and its standard uncertainty $u(R_c) = 0.002$ m. The standard uncertainties $u(h)$ and $u(V)$ in the file `hV1.csv` are calculated from sensors calibration and *in situ* experiments.

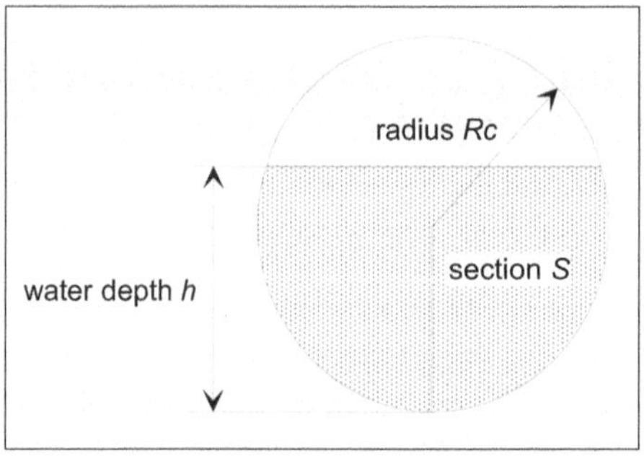

Figure 8.12 Scheme of the circular pipe. *Source*: Jean-Luc Bertrand-Krajewski (INSA Lyon).

Both the water level h and the mean flow velocity V are measured every 2 minutes during 24 hours. The file `hV1.csv` (separator ;) contains 720 values in the successive order of columns: h, $u(h)$, V and $u(V)$. The first 15 lines of `hV1.csv` are shown in Table 8.9.

This example shows successively

- The calculation of the discharge Q_c and of its uncertainty $u(Q_c)$ for each time step.
- The calculation of the daily volume V_d, its standard uncertainty $u(V_d)$ and its 95% coverage interval.

Step 1: Calculation of the discharge Q_c and of its uncertainty $u(Q_c)$ for each time step

One assumes that the quantities h, V and R_c are independent and normally distributed. Detailed calculations with both the MCM and the Type B method are given in Box 18. The results are shown graphically in Figure 8.13 and the first 15 lines of the numerical values are given in Table 8.10.

Table 8.9 First 15 lines of the file `hV1.csv`.

Date	*h* (m)	*u*(*h*) (m)	*V* (m/s)	*u*(*V*) (m/s)
01/01/2017 00:00	0.368	0.008	0.634	0.05
01/01/2017 00:02	0.368	0.008	0.632	0.05
01/01/2017 00:04	0.356	0.008	0.642	0.05
01/01/2017 00:06	0.356	0.008	0.642	0.05
01/01/2017 00:08	0.356	0.008	0.628	0.05
01/01/2017 00:10	0.349	0.008	0.634	0.05
01/01/2017 00:12	0.349	0.008	0.638	0.05
01/01/2017 00:14	0.349	0.008	0.628	0.05
01/01/2017 00:16	0.336	0.008	0.627	0.05
01/01/2017 00:18	0.336	0.008	0.634	0.05
01/01/2017 00:20	0.336	0.008	0.634	0.05
01/01/2017 00:22	0.349	0.008	0.629	0.05
01/01/2017 00:24	0.349	0.008	0.614	0.05
01/01/2017 00:26	0.356	0.008	0.614	0.05
01/01/2017 00:28	0.349	0.008	0.613	0.05

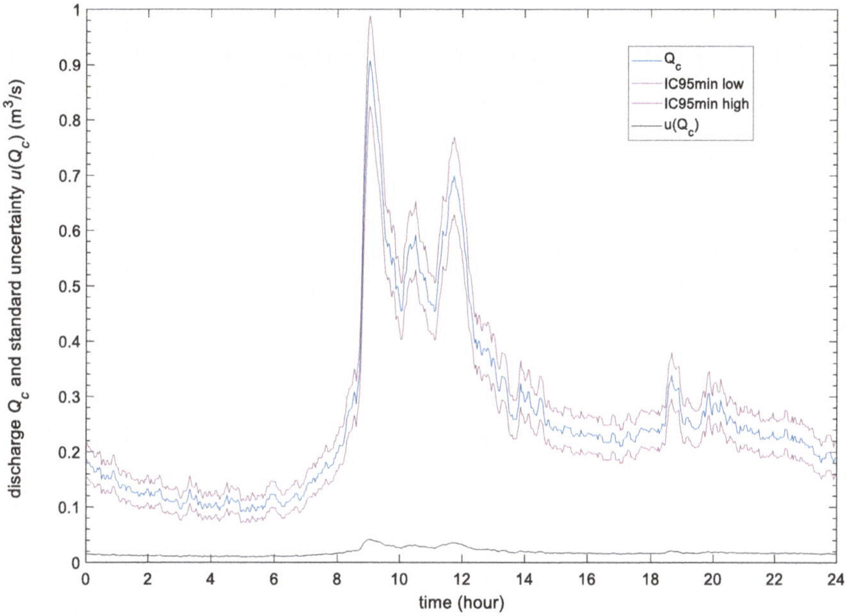

Figure 8.13 Plot of discharge Q_c over 24 hours, standard uncertainty $u(Q_c)$ and the 95% coverage interval. *Source*: Jean-Luc Bertrand-Krajewski (INSA Lyon).

Table 8.10 First 15 lines of the matrix QcMC (Q_c, $u(Q_c)$, *IC95min low* and *IC95min high*).

Q (m^3/s)	*u*(*Q*) (m^3/s)	*IC95min low* (m^3/s)	*IC95min high* (m^3/s)
0.1866	0.0157	0.1557	0.2173
0.1860	0.0157	0.1552	0.2166
0.1803	0.0151	0.1508	0.2099
0.1803	0.0151	0.1507	0.2099
0.1764	0.0150	0.1469	0.2058
0.1730	0.0147	0.1443	0.2019
0.1741	0.0147	0.1454	0.2029
0.1714	0.0146	0.1428	0.2002
0.1628	0.0140	0.1355	0.1903
0.1646	0.0140	0.1373	0.1922
0.1646	0.0140	0.1371	0.1921
0.1717	0.0147	0.1431	0.2006
0.1676	0.0146	0.1392	0.1965
0.1726	0.0150	0.1433	0.2021
0.1674	0.0146	0.1391	0.1964

BOX 18: CALCULATION OF THE DISCHARGE Q_C AND OF ITS UNCERTAINTY $u(Q_C)$ FOR EACH TIME STEP

(Matlab® codes and csv files available for download at https://doi.org/10.2166/9781789060102).
Import the data of the `hV1.csv` file:

```
data=dlmread('hV1.csv',';',1,1)
```

Then type

```
Z=data
A=[0.6 0.002]
alpha=0.95
Nmc=1e6
distrib=ones(3,1)
MatCor=eye(3)
chaineQc='(A(:,1).^2).*(acos(1-Z(:,1)./A(:,1))-(1-Z(:,1)./A(:,1))
.*sin(acos(1-Z(:,1)./A(:,1)))).*Z(:,2)'
```

and lastly, apply the MCM method for the entire time series of 720 time steps in one single instruction:

```
QcMC=uMCM(Z,A,chaineQc,alpha,MatCor,Nmc,distrib)
```

The results are shown graphically in Figure 8.13 and the first 15 lines of the matrix `Qcmc` are given in Table 8.10.
Apply now the Type B method for the entire time series of 720 values in one single instruction:

```
QcTB=uTypeB(Z,A,chaineQc,alpha,MatCor)
```

The Type B method runs much faster in such a case, also for the entire time series of 720 values. The maximum relative difference of the standard uncertainties $(u(Q_{cTB}) - u(Q_{cMC}))/u(Q_{cMC})$ is less than 0.25%: both methods deliver similar results.

Step 2: Calculation of daily volume V_d, standard uncertainty $u(V_d)$ and 95% coverage interval

With the time step $\Delta t = 120$ s, the daily volume $V_d = 22{,}763$ m^3. Applying the variograph method (see Section 8.2.6.2), the standard uncertainty is $u(V_d) = 620$ m^3 and the 95% coverage interval is [21547, 23979] m^3 (detailed calculations are given in Box 19).

8.3.3 Uncertainty in discharge calculation with the Manning-Strickler formula

It is frequent that the Manning-Strickler formula, due to its simplicity, is used to evaluate discharges in sewers where there is no backwater effect and where, as a first approximation, the discharge can be considered as locally uniform and permanent during each time step.

The Manning-Strickler formula (Equation (8.74)), in addition to the water level h (m) used to estimate the wet section S (m^2) and the hydraulic radius R_h (m), requires the values of both the roughness coefficient K (m$^{1/3}$/s) and the pipe invert slope I (m/m).

$$Q_{MS} = K\sqrt{I}SR_h^{\frac{2}{3}} \tag{8.74}$$

BOX 19: CALCULATION OF DAILY VOLUME V_d, STANDARD UNCERTAINTY $u(V_d)$ AND 95% COVERAGE INTERVAL

(Matlab® codes and csv files available for download at https://doi.org/10.2166/9781789060102).
Type

```
Dt=120
```

Calculate the daily volume V_d from the 720 values of the discharge Q_c given in the first column of `QcMC` calculated in Box 18:

```
Vd=sum(QcMC(:,1).*Dt)
```

which gives 22,763 m^3.
To use the Matlab® code `uTypeBsum` (see Box 10), select the first and second columns of `QcMC` containing, respectively, the values of the discharge Q_c and their standard uncertainties $u(Q_c)$, and multiply them by the time step Δt to get the corresponding values of the volume. In this example, one assumes that the uncertainty in the time step Δt is negligible.
Type

```
Vd=uTypeBsum(QcMC(:,1:2).*Dt)
```

The four components of the vector `Vd` correspond respectively to

- $V_d = 22{,}763$ m^3
- standard uncertainty with no autocorrelation $u(V_d)_{nc} = 62$ m^3, i.e. relative standard uncertainty of 0.3%.
- standard uncertainty with full autocorrelation $u(V_d)_{fc} = 1563$ m^3, i.e. relative standard uncertainty of 6.9%.
- standard uncertainty with partial autocorrelation $u(V_d)_{pc} = 620$ m^3, i.e. relative standard uncertainty of 2.7%.

Calculate the 95% coverage interval for the case of partial autocorrelation:

```
[Vd(1)-1.96*Vd(4), Vd(1)+1.96*Vd(4)]
```

One gets [21547, 23979] m^3.

Roughness and slope are critical quantities as the discharge Q_{MS} is directly proportional to K and to the square root of I. Instead of using (i) approximate values of K found in textbooks for the material of the sewer pipe, and (ii) values of I based on non-verified maps or GIS data, it may be better to estimate the value of the quantity $K\sqrt{I}$ from temporary field measurements.

Let us consider the case of a circular pipe with a radius $R_c = 0.8$ m, equipped with a water level sensor to estimate discharges during dry weather and most frequent storm events generating water levels lower than 0.5 m. The water sensor is installed permanently. For one week, a flow velocity sensor has been added temporarily and verified with tracing experiments (for tracing experiments in sewers, see e.g. Lepot *et al.*, 2014).

Two days of data recorded with a time step of 2 minutes are available in the file `manning1.csv`. The first 15 lines are shown in Table 8.11. The ranges of water level h and mean flow velocity V are, respectively, 0.009 to 0.51 m and 0.16 to 2.93 m/s. Standard uncertainties in h, V and R_c are, respectively, $u(h) = 0.003$ m, $u(V) = 0.03$ m/s and $u(R_c) = 0.002$ m.

Table 8.11 First 15 lines of the file `manning1.csv`.

Date UT	*h* (m)	*V* (m/s)
13/08/2018 00:00	0.104	1.50
13/08/2018 00:02	0.103	1.53
13/08/2018 00:04	0.103	1.53
13/08/2018 00:06	0.101	1.52
13/08/2018 00:08	0.098	1.51
13/08/2018 00:10	0.096	1.48
13/08/2018 00:12	0.093	1.46
13/08/2018 00:14	0.090	1.44
13/08/2018 00:16	0.088	1.42
13/08/2018 00:18	0.085	1.37
13/08/2018 00:20	0.083	1.36
13/08/2018 00:22	0.082	1.34
13/08/2018 00:24	0.079	1.30
13/08/2018 00:26	0.079	1.30
13/08/2018 00:28	0.079	1.30

Let us determine the values of the quantity $K\sqrt{I}$ and of its standard uncertainty. Using the independent quantities h, V and R_c, Equation (8.74) can be rewritten as

$$Q_{MS} = K\sqrt{I}\left[\arccos\left(1-\frac{h}{R_c}\right)-\left(1-\frac{h}{R_c}\right)\sin\left(\arccos\left(1-\frac{h}{R_c}\right)\right)\right]^{\frac{5}{3}}\left[2R_c\arccos\left(1-\frac{h}{R_c}\right)\right]^{-\frac{2}{3}} \tag{8.75}$$

One calculates directly the values of the quantity $K\sqrt{I}$ by combining Equations (8.75) and (8.73):

$$K\sqrt{I} = R_c^{-\frac{2}{3}}\left[\arccos\left(1-\frac{h}{R_c}\right)-\left(1-\frac{h}{R_c}\right)\sin\left(\arccos\left(1-\frac{h}{R_c}\right)\right)\right]^{-\frac{2}{3}}\left[2\arccos\left(1-\frac{h}{R_c}\right)\right]^{\frac{2}{3}}V \tag{8.76}$$

The set of 1440 values of the quantity $K\sqrt{I}$ calculated from the recorded data can be considered as a sample for the Type A method.

Calculations (detailed in Box 20) give $\overline{K\sqrt{I}} = 7.562\ \text{m}^{1/3}/\text{s}$ and $u\left(\overline{K\sqrt{I}}\right) = 0.034\ \text{m}^{1/3}/\text{s}$.

Based on these results, future values of the discharge in the circular pipe can be calculated by means of Equation (8.75) by using $K\sqrt{I}$ as an independent quantity. For example, let us calculate the discharge and its uncertainty for a water level $h = 0.3$ m (detailed calculations are given in Box 21). One gets $Q = 1.33\ \text{m}^3/\text{s}$, $u(Q) = 0.03\ \text{m}^3/\text{s}$, and the 95% coverage interval is [1.28, 1.39] m^3/s.

This example does not account for the uncertainties in the 1440 values of $K\sqrt{I}$ due to uncertainties in the quantities h, V and R_c, which would add some additional uncertainty in the estimation of $K\sqrt{I}$.

BOX 20: CALCULATION OF $K\sqrt{I}$ AND OF ITS STANDARD UNCERTAINTY

(Matlab® codes and csv files available for download at https://doi.org/10.2166/9781789060102).
Import the data of the `manning1.csv` file:

```
data=dlmread('manning1.csv',';',1,1)
```

Type

```
Rc=0.8
```

For convenience, type

```
h=data(:,1)
V=data(:,2)
```

According to Equation (8.76), type

```
KVI=Rc^(-2/3).*(acos(1-h/Rc)-(1-h/Rc).*sin(acos(1-h/Rc))).^(-2/3)
.*(2*acos(1-h/Rc)).^(2/3).*V
```

Then apply the Type A method with a 95% level of probability:

```
KVIbar=uTypeA(KVI, 0.95)
```

One gets $\overline{K\sqrt{I}} = 7.5621\ \mathrm{m}^{1/3}/\mathrm{s}$ and $u\left(K\sqrt{I}\right)$ is equal to $0.0338\ \mathrm{m}^{1/3}/\mathrm{s}$.

BOX 21: CALCULATION OF Q, $u(Q)$ AND 95% COVERAGE INTERVAL

(Matlab® codes and csv files available for download at https://doi.org/10.2166/9781789060102).
In addition to the variables defined in Box 20, type

```
h=0.3
uh=0.003
KVI=KVIbar(1)
uKVI=KVIbar(2)
Rc=0.8
uRc=0.002
Z=[h uh]
A=[KVI uKVI Rc uRc]
alpha=0.95
distrib=ones(3,1)
MatCor=eye(3)
Nmc=1e6
chaineQh='A(:,1).*(acos(1-Z(:,1)./A(:,2))-(1-Z(:,1)./A(:,2)).*sin(acos
(1-Z(:,1)./A(:,2)))).^(5/3).*(2.*A(:,2).*acos(1-Z(:,1)./A(:,2))).^(-2/3)'
```

and lastly

```
Q=uMCM(Z,A,chaineQh,alpha,MatCor,Nmc,distrib)
```

One gets $Q = 1.3342\ \mathrm{m}^3/\mathrm{s}$, $u(Q) = 0.0292\ \mathrm{m}^3/\mathrm{s}$ and the 95% coverage interval is [1.2771, 1.3916] m^3/s.

This could be done by means of Monte Carlo simulations. In addition, this example implicitly assumes that K is constant, whereas it may happen that it depends on the water level h. A relation $K = f(h)$ can be investigated along with, if significant, another relation $u(K) = f(h)$.

8.3.4 Uncertainty in velocity-area methods

The velocity-area method is one of the most common non-continuous methods to determine flow data in open channels. The example introduces the calculation of flow uncertainty of velocity-area methods using the approach given in ISO (2020). Furthermore, it demonstrates the sensitivity of flow uncertainty on simplifications and inaccuracies in the measurement process.

8.3.4.1 Case study

In a rectangular channel (width $B = 0.8$ m, flow depth $h = 0.5$ m), flow velocities $v_x(y_i, z_j)$ (m/s) are measured at 135 points (y_i, z_j) of a grid with $\Delta y = \Delta z = 0.05$ m during uniform steady state (Figure 8.14).

8.3.4.2 Uncertainty of flow and mean velocity

Flow Q (m^3/s) and mean flow velocity V (m/s) can be calculated from a sufficiently dense velocity field $v_x = f(y,z,t)$ in a representative flow cross section S (m^2). For uniform steady state flow conditions, v_x values do not depend on time t and the following applies:

$$Q = \int_S v_x(y, z) dS \tag{8.77}$$

For $m \times n$ elementary sections $S_{ij} = \Delta y_i \times \Delta z_j$ and point velocities $v_x(y_i, z_j)$ in the middle of each elementary section S_{ij}, the flow Q can be calculated as

$$Q = F_x F_z \sum_{i=1}^{m} \sum_{j=1}^{n} v_x(y_i, z_j) \Delta z_j \Delta y_i + Q_p \tag{8.78}$$

where $m = 15$ is the number of points per row, $n = 9$ is the number of points per vertical (Figure 8.14), F_y

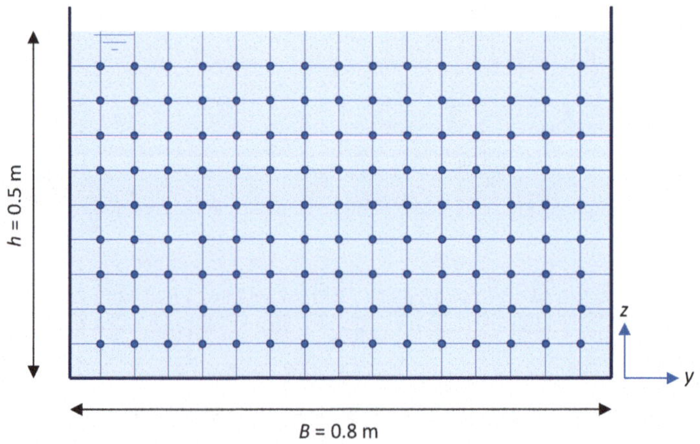

Figure 8.14 Scheme of the rectangular cross section with 135 points (y_i, z_j) where longitudinal flow velocity $v_x(y_i, z_j)$ is measured. *Source*: Mathias Uhl (Muenster University of Applied Sciences).

and F_z are factors relating the sums, respectively, in the y- and z-direction to an ideal integration of the real velocity profile, and Q_p (m^3/s) is the perimeter flow.

In the case of velocity grid measurements, F_y and F_z can be set to unity. The perimeter flow Q_p represents a part of the flow near the boundary of the cross section where measurements are impossible or very uncertain because of turbulence, velocity gradients or influences on the measurement sensor or just remaining areas between the real cross section and the measurement grid.

According to ISO (2020), the uncertainty of the flow $u(Q)$ is

$$\begin{aligned}u(Q)^2 = &\left[\sum_i^m \sum_j^n v_x(y_i z_j)\Delta y_i \Delta z_j\right][u^*(F_y)^2 + u^*(F_z)^2] \\ &+ \sum_i^m \sum_j^n [v_x(y_i z_l)\Delta y_i \Delta z_j]^2 u^*|[v_x(y_i z_j)]^2 \\ &+ \sum_i^m \left[\Delta y_i^2 \cdot \left\{\sum_j^n v_x(y_i z_j)\Delta z_j\right\}^2 \cdot u^*(\Delta y_i)^2\right] \\ &+ \sum_j^n \left[\Delta z_i^2 \cdot \left\{\sum_i^m v_x(y_i z_j)\Delta y_j\right\}^2 \cdot u^*(\Delta z_i)^2\right] + u(Q_p)^2\end{aligned} \tag{8.79}$$

with the mean velocity $V = Q/S$, its uncertainty $u(V)$ is

$$u(V)^2 = \left(\frac{1}{S^2}\right)u^2(Q) + \frac{Q^2}{S^4}u^2(S) \tag{8.80}$$

8.3.4.3 Flow cross section uncertainty

The assumed range of uncertainty of the channel width $B = 0.80$ m is $0.002 \leq u(B) \leq 0.005$ m and as well $0.002 \leq u(h) \leq 0.005$ m for the water level $h = 0.5$ m.

8.3.4.4 Measurement segment uncertainty Δy, Δz

The uncertainty of Δy and Δz was estimated in the range of $0.002 \leq u(\Delta y) = u(\Delta z) \leq 0.005$ m.

8.3.4.5 Flow velocity uncertainty $u(v_x)$

The uncertainty of the flow measurements consists of (i) the relative uncertainty u_k* of the measurement device as specified in the calibration certificate and (ii) a random relative component u_r* resulting from the unsteady nature of flow in channels. The uncertainty depends on the sensor type and the local velocity and turbulence. The uncertainty of the local velocity $u(v_x)$ is

$$u^*[v_x(y_i z_j)]^2 = u_k^*[v_x(y_i z_j)]^2 + u_r^*[v_x(y_i z_j)]^2 \tag{8.81}$$

This example uses $0.001 \leq u_k \leq 0.01$ m/s with $u_k^* = u_k/v_x$ and $0.01 \leq u_r^* \leq 0.02$ as a Type B estimate.

8.3.4.6 Measurement grid relative uncertainties $u^*(F_y)$ and $u^*(F_z)$

The density of the measurement grid influences the number and the representativeness of the measurements. ISO (2020) provides recommended default values in its Annex D: Table D.4 for the number of measurement points in a vertical and Table D.6 for the number of verticals in a cross section. The respective values in Table 8.13 are taken from these tables.

8.3.4.7 Perimeter flow uncertainty $u(Q_p)$

Near to the walls of the cross section, measurements of the flow velocity are either impossible or have large errors or uncertainties due to boundary effects, large gradients of the flow velocities or disturbed measurement signals. In those perimeter sections, the flow has to be estimated by extrapolated flow velocities. A simple approach had been used in this study. The free surface velocity $v_x(y, z=h)$ was set equal to the first measured value $v_x(y, z=h-\Delta z)$. At walls and bottom, flow velocities were assumed to be 50% of the nearest measured values, i.e. $v_x(y=0, z)=0.5 \times v_x(y=0+\Delta y, z)$, $v_x(y=B, z)=0.5 \times v_x(y=B-\Delta y, z)$ and $v_x(y, z=0)=0.5 \times v_x(y, z=0+\Delta z)$. The flow in the perimeter strip of 0.025 m width was calculated with a velocity at 0.0125 m derived from linear interpolation between the wall and bottom velocities and the next corresponding measured velocities.

It is very complicated or even impossible to accurately determine the perimeter flow uncertainty. Therefore ISO (2020) recommends a flat estimate for relative uncertainties of the perimeter flow in the range of $0.2 \le u^*(Q_p) \le 0.4$.

8.3.4.8 Scenarios using different measurement quality

Scenario 1 assumes the best practice in velocity measurement in a close measurement grid with a very accurate velocity sensor and careful positioning. The measurement grid consists of 135 points with $\Delta y = \Delta z = 0.05$ m. High sensor accuracy and precise handling is assumed and expressed by the values for uncertainty given in Table 8.13. Table 8.12 shows a screenshot of the Excel®-based calculations. The yellow marked perimeter section consists of a 0.025 m layer along the bottom and walls and at the surface. The measurement result of scenario 1 is $Q=0.2458$ m^3/s with the relative uncertainty $u^*(Q)= 0.0388$, and $V=0.6144$ m/s with the relative uncertainty $u^*(V)=0.0391$.

Scenario 2 assumes less accurate measurements in the same close measurement grid as scenario 1. Scenarios 3 and 4 use a less dense standard measurement grid with five measurements in eight verticals with measurement accuracy being high in scenario 3 and lower in scenario 4. Details on the parameters for all scenarios are given in Table 8.13.

Results in Table 8.14 show relative uncertainties for flow ranging in $0.0355 \le u^*(Q) \le 0.0653$ and with $0.0355 \le u^*(V) \le 0.0653$ slightly higher for the mean velocity. The low uncertainties confirm the state of velocity grid measurements as a standard for calibration of other measurement devices. The results of scenarios 2 and 3 show that accurate measurements can compensate for the uncertainties being introduced by larger measurement grids.

8.3.4.9 Scenarios using symmetry properties of the velocity field

Velocity grid measurement is undoubtedly one of the most accurate methods to determine flow and mean velocity. The high metrological effort can be reduced by using theoretically founded symmetry properties of fully developed flows. Necessary and often sufficient prerequisites are long and straight prismatic channels with low roughness and constant gradient without lateral junctions and installations. In these cases, the following symmetries of the velocity field can be observed and assumed:

- Prismatic channels with a free water surface: symmetry to the vertical centre axis.
- Prismatic channels with pressure flow: symmetry to the vertical and horizontal central axis.
- Circular cross sections with pressure flow: rotational symmetry.

Scenario 5 shows how the use of axial symmetry of velocity in the given rectangular cross section affects the uncertainty of Q and V. It is based on scenario 1 and assumes that velocity data are available only for the left side including the middle axis at $y=0.5 \times B=0.40$ m of the rectangular section. The cross section was split

scenario 1: best practice and close measurement grid

$v_x(y,z)$ [m/s]

z [m]	0.00	0.0125	0.05	0.10	0.15	0.20	0.25	0.30	0.35	0.40	0.45	0.50	0.55	0.60	0.65	0.70	0.75	0.7875	0.80	Δz [m]
0.50	0.303	0.378	0.605	0.671	0.746	0.711	0.795	0.797	0.801	0.762	0.771	0.793	0.775	0.728	0.692	0.676	0.633	0.396	0.317	0.025
0.45	0.303	0.378	0.605	0.671	0.746	0.711	0.795	0.797	0.801	0.762	0.771	0.793	0.775	0.728	0.692	0.676	0.633	0.396	0.317	0.05
0.40	0.330	0.413	0.660	0.717	0.736	0.727	0.778	0.778	0.764	0.829	0.804	0.842	0.802	0.709	0.727	0.689	0.676	0.423	0.338	0.05
0.35	0.309	0.386	0.618	0.665	0.747	0.717	0.769	0.784	0.745	0.756	0.768	0.775	0.729	0.686	0.730	0.706	0.649	0.406	0.324	0.05
0.30	0.307	0.384	0.615	0.679	0.676	0.677	0.778	0.781	0.788	0.788	0.784	0.803	0.761	0.683	0.692	0.671	0.639	0.399	0.319	0.05
0.25	0.298	0.373	0.597	0.690	0.681	0.680	0.738	0.764	0.769	0.783	0.724	0.727	0.688	0.651	0.644	0.648	0.637	0.398	0.319	0.05
0.20	0.303	0.378	0.605	0.650	0.627	0.637	0.681	0.690	0.736	0.717	0.686	0.730	0.678	0.635	0.646	0.624	0.571	0.357	0.286	0.05
0.15	0.262	0.327	0.523	0.568	0.600	0.569	0.610	0.652	0.653	0.651	0.620	0.650	0.607	0.559	0.587	0.535	0.539	0.337	0.270	0.05
0.10	0.227	0.284	0.454	0.495	0.523	0.493	0.564	0.584	0.548	0.590	0.578	0.591	0.549	0.511	0.509	0.469	0.455	0.284	0.227	0.05
0.05	0.189	0.236	0.378	0.406	0.416	0.409	0.458	0.451	0.443	0.451	0.459	0.438	0.435	0.406	0.423	0.386	0.385	0.240	0.192	0.05
0.025	0.118	0.148	0.236	0.254	0.260	0.255	0.286	0.282	0.277	0.282	0.287	0.274	0.272	0.254	0.264	0.241	0.240	0.186	0.120	0.025
0.00	0.094	0.118	0.189	0.203	0.208	0.204	0.229	0.225	0.222	0.225	0.230	0.219	0.218	0.203	0.211	0.193	0.192	0.120	0.096	
0.00	0.00	0.0125	0.05	0.10	0.15	0.20	0.25	0.30	0.35	0.40	0.45	0.50	0.55	0.60	0.65	0.70	0.75	0.7875	0.80	y [m]
		0.025	0.05	0.05	0.05	0.05	0.05	0.05	0.05	0.05	0.05	0.05	0.05	0.05	0.05	0.05	0.05	0.025		Δy [m]

ΔQ(y,z)		0.0125	0.05	0.10	0.15	0.20	0.25	0.30	0.35	0.40	0.45	0.50	0.55	0.60	0.65	0.70	0.75	0.7875	ΣQ
perimeter	[m³/s]	0.004	0.001	0.001	0.001	0.001	0.001	0.001	0.001	0.001	0.001	0.001	0.001	0.001	0.001	0.001	0.001	0.004	0.02734
measured	[m³/s]		0.013	0.014	0.014	0.014	0.015	0.016	0.016	0.016	0.015	0.016	0.015	0.014	0.014	0.014	0.013		0.21843
sum	[m³/s]	0.004	0.014	0.015	0.016	0.015	0.017	0.017	0.017	0.017	0.017	0.017	0.016	0.015	0.015	0.015	0.014	0.004	0.24577

Δy, Δz, u, u*	Δy	Δz	$u^*(F_y)$	$u^*(F_z)$	$u^*(\Delta y)$	$u^*(\Delta z)$	u_r^*	u_k	u(h)	u(B)	$u^*(Q_p)$	Q	u(Q)	u*(Q)	S	u(S)	u*(S)	V	u(V)	u*(V)	results
	m	m	-	-	-	-	-	m/s	m	m	-	[m³/s]	[m³/s]	[-]	[m²]	[m²]	[-]	[m/s]	[m/s]	[-]	
	0.050	0.050	0.030	0.005	0.040	0.040	0.010	0.001	0.002	0.002	0.200	0.2458	0.0095	0.0388	0.4000	0.0019	0.0047	0.6144	0.0240	0.0391	

Table 8.13 Grid size and uncertainty parameters of scenarios 1–4.

Scenario	Δy m	Δz m	$u^*(F_y)$ –	$u^*(F_z)$ –	$u^*(\Delta y)$ –	$u^*(\Delta z)$ –	u_r^* –	u_k m/s	$u(h)$ m	$u(B)$ m	$u^*(Q_p)$ –
1	0.050	0.050	0.030	0.005	0.040	0.040	0.010	0.001	0.002	0.002	0.200
2	0.050	0.050	0.030	0.005	0.100	0.100	0.020	0.010	0.005	0.005	0.400
3	0.100	0.100	0.030	0.005	0.040	0.040	0.010	0.001	0.002	0.002	0.200
4	0.100	0.100	0.057	0.005	0.050	0.050	0.020	0.010	0.005	0.005	0.400

Table 8.14 Flow, flow cross section, and mean velocity with uncertainties for scenarios 1–4.

Scenario	Q m^3/s	$u(Q)$ m^3/s	$u^*(Q)$ –	S m^2	$u(S)$ m^2	$u^*(S)$ –	V m/s	$u(V)$ m/s	$u^*(V)$ –
1	0.2458	0.0095	0.0388	0.4000	0.0019	0.0047	0.6144	0.0240	0.0391
2	0.2458	0.0164	0.0667	0.4000	0.0047	0.0118	0.6144	0.0416	0.0677
3	0.2512	0.0103	0.0411	0.4000	0.0019	0.0047	0.6279	0.0260	0.0414
4	0.2512	0.0186	0.0741	0.4000	0.0047	0.0118	0.6279	0.0471	0.0751

into the left section 5l ($0 \leq y < 0.5 \times B - 0.5 \times \Delta y$), the middle section 5m ($0.5 \times B - 0.5 \times \Delta y \leq y \leq 0.5 \times B + 0.5 \times \Delta y$) and right section 5r ($0.5 \times B + 0.5 \times \Delta y < y \leq B$).

The flow velocities on the right side v_r ($y > 0.5 \times B$, z) are assumed to be equal to the axially mirrored data v_l ($y < 0.5 \times B$, z) on the left side. In this study case, data v ($y > 0.5 \times B$, z) are available. Therefore, the uncertainty of the mirrored data v_r can be derived from the differences $\Delta v_r = v_r - v_l$. They range from $\Delta v_{r,min} = -0.055$ to $\Delta v_{r,max} = 0.065$ (mean -0.005, standard deviation -0.025, skewness 0.341). The uncertainty $u(v_r)$ was determined by a Type B estimate using a uniform density function (Section 8.2.3.3, Figure 8.1) which might slightly overestimate the uncertainty. It amounts to $u(v_r) = 0.0344$ m/s.

The parameters of the uncertainty calculation (Table 8.15) were set equal to scenario 1 (Table 8.13) for the sections 5l and 5m. For section 5r the uncertainty of the velocity is higher due to (i) the lower number of measurement verticals ($u^*(F_y) = 0.063$ according to ISO (2020)) and (ii) the uncertainty of the estimated velocities on the right side which can be considered by $u_k = u(v_r) = 0.0344$ m/s.

Table 8.16 shows equal results for Q, S and V in sections 5l and 5r but higher relative uncertainties $u^*(Q)$ and $u^*(V)$ in section 5r due to the higher uncertainties u_k. In section 5m, $u^*(Q)$ and $u^*(V)$ are higher than in the left section due to the higher influence of $u^*(\Delta y)$ and $u^*(\Delta z)$ on S which results in a high relative uncertainty $u^*(S)$. This small segment introduces additional uncertainty which in the end decreases

Table 8.15 Grid size and uncertainty parameters of scenario 5.

Section	Δy m	Δz m	$u^*(F_y)$ –	$u^*(F_z)$ –	$u^*(\Delta y)$ –	$u^*(\Delta z)$ –	u_r^* –	u_k m/s	$u(h)$ m	$u(B)$ m	$u^*(Q_p)$ –
5l, 5m	0.05	0.05	0.030	0.005	0.04	0.04	0.010	0.0010	0.002	0.002	0.2
5r	0.05	0.05	0.063	0.005	0.04	0.04	0.010	0.0344	0.002	0.002	0.2

Table 8.16 Flow, flow cross section, and mean velocity with uncertainties for scenario 5.

Section	Q m^3/s	$u(Q)$ m^3/s	$u^*(Q)$ –	S m^2	$u(S)$ m^2	$u^*(S)$ –	V m/s	$u(V)$ m/s	$u^*(V)$ –
5l	0.1146	0.0043	0.0380	0.1875	0.0013	0.0067	0.6111	0.0236	0.0385
5m	0.0171	0.0009	0.0550	0.0250	0.0010	0.0402	0.6822	0.0465	0.0681
5r	0.1146	0.0077	0.0676	0.1875	0.0013	0.0067	0.6111	0.0415	0.0679
5	0.2462	0.0089	0.0363	0.4000	0.0020	0.0051	0.6155	0.0666	0.1082
5l+m	0.1316	0.0049	0.0373	0.2125	0.0013	0.0062	0.6194	0.0234	0.0378
5r	0.1146	0.0077	0.0676	0.1875	0.0013	0.0067	0.6111	0.0415	0.0679
5	0.2462	0.0092	0.0372	0.4000	0.0018	0.0045	0.6155	0.0476	0.0774

slightly when summing up Q and S. A joint calculation for sections 5l and 5m, noted 5l+m, decreases $u^*(S)$ but increases $u^*(Q)$ due to the lower number of summands, and also slightly decreases $u^*(V)$.

Using symmetry properties of the velocity field in scenario 5 showed neglectable systematic deviations of Q, S and V compared to scenario 1 (Table 8.13) but higher uncertainties in V.

8.3.4.10 Scenarios using mean velocities estimates

Grid measurements of flow velocity fields are complex and unsuitable for continuous measurements of $v_x(y, z, t)$ in practice. Continuously-measuring devices capture a more or less well-defined part of the velocity field, and (i) extrapolate the non-measured part of the field with empirical or hydromechanical based algorithms or (ii) calculate the mean flow velocity V from the measured values using empirical factors or algorithms (see Chapter 3).

Exemplarily, two simple approaches are investigated with regard to their uncertainties by using the data set of scenario 1 in Table 8.12.

Scenario 6 uses the flow velocity V_{fs} in an area of the free water surface to calculate $V = e_{fs} \times V_{fs}$. The free surface velocity V_{fs} is calculated from $v(y = 0.25$ m, $z = 0.45$ m) to $v(y = 0.55$ m, $z = 0.45$ m) with an arithmetic mean 0.785 m/s, a median 0.793 m/s, standard deviation 0.0151 m/s, skewness −0.513, minimum 0.762 m/s and maximum 0.801 m/s. The Type B estimate using a uniform density distribution results in the uncertainty $u(V_{fs})$ and the uncertainty $u(e_{fs})$ for $e_{fs} = V/V_{fs}$ can be calculated by Equation (8.9) with results given in Table 8.17.

Scenario 7 is based on scenario 6 but uses the maximum flow velocity V_{max} to calculate the mean flow velocity $V = e_{max} \times V_{max}$. V_{max} is taken from the area of maximum velocities $v(y = 0.35$ m, $z = 0.45$ m), to v ($y = 0.45$ m, $z = 0.35$ m) with an arithmetic mean 0.7779 m/s, a median 0.7683 m/s, standard deviation 0.0274 m/s, skewness +0.8795, minimum 0.7451 m/s and maximum 0.8294 m/s. Results with uncertainties are given in Table 8.18.

Table 8.17 Mean velocity, mean surface velocity, and their ratio with uncertainties for scenario 6.

V m/s	$u(V)$ m/s	V_{fs} m/s	$V_{fs,min}$ m/s	$V_{fs,max}$ m/s	$u(V_{fs})$ m/s	e_{fs} –	$u(e_{fs})$ –	$u^*(e_{fs})$ –
0.6144	0.0240	0.7849	0.7625	0.8008	0.0111	0.7828	0.0325	0.0415

Table 8.18 Mean velocity, maximum velocity, and their ratio with uncertainties for scenario 7.

V m/s	$u(V)$ m/s	V_{max} m/s	$V_{max,min}$ m/s	$V_{max,max}$ m/s	$u(V_{max})$ m/s	e_{max} –	$u(e_{max})$ –	$u^*(e_{max})$ –
0.6144	0.0240	0.7779	0.7451	0.8294	0.0244	0.7898	0.0395	0.0501

Table 8.19 Coverage intervals of e_{fs} and e_{max}.

	Scenario 6				Scenario 7			
k	e_{fs}	$k \times u(e_{fs})$	$e_{fs} - k \times u(e_{fs})$	$e_{fs} + k \times u(e_{fs})$	e_{max}	$k \times u(e_{max})$	$e_{max} - k \times u(e_{max})$	$e_{max} + k \times u(e_{max})$
1	0.7828	0.0325	0.7503	0.8153	0.7898	0.0395	0.7503	0.8294
1.96	0.7828	0.0637	0.7191	0.8465	0.7898	0.0775	0.7123	0.8673
2.58	0.7828	0.0839	0.6989	0.8667	0.7898	0.1020	0.6878	0.8918

The results of both scenarios 6 and 7 are rather similar and might indicate a reasonable accuracy when looking at $u^*(e_{fs})$ and $u^*(e_{max})$ only. But note that the values of e_{fs} and e_{max} are one of the potential realizations for these particular cases. Repeated tests of the same configuration would give other results. The range for e to be expected can be estimated by the coverage interval according to Equation (8.17) as $[e - k \times u(e), e + k \times u(e)]$ with k from Table 8.3. Table 8.19 shows the coverage intervals for e_{fs} and e_{max}.

The spreads between the low and high limits are 1.18 for e_{fs} and 1.22 for e_{max} at the 95% confidence level ($k = 1.96$), and 1.24 for e_{fs} and 1.30 for e_{max} at the 99% confidence level ($k = 2.58$). Note that these results were derived from one clearly defined steady state flow experiment. The rather wide ranges should be reason enough for equipment suppliers to carefully derive the values for e_{fs} or e_{max} for a broad variety of flow conditions and quantify their dependencies on influencing factors.

8.4 SENSOR UNCERTAINTY AND *IN SITU* MEASUREMENT UNCERTAINTY

8.4.1 Definitions and explanations

In uncertainty assessment, it is of crucial importance to account for two main and independent sources of uncertainty in measured data: (i) sensor uncertainty and (ii) *in situ* measurement uncertainty.

8.4.1.1 Sensor uncertainty

All sensors used in UDSM should be calibrated according to rigorous calibration procedures and protocols (see Section 7.6 on sensor calibration). It is very important to note that the calibration should involve the entire chain of sensors and instruments, from the transducer to the final data storage in data loggers, SCADA systems or databases, i.e. to the final state of the data as they are later used by operators, researchers, regulators, etc.

Calibration of the transducer or the sensor only (uncertainties reported by manufacturers are frequently given only for such conditions) will lead to systematic underestimation of uncertainties. Indeed, other sources of uncertainties are due to data transmission, signal amplification, analogue-digital conversions, format conversion, re-scaling, display rounding, etc.

In addition, uncertainties reported by sensor manufacturers are obtained under specific conditions which may differ significantly for the *in situ* conditions. Therefore, it is the responsibility of the user to systematically calibrate all sensors under their real conditions of use, whatever manufacturers indicate.

8.4.1.2 In situ measurement uncertainty

This source of uncertainty may be summarized as 'the uncertainty due to the *in situ* conditions of measurement for a given sensor'. Its evaluation is mainly based on detailed *in situ* observations and expertise.

Let us consider water level measurement in a sewer system. In addition to the sensor standard uncertainty (e.g. 1 mm for an ultrasonic sensor determined from a calibration under static controlled conditions), there are other sources of *in situ* uncertainty which affect the measured value of the water level:

- The uncertainty in the position of the sensor in the pipe.
- The waves and oscillations of the free surface.
- The horizontality of the free surface through the width of the sewer.
- Local flow depending on changes of water level due to hydraulic effects.
- Etc.

Visual observations during dry weather in a man entry sewer showed that the free surface wave and oscillations had an amplitude up to 15 mm, which may be equivalent to a standard uncertainty of 7.5 mm. In the same sewer under wet weather conditions, the amplitude may reach 30 mm and even more. The resulting total uncertainty $u(h)$ in water level measurement h may be estimated from both the sensor uncertainty $u_s(h)$ and the *in situ* measurement uncertainty $u_i(h)$ by the law of propagation of uncertainty:

$$u(h) = \sqrt{u_s(h)^2 + u_i(h)^2} = \sqrt{1^2 + 7.5^2} = 7.6\,\text{mm} \tag{8.82}$$

In this case, the *in situ* uncertainty is clearly the dominant contribution. It may be very different if the same sensor is used e.g. in a Venturi flume with an upstream channel of sufficient length to stabilize the free surface waves and oscillations.

In situ uncertainties should always be investigated case by case to evaluate their significance. Experience has revealed that they are too frequently ignored. Their importance may be great, especially for velocity sensors (Lepot *et al.*, 2014) and some quality sensors like e.g. turbidity sensors as suspended solids are not always evenly distributed through the entire cross section.

8.4.2 Examples/orders of magnitude for some common sensors and methods

Uncertainty assessment (UA) must always be done on a case-by-case basis, accounting for local conditions: sensors used, associated calibration and verification protocols and data, operating conditions including maintenance, and operators. However, to provide orders of magnitude for beginners in UA and to facilitate comparisons, Table 8.20 indicates typical relative standard uncertainties $u^*(x) = u(x)/x$ for common sensors and methods used to measure water level, flow velocity and discharge in sewer systems. These relative standard uncertainties correspond to *in situ* best practice usual operational conditions.

Table 8.20 Order of magnitude of typical relative standard uncertainties $u^*(x)$ (in %) for most common sensors used to measure water level, velocity and discharge in sewer systems (adapted from DWA, 2011).

Percentage of the full measurement range	6	10	20	25	30	50	75	100
Water level *								
Ultra-sound sensor 0–2 m	1.398	0.893	0.388	0.311	0.311	0.194	0.116	0.116
Pressure sensor 0–2 m	2.407	1.398	0.699	0.582	0.505	0.311	0.194	0.116
Pressure sensor 0–6 m	1.902	1.203	0.582	0.505	0.388	0.194	0.194	0.116
Discharge								
Venturi flume	4.309	3.998	3.610	3.610	3.494	3.416	3.300	3.300
Electromagnetic sensor - full pipe	2.019	1.009	1.009	1.009	1.009	1.009	1.009	1.009
Electromagnetic sensor - partly filled pipe	2.989	2.989	2.019	2.019	2.019	2.019	2.989 **	2.989 **
Rectangular weir	2.484	2.096	1.786	1.708	1.708	1.514	1.514	1.281
Triangular weir	2.484	2.407	2.213	2.096	2.096	2.019	1.902	1.902
Free orifice outflow	na	na	na	1.281	1.203	1.203	1.087	1.087
Combined measurements of water level *h* and maximum velocity *v max* in partly filled pipe, diameter from 600 to 1400 mm, $h \geq 8$ cm, and	6.599	6.599	6.483	6.483	6.483	6.289	6.095	6.483 **
Combined measurements of water level *h* and velocity $v(h)$ in partly filled circular pipe, diameter from 600 to 1400 mm, $h \geq 8$ cm, and	3.998	3.998	2.989	2.989	2.989	2.989	2.989	3.998 **
Tracing experipment with continuous injection	3.300	3.300	3.300	3.300	3.300	3.300	3.300	3.300

*Without accounting for uncertainty components from height measurement and surface waves (see *in situ* uncertainty in Section 8.4.1.2).
**In the upper part of the measurement range, uncertainty increases due to hydrodynamic and measurement effects.

Example: for a water level ultra-sound sensor with a measurement range of 0–2 m measuring a water level $h = 1$ m: this corresponds to 50% of the measurement range. Table 8.20 indicates that the relative standard uncertainty is $u^*(h) = u(h)/h = 0.194\%$. Consequently, the standard uncertainty is $u(h) = 1.94$ mm and the corresponding 95% coverage interval for $h = 1$ m is $[h - 1.96 \times u(h),\ h + 1.96 \times u(h)] =$ [0.996, 1.004] m.

8.5 SUMMARY AND TRANSITION

This chapter explained in detail and illustrated with various examples the three methods one can use to estimate uncertainties: (i) the Type A method (repeated measurements), (ii) the Type B method (law of propagation of uncertainties), and (iii) the Monte Carlo method (propagation of distributions). It is essential that uncertainty assessment becomes a routine practice in urban drainage and stormwater management. Maintenance, periodic calibration and verification of sensors (Chapter 7), and uncertainty assessment (this chapter) are key elements for the next step in data processing: data analysis, quality assessment and validation (Chapter 9).

ACKNOWLEDGEMENTS

The authors thank Manuel Froidevaux (High School of Engineering and Architecture, Fribourg, Switzerland), Qingchuan Zhu (INSA Lyon, France) and Roberto Pintos (INSA Lyon, France) for their readings and corrections.

REFERENCES

Bertrand-Krajewski J.-L. & Bardin J.-P. (2001). Estimation des incertitudes de mesure sur les débits et les charges polluantes en réseau d'assainissement : application au cas d'un bassin de retenue-décantation en réseau séparatif pluvial [Estimation of uncertainties on discharges and concentrations measurements in a storage and settling tank during rainfall events]. *La Houille Blanche*, **6/7**, 99–108. doi: 10.1051/lhb/2001078. (in French).

Bertrand-Krajewski J.-L., Laplace D., Joannis C. & Chebbo G. (2000). *Mesures en hydrologie urbaine et assainissement [Measurements in urban hydrology and sewer systems]*. Lavoisier – Technique et Documentation, Paris (France), juin 2000, 794 p. ISBN 2-7430-0380-4. (in French).

CEN (1999). *ENV 13005:1999 Guide to the expression of uncertainty in measurement.* CEN, Brussels (Belgium), May 1999, 113 p.

DWA (2011). *Merkblatt DWA-M 181 – Messung von Wasserstand und Durchfluss in Entwässerungssystem [Leaflet DWA-M 181 - Measurement of water level and flow in drainage systems].* DWA – Deutsche Vereinigung für Wasserwirtschaft, Hennef (Germany), Abwasser und Abfall e.V., 101 p. ISBN 978-3-941897-94-6. (in German).

Gentle J. E. (2003). *Random Number Generation and Monte Carlo Methods, 2nd edn.* Springer, New York (USA), 382 p. ISBN 978-1441918086.

Gy P. (1988). *Hétérogénéité, échantillonnage, homogénéisation – Ensemble cohérent de théories [Heterogeneity, sampling, homogenization – Coherent set of theories].* Masson, Paris (France), 607 p. ISBN 2-225-81313-2. (in French).

Gy P. (2012). *Sampling of Heterogeneous and Dynamic Material Systems: Theories of Heterogeneity, Sampling and Homogenizing*. Elsevier Science Publishers, Amsterdam (The Netherlands), 684 p. ISBN 978-0444556066.

ISO (1993). *ISO/IEC Guide 98:1993 Guide to the expression of uncertainty in measurement (GUM).* ISO, Geneva (Switzerland), January 1993, 105 p.

ISO (2008a). *ISO/IEC Guide 98-3:2008(E) Uncertainty of measurement - Part 3: Guide to the expression of uncertainty in measurement (GUM: 1995).* ISO, Geneva (Switzerland), December 2008, 120 p.

ISO (2008b). *ISO/IEC Guide 98-3/suppl.1:2008(E) Uncertainty of measurement - Part 3: Guide to the expression of uncertainty in measurement (GUM: 1995) Supplement 1: Propagation of distributions using a Monte Carlo method.* ISO, Geneva (Switzerland), December 2008, 88 p.

ISO (2009a). *ISO/IEC Guide 98-1:2009(E) Uncertainty of measurement – Part 1: Introduction to the expression of the uncertainty in measurement.* ISO, Geneva (Switzerland), September 2009, 21 p.

ISO (2009b). *ISO/IEC Guide 98-3/S1/AC1:2009(E) Uncertainty of measurement - Part 3: Guide to the expression of uncertainty in measurement (GUM: 1995), Supplement 1: Propagation of distributions using a Monte Carlo method, technical corrigendum 1.* ISO, Geneva (Switzerland), May 2009, 2 p.

ISO (2017). *ISO 1438:2017 Hydrometry – Open channel flow measurement using thin-plate weirs.* ISO, Geneva (Switzerland), April 2017, 60 p.

ISO (2020). *ISO/FDIS 25377:2020 Hydrometric uncertainty guidance (HUG).* ISO, Geneva (Switzerland), August 2020, final draft.

Lencastre A. (1999). *Hydraulique générale [General hydraulics].* Eyrolles, Paris (France), 648 p. ISBN 978-2-212-01894-3. (in French).

Lepot M., Momplot A., Lipeme Kouyi G. & Bertrand-Krajewski J.-L. (2014). Rhodamine WT tracer experiments to check flow measurements in sewers. *Flow Measurement and Instrumentation*, **40**, 28–38. doi: 10.1016/j.flowmeasinst.2014.08.010.

Nelsen R. B. (2010). *An Introduction to Copulas, 2nd edn.* Springer, New York (USA), 272 p. ISBN 978-1441921093.

OJEU (2014). Directive 2014/32/EU of the European Parliament and of the Council of 26 February 2014 on the harmonisation of the laws of the Member States relating to the making available on the market of measuring instruments (recast). *Official Journal of the European Union*, **29/03/2014**, L 96/149–L 96/250.

Press W. H., Teukolsky S. A., Vetterling W. T. & Flannery B. P. (2007). *Numerical Recipes – The Art of Scientific Computing, 3rd edn.* Cambridge University Press, New York (USA), 1256 p. ISBN 978-0521880688.

Robert C. & Casella G. (2005). *Monte Carlo Statistical Methods, 2nd edn.* Springer, New York (USA), 649 p. ISBN 978-0387212395.

Thomopoulos N. T. (2018). *Probability distributions: with truncated, log and bivariate extensions.* Springer International Publishing, Cham (Switzerland), 184 p. ISBN 978-3319760414.

Chapter 9

Data validation and data quality assessment

Francois H. L. R. Clemens-Meyer[1,2,3], *Mathieu Lepot*[1,4], *Frank Blumensaat*[5], *Dominik Leutnant*[6] *and Guenter Gruber*[7]

[1]*Delft University of Technology, Faculty of Civil Engineering and Geosciences, Water Management Department, Delft, The Netherlands*
[2]*Norwegian University of Science & Technology, Faculty of Engineering, Dept. Civil & Environmental Engineering, Trondheim, Norway*
[3]*Deltares, Unit Hydraulic Engineering, Delft, The Netherlands*
[4]*Un poids une mesure, Lyon, France*
[5]*Eawag, Swiss Federal Institute of Aquatic Science and Technology, Dübendorf, Switzerland*
[6]*Emschergenossenschaft, Essen, Germany*
[7]*Graz University of Technology, Institut für Siedlungwasserwirtschaft und Landschaftwasserbau, Graz, Austria*

ABSTRACT

Once data have been recorded, data validation procedures have to be conducted to assess the quality of the data, i.e. give a confidence grade. Furthermore, gaps may occur in time series and, depending on the purposes, these can be given values by application of e.g. interpolation. Since both aspects are strongly correlated, this chapter gives an overview on the main data validation and data curation/imputation methods. Instead of offering exhaustive details on existing methods, this chapter aims at providing concepts for most popular techniques, a discussion of their advantages and disadvantages in the light of different cases of application, and some thoughts on potential impacts of the choices that must be made. Despite involving mathematical methods, data validation remains a largely subjective process: every data user must be aware of those subjectivities.

Keywords: Data curation/imputation, data quality assessment, data validation, interpolation.

doi: 10.2166/9781789060119_0327

SYMBOLS

(Some symbols are used for different parameters; it should be clear from the context what is meant in a specific case.)

a	fitted parameter in the linear regression
A	cross-sectional area (m^2)
b	fitted parameter in the linear regression
c	constant in an ARMA (Auto Regressive Moving Average) model
d	duration (month)
d_{RC}	maximum delay recommended between two verifications or calibrations (month)
d_{RM}	recommended duration between two maintenance procedures (month)
D	sewer pipe diameter (m)
D_j	Cook's distance for point j
G	test value for the Grubbs test
$G_{Q,t+\Delta t}$	gradient between two discharge values ($m^3/s/min$)
$G_{v,t+\Delta t}$	gradient between two velocity values (m/s/min)
$G_{V1,t+\Delta t}$	gradient between two values (V_1) (various units)
$G_{WL,t+\Delta t}$	gradient between two water level values (m/min)
$Gradient_{min}$	minimal gradient (various units)
$Gradient_{max}$	maximal gradient (various units)
h	water level (m)
h_c	hydraulic gradient (m/m)
i	counter
I	slope of a sewer pipe (m/m)
k_{st}	Manning-Strickler roughness coefficient ($m^{1/3}/s$)
K	quantity in the Mann-Whitney test
l_u	length of the wetted perimeter (m)
m	number of elements in a time series
MSE	mean squared error
n	number of elements in a time series
N	number of data points in a time series for the trend test
$N(\Delta t, T)$	number of observations in T
N_A	number of data points available in a data set
N_D	number of data points labelled as 'Doubtful'
$N_{D\text{-}D}$	number of data points labelled as 'Doubtful' after the final validation
$N_{D\text{-}G}$	number of data points labelled as 'Good' after the final validation
$N_{D\text{-}U}$	number of data points labelled as 'Unsuitable' after the final validation
$N_{equi}(\Delta t, T)$	equivalent number of observations in T, eliminating redundant information
N_E	expected number of data points in a data set
N_G	number of data points labelled as 'Good'
N_M	number of measured data points in a data set
N_T	number of tests applied to a data set
N_U	number of data points labelled as 'Unsuitable'
p	probability value (p-value) or ARMA model first parameter

q	ARMA model second parameter
Q	discharge (m^3/s)
Q_t	discharge value recorded at the date t (m^3/s)
$Q_{t+\Delta t}$	discharge value recorded at the date $t + \Delta t$ (m^3/s)
r	residues in the linear regression
R_{hyd}	hydraulic radius (m), defined as A/l_u
$R(y)$	rank of element y in a series (Mann-Whitney test)
s	standard deviation of V_t during a time window w (various units)
t	time (min, s) or Student t value
t_i	the number of subjects having the rank i (Mann-Whitney test)
T	time series, i.e. pairs of (t_i, x_i) in a time window
T_r	magnitude of the trend
$u(V_1)$	standard uncertainty of the value V_1 (various units)
$u(V_2)$	standard uncertainty of the value V_2 (various units)
$u(V_3)$	standard uncertainty of the value V_3 (various units)
$u(V_{1,t})$	standard uncertainty of the value V_1 recorded at the date t (various units)
u_{MAX}	maximal acceptable standard uncertainty (various units)
v_t	velocity recorded at the date t (m/s)
$v_{t+\Delta t}$	velocity recorded at the date $t + \Delta t$ (m/s)
$\bar{V}$	mean value of V_t during a time window w (various units)
V_1	value 1 (various units)
$V_{1,t}$	value 1 recorded at the date t (various units)
$V_{1,t+\Delta t}$	value 1 recorded at the date $t + \Delta t$ (various units)
V_2	value 2 (various units)
V_3	value 3 (various units)
$V_{I,i}$	interpolated value at the step i (various units)
$V_{LL,CR}$	lower limit for the calibration range test (various units)
$V_{LL,ER}$	lower limit for the expertise range test (various units)
$V_{LL,MR}$	lower limit for the measuring range test (various units)
$V_{LL,PR}$	lower limit for the physical range test (various units)
V_{max}	maximum value of V_t in a time window w (various units)
V_{min}	minimum value of V_t in a time window w (various units)
V_t	value recorded at the date t (various units)
$V_{t+\Delta t}$	value recorded at the date $t + \Delta t$ (various units)
$V_{UL,CR}$	upper limit for the calibration range test (various units)
$V_{UL,ER}$	upper limit for the expertise range test (various units)
$V_{UL,MR}$	upper limit for the measuring range test (various units)
$V_{UL,PR}$	upper limit for the physical range test (various units)
w	time window
WL_t	water level recorded at the date t (m)
$WL_{t+\Delta t}$	water level recorded at the date $t + \Delta t$ (m)
$\bar{x}$	mean value of x_i
x_i	observed values in the linear regression
X_k	element number k in a time series X

$\hat{y}_i$	i^{th} value of y for the fitted linear function
$\hat{y}_{i(j)}$	i^{th} value of y for the fitted linear function leaving out the jth observation in the regression
z	a time series or test value in the Mann-Whitney test
z_q	quantiles of the time series z
z_{max}	maximum threshold value in the Mann-Whitney test
z_{min}	minimum threshold value in the Mann-Whitney test
Z	Z-value in the Z-test for outliers
Z_{max}	threshold in the Z-test for outliers
α	p-value for Type I error, level of confidence
β	p-value for Type II error
γ_i	polynomial coefficient in the AutoRegressive part of an ARMA model
γ_x	weighing function for the autocorrelation function
Δt	time step between two consecutive measurements (min)
$\varepsilon(i)$	residuals at the step i (various units)
ε_i	noise term at step i in an ARMA model
θ_i	polynomial coefficient in the Moving Average part of an ARMA model
ρ	autocorrelation function, Spearman's test value, density (kg/m^3)
ρ_p	autocorrelation function for the process (for the window T)
σ_a	standard deviation of a
σ_b	standard deviation of b
σ_m	standard deviation in the measuring data (various units)
σ_P	standard deviation of the process (various units)
σ_r	standard deviation of the residues (various units)
$\xi(\alpha/2)$	quantile for $\alpha/2$
$\psi(\Delta t, T, T_r)$	quantile value as defined by Equations (9.23) and (9.24)

Motivation anecdote 'Disturbing lamppost'

After installing a Doppler flow meter, on some days a very clear signal was produced and on some days very regular outliers occurred. After analysing a few weeks of data, it became apparent that the outliers only occurred during working hours. This led to the discovery of some industrial discharge of wastewater that interfered with the measuring equipment. Once this was acknowledged the outliers could be safely imputed (in this case, by taking the average of the two adjacent values).

An alternative measure could have been to install a measuring device that could handle the specific type of wastewater without problems. The imputed data were given the tag 'imputed' in the meta-data. A similar issue occurred with a water level sensor: twice a day an enormous outlier occurred, this turned out to be caused by a defect in a lamppost located near to the monitoring location.

Francois Clemens-Meyer

9.1 INTRODUCTION

Data acquired from individual sensors and monitoring stations are prone to systematic and random errors. There are many causes varying from instrumental/device errors, human failure, software bugs, incorrect installations, discontinuities in data communication or power supply, electromagnetic interferences, etc.

This implies that raw data obtained from any monitoring system are not 100% flawless, making a 'blind' use of them potentially risky. Avoiding misleading decisions based on faulty, non-verified data is perhaps the most important reason why data should be carefully validated in any case (see the motivation anecdote). Other reasons to conduct data validation are e.g. avoiding system/catchment misunderstanding, and continuous maintenance and update of the monitoring system.

In addition, validating data on a regular and frequent basis, preferably in (almost) real-time modus, can reveal underlying causes of incorrect or missing data, and hence allow an early-on action to prevent undetected faulty recordings, and improve the maintenance protocols and tasks.

Furthermore, it can help to:

- Improve design and operation protocols.
- Detect failures of sensors and data communication.
- Identify errors which were man-made during installation and maintenance actions.
- In case of malfunctioning elements, preserve potential recourse claims involved.
- Detect and understand abnormal events that occurred at the monitoring location.

In the course of data validation, confidence grades are assigned to the subjected data, to ensure sufficient data quality as required for their purpose. In other words, data validation is a goal-driven process: required data quality changes according to the purpose of the subsequent data analysis. The level of quality strived for is different for, e.g. calculation of annual fluxes to comply with regulation obligations and real-time control of a complex system or process; data users may accept a lesser data quality for the first goal. While continuing with those two goals, the delay between records and validation is another key factor to take into account for the validation methods. If for annual fluxes data can be validated on a weekly or monthly basis, real-time control requires online data validation. The required methods depend on the purpose the data will be used for *and* the timeliness in which the validation can be accomplished after the data had been recorded. Passively measuring and collecting data without clear objectives and/or questions is not only inefficient but also makes the data validation difficult (Lindenmayer & Likens, 2018).

Prior to stepping into data validation and quality assessment procedures, some general conventions are introduced:

- First regarding data themselves:
  - A data point is a value recorded by a monitoring station from a given sensor.
  - This value can be raw (i.e. the raw data recorded by the system), 'processed' once the calibration correction has been performed (see Chapter 7), or have 'basic' and 'classical statistic' validation techniques applied once pre-validation and validation processes have been conducted.
- Then with respect to methods and procedures detailed in this chapter:
  - Data pre-validation is done by application of a sequence of basic procedures applied on corrected data aiming at automatically pinpointing data points which can be erroneous.

- o Data validation is done by application of a sequence of more or less advanced procedures (including manual checks by experts) on pre-validated data.
- o Data quality assessment is the output of those two steps: a 'validated' data point is then flagged with a colour (e.g. traffic light colour – green, orange or red) or a label (e.g. G for good, D for doubtful or U for Unsuitable, i.e. poor data quality – not fitting for the given use or purpose).

Data validation is about judging data quality in relation to the purpose the data are being meant to be used for. The quality of data points can be judged by a number of criteria:

- Plausibility: data points seem consistent with the expected conditions.
- Consistency: there are no internal inconsistencies in the data, e.g. no data beyond the physical defined interval of possible values.
- Accuracy: data points are too inaccurate and, therefore, meaningless.
- Auditability: this refers to the ability for users of the data set to obtain knowledge on the 'history' of the data, i.e. information on e.g. correction, interpolations, etc. being done on the data and the availability of meta-data on e.g. calibration and maintenance of sensors.
- Synchronicity: time stamps of measured data should be correct in relation to different global time systems, e.g. UTC (Coordinated Universal Time) and, again depending on the purpose the data is collected for, synchronized with associated sensor applications in the same network.

It is recommended to validate data as soon as possible after the measurements have been taken, for which an interval of one week has proven itself in practice, since many available meta-data such as the prevailing weather of the last seven days are mostly still mentally present.

In this sense, data validation is mostly done by computer software (see e.g. Mourad & Bertrand-Krajewski, 2002) largely since the amount of data gathered is normally too huge for manual validation. To date a 100% automatized data validation does not seem possible. What can be achieved, however, is a subdivision in data quality: 'fit for use', 'questionable quality' or 'unfit for use', i.e. in other words 'Good', 'Doubtful' and 'Unsuitable'. Since standardized and general applicable automated procedures are as yet unavailable, the assignment of those confidence levels to data points remains highly subjective with respect to the different methods discussed in this chapter, machine learning training data sets, annotating or labelling. The main challenge is to automate this subdivision in such a manner that false negative and false positive outcomes are minimal, while at the same time keeping the category 'questionable quality' as small as possible. The latter category represents data that may be of use when looked into in more detail, combining domain and process knowledge with familiarity with the system studied, the set-up applied and general engineering experience. Another very important source of information in this respect are the meta-data, such as logbooks (see also Chapters 5, 6, 7 and 10) in which information can be found on maintenance activities and calibration information for each measuring device. For this reason, it is of utmost importance during operation of the measuring systems that this information is logged by the operator with the greatest care and very promptly.

The main objective of this chapter is to make practitioners aware of the techniques that have been widely demonstrated (Figure 9.1) to be useful and work in practice for sensor signal quality within the urban hydrology context.

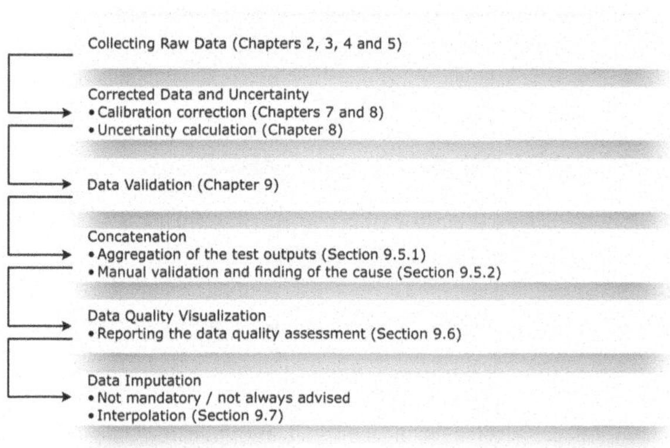

Figure 9.1 Flow chart of the data validation and quality assessment procedures. *Source*: Mathieu Lepot (TU Delft/Un poids une mesure).

This chapter reflects on data validation methods with a focus on urban drainage and stormwater management (UDSM) applications. It is by no means intended to be exhaustive on the subject, as the underlying methods find their roots in a vast and comprehensive research field in mathematics and are widely applied while generic, exhaustive texts are readily available (e.g. ESS, 2018). After defining the characteristics of 'good data', both basic and some 'classical' data validation routines with respect to their purposes are presented.

After a brief review of the different approaches (Section 9.2), this chapter is devoted to the description of the principles (Figure 9.1) of data validation of corrected data (i.e. after implementing calibration corrections on the raw data):

- Pre-validation tests with basic (Section 9.3) and advanced (Section 9.4) data analytical techniques.
- Once each data point has been flagged for each test (according to the results of each test, Section 9.5.1), those flags or labels have to be concatenated (Section 9.5.2) in order to label the data point (Section 9.5.4).
- The data quality representation (Section 9.6.1) and monitoring system analysis (Section 9.6.2) for communication purposes.
- Section 9.7 aims at introducing some methods for data imputation, i.e. replace unsuitable data to achieve some goals (e.g. calculated volumes or fluxes that require complete and equidistant time series). This step, mandatory for some applications, is not really recommended when not needed to avoid working with artificial (interpolated or imputed) data.
- Emerging techniques and methods are briefly introduced in Section 9.8.

This chapter does not aim at offering a complete guideline nor protocol for data validation: it is meant to be an introduction to data validation in itself, a review of the main existing methods (including their advantages and disadvantages) and a list of warnings regarding validation (a mandatory step in UDSM monitoring, but rather prone to bias).

Key messages on data validation

- KM 9.1: *Data validation is mandatory* – never use the data without a careful check.
- KM 9.2: Data validation based on the separation of concerns: *two steps* – (i) pre-validation (unified basic checks), (ii) goal-driven validation.
- KM 9.3: *Purpose dependency:* the results of the data validation depends on the anticipated use of the data.
- KM 9.4: *Subjectivity and reproducibility*: despite there being numerous methods and protocols, data validation remains a subjective process. Keep track of tasks performed.

9.2 CONCEPTS APPLIED IN DATA VALIDATION

9.2.1 What is data validation

Data validation is a process that determines if available data satisfy quality objectives (which have been *a priori* agreed upon) and requirements defined by the anticipated use of the data, here in the context of urban drainage and stormwater management. The process results in adding a quality indicator to each individual data point based on objective criteria as far as possible.

This quality indicator ideally reflects both the *correctness* and the *usefulness* of the data point. Whereas the correctness of a data point can be attributed to the physical meaning, the latter aspect indicates that there is no 'absolute' metric for the quality of a data point. To a certain extent, the evaluation of whether a data point is of high or poor quality depends on the purpose for which the data are to be used. Speaking in these terms, the process of data validation combines (i) an objective, physically-based assessment and (ii) a somewhat subjective perception of how confident the user can be that the measured data point reflects 'reality'.

Example: In real-time control (RTC) applications, there is very little time between obtaining data and using them, which implies that time for an extended validation of the data is limited at best. In such cases, a minimal (if any) validation is performed, e.g.:

- Is the data point there?
- Is the data point within the expected range?

If both questions are answered positively, the data can be used for feeding the RTC algorithm; if one question is negatively answered, the data point is omitted and a default action (in terms of RTC) is taken. In such a situation, it is good practice to store data and the outcome of the two tests mentioned as it allows for a posterior evaluation of the quality of the monitoring, and it may allow the future use of the data for other applications.

Therefore, prior to setting up a protocol for data validation in a given case, case-specific quality levels have to be agreed upon along with a method of organizing the meta-data (see Chapters 5 and 10 on this subject along with Section 9.2.5) that are produced by the validation protocol. This furthermore implies that, when starting a monitoring project, designing the data structure (see Chapter 5) essentially requires considering the envisioned process for the data validation.

9.2.2 How to quantify the quality of data

There are various ways of assigning confidence levels (i.e. quality flags) to individual data points as an indicator for data quality. Practically, the quality assessment of a data point may range from a very basic 0/1 flagging or a more distinct traffic-light labelling 'Good', 'Doubtful', 'Unsuitable' (Table 9.1) to a very refined system in which a wide range of specific qualifications can be added, e.g. attributed to a specific anomaly type (see Table 9.2).

Do's

- *'Only recordings that have a value can be assessed regarding their quality. Keep a record of the fact that there was a missed recording for as long as possible. Do not mix data quality assessment and data curation.'*
- *'The interpretation of data regarding its quality can substantially be qualified through meta-data information. Carefully document meta-data and associate them with data.'*
- *'Prior to assessing data quality, a thorough reflection is mandatory to ensure: (i) are the performed tests useful to reflect likely dubious behaviour of data? and (ii) can all available data be used to conduct individual tests?'*

Differentiating data into just two states, good and poor quality (dichotomous flagging) may be unambiguous and well-achievable for a machine, but insufficient for differentiation. For this reason, often three levels of confidence, e.g. good-doubtful-unsuitable, are assigned, allowing for a more distinguishing assessment. Still, the aspect when a data point is labelled *doubtful* can be somewhat subjective. One labeller may consider an obvious outlier as doubtful whereas the other labeller clearly labels it as *unsuitable*. Clear mind models or 'gold standards' need to be established to avoid subjectivity and allow cross-comparison within one data set. The term 'gold(en) standard' stands for an external criterion representing a kind of benchmark that is the best available under reasonable conditions.

Table 9.1 Example of a traffic-light-system for a gross quality assessment of a data point.

Primal label	Shortened as	Colour	Description
'Good'	'G'	Green	Data point passed all validation tests
'Doubtful'	'D'	Orange	Data point is physically valid but somewhat questionable when evaluated in a wider context
'Unsuitable'	'U'	Red	Data point is physically invalid or is definable erroneous so that it cannot be used
'Missing'	'M'	White or black	Missing data point

Table 9.2 Didactical example of advanced refinement of the quality assessment of a data point. Further examples are given in Leigh *et al.* (2019).

Minor label	Meaning
A	Sensor failure
C	Trend
D	Outlier
E	Constant offset
F	Time shift
G	Value < lower bound of the valid range
H	Value > upper bound of the valid range
I	Low variability, persistently constant value, freeze
J	Imputed by application method x or y
K	Wrong data format
M	Missing time stamp; no data point available

Defining a 'gold(en) standard' is a matter of consensus or opinion, not some kind of statistical property. Whereas dichotomous flagging can be accomplished by a machine, tripartite scoring mostly involves human assessment, i.e. expert knowledge. The general idea is to add relevant information to enhance the probability of finding the cause of a poor data quality.

Automatized flagging of individual signals (no additional information) results in a 0/1 assessment. Adding further information, i.e. extending it to a multi-signal analysis, allows tripartite scoring through a machine.

One can argue about whether or not to include missing data points ('M') in the data quality assessment. Strictly speaking, in a case where there is no measurement recorded, i.e. no data point available, the quality cannot be evaluated. On the other hand, the indication and qualification of gaps in time series at which a data point would have been expected, due to sensor failures, data communication outages, or erroneous data formats allows for characterizing time series regarding their consistency and completeness. The information on amount and distribution of periods at which no data is available may be decisive for the subsequent use of the data, but also for the data validation itself (Section 9.3.6).

Data points labelled as 'Unsuitable' or 'Doubtful' can further be qualified according to the (likely) cause of the less-than-ideal quality. A didactical example of such refined data quality labelling is given in Table 9.2. Note that qualification of quality labels can be supported through operational information, often referred to as meta-data. Meta-data, i.e. additional information on the sensor performance, operation of periphery devices, maintenance actions, and changes to the monitoring environment, are essential to interpret field data correctly (Section 9.2.5).

Authors suggest outputs of those tests: G, D or U. Those are only a suggestion and may or should change according to the monitoring purposes, legal regulations and the expected data quality. However, those suggestions are based on rather long experiences and we advise slight adaptations without completely changing the tests and their outputs.

Data available?

- CL 9.1: *Which?* – Which meta-data are available? Catchment, sewer, sensor, maintenance data.
- CL 9.2: *How?* – How can we make use of this information? Run-off model to correlate catchment, rain and discharge data.
- CL 9.3: *Missing data?* – Is there any data easily acquirable that could be used to conduct additional and relevant tests?

Performed tests

- CL 9.4: *Cover* – Do the applied tests cover any likely behaviour of my data?
- CL 9.5: *Complex situations* – Is (are) there any situation(s) that could bias the output of a few applied tests? Such as backflow effect, complex hydraulic geometry, etc.
- CL 9.6: *Full use* – Do the applied tests make full use of available data?

9.2.3 Subjectivity

The subjectivity in the process of data quality assessment is basically present in discriminating between data in categories 'Good', 'Doubtful' and 'Unsuitable' as defined in Table 9.1. Without going into the discussion of what 'truth' is and whether or not it can be known, 'Good' data is equivalent to 'passed all validation tests'.

This implies that the range of tests a data point is subjected to has a stringency convincing the data user that it is fit for its purpose in the case where the data point passes all these tests. But it does *not* automatically imply that it therefore necessarily reflects the 'truth'.

At the same time, one is striving for a data yield as high as possible, implying that the range of tests should produce a small portion of false positives and false negatives. In other words, the number of data labelled as 'Doubtful' should be minimized, as this fraction of data points requires further attention to investigate the cause of the imperfectness. This can be a very tedious job requiring domain knowledge, and in many cases also knowledge and understanding of the actual situation in the system at hand (e.g. documented as meta-data). Depending on the level of expertise and the solidity of the given information, different answers can be expected when asking a group of experts about the quality of a data point. A certain amount of subjectivity is introduced.

For instance, in the case where rehabilitation works are ongoing, this may result in abnormal sensor readings that may be classified as being 'Doubtful', while recorded values actually represent the (disturbed) process in reality. In the case where one is aware of such an event, the data may be useable after all; otherwise, the data may remain classified 'Doubtful'.

9.2.4 Automation of data validation

The example in the preceding paragraph nicely illustrates that it is likely that data validation cannot be 100% left to computerized algorithms. One way or the other there seems always to be a need for an expert judgement regarding the quality/useability of the data obtained. Having said that, it has to be added immediately that for practical purposes the application of software, i.e. some degree of automation of data validation, is very favourable.

In principle, data validation can be done manually, which implies that trained individuals have to study the raw data obtained and judge whether or not the data obtained are fit for purpose. Manual data validation, however, has some serious drawbacks:

- It is very labour-intensive and therefore expensive.
- The criteria for accepting/rejecting data points are subjective and will result into a non-reproducible assessment.
- For some purposes, e.g. RTC applications, the processing time is simply too long to be practically applicable.

For these reasons, a certain degree of automated data validation is applied in practice. This may at least relieve the workload, although applied schemes seem to show a variation of success. For example, the validation scheme as proposed by Upton & Rahimi (2003) for validating data from tipping bucket rain gauges proved to be very efficient: up to 90% of the anomalies proved to be correctly identified after manually checking. However, when applying the same procedure to a grossly similar case, Schilperoort (2011) found a percentage of only 60% of correctly identified anomalies. This reduced yield in the latter case was caused by the huge amount of data missing due to data communication issues and the lack of meta-data, the latter underlining the importance of keeping track of such additional information.

9.2.5 Meta-data

Meta-data is essential to interpret field data correctly. When trying to identify causes of data being classified as 'Doubtful' or 'Unsuitable', the presence of additional information, i.e. meta-data, is vital. Ideally, this information on sensor operation and maintenance actions is in standardized logbooks. Meta-data should be collected systematically, i.e. formalized in individual categories, and continuously over time.

Meta-data information can comprise (non-exhaustive list):

- Sensor maintenance actions.
- Antecedent and last calibration results.
- Access to plans for and reports on construction works.
- Data from adjacent and/or related monitoring sites, e.g. rain gauges to discriminate between dry and wet weather or the reading from a sensor showing overlap in its readings – see also Section 6.2 on macro design.
- Weather reports, e.g. thunderstorms may cause loss of communication or damage caused by an electromagnetic pulse (EMP).
- Development over time of the sensor performance; sometimes a problem may repeat, so using the sensor history may hint at a (external) cause that induces the problem.
- Data on the performance of similar sensors (brand, production batch, etc.), this may reveal some inherent issues with the device(s) used. This information may be used to upgrade the system when replacing parts.

It is a managerial decision to what extent of detail one should go in gathering meta-data when operating a long-term observation campaign, as these administrative tasks tend to expand (certainly in large bureaucratic organizations). It is essential – not just against the background of an increasing degree of automation of data handling – to foresee gathering information on the performance of the monitoring system as a whole and in a systematic manner and for well-defined purposes only (Chapter 6). This implies man-hours spent on maintenance, data analysis and validation are monitored as well. Such systems allow fine-tuning of the design and the operation of the monitoring system to ensure a certain level of data quality. The creation of standard operational procedures (SOPs) for the documentation of all meta-data to be recorded is recommended to ensure that the meta-data are documented as uniformly as possible and to reduce subjective elements as much as possible.

9.3 BASIC CHECKS

There are numerous data validation methods, from very simple to very advanced ones. This section provides an overview of the existing ones, their advantages, disadvantages, and limitations. The methods are divided in two categories, the basic checks (Section 9.3) and the advanced validation ones (Section 9.4). The choice of those tests is strongly dependent on what purpose(s) the data are collected for, the available skillset and knowledge of persons in charge, and the delay between measurement and validation. All the methods presented in this chapter are applied on data from calibrated sensors. Contrary to the methods discussed in Section 9.4, the basic methods can easily be automated.

Thresholds

Most of the tests presented hereafter are based on thresholds. The output of each test is directly dependent of the selected threshold(s). Careful attention must be paid to the threshold selection: the output can be too pessimistic or too optimistic.

This warning is valid for everyone: from data provider, data curator to the data user. Always keep in mind a famous quote from W.S. Churchill: 'I only believe in statistics that I doctored myself'.

9.3.1 Test on plausibility

Plausibility tests using numerical criteria or based on common sense usually do not require significant resources and are hence suitable for low-computation online validation.

9.3.1.1 Physical range

This is a first test, only based on physical boundaries of the measured phenomena: a water level at free surface flow mode cannot be negative or higher than the diameter of a circular pipe, the temperature of liquid water cannot be below 0 or above 100 degrees Celsius (at atmospheric pressure), rain intensity cannot be negative, etc. If the data values are outside the physical range, they should be labelled as 'Doubtful' or 'Unsuitable' for this test. A value $V_{1,t}$ recorded at the date t successfully passes this test if Equation (9.1) is verified.

$$V_{LL,PR} \leq V_{1,t} \leq V_{UL,PR} \tag{9.1}$$

where $V_{LL,PR}$ and $V_{UL,PR}$ are, respectively, the lower and upper limits of the physical range.

Example: In a circular pipe or 1000 mm of diameter, the water level values have the following threshold: $V_{LL,PR} = 0$ mm and $V_{UL,PR} = 1000$ mm. If a recorded value ($V_{1,t}$) is negative or higher than 1000 m, Equation (9.1) is not verified and, therefore, this value is flagged as 'Doubtful' or 'Unsuitable' with respect to this test on physical range (Figure 9.2).

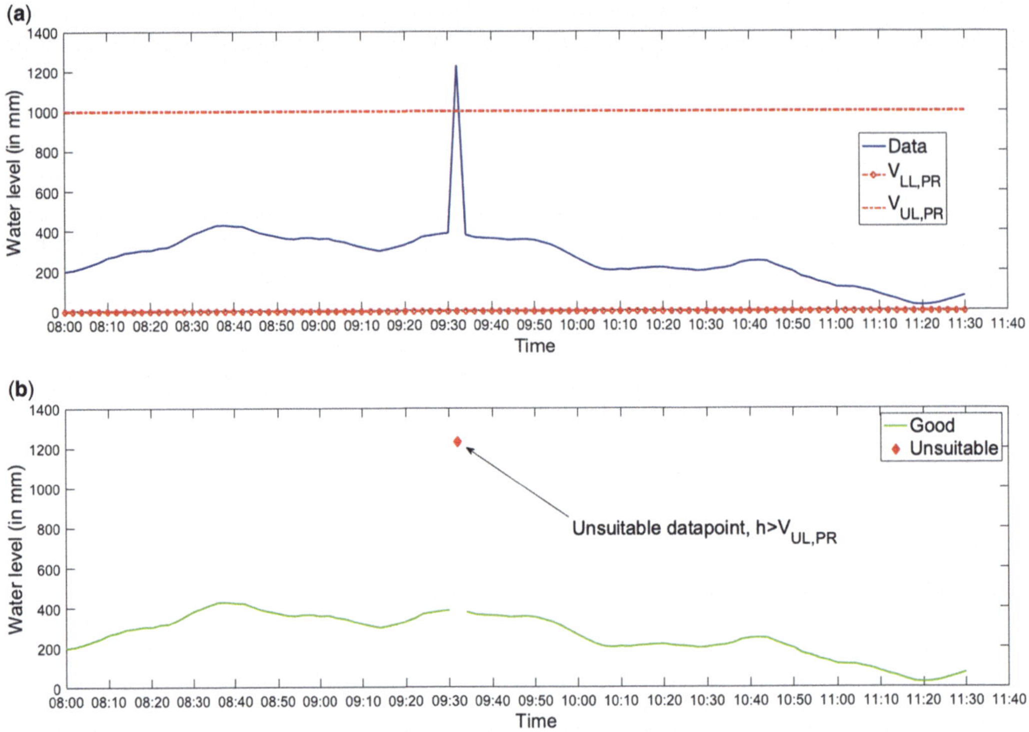

Figure 9.2 (a) water level data (blue) and physical range thresholds (red); (b) 'Good' (green) and 'Unsuitable' (red) data according to the physical range test. *Source*: Mathieu Lepot (TU Delft/Un poids une mesure).

Limitations

The statement made in Section 9.3.1.1 for a single data point is certainly correct. However, if pressurized flow may occur at the measurement location, piezometric water level sensors may give a water level greater than the pipe diameter (i.e. the flow pressure at the measurement section). Even if this test is rather easy to set up, it requires some expertise and knowledge about (un)likely conditions at the measurement point.

9.3.1.2 Measuring range

This test is rather similar to the previous one, but based on the measuring range of each sensor. Sensors are designed to measure and work over certain ranges of measurement or environmental conditions. If the recorded value is outside the measuring range or has been recorded in unusual conditions, it should be labelled as 'Doubtful' or 'Unsuitable' for this test (Equation (9.2)).

$$V_{LL,MR} \leq V_{1,t} \leq V_{UL,MR} \tag{9.2}$$

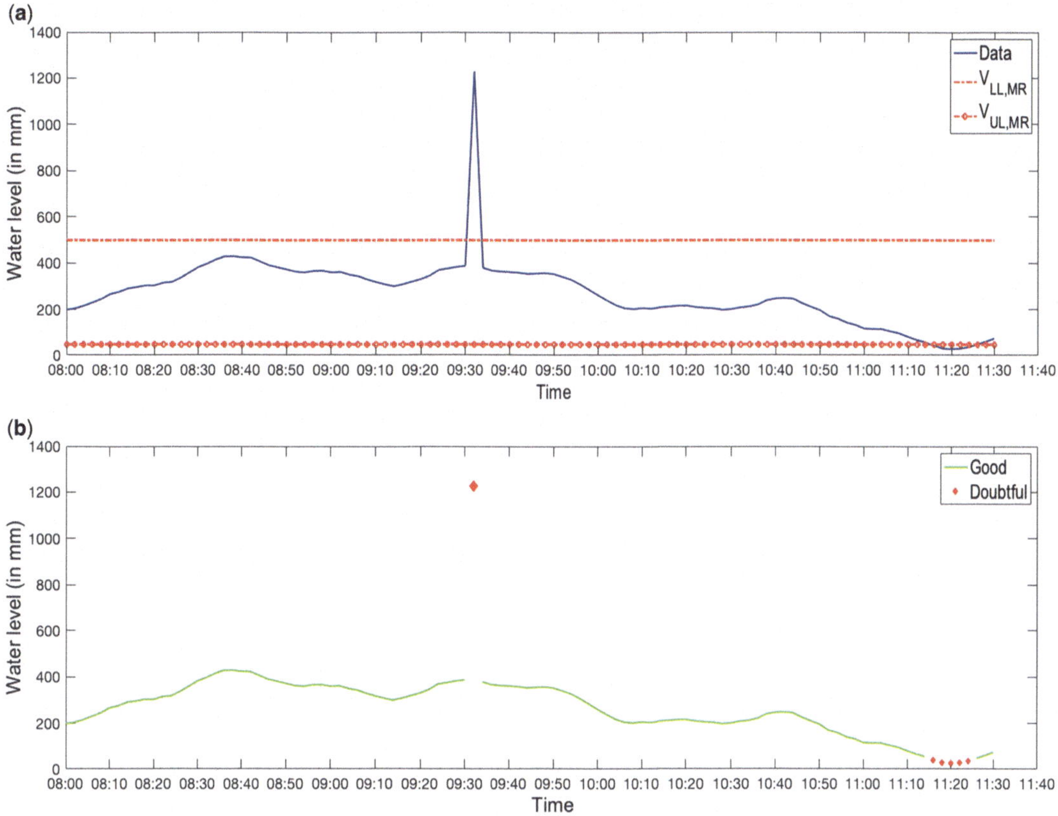

Figure 9.3 (a) water level data (blue) and measuring range thresholds (red); (b) 'Good' (green) and 'Doubtful' (red) data according to the measuring range test. *Source*: Mathieu Lepot (TU Delft/Un poids une mesure).

where $V_{LL,MR}$ and $V_{UL,MR}$ are, respectively, the lower and upper limits of the measuring range, i.e. they are sensor dependent. To avoid 'not good data' for this test, the measuring ranges of the different devices have to overlap.

Example: In the same pipe as in the previous example, the water level sensor has a measuring range between 50 mm and 500 mm (according to its specifications). The water level values recorded by this sensor have the following threshold: $V_{LL,MR} = 50$ mm and $V_{UL,MR} = 500$ mm. If a recorded value ($V_{1,t}$) is lower than 50 mm (e.g. 30 mm) or higher than 500 m, Equation (9.2) is not verified and, therefore, this value is flagged as 'Doubtful' with respect to this test on measuring range (Figure 9.3).

9.3.1.3 Calibration range

This test is quite similar to the previous ones. A sensor is calibrated over a given range, from the minimum to the maximum values of calibration standards. For this test also, if the value is outside the calibration range, it should be labelled as 'Doubtful' or 'Unsuitable' for this test (Equation (9.3)).

$$V_{LL,CR} \leq V_{1,t} \leq V_{UL,CR} \tag{9.3}$$

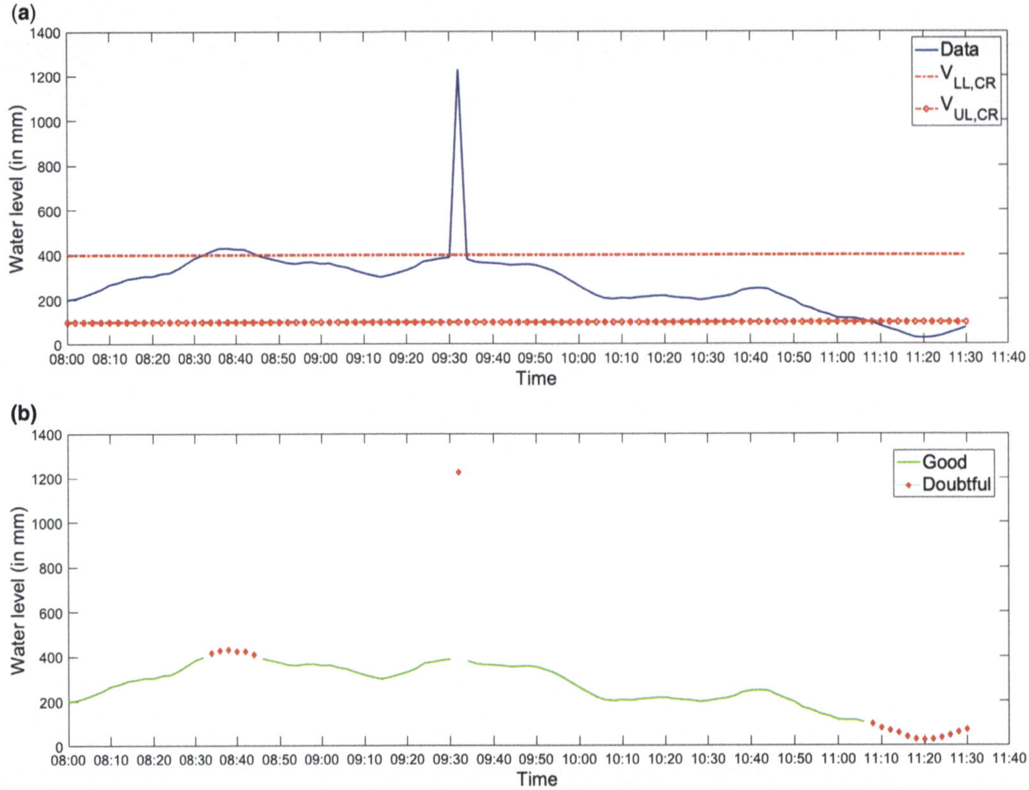

Figure 9.4 (a) water level data (blue) and calibration range thresholds (red); (b) 'Good' (green) and 'Doubtful' (red) data according to the calibration range test. *Source*: Mathieu Lepot (TU Delft/Un poids une mesure).

where $V_{LL,CR}$ and $V_{UL,CR}$ are, respectively, the lower and upper limits of the calibration range, i.e. calibration standards dependent. To avoid 'not good data' for this test, the calibration standard values should cover the full range of expected conditions.

Example: The water sensor used in the previous example has been calibrated from 100 ($V_{LL,CR}$) to 400 ($V_{UL,CR}$) mm. If a recorded value ($V_{1,t}$) is outside those boundaries, it should be flagged as 'Doubtful' with respect to this test on calibration range (Figure 9.4).

9.3.1.4 Expertise range

This range test requires domain knowledge and knowledge of the system under study. Despite all the previous checks, experts may judge that a value is doubtful if measured under certain conditions, generally narrower than the ones given in the design specifications of a sensor (Equation (9.4)). As an example, a Doppler probe can measure water levels in a range from 0 to 0.7 m, but experts may consider that data cannot be fully trusted outside 0.1 to 0.4 m due to the intrinsic limitation of the probe and acoustic attenuation of the signal.

$$V_{LL,ER} \leq V_{1,t} \leq V_{UL,ER} \quad (9.4)$$

where $V_{LL,ER}$ and $V_{UL,ER}$ are, respectively, the lower and upper limits of the expertise range.

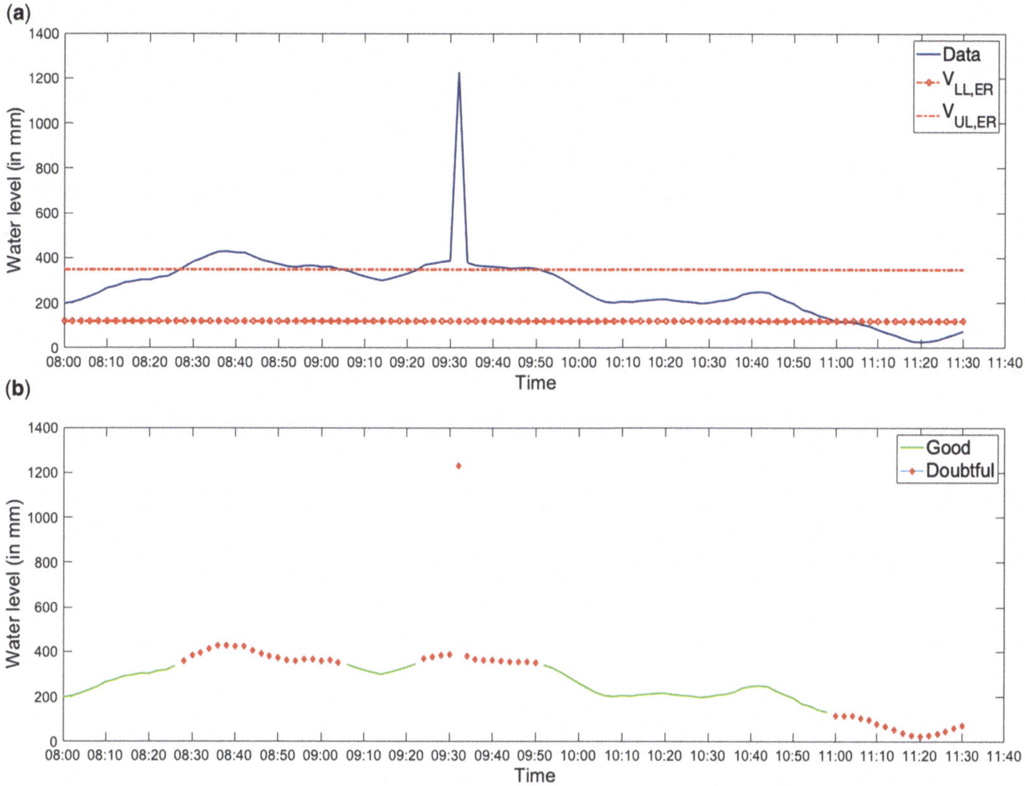

Figure 9.5 (a) water level data (blue) and expertise range thresholds (red); (b) 'Good' (green) and 'Doubtful' (red) data according to the expertise range test. *Source*: Mathieu Lepot (TU Delft/Un poids une mesure).

Example: Experience shows that the recorded values given by sensor are doubtful below 120 mm ($V_{LL,ER}$) and above 350 mm ($V_{UL,ER}$). If a recorded value ($V_{1,t}$) is outside those boundaries (e.g. 110 mm or 360 mm), it should be flagged as 'Doubtful' with respect to this test on expertise range (Figure 9.5).

9.3.1.5 Gradient range

Time series give information on phenomenon dynamics. With some expertise, the usual dynamics of the phenomena are known and can be used to validate or not the data. Time series showing no or too sudden dynamics can be considered as doubtful. Given a value V_1 recorded at two different dates (t and $t + \Delta t$), the value $V_{1,t+\Delta t}$ could be considered as doubtful if one of the Equation (9.5) is verified.

$$\begin{cases} G_{V1,t+\Delta t} = \dfrac{V_{1,t+\Delta t} - V_{1,t}}{\Delta t} > Gradient_{MAX} \\ V_{1,t+\Delta t} = V_{1,t} \\ G_{V1,t+\Delta t} = \dfrac{V_{1,t+\Delta t} - V_{1,t}}{\Delta t} < Gradient_{MIN} \end{cases} \tag{9.5}$$

where $Gradient_{MAX}$ and $Gradient_{MIN}$ are, respectively, the maximum and minimum likely gradients for the given phenomenon.

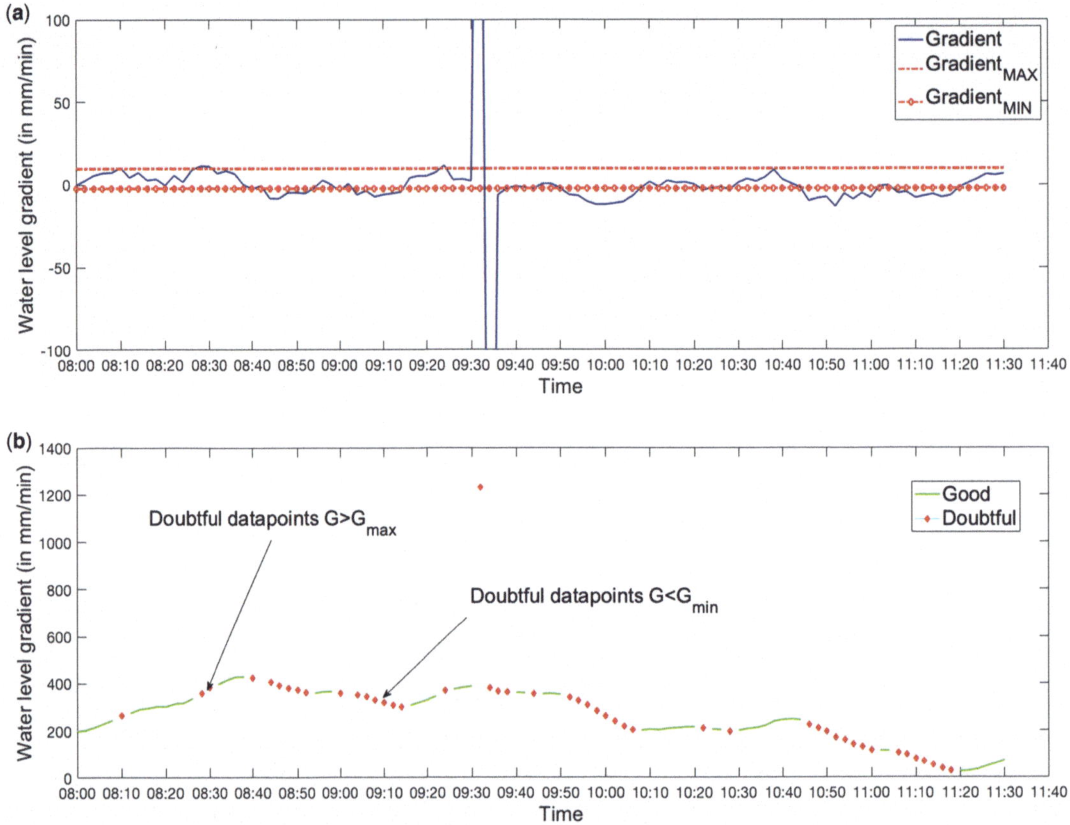

Figure 9.6 (a) water level data (blue) and gradient range thresholds (red); (b) 'Good' (green) and 'Doubtful' (red) data according to the gradient range test. *Source*: Mathieu Lepot (TU Delft/Un poids une mesure).

Example: Experience shows that the hydraulic dynamics of the catchment barely ever exceed 10 mm/min for the rising part of a storm event and are, in most cases, below 2 mm/min for the declining part (Figure 9.6).

Assuming a Δt equal to 2 min, the difference ($V_{1,t+\Delta t} - V_{1,t}$) should not be higher than 20 mm when the flow rises or be lower than 4 mm when the flow decreases. Otherwise, the value should be flagged as 'Doubtful'. As an example, the following couples ($V_{1,t+\Delta t}$, $V_{1,t}$) will flag $V_{1,t+\Delta t}$ as 'Doubtful': (160,190), (50,45) and respectively (70,70) – for these couples the gradients are, respectively, 15, −2.5 and 0 mm/min.

9.3.2 Test on consistency

Consistent data are logical and do not contradict themselves. Inconsistencies are usually caused by gross errors (DWA, 2011).

9.3.2.1 Comparison between redundant recordings (signal redundancy)

If, as advised, a monitoring station has redundant sensors to measure the same type of information (e.g., water level, velocity, etc.), each value can be compared to the other ones in order to identify if one or a

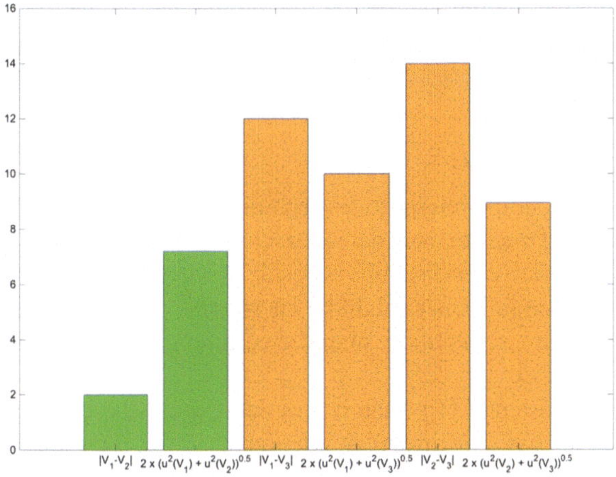

Figure 9.7 Comparison between absolute difference and their uncertainties (with 95.4% confidence level). *Source*: Mathieu Lepot (TU Delft/Un poids une mesure).

few of them are too different from the other ones. Given three measured values V_1, V_2 and V_3 and their associated standard uncertainties $u(V_1)$, $u(V_2)$ and $u(V_3)$, the value V_1 can be considered as doubtful if it is significantly different from the other ones, i.e. if V_1 satisfies the subsequent three Equation (9.6).

$$\begin{cases} |V_1 - V_2| < 2\sqrt{u^2(V_1) + u^2(V_2)} \\ |V_1 - V_3| \geq 2\sqrt{u^2(V_1) + u^2(V_3)} \\ |V_3 - V_2| \geq 2\sqrt{u^2(V_3) + u^2(V_2)} \end{cases} \tag{9.6}$$

This test requires at least three values. If there are only two, the test is just able to say that both values are significantly different, without pinpointing which one might be wrong. This test is applicable on recorded values or calculated values, such as water levels, velocities and discharges calculated from those two.

Example: At the same monitoring location, and while using the uncertainty calculation methods presented in Chapter 8, three water levels are recorded with known uncertainties: $V_1 = 50$ mm and $u(V_1) = 3$ mm, $V_2 =$ 48 mm and $u(V_2) = 2$ mm and $V_3 = 62$ mm and $u(V_3) = 4$ mm. The three Equation (9.6) are verified: V_3 is flagged as 'Doubtful' while V_1 and V_2 pass the consistency test, i.e. are flagged as 'Good' (Figure 9.7).

9.3.2.2 Dynamic consistency

As an example, the dynamic behaviour of water level, velocity and discharge should be consistent: under standard condition, e.g. when no downstream effects occur, if the water level increases, the velocity increases and the discharge too. Consistencies between gradients could be checked to identify potentially doubtful data. While reusing the same notation as in Equation (9.5) for this example, i.e. $G_{WL,t+\Delta t}$, $G_{v,t+\Delta t}$ and $G_{Q,t+\Delta t}$ being the gradients for water level, velocity and discharge, the velocity can be considered as doubtful if Equation (9.7) are verified.

$$\begin{cases} (G_{WL,t+\Delta t} > 0 \text{ and } G_{Q,t+\Delta t} > 0) & \text{or} & (G_{WL,t+\Delta t} < 0 \text{ and } G_{Q,t+\Delta t} < 0) \\ (G_{WL,t+\Delta t} > 0 \text{ and } G_{v,t+\Delta t} < 0) & \text{or} & (G_{WL,t+\Delta t} < 0 \text{ and } G_{v,t+\Delta t} > 0)\ . \\ (G_{v,t+\Delta t} < 0 \text{ and } G_{Q,t+\Delta t} > 0) & \text{or} & (G_{v,t+\Delta t} > 0 \text{ and } G_{Q,t+\Delta t} < 0) \end{cases} \tag{9.7}$$

The potential combinations of such tests are endless and too site specific to draft an exhaustive list here.

Example: Figure 9.8 presents an example of such a test. The gradients for water level, velocity and discharge data are plotted at the top (Figure 9.8a). Based on sign analysis (Equation (9.7)), data are then labelled as 'Good' or 'Doubtful' according to this test (Figure 9.8b).

9.3.2.3 Time stamp consistency

Measurement data always have a time reference, as each individual measurement point has been observed and recorded at a specific time. If measurement data are recorded at a regular time interval (e.g. each minute), the distance between two consecutive time stamps is equal. However, depending on the quality of hard- and software installed, an expected equidistance might be interrupted, resulting in irregular time series causing loss of information. Irregular time series show unexpected time gaps or even different measurement data assigned to an identical time index.

Testing the time stamp consistency of measurement data requires knowledge of whether the signal is expected to be equidistant or have an irregular interval, and this must be communicated before the measurement is under operation. Estimating the correct periodicity after data has been collected would otherwise require statistical tests to be applied.

Nowadays, monitoring stations tend to measure and record at fixed and regular time intervals. However, irregularly-recording measurement stations are still maintained, e.g. to save battery life when remotely installed. Regular time changes due to daylight saving taking place twice a year can be a further cause of time stamp inconsistencies. If possible, these should be avoided by, for example, storing the measured data uniformly with a global time system, e.g. UTC (Coordinated Universal Time).

9.3.3 Test on accuracy

If a value is too inaccurate, i.e. if its standard uncertainty is higher than a given threshold adapted to its future use, it should be labelled as 'Doubtful' or 'Unsuitable' (Equation (9.8)).

$$u(V_{1,t}) \leq u_{MAX} \tag{9.8}$$

where $u(V_{1,t})$ is the standard uncertainty in the value $V_{1,t}$ and u_{MAX} is the threshold of the uncertainty. This test can be extended to two thresholds, one for 'Doubtful' and another one for 'Unsuitable'. This rather basic test is sensitive to the selected threshold, which is sensor specific and could either be absolute or relative.

Given certain standards, by law or for the final use, a value can be labelled as D or B if there is no uncertainty associated.

Example: At the same monitoring location, a value $V_{1,t} = 67$ mm is recorded by a water level sensor. Once the uncertainty calculations are done, $u(V_{1,t})$ is equal to 3 mm (see Chapter 8). Given a u_{MAX} of, e.g., 5 mm, $V_{1,t}$ is flagged as 'Good' (Figure 9.9).

9.3.4 Test on auditability

Although the term auditability is mostly used in accountancy, the principle of trackability of what 'happened' to a measured parameter value can be transferred to monitoring projects. In the end, the product that a monitoring project has to deliver is data of a known and well described quality. To be able to implement the underlying principle of quality control – 'collect data on the manner in which the procedures and protocols in an organization are applied and learn from evaluating them' – to monitoring projects, the following aspects are to be considered:

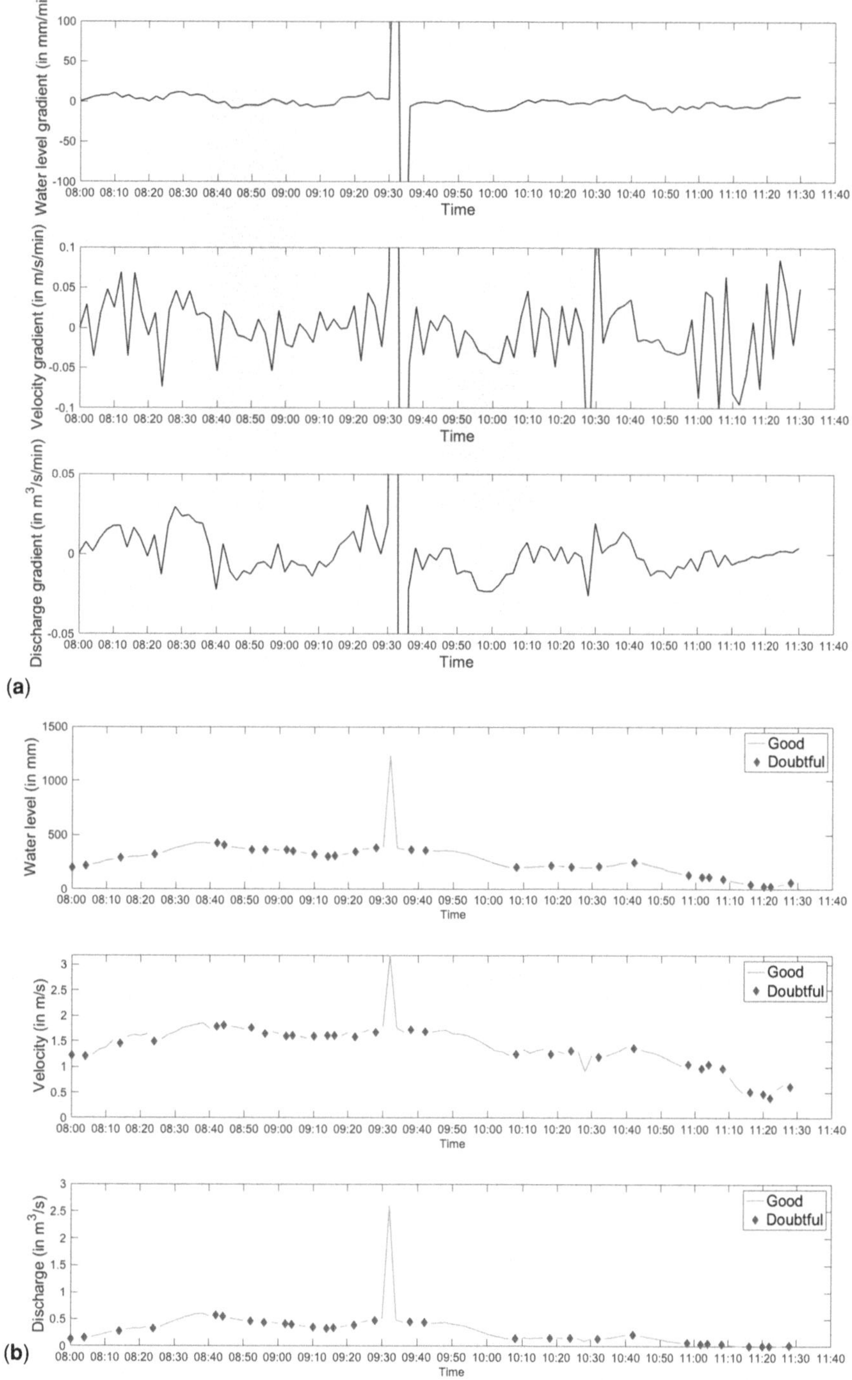

Figure 9.8 (a) gradient for water level, velocity and discharge time series; (b) output of the test based on dynamic consistencies. *Source*: Mathieu Lepot (TU Delft/Un poids une mesure).

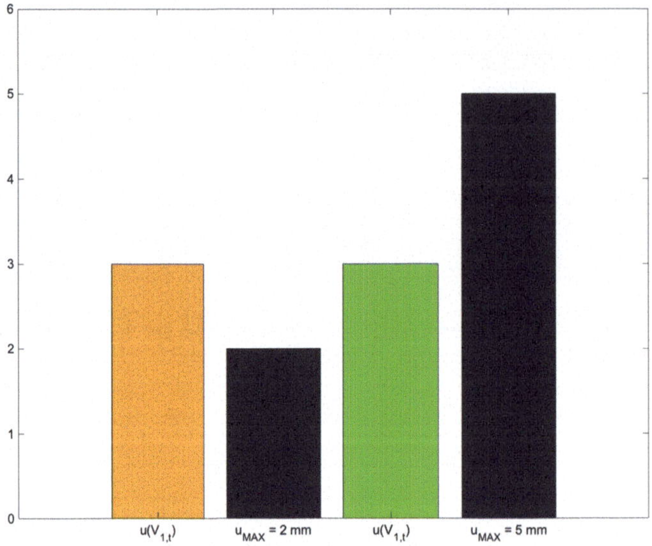

Figure 9.9 Comparison between absolute difference and their uncertainties (with 95.4% confidence level). *Source*: Mathieu Lepot (TU Delft/Un poids une mesure).

- Formulate clear procedures and protocols with respect to data storage, postprocessing and authorization levels of personnel working with or on the data.
- Make sure meta-data is kept accurate and up to date.
- Be sure to have back-ups of the 'raw' data at all times, this ensures being able to 'redo' postprocessing when in the course of time protocols or procedures applied earlier turn out to have flaws (e.g. some bug in a piece of computer code).

Testing on auditability may well be part of a regular/general quality systems check in an organization which not only tests the applicability and application of all protocols and procedures but also the motivation and awareness of people working with them.

9.3.4.1 Calibration

As introduced in Chapter 7 on calibration methods, sensors need to be calibrated regularly. If, for some reason, the data have not been corrected for calibration (no calibration has been done, the data of the calibration correction are not recorded or stored), i.e. the data are raw, those data points should be considered as 'Unsuitable'.

Example: If there is no calibration correction (Chapters 7 and 8), which can be identified by either the absence of corrected data, of calibration correction parameter or (but not always) no difference between the corrected and the raw values, this value has to be flagged as 'Unsuitable' according to the calibration test.

9.3.4.2 Latest calibration

The duration since last verification or calibration might be used for data validation. Given a sensor that requires a monthly calibration, the data recorded between 1 and 2 months after the latest calibration could be considered as 'Doubtful' for a longer delay. Given the maximum delay recommended between

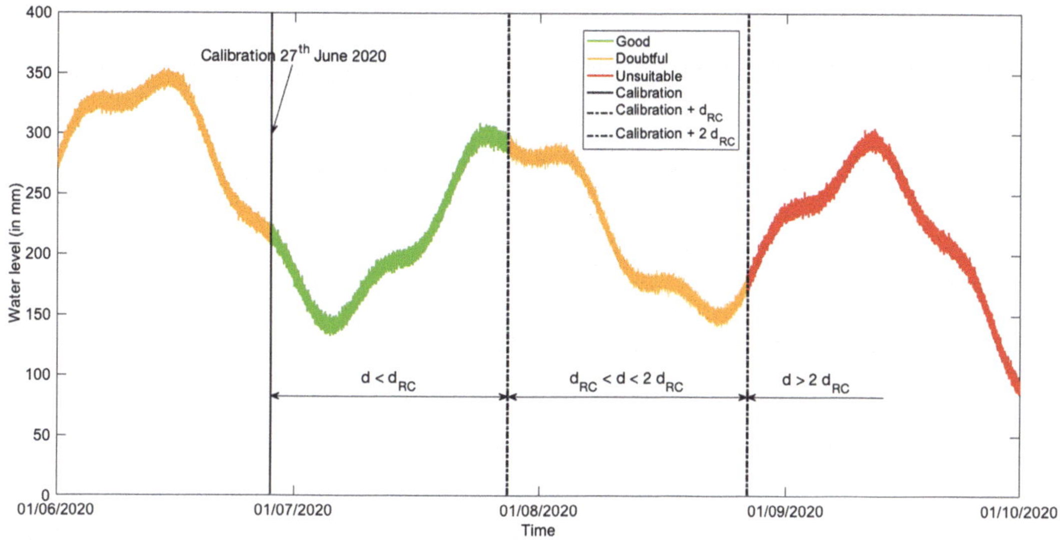

Figure 9.10 Test based on the duration since the last calibration. *Source*: Mathieu Lepot (TU Delft/Un poids une mesure).

two verifications or calibrations (d_{RC}), the duration since the last maintenance (d) should respect one of Equation (9.9):

$$\begin{aligned} &d \leq d_{RC} \\ &d_{RC} < d \leq 2 \times d_{RC} \\ &d > 2 \times d_{RC} \end{aligned} \qquad (9.9)$$

The first test leads to data flagged as 'Good', the second one to 'Doubtful' and the third one to 'Unsuitable'. This refers to the ability for users of the data set to obtain knowledge on the 'history' of the data, i.e. information like corrections, interpolations, etc. being done on the data and the availability of meta-data on calibration and maintenance of sensors.

Example: Manufacturer or user expertise on a water level sensor advises a verification (and a potential re-calibration) every month. If the value has been recorded within a month since the last verification (or re-calibration), it is flagged as 'Good'. If this duration is longer than a month but shorter than 2 months, the value is flagged as 'Doubtful'. The flag is 'Unsuitable' if this duration d exceeds 3 months (Figure 9.10).

9.3.4.3 Maintenance

Sensors and data acquisition systems require maintenance. During maintenance or calibration, manually or automatically recorded data points should be considered as 'Unsuitable'. If there is no maintenance operation log in the system (automatic or logbook), this test cannot take place.

Example: A water level sensor has been cleaned between 2:00 and 2:30 pm. All the data recorded between those hours are flagged as 'Unsuitable' because the measurements have been disturbed by the cleaning actions. Outside this time slot, the values are considered as 'Good' according to this test.

9.3.4.4 Last maintenance

Maintenance has to be done on a periodic basis. According to manufacturer or expert recommendations, the delay between two maintenance activities should not exceed a certain duration d_{RM} for a given sensor. Therefore, if the data point has been recorded within this delay, it could be considered as 'Good', between this and twice this duration as 'Doubtful', and beyond twice this duration as 'Unsuitable' (Equation (9.10)).

$$\begin{aligned} &d \leq d_{RM} \\ &d_{RM} < d \leq 2 \times d_{RM} \\ &d > 2 \times d_{RM} \end{aligned} \tag{9.10}$$

Those durations and thresholds can be shortened based on expert judgements and site knowledge. It is not recommended to extend both values (durations and thresholds).

Example: A velocity sensor has to be cleaned every two months according to the manufacturer recommendations or your expertise. If the value has been recorded within 2 months since the last cleaning operation it is flagged a 'Good'. If this duration is longer than 2 months but shorter than 4 months, the value is flagged as 'Doubtful'. The flag is 'Unsuitable' if this duration d exceeds 4 months.

9.3.5 Test on synchronicity

Depending on their quality, quartz timers of measuring systems tend to drift (e.g. Leutnant *et al.*, 2015). In order to guarantee the accuracy of time stamps, measurement devices should be regularly synchronized, either automatically with an available time-server or manually in the case where an automated synchronization is not possible. As in urban drainage normally dynamic processes are studied involving an interest in the relation between e.g. rainfall and discharge, the mutual synchronicity between time series obtained from a monitoring network is of key importance. Depending on the goal of the analysis, a certain time shift between time series may be tolerable. However, preferably all individual series will share the same time basis and have an equal time interval between readings (temporal equidistance). The former is particularly important to avoid complications when analysing interrelations between different time series. An important, yet trivial, issue to address related to synchronicity is to make sure that, when daylight saving time is taken into account, all sensors in the network switch at the same moment in time, which in practice is not always easily achieved. Interpretation of time series can be significantly hampered by a disparate adjustment of the time stamps.

Deficiency analysis can be accomplished by a fragment-wise application of cross-correlation and/or the method of least squares. Sensor networks with wireless data transmission inherently ensure synchronicity between sensors as they synchronize regularly with an external time reference, such as the clock of the central computer used for data acquisition, a time-server via Network Time Protocol (NTP), through GPS signalling, or via DCF77. With DCF77, the legal time is transmitted from Frankfurt, Germany, across Europe according to the standards ISO 8601 or DIN EN 28601. DCF77 is registered on the international frequency list of the ITU (International Telecommunication Union) as 'Fixed Service' with the carrier frequency 77.5 kHz and the bandwidth 2.4 kHz.

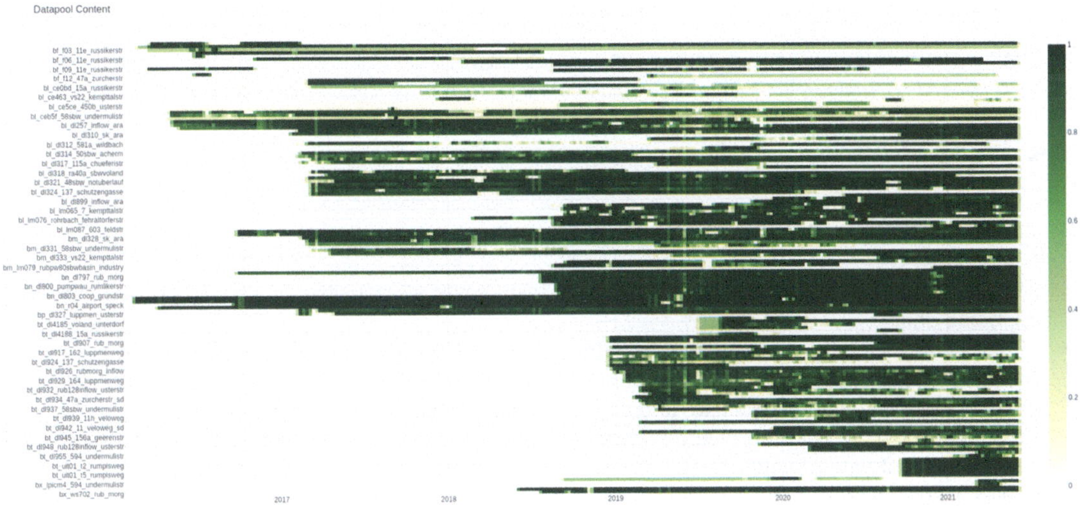

Figure 9.11 Heatmap that illustrates the consistency (here: degree of completeness) of various urban hydrology monitoring series. *Source*: Eawag. Web-based dashboard produced with the plotly app. Data: www.uwo-opendata.eawag.ch; Dashboard: Christian Förster (Eawag).

9.3.6 Test on completeness (degree of incompleteness)

To assess the usefulness of a set of measurements, it may be interesting to see how complete a time series is in terms of missing recordings (Figure 9.11). Depending on the sensors, the data communication and the operational environment at the monitoring location, the measuring system may be subject to outages at which an instance is not measured, not recorded or not transmitted. A missing recording, i.e. a gap in the time series, is the consequence. Considered over a longer period, the number of missing records may accumulate to a considerable amount. However, quite a few advanced data validation methods rely on gapless data series – they would not work with a single inconsistency in the time vector.

But depending on the distribution of missing recordings, e.g. small but frequently occurring gaps or few but large gaps, the missing data points may be curated or not (Section 9.7). Therefore, a *posterior* assessment of a data series with regard to degree of completeness and distribution of missing records is relevant. Even better, real-time data validation triggers an alarm in the case where recordings do not arrive at central servers, allowing *ad hoc* maintenance to be carried out. Assessment can be accomplished by analysing the plot of the gap distribution, i.e. number of days (% of total monitoring period) for which a certain accumulated gap length per day/hour is exceeded. Log(x) may be more favourable for a visual inspection. This test delivers two quality flags: 'Good' or 'Missing'. Note that the degree of completeness of a data set can change throughout the validation process, as some data that is identified as implausible may need to be excluded.

9.3.7 Summary of main basic tests available for data pre-validation

Table 9.3 proposes a non-exhaustive list of basic tests. Those tests are threshold dependent: the test outputs are sensitive to the chosen thresholds. The output can be 'Good', 'Doubtful', 'Unsuitable' or

'Missing' depending on the purposes, i.e. the required selectivity. Those tests can be combined to create new ones.

Selection of basic tests

- I 9.1: *Coverage* – Did the selected and applied tests cover all plausible 'Doubtful' behaviour of the corrected data?
- I 9.2: *Update* – Update, add but never withdraw tests from the list. 'Doubtful' behaviour can occur for numerous reasons. The list proposed (Table 9.3) is rather extended but is not exhaustive. Experiences will lead to the creation of new site- and sensor-specific tests.
- I 9.3: *Thresholds* – With the given warning at the beginning of this section, the authors advise a sensitivity analysis is conducted on the selected thresholds to ensure robustness in data quality assessment.
- I 9.4: *Redundancy* – Be aware that a few tests can identify data as 'Unsuitable' (U) or 'Doubtful' (D) several times but for the same reasons. Please check on this possibility and pay careful attention during the concatenation (see Section 9.5).

9.4 APPLIED CLASSICAL METHODS

This section presents some more advanced methods than Section 9.3. Those methods still aim at flagging a data point according to different tests. The main difference to the previous section comes from the complexity of those methods, either to perform the test or to interpret the result that leads to a certain flag. The methods presented hereafter remain strongly recommended, but they require some mathematical skills and, more importantly, a close evaluation of the results of those tests. At the same time, the examples given and the codes supplied are to be regarded as material to be used for illustration, and/or educational purposes as 'real world' applications are far more complicated and more advanced. As data validation is an activity in many fields of (scientific and industrial) application, the development of mathematical methods and their implementation into software is a field of science in itself. The interested reader is referred to the vast library of literature on the subject (e.g. Hamilton, 1994).

As with most classical statistically based methods, time series analysis implicitly assumes certain characteristics of the time series (stationarity, absence of autocorrelation and certain assumptions regarding the distribution of the data). These requirements are normally not (all) met in time series obtained in UDSM monitoring. In addition it is not always straightforward to manipulate the series in such a manner that the methods become applicable in a strict mathematical sense. Nevertheless, when taking into account these limitations, a naïve application of these methods can be effective when the results are used as a pre-filter for reducing the workload that comes with visual inspection by an expert. The latter aspect may be relaxed in the future by application of pattern recognition technology or other implementations of machine learning (ML).

Hereafter the detection of outliers, step and linear trends is described in a simple manner, just to illustrate the basics of these methods and their application. In practice much more complicated algorithms need to be applied but discussing them in detail is outside of the scope of this section.

Table 9.3 Possible basic tests for data validation. Output: G, Good, D, Doubtful, U, Unsuitable and M, Missing.

Category (Section)	Name (Section)	Output	Advantages	Disadvantages
Plausibility (9.3.1)	Physical range (9.3.1.1)	G U	Easy to set up Based on common sense	Basic
	Measuring range (9.3.1.2)	G D	Easy to set up Limits the sensor choice	Requires overlapping ranges
	Calibration range (9.3.1.3)	G D	Easy to set up	Depends on calibration standard ranges
	Expertise range (9.3.1.4)	G D	Easy to set up	Requires some expertise, site and sensor specific
	Gradient range (9.3.1.5)	G D	Easy to set up	Sensitive to noisy data and gaps
Consistency (9.3.2)	Redundancy (9.3.2.1)	G U	Easy to set up	Requires three measurements
	Dynamics (9.3.2.2)	G U	Easy to set up	Requires expertise, sensitive to special conditions
	Time step (9.3.2.3)	G U	Easy to set up	Requires time stamp
Accuracy (9.3.3)	Accuracy (9.3.3)	G U	Easy to set up once uncertainty is known	Sensitive to the threshold
Auditability (9.3.4)	Calibration (9.3.4.1)	G U	Easy to set up	Requires proper site book to record maintenance
	Last calibration (9.3.4.2)	G D U	Easy to set up	Requires proper site book and calibration data
	Maintenance (9.3.4.3)	G U	Powerful tool	Not easy to set up
Synchronicity (9.3.5)	Synchronicity (9.3.5)	G D	Easy to set up	Requires time stamp and a reference
Completeness (9.3.6)	Completeness (9.3.6)	G M	Easy to set up	'Good' data just means data point exists

A basic working sequence could be:

- Step 1: Perform basic validation methods.
- Step 2: Perform detection of outliers.
- Step 3: Detect step trends.
- Step 4: Detect linear trends.
- Step 5: Apply advanced methods.
- Step 6: Try to find the cause of data that do not pass the checks.
- Step 7: Decide what to do with discarded data points and how to proceed with data analysis.

Steps 2, 3 and 4 will be discussed in some detail while the more advanced methods are discussed in a more superficial manner, with reference to the emerging literature on e.g. ML techniques.

9.4.1 Detection of outliers

Outliers are data points that deviate significantly from the data points in their close vicinity (in time or space, the discussion here is limited to the time dimension). Outliers can occur due to e.g. (i) human error, (ii) some unforeseen process, e.g. clogging of a sensor, maintenance activities interfering with a sensor functioning, or (iii) erroneous sensor readings that are not filtered through on-board processing at the sensor and/or after applying basic validation (Section 9.3). Therefore, it should be kept in mind that data points identified as 'outliers' are not necessarily incorrect (e.g. when situation *ii* has occurred).

Outlier detection has become almost a science in itself. Many methods have been developed and applied in a wide range of application fields, and depending on system and signal characteristics, one or another may be preferential. However, identical methods utilized in different application fields are likely to be parameterized differently. Given our application field, UDSM, chosen methods and thresholds, confidence levels, etc. may have to be made adaptable between, e.g., storm and dry weather conditions.

In UDSM the main cause of outliers is likely to be found in malfunctioning equipment, so methods selected for outlier detection are logically chosen to 'catch' specific behaviour with this type of cause.

The interested reader is referred to literature, e.g. Barnett & Lewis (1996) provide a very comprehensive book on outliers in data sets and methods for detecting them, and Iglewicz & Hoaglin (1993) provide an exhaustive text on the fundamentals and application of a wide range of techniques for outlier detection. Here the discussion is limited to only a few of them that are found to be useful for time series in the UDSM context. As stated before, for most of these techniques some implicit assumptions are made:

- The data are equidistant in time (hence the importance of validating time stamps and synchronicity).
- There are no data gaps (no missing data).
- Time series are (piecewise) stationary.
- Uncertainties are assumed to be normally distributed.

A logical next step to take, once an outlier is detected, is to decide to either remove, correct or keep the data point. Simply removing the data point may hamper the application of analysing tools, correcting implies the need to 'make up' information while keeping it implies knowingly using wrong data. The following steps are distinguished:

- Detection of outliers (discussed hereafter).
- Deciding what to do with them (Section 9.7).
- Data curation (Section 9.7).

9.4.1.1 Z-test for outliers

A first, very basic and simple to implement test for the presence of outliers is the so-called Z-test, in which for each data point V_t a value is added according to Equation (9.11):

$$Z_t = \frac{V_t - \bar{V}}{s} \tag{9.11}$$

where $\bar{V}$ and s are, respectively, the mean value and the standard deviation of a shifting time window $w(V_t, \ldots, V_{t+n\Delta t})$.

This indicates the quotient between the absolute difference between the recorded value V_t and the average value in a time window w covering a shifting subset of the time series $(V_t, \ldots, V_{t+n\Delta t})$, and the corresponding standard deviation. The window size to be applied depends on the system characteristics and needs to be individually estimated. The test statistics of the Z-test are defined for the hypothesis H_0

(no outliers in the data set) against H_1 (at least one outlier is present). A comprehensive text on hypothesis testing is given in Wilcox (2016).

In the Z-test, a Gaussian probability distribution is implicitly assumed, which implies that when $Z_t > 2.5$ there is a $< 1\%$ probability that the corresponding reading V_t is not an outlier. Choosing an adequate threshold is a matter of preference. When setting the threshold very low (e.g. 0.5), this will result in many 'false alarms' leading to an increased effort ('manual labour') to decide whether to keep the data entry or not. On the other hand, when setting the threshold too high (e.g. >4), the risk of missing outliers increases, increasing the risk of obtaining incorrect information.

Another practical issue is that when using small time series (or a small shifting window), the Z value obtained can be misleading as the maximum value is limited to $(n-1)/n^{0.5}$, e.g. for $n = 10$ the maximum value is 2.84. When a threshold of 3 is applied, no outlier will be detected. Therefore, one is well-advised to carry out some test runs on available data sets and evaluate the effectiveness of a chosen threshold. In this process, information is obtained on the amount of time and means needed to manually process the indicated outliers against the improvement of the information which will be obtained.

9.4.1.2 Grubbs test

In the two-sided Grubbs test (Grubbs, 1969), the underlying hypothesis is the same as for the Z-test, the test value G is defined by Equation (9.12):

$$G = \frac{\max|V_t - \bar{V}|}{s} \tag{9.12}$$

Testing whether the minimum is an outlier is tested by Equation (9.13):

$$G = \frac{\bar{V} - V_{min}}{s} \tag{9.13}$$

And, correspondingly, for the maximum value (Equation (9.14)):

$$G = \frac{V_{max} - \bar{V}}{s} \tag{9.14}$$

At a significance level α the test statistics (the hypothesis H_0, i.e. no outlier) is rejected if:

$$G > \frac{n-1}{\sqrt{n}} \sqrt{\frac{\left[t_{\frac{\alpha}{2n}}, n-2\right]^2}{n-2\left[t_{\frac{\alpha}{2n}}, n-2\right]^2}} \tag{9.15}$$

in which $t_{\alpha/2n,n-2}$ is the critical value of the Student t distribution with $n-2$ degrees of freedom. For the one-sided test (for maximum and minimum values) the significance level is α/n – functions for Student t values in Microsoft Excel® and Matlab® are given in Table 8.2).

9.4.1.3 Cook's distance

Another, often applied, metric to decide whether or not a specific measuring result is an outlier is known as the Cook's distance. Basically, the Cook's distance (Equation (9.16)) is a metric for the 'influence' an individual data point has on the 'fit' of a regression model for the time series of (monitoring) data. The test is based on the linear regression (see also Section 9.4.2.1):

$$D_j = \frac{\sum_{i=1}^{i=n} (\hat{y}_i - \hat{y}_{i(j)})^2}{MSE} \tag{9.16}$$

The subscript $i(j)$ implies that when $i = j$, the element is omitted. MSE is defined by Equation (9.17):

$$MSE = \sum_{i=1}^{i=n} (\hat{y}_i - y_i)^2/(n-1) \tag{9.17}$$

A generally accepted criterion for detecting outliers is that an individual observation has a Cook's distance $D_i > 3\bar{D}$ with $\bar{D}$ the mean value for D_j $(j = 1{:}n)$, with n the number of observations in the time series.

9.4.1.4 Example of Cook's distance, Z and Grubbs tests

An example of a hydrograph is shown in Figure 9.12, where 10 apparent outliers are introduced. To illustrate the effect of choosing thresholds in the Z-test, the Grubbs test and the Cook's distance, this hydrograph is used as a test. Figure 9.13 shows the number of outliers detected using the Z-test as a function of the Z_{max} value.

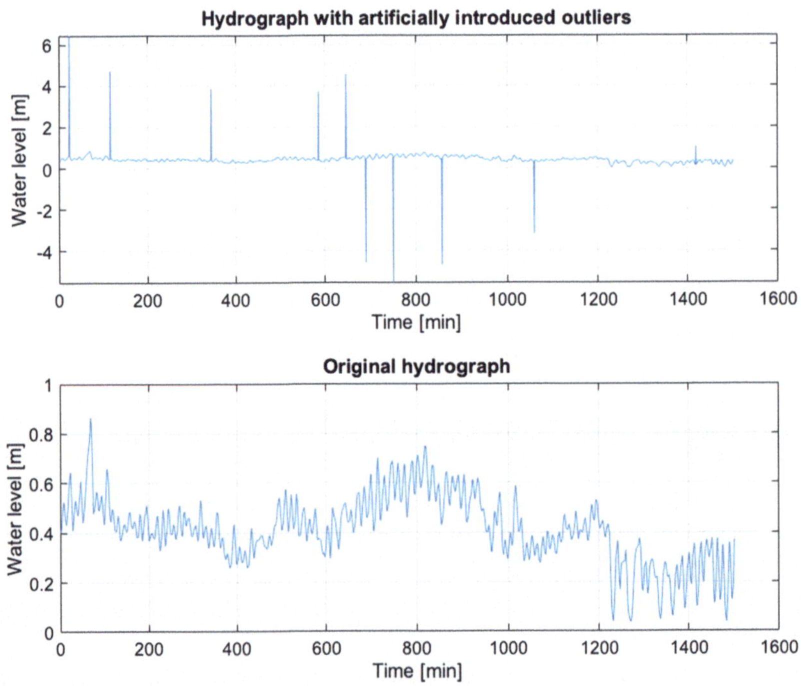

Figure 9.12 Example of a hydrograph with 10 artificial outliers. *Source*: Francois Clemens-Meyer (Deltares/TU Delft/NTNU).

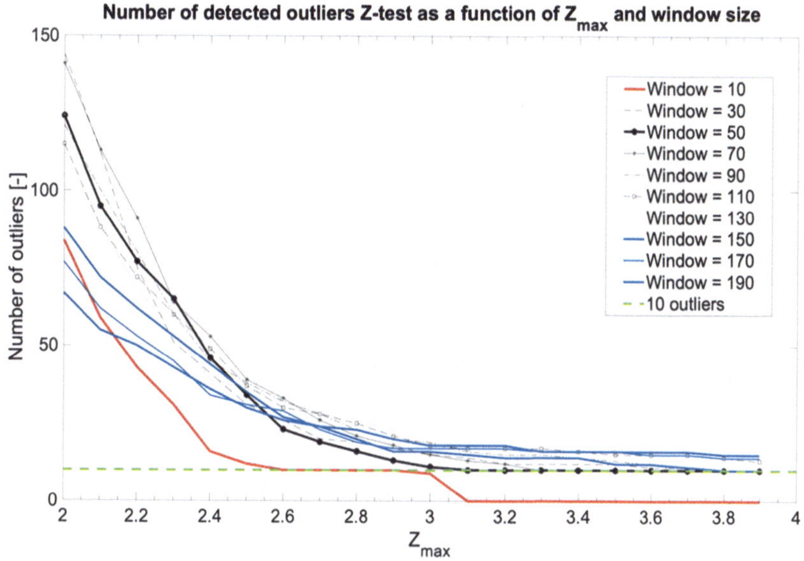

Figure 9.13 Number of outliers detected by the Z-test as a function of window size and Z_{max}. *Source*: Francois Clemens-Meyer (Deltares/TU Delft/NTNU).

As can be seen in Figure 9.13, the correct result (10 outliers) is only achieved for a limited number of combinations of window size and Z_{max}, a window size of 50 minutes seems to be a robust choice, as the outcome is constant for Z_{max} values >3.1. On the other hand, for window sizes >110 the number of detected outliers seems to be too high regardless of the Z_{max} value chosen.

Figures 9.14, 9.15 and 9.16 show some detailed results from some combinations of window size and Z_{max}.

It has to be emphasized that these graphs are only valid for the examples shown. The settings of the test parameters depend on the signal used, the defined rigidness in terms of false positives and false negatives,

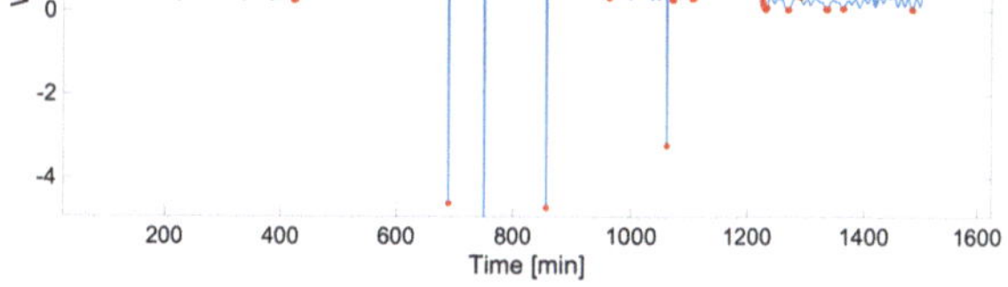

Figure 9.14 Detected outliers for window size = 70 and Z_{max} = 2.0. *Source*: Francois Clemens-Meyer (Deltares/TU Delft/NTNU).

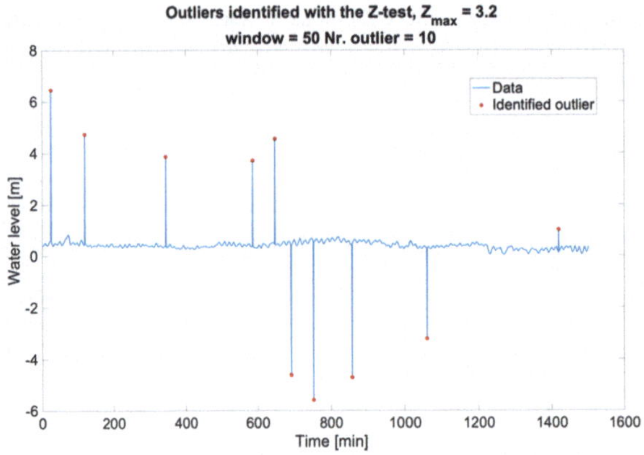

Figure 9.15 Detected outliers for window size = 50 and Z_{max} = 3.2. *Source*: Francois Clemens-Meyer (Deltares/TU Delft/NTNU).

and the subjective preferences of the user of the data. Nevertheless, as a rule of thumb, a Z_{max} value of 2.5 seems a good starting point, with respect to window size and the range of values between the relevant characteristic timescales of the processes (Section 6.3).

Figure 9.16 shows the performance of the Grubbs test on the hydrograph shown in Figure 9.12. The Grubbs test performs well for this example for a window size of 70 and for a range of significance levels ($\alpha > 0.015$). The test is sensitive to the window size, as for a window size of 90, the test overpredicts for any confidence level. Figure 9.17 shows some detailed results for the Grubbs test.

Finally, the hydrograph shown in Figure 9.12 was subjected to the Cook's distance test. Figure 9.18 shows the dependency of the result on the threshold. The Cook's distance test proves to produce reliable results for a wide range of threshold values. Threshold values between 2 to 7 times the mean values result in the correct identification of all outliers. Figure 9.19 shows an example of the detailed results of the Cook's distance test.

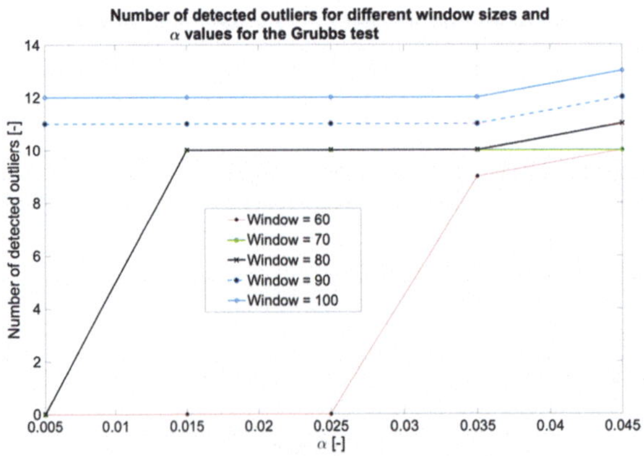

Figure 9.16 Performance of the Grubbs test as a function of window size and significance. *Source*: Francois Clemens-Meyer (Deltares/TU Delft/NTNU).

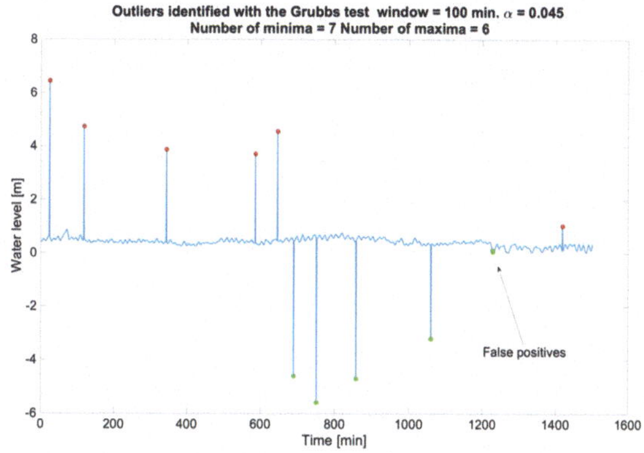

Figure 9.17 Example of the introduction of 3 local minima (very close together) falsely identified as outliers by the Grubbs test. *Source*: Francois Clemens-Meyer (Deltares/TU Delft/NTNU).

9.4.2 Detecting trends and sensor drifts

A trend in a time series is basically a variation of the process mean values in time and/or space.

A linear trend in a time series may point to either, (i) a change in the process under study, or (ii) zero drift of a sensor. Step trends (a sudden change in the mean value) may hint at (i) a change in reference level of a sensor (e.g. due to a wrong reinstallation after maintenance), or (ii) a change in the system studied (e.g. a sudden blockage of a conduit in a sewer system due to collapse or a closure during construction activities).

Methods to detect such trends are numerous (e.g. Gray, 2007). It is noted however that the detectability of (linear) trends depends on the variability of the process monitored, the uncertainty in the measuring system applied and the sampling frequency.

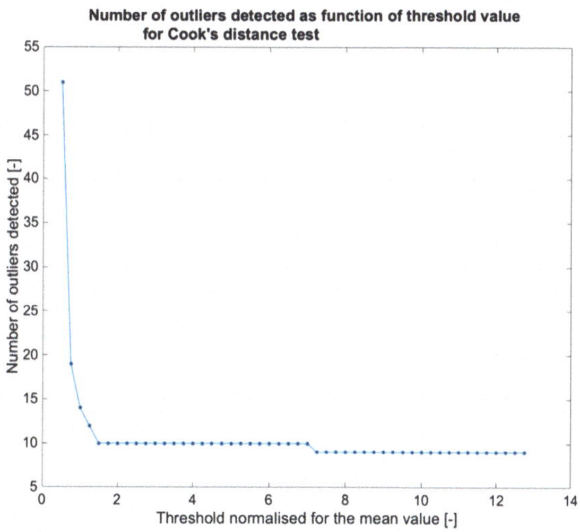

Figure 9.18 Performance of Cook's distance test. *Source*: Francois Clemens-Meyer (Deltares/TU Delft/NTNU).

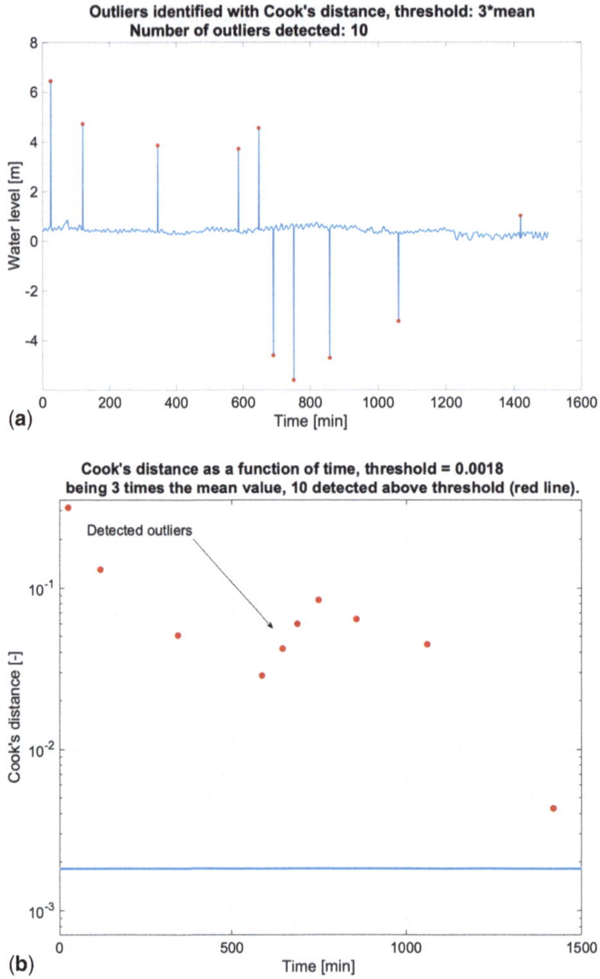

Figure 9.19 Results of Cook's distance test for the example hydrograph in Figure 9.12. (a) hydrograph with detected outliers; (b) values of Cook's distance for each individual data point in the series. *Source*: Francois Clemens-Meyer (Deltares/TU Delft/NTNU).

Lettenmaier (1976) approaches trend detectability in terms of statistical tests, with a null hypothesis H_0 stating 'no trend is present' against the hypothesis H_1 'a trend did occur'. Based on the data, either one of these hypotheses is rejected or accepted.

In statistical testing on accepting or rejecting a hypothesis, the relation between type I and type II errors and confidence and power are shown in Table 9.4. Emmert-Streib & Dehmer (2019) provide a review on hypothesis testing in general along with methods applied.

Obviously, the values for both α and β should be as small as feasible. As these values, apart from choices made like measuring frequency and uncertainty in the measured values, depend largely on the process studied, the settings chosen for acceptance limits of the test outcome (like a maximum value for α) can therefore never be regarded as generic.

Table 9.4 Accepting or rejecting a hypothesis.

	Test indication H_0	Test indication H_1
'Real' state H_0	Confidence = $(1-\alpha)$	Type I error $p = \alpha$
'Real' state H_1	Type II error $p = \beta$	Power = $(1-\beta)$

The discussion here is limited to two types of trends that occur frequently in UDSM, namely:

- The step trend (typically occurring after e.g. misplacing a water level sensor after maintenance).
- A linear trend, often attributable to sensor drift.

Most algorithms for trend detection are sensitive to the presence of outliers, therefore one is well advised to first analyse for outliers prior to analysing for the presence of trends.

9.4.2.1 Linear regression

A very simple method to detect a linear trend is by fitting a linear function through the measuring data using the relation: $y = ax + b$.

The values of a and b can be obtained using the method of maximum likelihood estimates, which, when assuming a Gaussian distribution for the residues, boils down to the classical ordinary least squares method.

The least squares estimators for a and b are:

$$\left.\begin{aligned} \hat{a} &= \frac{\sum_{i=1}^{i=n}(x_i - \bar{x})(y_i - \bar{y})}{\sum_{i=1}^{i=n}(x_i - \bar{x})^2} \\ \hat{b} &= \bar{y} - \hat{a}\bar{x} \end{aligned}\right\} \tag{9.18}$$

Giving the 'fitted' relation:

$$\hat{y} = \hat{a}x + \hat{b} \tag{9.19}$$

the residuals (difference between measured values and fitted results) are defined as:

$$r = y - \hat{y} \tag{9.20}$$

Assuming that the variance of the residuals is constant, their variance is estimated by:

$$\sigma_r^2 = \frac{\sum_{i=1}^{i=n} r_i^2}{n-2} \tag{9.21}$$

The standard deviations in the estimated parameter values follow from Equation (9.21) (neglecting covariance terms for the sake of simplicity – see Section 7.6.4.2 for more detail):

$$\left.\begin{aligned} \sigma_{\hat{a}} &= \sigma_r \sqrt{\frac{1}{\sum_{i=1}^{i=n}(x_i - \bar{x})^2}} \\ \sigma_{\hat{b}} &= \sigma_r \sqrt{\frac{1}{n} + \frac{\bar{x}^2}{\sum_{i=1}^{i=n}(x_i - \bar{x})^2}} \end{aligned}\right\} \tag{9.22}$$

When $|\hat{a}| > 0$, this may indicate there is a linear trend in the time interval studied. Or, more formally, the hypothesis tested is 'H$_0$: there is no linear trend present' against 'H$_1$: there is a linear trend present'. This essentially boils down to deciding whether or not the value of $|\hat{a}|$ deviates significantly from zero. For this, the Spearman's ρ test is used, as will be discussed later.

9.4.2.2 Relation between trend, window length, process characteristics, confidence and power of trend detection tests

Conover (1999) identifies the Mann-Whitney test and the Spearman's ρ test as best suited for the trends mentioned. Lettenmaier (1976) has shown that the detectability of a trend depends on the following parameters for:

- Detecting a step trend (Equation (9.23)):

$$\psi(\Delta t, T, T_r) = \frac{T_r}{2\sigma_x}\sqrt{N_{equi}(\Delta t, T)} \tag{9.23}$$

- Detecting a linear trend (Equation (9.24)):

$$\psi(\Delta t, T, T_r) = \frac{T_r}{N_{equi}(\Delta t, T)\sqrt{12}\sigma_x}\sqrt{N_{equi}(\Delta t, T)[N_{equi}+1][N_{equi}-1]} \tag{9.24}$$

The value N_{equi} is the equivalent number of samples in a series of N data points corrected for the mutual correlation between these N points (implying these hold redundant information).

N_{equi} depends on the autocorrelation function of the time series under study, as defined by Equation (9.25) (Bayley & Hammersley, 1946):

$$N_{equi}(\Delta t, T) = \frac{N(\Delta t, T)}{1 + 2\sum_{i=1}^{i=N(\Delta t, T)}\left[\left(1 - \frac{i}{N(\Delta t, T)}\right)\rho_x(i\Delta t)\right]} \tag{9.25}$$

in which $\rho_x(i\Delta t)$ is defined as:

$$\rho_x(i\Delta t) = \frac{\gamma_x(i\Delta t)}{\sigma_p^2 + \sigma_m^2} \tag{9.26}$$

and $\gamma_x(i\Delta t)$ is defined by:

$$\gamma_x(i\Delta t) = \begin{cases} \sigma_p^2 + \sigma_m^2 & i = 1 \\ \sigma_p^2 \rho_p(i\Delta t) & i > 1 \end{cases} \tag{9.27}$$

When a time series is absolutely uncorrelated (i.e. a random sequence), then $N_{equi}(\Delta t, T)$ is equal to N. When all points are 100% correlated, then $N_{equi}(\Delta t, T)$ is equal to 1 (the first value in the series is the perfect predictor for all the rest). The power and confidence of the tests are related as Equation (9.28):

$$1 - \beta = \phi\left[\psi(\Delta t, T, T_r) - \xi\left(\frac{\alpha}{2}\right)\right] \tag{9.28}$$

This implies that, given a value for α (type I error, or significance), a time series and the measuring uncertainty, the limits of detecting a step and/or linear trend are defined when choosing a certain window size $N(\Delta t, T)$ in terms of confidence and power of the test.

Note that in the obtained values for power and confidence of the tests, the effects of (auto) correlation and measuring uncertainty are accounted for. This allows for determining the characteristics of trends that are detectable given the measuring frequency, measuring uncertainty and window length.

Figure 9.20 shows some results of the application of the relations between the variables in Equations (9.23–9.28) (Matlab instructions can be found in `lin_step_power.m`). When striving for equal probability for type I and type II errors, one would like to have equal levels for confidence and power of a test. As can be seen in Figure 9.20 for the linear trend in the example time series, this is not a feasible option within the range of window sizes and T_r/σ_p values. For the step trend, however, it is feasible for all window sizes provided $T_r/\sigma_p > 2$. The 'optimum' window size for both step and linear trend is approx. 250 minutes, as it produces the highest values for the power over the whole range of T_r/σ_p. So, in this case, one would start with analysing the series with a window size of 250 min. Further it has to be emphasized that the detectability of trends is largely decided upon in the macro design of the monitoring network (Section 6.2).

9.4.2.3 Mann-Whitney test for step trend detection

The Mann-Whitney U test (or Wilcoxon rank-sum test) basically tests whether there is a difference in level (median) of two partitions y and z ($y = (x_1, x_2,\ldots, x_m)$, $z = (x_{m+1},\ldots, x_n)$) of a vector (i.e. a time series) $X = (x_1, x_2,\ldots, x_n)$. The hypothesis H_0 is that $p(y < z) = 0.5$ (no step trend) against H_1 $p(y < z) < 0.5$ (a step trend is present and y has a lower overall value than z) and H_2 $p(y < z) > 0.5$ (a step trend is present

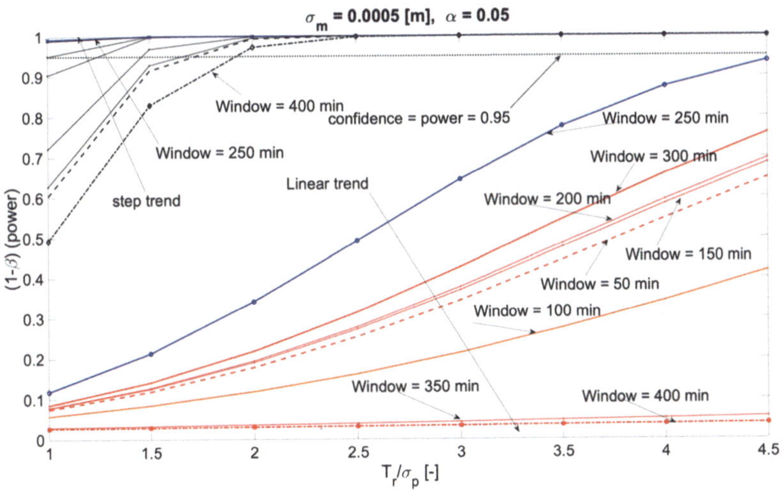

Figure 9.20 Results for the relation between confidence and power for the example hydrograph using different sizes of shifting windows and trend ratios. *Source*: Francois Clemens-Meyer (Deltares/TU Delft/NTNU).

and y has a higher overall value than z). The test value is defined in Equation (9.29):

$$z = \frac{\sum_{i=1}^{i=n} R(y_i) - \frac{m(n+1)}{2} + K}{s_w}$$

in which

$$K = \begin{cases} \sum_{i=1}^{i=n} R(y_i) - \frac{m(n+1)}{2} > 0.5 : K = 0.5 \\ \sum_{i=1}^{i=n} R(y_i) - \frac{m(n+1)}{2} = 0.5 : K = 0 \\ \sum_{i=1}^{i=n} R(y_i) - \frac{m(n+1)}{2} < 0.5 :: K = -0.5 \end{cases} \tag{9.29}$$

where $R(y_i)$ is the rank of y_i in the vector X, and s_w is defined by Equation (9.30):

$$s_w = \sqrt{\frac{m(n-m)(n+1)}{12} - \frac{m(n-m)\sum_{i=1}^{k}(t_i^3 - t_i)}{12n(n-1)}} \tag{9.30}$$

where t_i is the number of subjects having the rank i, and k is the number of (distinct) ranks in the data. H_0 is rejected in favour of H_1 or H_2 (step trend is present) when $z > z_{1-\alpha/2}$ or $z < z_{\alpha/2}$, respectively, in which z_q is the quantile for $\alpha/2$ (for $\alpha = 0.05$, $z_{\alpha/2} = -1.96$ and $z_{1-\alpha/2} = 1.96$). Common statistical software suites provide a standard function for this test. A simple implementation is provided in the Matlab® code `step_trend.m` (available for download at https://doi.org/10.2166/9781789060102) by defining a shifting window and applying it to the time series at a chosen value for α.

9.4.2.4 Spearman's ρ test for linear trend detection

Spearman's ρ test is used to decide whether or not a detected linear trend, as described in Section 9.4.1.1, is to be regarded as significant or not. The test parameter is defined as:

$$\rho = \frac{\frac{1}{n}\sum_{i=1}^{i=n}\left[i - \frac{n+1}{2}\right][R(x_i) - \overline{R(x)}]}{\sqrt{\sum_{i=1}^{i=n}\left[i - \frac{n+1}{2}\right]^2 \sum_{i=1}^{i=n}[R(x_i) - \overline{R(x)}]^2}} \tag{9.31}$$

If the value of ρ is negative, the trend is descending; if ρ is positive, the trend is ascending. At the same time, the value of ρ is used as a statistical test variable, assuming a normal distribution. This implies that when choosing a p-value for the test, α, the hypothesis that no linear trend is present is rejected when either $\rho n^{0.5} < z_{\alpha/2}$ or $\rho n^{0.5} > z_{1-\alpha/2}$ ($p < \alpha/2$ or $p > 1-\alpha/2$). A simple implementation in Matlab® is `[r,p] = corr(x,y, 'Type', 'Spearman')` here 'r' is the value for ρ and p is the p-value (to be related to α in the preceding text).

9.4.2.5 Examples of trend detection

Hereafter a rather naïve working sequence will be demonstrated which is not very sophisticated but will show the limitations of the methods applied and the need for some human supervision (although self-learning software may be expected to, at least partially, take over this task, see Section 9.8).

An important fact to consider it that the shifting time window over which the trend analysis is applied is crucial in recognizing any trend. A first logical step to take is to identify time intervals that behave in a more or less similar manner, at times where changes occur. An approach in relation to the former is to find a piecewise linear fit to the original signal. A simple implementation is shown in the Matlab® codes `piece_lin_fit.m`, `step_trend.m` and `linear_trend.m` (available for download at https://doi.org/10.2166/9781789060102).

When studying the lower graph in Figure 9.21, it can be seen that the critical z values are only surpassed below the z_{min} value (with $\alpha = 0.05$, this is -1.96). The Mann-Whiney test is applied in such a manner that the z values for both hypotheses are tested (z_1 tests hypothesis H_1): this implies that when $z_1 > 1.96$, the hypothesis that no trend is present is rejected in favour of a trend in which a sudden increase occurs. Regarding z_2, the situation is likewise: when $z_2 < -1.96$, the H_0 hypothesis is rejected in favour of the presence of a sudden decrease. In the graph, there are three time windows (indicated as 'A', 'B' and 'D') and one cluster of short windows (indicated 'C') that, given the chosen α, are marked to reject the hypothesis that no trend is present. As can be seen, the example used is composed of a range of (linear and step) trends. As we are 'hunting' for step trends using the Mann-Whitney test, we focus on the time windows indicated. In window A, it is clear from visual inspection that this would qualify a linear trend, indeed the test outcomes are not very convincing, so possibly choosing a somewhat smaller value for α would have eliminated this candidate. Window B shows that here the discrimination between step trend and linear trend seems to function well, which cannot be said for window D. Indeed, from visual inspection, it is clear that the signal is more or less constant, which is reflected in a somewhat ambiguous test result. In Figure 9.13 the maximum relative trend size as a function of window size is shown. As can be seen, the maximum relative trend that occurs in the example hydrograph is less than 2, and for most window sizes approx. 1.6–1.7. From Figure 9.20, it is concluded that for this range of trends, the power of the tests in not very high (not for step trends and certainly not for linear trends), which hints at ambiguous test results (Figure 9.22).

With respect to the cluster of windows C (Figure 9.21) overall, apart from the sub window at the far right, there is clear evidence that a sudden decrease in the water level occurs. Again, here a smaller value for α

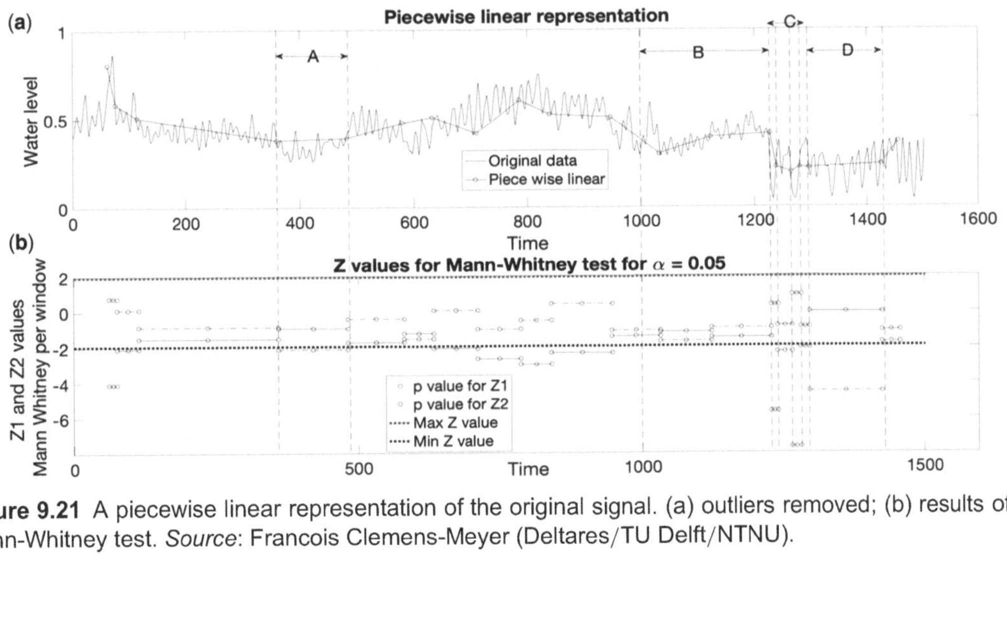

Figure 9.21 A piecewise linear representation of the original signal. (a) outliers removed; (b) results of the Mann-Whitney test. *Source*: Francois Clemens-Meyer (Deltares/TU Delft/NTNU).

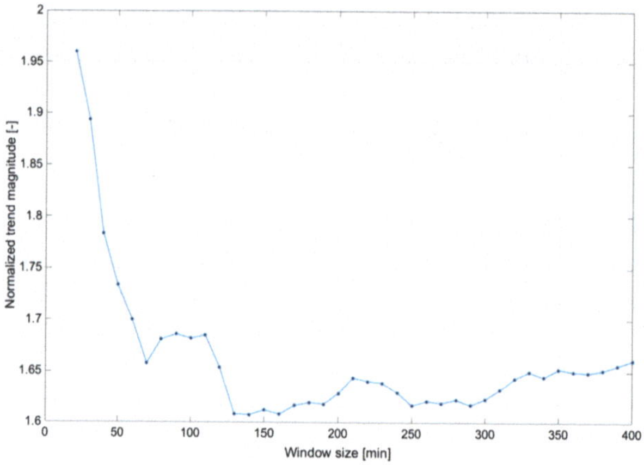

Figure 9.22 Maximum relative trend as a function of window size. *Source*: Francois Clemens-Meyer (Deltares/TU Delft/NTNU).

would have eliminated this candidate. After eliminating A, D and the far-right sub window C, a step trend in the time window between $t = 1226$ and $t = 1294$ minutes is recognized. In this respect one has to realize that when windows become small (i.e. <20 observations), implicit assumptions (most important normality) underlying many of the statistical methods applied are no longer valid. Filtering on window size prior to conducting the analysis is thus strongly advised.

After identifying the step trend, linear trends can be detected using the Spearman's ρ test applied on the piecewise linearization. In Figure 9.23, the results of the Spearman's ρ test are shown. It can be seen that at a significance level of 0.05, six time windows (noted A-F) are identified to contain a significant linear trend, the sign of the ρ values indeed corresponds with the visually observed trends (either ascending or descending). With respect to window F, it contains less than 20 data points and is therefore to be treated as an artefact of the test result. Notice that the time window B in Figure 9.23 corresponds with the time window B in Figure 9.21, which implies that both the Mann-Whitney test and the Spearman's ρ test detect, respectively, a step and a linear trend that has statistical significance. After visual inspection, one has to conclude that a linear trend is more obvious than a step trend in this case. In spite of the strictly taken non-compliance of the underlying data with the pre-assumptions set for the statistical test presented, these tests prove to be of value when validating time series, even though some manual inspection is needed as demonstrated in the examples.

9.4.3 Detecting abnormal processes

9.4.3.1 Using spline function

Villez & Habermacher (2016) propose a method to detect anomalies in processes. The methodology is based on *shaped-constrained splines* and is applicable for any univariate or multivariate time series. Without entering into all the details of this method, which is rather more complex than the previous ones, the overall idea is to identify abnormal trends or behaviours in time series, while fitting spline functions into different parts of time series.

As a basic example, let us look into the evolution of water levels, velocities and discharges in a sewer pipe once a rainfall peak is passed. Those three values are supposed to decrease while following a convex shape.

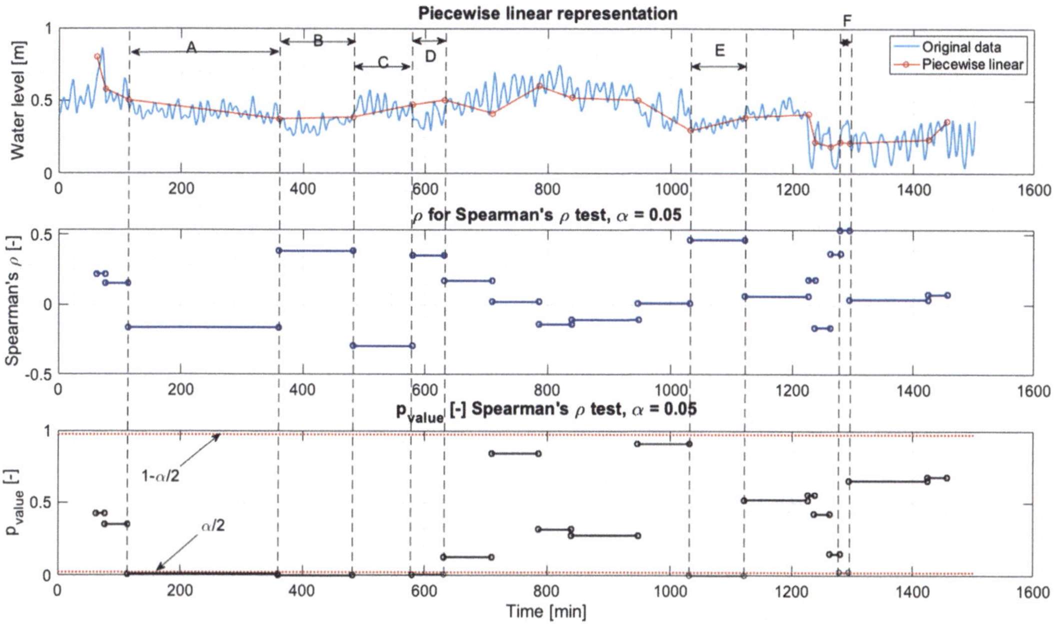

Figure 9.23 Results for Spearman's ρ test. *Source*: Francois Clemens-Meyer (Deltares/TU Delft/NTNU).

If, for a given period, one time series show a concave shape, the values obtained over this period should be considered as 'Doubtful' or 'Unsuitable'.

9.4.3.2 Detecting abnormal episodes based on conditional dependencies

Conditional dependencies identify if inconsistencies in data emerge as violations of these dependencies. That is, for instance, including topological information on a flow path network in the form of rules and using partially redundant information from up- or downstream located sensors. This allows detection of abnormal measurements (Figure 9.24).

In Figure 9.25, sensors F03 and F06 are installed at the same location. Sensor F04 is located 1 km downstream of F03 and F06. F04 should always show higher flows than F03 and F06. This can be questionable to a certain extent for dry weather periods. The two labelled anomalies in F04 appear questionable. A cross-comparison with correctly aligned adjacent measurements leads to the conclusion that the early anomaly obviously occurs due to hydraulic disturbances, while the later anomaly is obviously a non-natural artefact.

9.4.3.3 Detecting abnormal episodes based on the hydraulic gradient h_c

The consistency of flow observations can be verified by means of the hydraulic gradient h_c based on the Manning-Strickler relation. This method can be applied for flow ranges in which no disturbance due to minimal water levels and/or backwater effects is expected. The hydraulic gradient h_c is calculated according to Equations (9.32) and (9.33).

$$h_c = k_{st} * \sqrt{I} = \frac{Q}{(A * R_{hyd}{}^{2/3})} \tag{9.32}$$

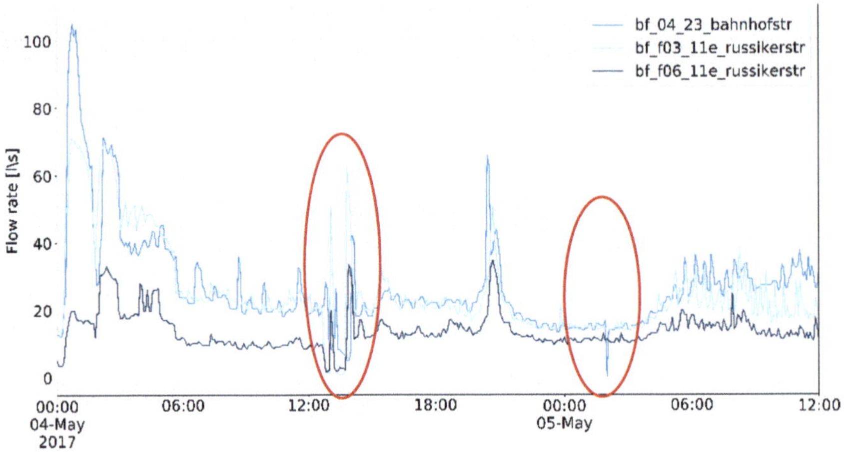

Figure 9.24 Example of the cross-comparison of hydrographs from three different sensors in the same drainage network. *Source*: Frank Blumensaat and Andy Disch (Eawag).

with

$$R_{hyd} = \frac{A}{l_u} \tag{9.33}$$

h_c is calculated for each recording, i.e. each time step. Subsequently a monthly/weekly average is determined. Inconsistencies can be detected by deviations from the mean value over the course of time.

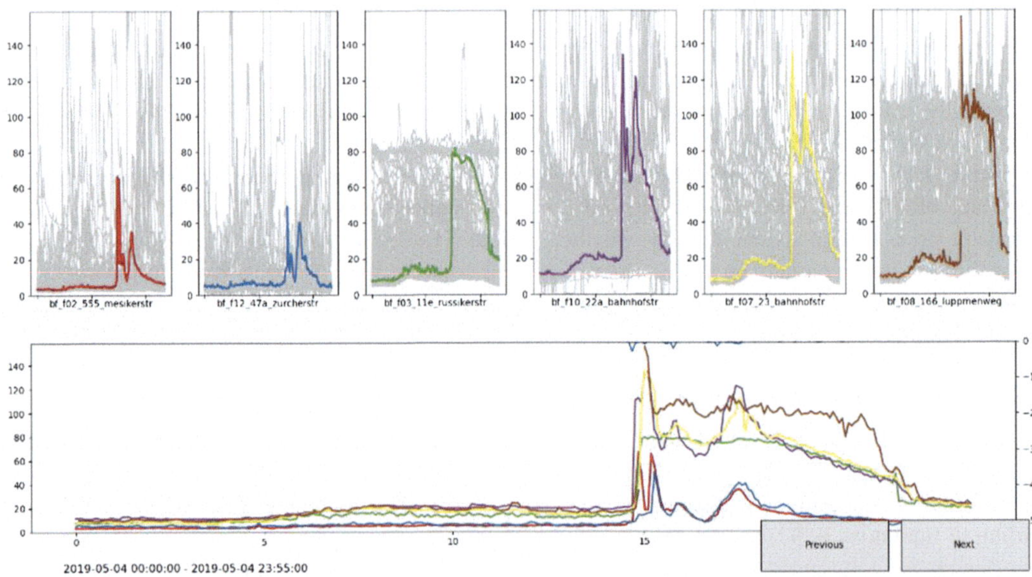

Figure 9.25 Example of multi-signal cross-comparison (here on a daily basis). Rules apply depending on the topological relation on the flow path network. *Source*: Frank Blumensaat and Andy Disch (Eawag).

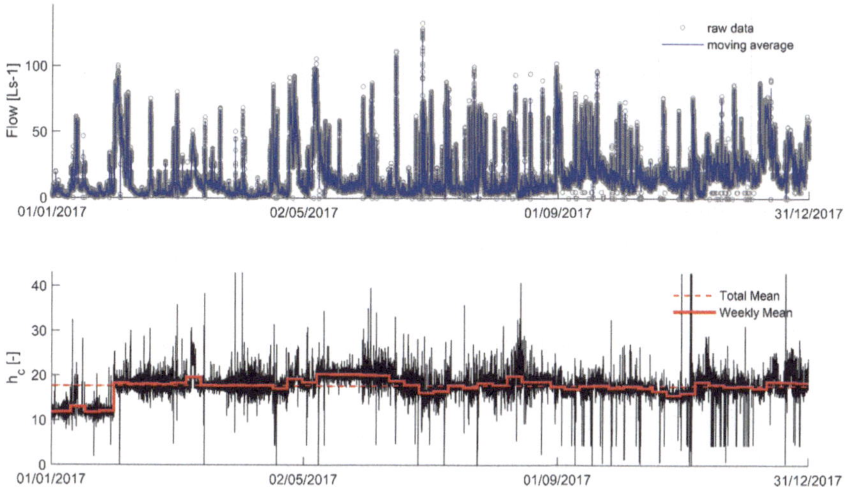

Figure 9.26 Hydrograph obtained through a wedge sensor that measures flow velocity (US backscatter and cross-correlation) and water level (pressure gauge). *Source*: Frank Blumensaat (Eawag).

Weeks/months for which h_c values considerably deviate from the mean are to be questioned; periods should be excluded from further usage, e.g. for calibration, or should be further analysed. The application of the hydraulic gradient test is exemplified in Figure 9.26. Here, in February 2017, the sensor was cleaned and re-configured without showing an impact on the resulting flow signal. The jump in the h_c value, however, reveals the hidden anomaly due to sensor maintenance (Figure 9.26).

9.4.3.4 Detecting abnormal flow conditions based on the Q(h) relation

Flow observations should be checked for plausibility by comparing measured data with the theoretical $Q(h)$ relationship, i.e. with the part-full circular pipe flow curve (Figure 9.27). For this purpose, the theoretical $Q(h)$ relationship according to Manning-Strickler (Equations (9.34) to (9.36)) is calculated for a given pipe diameter D.

$$Q = k_{st} \times A \times R_{hyd}^{\frac{2}{3}} \times \sqrt{I} \tag{9.34}$$

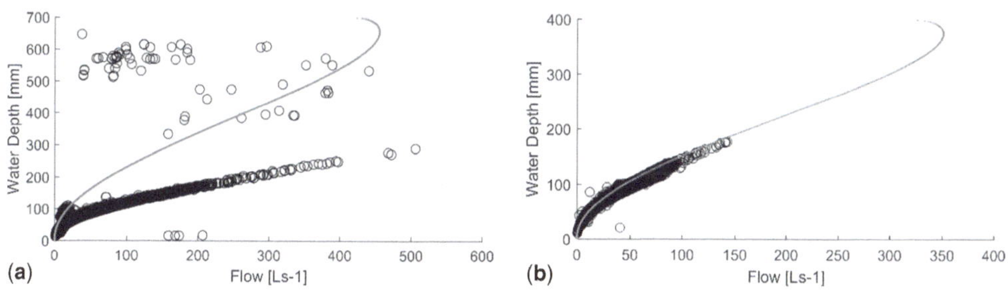

Figure 9.27 $Q(h)$ relationships plotted against the theoretical Manning-Strickler relation (parable test). Examples for a poor (a) diameter 700 mm and good (b) diameter 400 mm matches. *Source*: Frank Blumensaat (Eawag).

with

$$A = D^2 \frac{4 \times \arcsin\left(\sqrt{\frac{h}{D}}\right) - \sin\left(4 \times \arcsin\left(\sqrt{\frac{h}{D}}\right)\right)}{8} \tag{9.35}$$

$$l_u = 2 \times D \times \arcsin\left(\sqrt{\frac{h}{D}}\right) \tag{9.36}$$

9.4.4 Validation between correlated monitoring points (time series, ARMA models)

In many cases, time series obtained from two or more measuring devices in an urban drainage system show a mutual correlation structure. For example, water level sensors in a wastewater system will reflect more or less the same daily pattern in the recorded water levels, or rain measurements with discharge time series. In many cases an auto regressive moving average (ARMA) model can be used to obtain a description of a time series in the form of a polynomial function, which, to a certain extent, can provide a basic tool for forecasting or (re)constructing a missing value in the time series (see also Section 9.7 on data curation). The theory of ARMA models and applications is comprehensively discussed in e.g. Choi (1992).

The general equation for an ARMA(p,q) model is given in Equation (9.37):

$$V_t = c + \varepsilon_t + \sum_{i=1}^{i=p} \gamma_i V_{t-i} + \sum_{j=1}^{j=q} \theta_i \varepsilon_{t-j} \tag{9.37}$$

The first summation over p represents the auto regressive (AR) part while the second summation over q represents the moving average (MA) part of the model. c is a constant, ε_i represents white noise in step i and V_i is the dependent variable at step i, γ_i and θ_i are the polynomial coefficients for, respectively, the AR and MA parts of the model. The coefficients of the model, for given values of p and q, can be found by e.g. using the maximum likelihood estimate method. In most software packages like Matlab®, Python®, R®, etc. fast routines for ARMA fitting are available as standard.

There is no general manner or protocol for determining the values for p and q. A first requisite is to achieve (piecewise) stationarity of the time series. Using transformation techniques, stationarity of the series can be achieved. Apart from removal of trends and/or periodic signals in the series, differentiating is a popular and effective manner to achieve stationarity. Transforming the time series into a series of incremental differences between successive parameter values often achieves the sought after stationarity. The goodness of fit between the ARMA model and the original data can be expressed in a range of metrics, of which the Akaike information criterium (AIC) is the most commonly applied in this context as it not only takes into account the 'goodness of fit' of the model, but also penalizes for overfitting.

AIC is defined as:

$$\text{AIC} = \ln(\sigma_r^2) + \frac{2(p+q)}{n} \tag{9.38}$$

where n is number of elements in the time series and σ_r is the standard deviation of the residues.

A simple stepwise approach is as follows:

(1) Plot the autocorrelation function (ACF), and the autocorrelation function of the differentiated series (DACF) (Matlab® commands: `autocorr(x1)` and `autocorr(diff(x1))` respectively, where the differentiated time series of x1 is defined as $x_{d,i}, = x_{i+1} - x_i$).
(2) A first estimate for p is indicated by the DACF, where the DACF becomes (almost) zero, defines this first estimate.
(3) A first estimate for q is obtained from the ACF where it starts to tail off to zero.
(4) Estimate the model parameters: in most standard available applications, this is normally done by the application of e.g. the maximum likelihood estimates method.

In Figures 9.28 and 9.29 an example is shown. The Matlab® script `arma_space.m` can be downloaded here https://doi.org/10.2166/9781789060102. Based on the DACF, the value for p is expected to be approx. 5 while the q value is less easy to deduce, as the ACF tails off to zero at moderate time lags. A value of $q = 1$ or 2 would be a first guess.

Using the AIC metric, an ARMA ($p = 5$, $q = 2$) model using the differentiated time series was found to produce the best fit to the data over the first 60 minutes. An ARMA model on a differentiated time series is

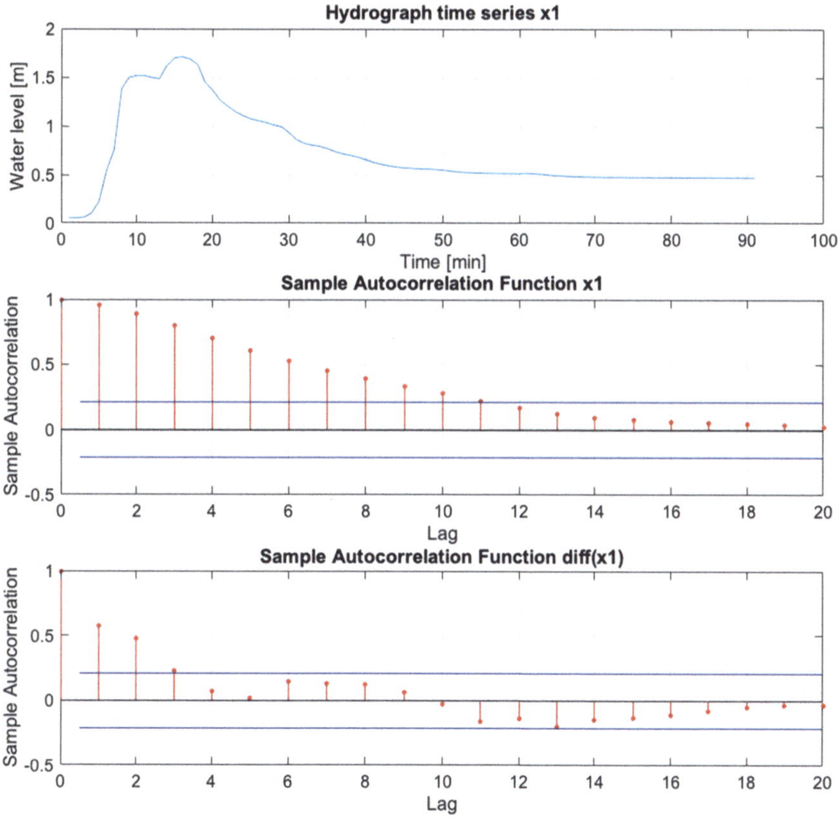

Figure 9.28 Time series and the corresponding ACF and DACF. *Source*: Francois Clemens-Meyer (Deltares/TU Delft/NTNU).

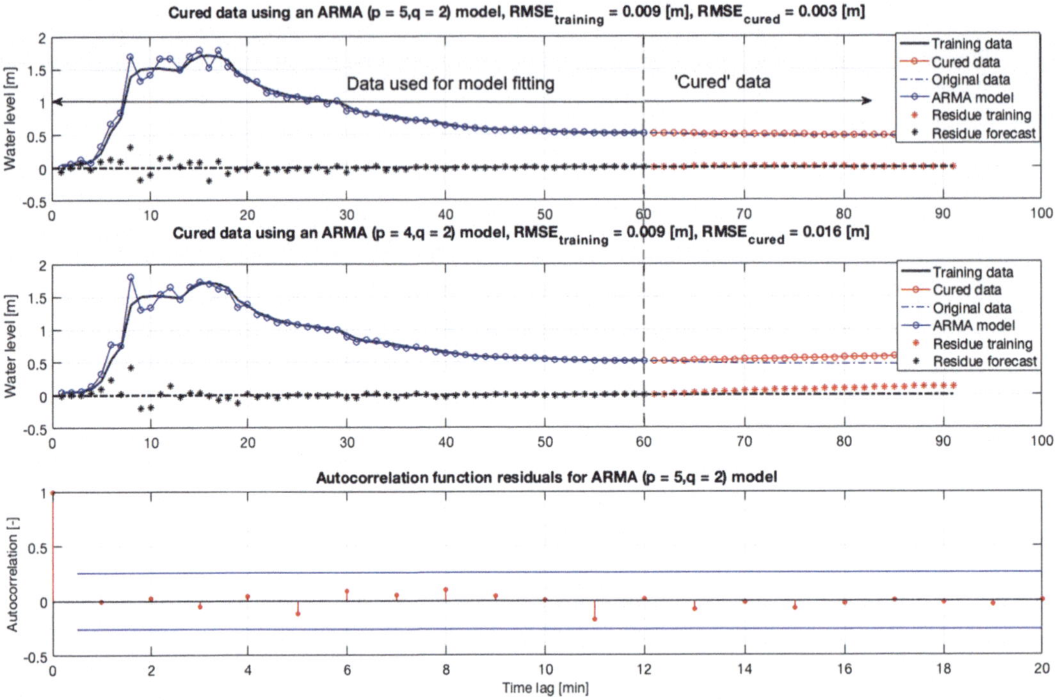

Figure 9.29 Example of the application of an ARMA model to forecast data. (ARIMA(*p*,*d*,*q*), in which *d* stands for differentiating (in this case $d = 1$: differentiating once). *Source*: Francois Clemens-Meyer (Deltares/TU Delft/NTNU).

also known as an ARIMA model (most software packages like Matlab®, R®, etc. contain standard functions for this). This model was used to forecast 30 minutes of additional data. Actually these data were measured as well, allowing for a comparison between forecasted and observed results. In Figure 9.29 the upper graph shows the results of the ARMA model that showed the best fit to the training data, which results in forecasted data with an RMSE (root mean squared error, chosen here since it offers a more 'intuitive' understanding compared to a value for the AIC, the model selection, however, has been done based on AIC) of 3×10^{-3} m. The lower graph shows the results for a slightly different ARMA model ($p = 4$, $q = 2$) that results in a significantly different forecast result (RMSE = 0.015 m), although the quality of the model-fit to the data was only incrementally different from the ARMA ($p = 5$, $q = 2$) model. Of course, these forecasts can be refined with confidence intervals, but the message from the example is clear: forecasting is a possibility but one is advised to apply it for short time windows only and test a range of model configurations (i.e. p and q values), as a 'good' fit to the training data does not guarantee the 'best model forecast'. The latter statement holds for any other type of model (be it process based, or a statistical model), the validity beyond the calibration domain cannot be taken for granted.

An alternative application is to use similarity between time series. Figure 9.30 shows an example: two hydrographs from locations 1 and 2 are shown together with their difference (top graph in Figure 9.30). The middle and bottom graphs show, respectively, the autocorrelation function for the difference between the time series (i.e. $x_3 = x_1 - x_2$) and the differentiated difference (i.e. `diff`(x_3)).

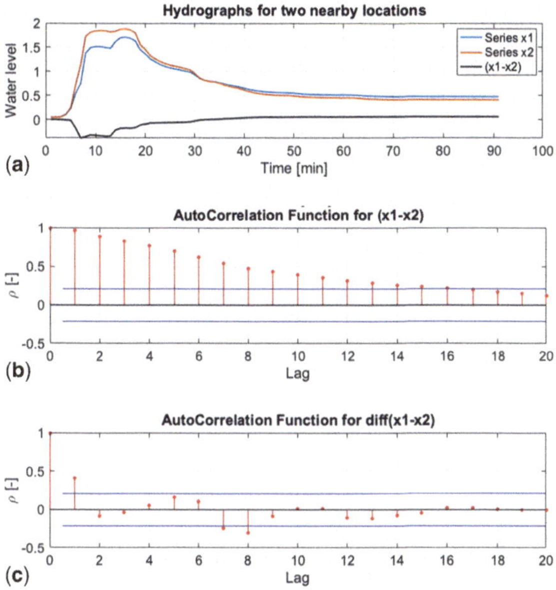

Figure 9.30 (a) two hydrographs and their difference; (b) ACF; (c) DACF. *Source*: Francois Clemens-Meyer (Deltares/TU Delft/NTNU).

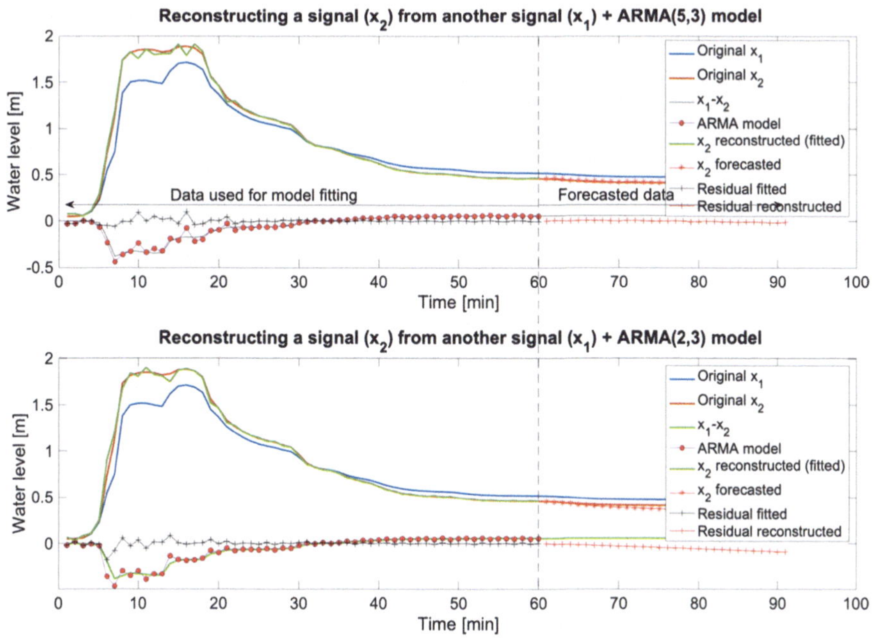

Figure 9.31 Results of forecasting of one signal based on an ARMA model for the signals differences combined with one signal available. *Source*: Francois Clemens-Meyer (Deltares/TU Delft/NTNU).

In this case a first estimate for p would be 2, while for the value of q the same reasoning is followed as before, as the ACF start tailing off to zero in the first few lags. Thus a first estimate for q would be 1 or 2. Again the best model is decided upon based on the AIC values.

Notice that the p value is found to be 5, in contrast to the indication from the DACF (see Figure 9.30). This illustrates that the indication obtained from the DACF does not necessarily correspond with the 'best' model using AIC as a metric. It is therefore, again, suggested to test a range of values for p and q.

Figure 9.31 shows some results: the upper graph for the best model configuration found, and the lower graph for a slightly different model configuration. Again, it is seen that a slightly different model results in significantly different results again. The advantage of the latter approach based on differences between two observation locations is that only one model needs to be maintained enabling the possible curation of two sensors' outputs. A regular update of the model configuration is suggested, as the recorded processes may change over time in terms of level of stationarity. In that respect, when it is found that the best fitting model changes over time in terms of AIC result or even in variation for the order (p, q) of the model, it can be used as an indication for changes in the system observed.

9.5 MAKING QUALITY FLAGS OPERATIONABLE

9.5.1 Concatenation of quality flags

Each individual recording can be checked and assessed through different tests. In the previous sections, these tests have been described, producing a more or less differentiated output for each individual test and for each recording. For each value V_t, i.e. each sensor and each time stamp, several outputs are available to further specify data quality according to the N_T tests presented above (see Table 9.5).

In order to assess a complete data set, the quality labels for individual recordings need to be concatenated: manually validated by a trained staff member to split the values labelled with a D into the G or the U categories.

There are several methods to achieve such concatenations, i.e. to perform a dichotomous flagging or label data points as 'Good' or 'Unsuitable', while further differentiating the quality of recordings labelled 'Doubtful':

- Method 1 (worst case): assign the final quality as the common minimum, i.e. the lowest quality.
- Method 2 (arithmetic mean): calculate an average, while assigning a numerical value to qualitative flags, e.g. 1 for 'Good', 0.5 for 'Doubtful' and 0 for 'Unsuitable' and assigning thresholds.
- Method 3 (median): based on the same principle as Method 2, but it calculates the median of the cardinal qualities.

Each of those methods has pros and cons: the first one being rather pessimistic, the second one being sensitive the grade attribution and potential weights in the average calculation, and the last one being

Table 9.5 Possible basic tests for data validation. Output: G, Good, D, Doubtful and U, Unsuitable.

Value	Test 1	Test 2	Test 3	Test 4	...	Test N_T
V_t	G	G	D	U	...	D

sensitive to a series of failed tests. Based on the output in the didactical example given in Table 9.5, the output of the concatenation will be:

- Method 1: 'Unsuitable', due to Test 4 or Test N_T (5 in this case).
- Method 2: 'Doubtful' while assigning the same weight for each test (average score of 0.6 $(1 + 1 + 0.5 + 0 + 0.5)/5$).
- Method 3: 'Doubtful', the median is equal to 0.5.

The *advantage* of the concatenation is its flexibility: the tests are taken into account, the weights assigned to each output can be changed according to the different purposes the data have been recorded for.

The *disadvantages* may be that (i) one reduces the overall amount of useful data when applying a stringent method (Method 1), or (ii) one introduces a bias when transforming an ordinal scale (G > D > U) into a cardinal, i.e. ratio scale (Method 2).

Those concatenations can be done automatically but are prone to subjectivity regarding the selected tests, thresholds, weights and the retained method to concatenate the outputs. However, this step is mandatory to simplify a subsequent manual validation. At the end of the automatic concatenations, a value can have one or several labels about its quality, one for each purpose.

A suggested concatenation method

- I 9.5: *Test* – Test the three proposed methods and compare the result to select the most appropriate for your needs and uses.
- I 9.6: *Sensitivity* – Try different values and weights if you use methods based on mean and median.
- I 9.7: *Update* – Update and design new concatenation methods if you are not satisfied with the proposed methods.
- I 9.8: *Report* – Always report the methods used in the meta-data, the values and the weights used for the concatenation.

9.5.2 Finding causes of unreliable data being rejected

Validating measurements generally aims to distinguish between dubious and plausible data. Unreliable data may lead to wrong findings and consequently have to be excluded before further processing. This, however, often requires manual intervention to explain *why* some data need to be labelled 'Unsuitable' or 'Doubtful'.

Labelling data as discussed in Section 9.2 supports this process, but it needs to be backed with domain knowledge of UDSM systems behaviour, monitoring techniques, signal processing and data transformation. Losses in data quality may occur during the entire data collection process due to: (i) inappropriate selection of sensor and/or location; (ii) inadequate sensor installation; (iii) sensor specific issues, e.g. failing barometric compensation or electrolytes in the wastewater; (iv) data-logger related issues, e.g. synchronicity, interval, smoothing algorithms, power supply, data transfer; (v) data storing techniques (database management); and (vi) the validation process itself (software, data import routines). Persons in charge of investigating doubtful data are required to be qualified accordingly or seek professional (and mental) support.

Depending on (i) the complexity of the monitoring set-up, (ii) the environmental conditions when the value has been recorded and (iii) the existence of a site-book, investigations may require several experts:

- Persons in charge of the design, construction, maintenance of the monitoring set-up.
- Experts in metrology.
- Experts in IT, electronics and signal processing.
- Local experts with in-depth knowledge about system and locations, e.g. hydraulic conditions in a specific pipe.
- Sensor manufacturers and data acquisition suppliers.

The process of *manual validation* can be done in several steps. The main way suggested for processing is summarized as follows: Be very strict in assigning the labels 'Good' or 'Unsuitable' and any manual modification (from 'Doubtful' to 'Good', 'Unsuitable' – and sometimes 'Doubtful', if the manual validation did not result in assigning another label) must be commented and recorded in order to keep track of the conducted investigations, e.g.:

- The values of water level and velocities, flagged as 'Doubtful' because of inconsistencies between them, have been finally flagged as 'Good'. Reason: recession phase during a storm event, and strong hysteresis (see Chapter 3 or the example in Section 9.4.3.4).
- Discharge values that are flagged as 'Doubtful' were linked to a change of position of the measuring device and finally found to be 'Unsuitable'. Reasons: error in the probe positioning, and the observed bias was confirmed by tracer experiments, as outlined in Chapter 3.

The final flagging and the reasoning that may have led to manual modification should be recorded as meta-data in the database (see Chapters 5 and 10).

Manual validation requires time, expertise and common sense for complicated cases: meta-data on maintenance, sensors, storm events etc., collected in a site book, are essential to conduct such investigations properly. There is no 'silver bullet' to investigate the reasons for 'Doubtful' flags. However, a few basic questions could help to draft hypotheses regarding doubtful data points:

- Are data points flagged as 'Doubtful' assigned for a single or several time step(s)?
- Does this phenomenon occur for a single, several or every time series?
- If several time series are flagged as 'Doubtful', is there any correlation between them? Do they occur through the same computer, the same data acquisition set-up, a specific software? Do data stem from the same measurement location?
- Can a change in data quality be associated with a specific event, such as an intense storm, a maintenance operation, the installation of a nearby sensor?

The list of possible causes can be long. Unfortunately, there is no known generic method or protocol that, when applied, can ensure that in all cases the underlaying cause for poor data quality, or missing data, can be determined. Common sense, domain expertise, site and maintenance logs, and a complete documentation of the monitoring station are necessary to increase chances of problem identification.

9.6 COMMUNICATING DATA QUALITY

9.6.1 Presenting validated data

Once the data set has been validated, one of the key methods to get a user-friendly overview on its overall quality consists of plotting the data quality. Such plots will help the data providers and data users to easily assess the quality of the monitoring set-up, the recorded data set and implicitly all the procedures along the monitoring processes. Here it is of importance that the quality level of the data is indicated in a manner that is unambiguous and fit for purpose. In this respect a difference has to be made between 'managerial' information and 'operational' information. For example, from a managerial point of view, indicators related to the overall performance of the monitoring system as discussed in Section 9.6.2 are of interest while for the operator of the same network detailed information on the level of individual sensors is sought after. Especially in long term monitoring projects, graphs visualizing the data quality provide an easy access to data quality on different levels.

9.6.1.1 Types of quality and availability plots

There are numerous ways to produce representations of data quality, mainly based on colour scale:

- 'Shades of grey' style (e.g. in Figure 9.32).
- Heatmap dashboard illustrating data consistency, i.e. availability, completeness and interpretability (e.g. in Figure 9.33).

From Figures 9.32 and 9.33, the availability and quality of data are easily recognized. Using this kind of charts allows for a quick identification of whether or not enough data are available for a given purpose. In Figure 9.32, for each sensor (Lev1-Lev27 and R1, R2) for each day (24 h), different shades of grey indicate the quality of the available data. The white columns in Figure 9.33 highlight likely failure of the whole system, since no data have been recorded by any sensor.

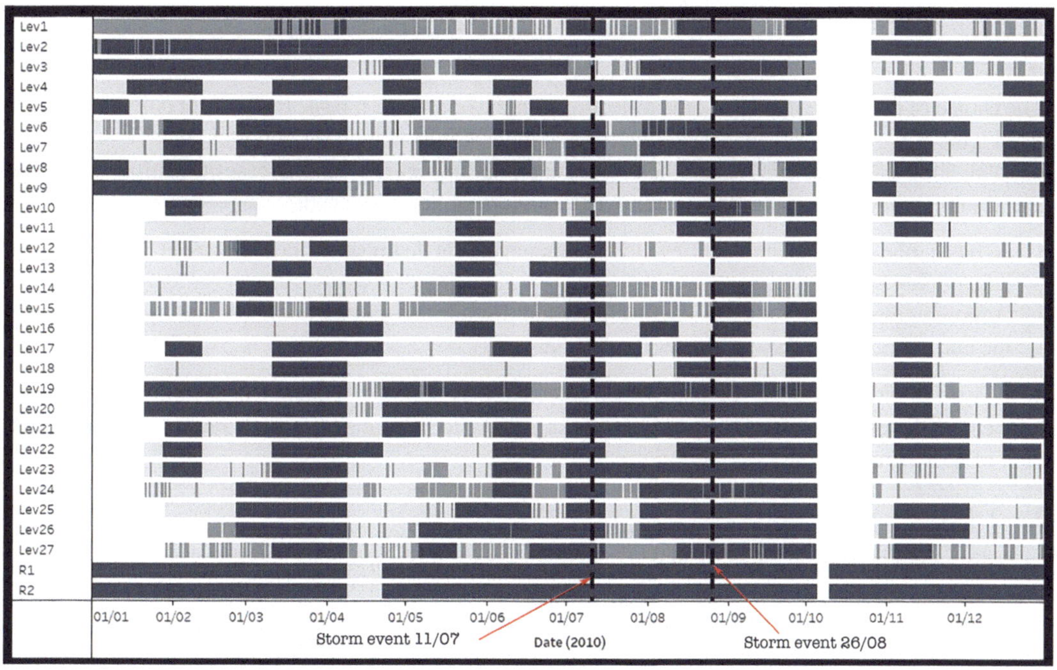

Figure 9.32 Example of a chart granting a 'quick' impression of the quality of a data set, discriminating between a range of labels indicating a range of possible 'issues' with data. *Source*: van Bijnen (2018).

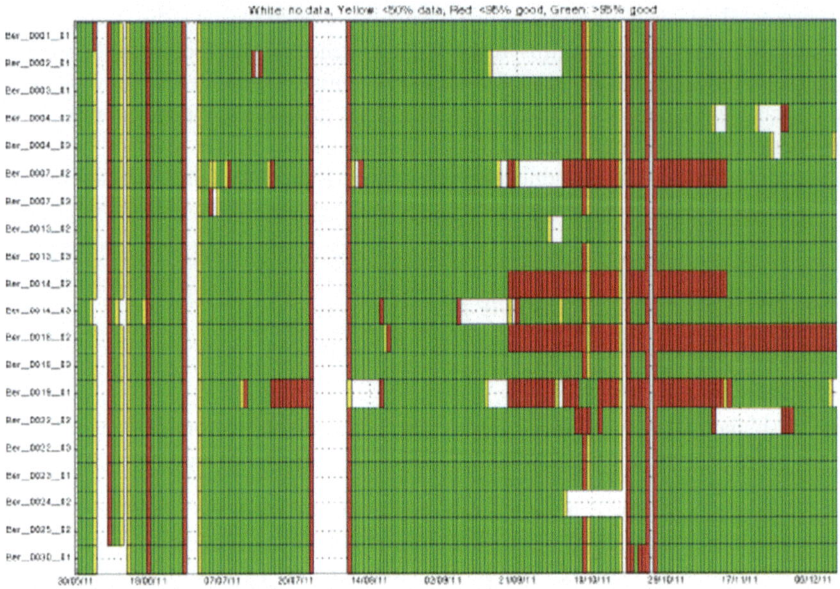

Figure 9.33 Example of a chart showing data quality in 'traffic-light' coding: green is >95% Good data, red is <95% Good data, orange indicates that <50% of data is available and needs a manual check prior to application. A blank cell indicates no data are available at all. Each row represents one sensor, each column represents a 24 h window. *Source*: van Bijnen (2018).

The choice of map style requires some reflection. On the one hand, if basic traffic light colours (or similar plots – Figure 9.33) are relatively easy to interpret, they lack information on important data or facts. On the other hand, adding too many layers on the colour map (e.g. in Figure 9.32) increases the information given by the graphic while decreasing its readability. Both plots can be produced:

- A basic one to give an overview to the management team and draft the main outlines on the data quality.
- A more complex one for technical meetings and discussions aiming at understanding and improving the current performances of the monitoring system.

Once the type of plots and the legend have been selected, they should not change over time to allow a quick assessment of the evolution of data quality, without having to learn a new way of reading for the updated version of those graphics.

9.6.1.2 Data to plot

Plots as discussed in Section 9.6.1.1 can be created with several types of data (results of the tests, automatic and final quality grades).

Creating those graphics with the results of individual tests (like consistency, gradient, etc.) can help to identify tests that all fail or succeed and, later on, can help to adjust the tuning of those tests or identify permanent errors between sensors.

Comparing graphics on global data quality grade, before and after manual validation, can help to highlight changes in the manual validation, between different data validators or over time.

The possibilities are endless and existing tools to generate plots do not really limit the number of graphics that can be produced. However, multiplying the number of plots requires more time to perform such analysis and may lead to unseen problems. The tendency to produce more and more graphics is not always suitable: it might be useful at the beginning, but only the relevant plots should be drawn once the experience is sufficient to identify the meaningless ones.

9.6.1.3 Use of those graphics

Graphs as shown in Figures 9.32 and 9.33 highlight 'Good' and 'Unsuitable' data: column(s) with 'Missing' or 'Unsuitable' data will most likely indicate an error in the data acquisition system and row(s) presenting the same characteristics will indicate a problem with one or more sensors (with various plausible causes). Adding on the timeline every single event that took place on site (storm event, maintenance, calibration, probe change, etc.) will give key information to understand what might have been the cause leading to either an 'Unsuitable' or a 'Doubtful' flag.

The main utilities of such plots are to:

- Understand what occurred on the system.
- Improve the quality of the data while proposing and testing solutions of the causes.
- Communicate the overall quality of the data to the final users (modellers, managers) or financers, while, in most cases, regularly delivering some statistics on data quality.

9.6.2 Using statistics as indicator of the overall monitoring system quality

Even if graphics are rather user-friendly tools to communicate data quality, basic statistics offer some additional highlights, especially for reporting and conducting good asset management of the monitoring networks.

9.6.2.1 Additional information given by statistics

Statistics involved in this part are really basic: they mainly consist of calculating the percentage of each flag (typically, 'Good', 'Doubtful' or 'Unsuitable') for the recorded data set. Those percentages can be and should be calculated for different subsets of the data sets: per sensor, per group of sensors (e.g. inside or outside of the sewer, sensors connected to the same hardware, same type – such as rain gauges, water level probes), per monitoring station, per catchment, etc.

Those statistics will help the data user to highlight and quantify the first impression derived from the previous graphics. Weekly, monthly or yearly values will highlight the data quality trends over longer durations.

9.6.2.2 Suggested indicators

Various indicators could be calculated with the validated data set:

- Percentage of available data (Equation (9.39a) or Equation (9.39b))

$$100 \times \frac{N_G + N_D + N_U}{N_E} \tag{9.39a}$$

$$100 \times \frac{N_E - N_M}{N_E} \tag{9.39b}$$

- Percentage of 'Good' data (Equation (9.40))

$$100 \times \frac{N_G}{N_A} \tag{9.40}$$

- Percentage of 'Doubtful' data (Equation (9.41))

$$100 \times \frac{N_D}{N_A} \tag{9.41}$$

- Percentage of 'Unsuitable' data (Equation (9.42))

$$100 \times \frac{N_U}{N_A} \tag{9.42}$$

- Percentage of 'Doubtful' data finally considered as 'Good' after the manual validation (Equation (9.43))

$$100 \times \frac{N_{D\rightarrow G}}{N_D} \tag{9.43}$$

- Percentage of 'Doubtful' data finally considered as 'Unsuitable' after the manual validation (Equation (9.44))

$$100 \times \frac{N_{D\rightarrow U}}{N_D} \tag{9.44}$$

- Percentage of data that remain 'Doubtful' after the manual validation (Equation (9.45))

$$100 \times \frac{N_{D\rightarrow D}}{N_D} \tag{9.45}$$

Those indicators should be calculated for several subsets of the entire data set:

- For individual sensors.
- For groups of sensors, e.g. inside/outside pipe, water level/velocity/discharge probes, by manufacturer/sensor connected to the same data acquisition hardware.
- For each monitoring station.
- For each catchment.

This type of indicator can be used when judging the performance of a monitoring network. Especially for long-term monitoring activities, keeping track of the monitoring system performance can provide crucial information for, amongst others, improving the data yield.

9.6.2.3 Indicators and asset management of the monitoring system

Data quality indicators and their evolution over time offer multiple opportunities to conduct asset management of a monitoring system, especially while using the meta-data associated with the validated data set. Any positive or negative change in the data quality indicators can lead to confirmation of good decisions/practices or to new decisions to improve the monitoring system. Since an exhaustive list of cases/situations/conclusions is nearly impossible to draft, a few examples are listed below:

- The percentage of data labelled as 'Doubtful' increases a few weeks after a cleaning procedure (meta-data) of a water sensor. This behaviour occurs several times. The delay between two cleaning procedures should be shortened.
- The percentage of data labelled as 'Good' is higher for a pressure sensor than for a US water level sensor at the same location. Pressure sensors seem to be more suitable at this location.
- The percentage of available data for a group of sensors (connected to the same hardware) dropped after its replacement. The new hardware and its installation need to be checked.
- The percentage of data labelled as 'Good' increases after the refurbishment of a monitoring station. The new design and set-up are better than the previous ones: the future replacement of existing monitoring stations should be done the same way.
- The quality of data decreased after some changes in the maintenance/calibration protocols. Those protocols and their realizations need to be carefully checked and compared with the previous ones in order to identify potential issues in the new protocols.

The list of examples is virtually endless. The general rule consists of having a deep look into data quality indicators, keeping track of any change and correlating those behaviours with other data and meta-data, i.e. rain event, maintenance events, hardware or software upgrades, etc.

9.7 DATA CURATION

When deciding on the methods, protocols and their thresholds and further settings, a decision has to be made on what to do with missing or discarded data (i.e. data in category 'Doubtful', 'Unsuitable' and 'Missing'). In many cases missing a few data does not influence the decisions ultimately taken based on them. For instance, when it comes to the evaluation of the environmental performance of a combined sewer system, three or four missing data points over a period of a year is not an issue. However, in the case where six months of data are missing in the same situation, this may make the data set useless.

At the other end of the spectrum, when e.g. model calibration is the main objective of the monitoring project, a time shift of just 1 minute can be enough to make data completely worthless. In the former case no action is needed, other than reporting that out of the 20,000 recorded data points 4 were missing. In the latter case a decision has to be made; either discard the data set altogether and wait for better times, or 'repair' the data obtained and 'make them useable'. From a strictly scientific perspective, the latter is considered a death sin, as information is added that was not actually measured and as such cannot be accepted as an objective basis to work upon.

However, when relaxing this point of view a little bit, it can be argued that when there is convincing evidence that an observed time shift is due to a cause that has been identified, (e.g. a documented application of an incorrect reference level), a correction of the data can be accepted and imposed under the condition that this is clearly stated, hence the label 'Doubtful' for cured data.

In any case, it is strongly suggested to always keep and maintain two data sets: (i) the original time series, with gaps and (ii) the 'cured' time series, with interpolated values. This solution offers some advantages: (i) original data are not overwritten by interpolated data, (ii) the two data sets may serve different purposes and needs, and (iii) alternative interpolation methods can be tested and performed afterwards.

Can we trust or rely on interpolated values? Some points have to be clearly stated: (i) an interpolated value is, as repeatedly stated, a virtual value. Unusual, but real, phenomena may have occurred during the gaps, (ii) interpolation methods are likely not to be able to reconstruct such a phenomenon.

How can the uncertainties of interpolated values be quantified? In line with one of the main messages of this book, the standard uncertainty associated with each value should be estimated. The standard uncertainty of an interpolated value has to take into account two sources: the uncertainty of the measurement (like this value has been normally recorded) and an additional uncertainty due to the interpolation process itself.

Imputation of data in time series is a subject not restricted to the field of UDSM. The study of time series and all their aspects are comprehensively discussed and explained in textbooks (e.g. Hamilton, 1994). Wongoutong (2020) provides a state-of-the-art review on methods applied for imputation of missing values in time series.

9.7.1 What to do with outliers, trends or data gaps in general?

Once gaps, outlier(s) or trends are detected in a data set, the question arises over what to do with them. A first omni-important action to take is to try to find out what caused the outlier, trend or missing data. In many cases malfunctioning, or wrongly installed equipment proves to be the cause. On the other hand, in many cases the cause remains unknown, and could therefore represent a 'real' value and as such hold information on the system studied that might be important. In such cases it is worth trying to find out whether there is some temporal pattern in the occurrence of outliers at the given location for the specific sensor. The example presented in the beginning of the chapter (disturbing lamppost) illustrates the importance of meta-data. In this case the outliers were first marked as 'outlier', which is as such a warning for the use for further analysis. After unambiguously determining the cause and interpolation, the meta-data was changed into 'imputed'. Of course, this type of protocol has to be designed and applied to the specific demands of a given project.

Basically, applying the following sequence with respect to applying data imputation is suggested:

(1) Try to use the data in a piecewise manner, that is, use those time windows in which 'Good' data is present without gaps.
(2) If 1/does not apply, look for the time windows with as much 'Good' data as possible.
(3) If 2/does not result in enough data for analysis, a first data imputation can be made.
(4) A first step is to impute data for single point outliers (see Section 9.7.2).
(5) If data gaps are present (more than a few time steps), data reconstruction may be considered, e.g. from known correlations with other measuring locations or from a model running in parallel.

The latter situation is the most difficult one as it is not simple to set a limit on the allowable amount of missing data. Essentially this boils down to carrying out a risk assessment in terms of making a wrong decision. For pure scientific purposes, the situation is relatively simple: when no data are available, no analysis can be made, so the challenge is to obtain more data and improve the quality until the data are usable. For practical applications, it is more complicated: depending on the purpose for which the data are being collected, more or less missing data can be acceptable, or imputation is seen as 'normal and accepted practice'. For example, if a system has been monitored on dry weather flow patterns for a few years, missing a few days of data per year does not pose a serious impact on the uncertainty on e.g. the total volume discharged per year (one could impute the expected behaviour from historical data). On the other hand, when data gaps occur in a real time controlled system, this has a direct impact on the effectivity of the system. Therefore, again, how to handle missing data is largely a matter of subjective decision making. Regardless of the circumstances, however, the fact that a data value stems from imputation has to be explicitly clear from the meta-data.

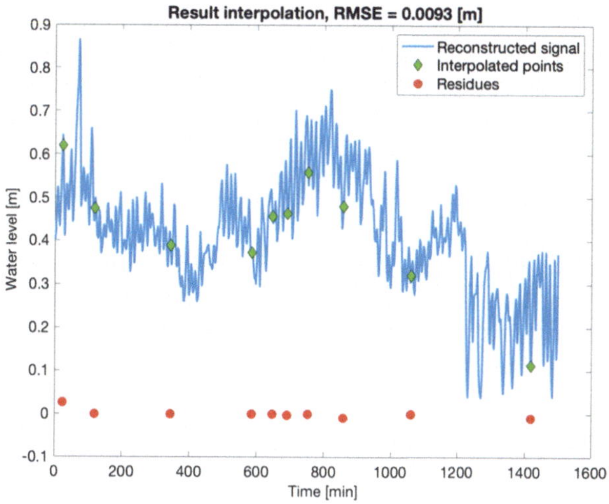

Figure 9.34 Example of interpolated data points. *Source*: Francois Clemens-Meyer (Deltares/TU Delft/NTNU).

9.7.2 Imputation of small data gaps

When 'curing' a single outlier, linear midpoint interpolation (Equation (9.46)) is the easiest and most straightforward method (Figure 9.34):

$$V_t = 0.5(V_{t-\Delta t} + V_{t+\Delta t}) \tag{9.46}$$

Figure 9.34 shows the results of the midpoint interpolation of artificially removed data points. The RMSE of the residues is approx. 9 mm, which is in the same order of magnitude of the confidence range as normally expected for water level sensors. This implies that the curation through interpolation in this case has no noticeable effect on the confidence interval of the data point (see Lepot *et al.*, 2017).

Al Janabi (2005) suggests the following interpolation values for up to two successive missing values, based on the previous and following value:

One point missing (Equation (9.47)):

$$V_t = \sqrt{V_{t-\Delta t} \times V_{t+\Delta t}} \tag{9.47}$$

For two successive points missing (Equation (9.48)):

$$V_t = \sqrt[3]{V_{t-\Delta t}^2 \times V_{t+2\times\Delta t}} \tag{9.48a}$$

$$V_{t+\Delta t} = \sqrt[3]{V_{t-\Delta t} \times V_{t+2\times\Delta t}^2} \tag{9.48b}$$

When filling larger data gaps, a more advanced method is to use a (calibrated) model to fill in gaps, e.g. either a deterministic model, or a conceptual model like an ARMA model, spline fitting method or ML applications.

Gaps may, and will, occur in data sets. There are numerous methods to fill the gaps (if needed) and, in any case, those interpolated values should be labelled as 'D' or 'U' (see Table 9.1), depending on the goals the data have been recorded for. Gaps vary in: (i) size, a single value missing or more; (ii) continuity, a single gap or a series of several gaps; and (iii) the number of impacted time series, typically if a gap is due to a sensor failure or if gaps are due to a data acquisition system failure (a few time series present gaps at the same time). Lepot *et al.* (2017) wrapped up the state of the art regarding interpolation in time series. The next section presents a brief summary of this review, while being restrained to the main method, adding a few examples and advice for practitioners.

9.7.3 Imputation of larger data gaps

Applied methods for filling larger data gaps (i.e. >2 consecutive missing records) are divided into two categories: the deterministic and the stochastic ones. This distinction, based on the existence or not of residuals (differences between prediction at known location and observations) in the interpolation function, deals also with uncertainty.

9.7.3.1 Deterministic methods

The easiest method, not highly recommended for large gaps in dynamic time series, is the nearest-neighbour interpolation: the interpolated values are equal to the closest recorded ones. Other straight forward to implement methods are LOCF (last observation carried forward) or NOCB (next observation carried backward). Such methods are fast and simple, and find their main application in RTC systems.

Example: There is a gap in water level data with missing values between 10:51 am and 10:59 am. At 10:50 and 11:00, the recorded water levels are respectively 10 cm and 12 cm. With this method, interpolated water levels are 10 cm until 10:55 and 12 cm afterwards (Figure 9.35).

In order to avoid this discontinuous behaviour, smoother functions can be used to interpolate data e.g. with linear or polynomial interpolation methods.

Example: While re-using the same example with a linear interpolation, the interpolated values will slowly vary from 10 to 12 cm, while reaching 11 cm at 10:55 am (Figure 9.35).

Phenomena in urban hydrology are rarely linear: polynomial interpolation or application of ARMA models (see Section 9.4.4) will better mimic the expected shape of the time series. Methods based on

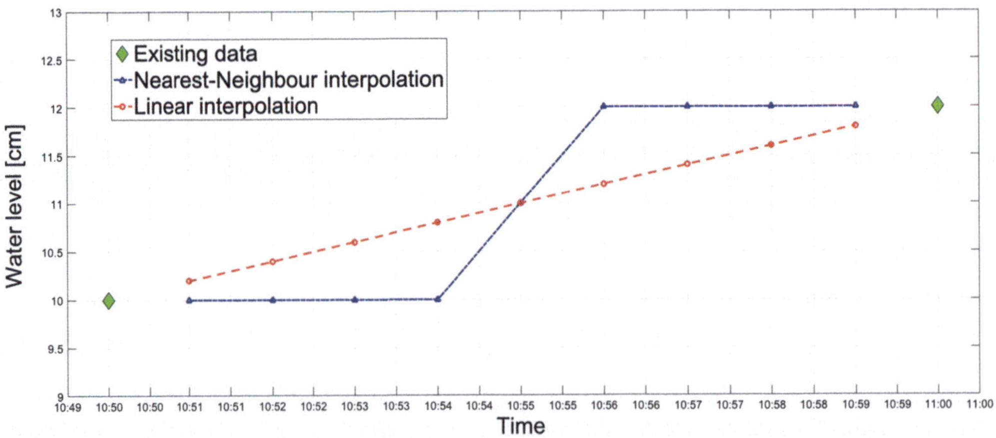

Figure 9.35 Existing and interpolated data. *Source*: Mathieu Lepot (TU Delft/Un poids une mesure).

distance-weighting could be also used: an average, weighted by the distance between the interpolated and recorded values.

Other functions can be used to achieve such interpolation, especially equations that reflect the phenomenon: dispersion of pollutant, correlation between water level, velocity and discharge (like the Manning-Strickler equation), run-off models, etc. Those methods are less based on mathematics and more on the physical processes that occur at the measuring location. A few of those approaches are nearly impossible to perform if the entire data acquisition system collapses.

9.7.3.2 Stochastic methods

Machine learning approaches such as neural networks, kernel methods and kriging are also available for data interpolation. Methods based on data dynamics seem to be more appropriate in urban hydrology and deserve more detail.

The k-Nearest Neighbours (k-NN) takes into account the cyclic variations of a time series (e.g. discharge during a dry business day). Assuming or knowing the dynamics are similar from day to day, the gap can be fulfilled with data from another day. There are several metrics to identify which part of the recorded values is most suitable for the interpolation e.g. city block, Euclidean or Chebychev. Box-Jenkins models are suitable for polycyclic data, including seasonality and daily or weekly patterns. Pratama *et al.* (2016) provide a review on handling missing data in time series.

9.7.3.3 Uncertainty assessment

The uncertainty of interpolated data has two components: (i) the uncertainty of measurement if this value has been normally recorded and (ii) the uncertainty from the interpolation process itself. If the first component has been detailed in Chapter 8, the second one requires a few tips given in this section.

Given two known values (V_t and $V_{t+\Delta t}$), the standard uncertainty $u(V_{I,i})$ of an interpolated value $V_{I,i}$ is calculated according to Equation (9.49) (Lepot *et al.*, 2017):

$$u(V_{I,i}) = \sqrt{\frac{1}{2}\sigma_M^2\left(\begin{array}{c}3 + |\rho(V_t, V_{t+\Delta t})| - 2|\rho(V_t, V_{I,i})| - 2|\rho(V_{t+\Delta t}, V_{I,i})| \\ -\dfrac{(\rho(V_{t+\Delta t}, V_{I,i}) - \rho(V_{t+\Delta t}, V_{I,i}))^2}{1 - \rho(V_t, V_{t+\Delta t})}\end{array}\right) + \sigma_P^2} \tag{9.49}$$

where σ_P is the process variance, ρ is the autocorrelation function and σ_M is the measurement error.

Figure 9.36 shows several methods to assess uncertainties of interpolated data: the Law of Propagation of Uncertainty (LPU, top left), Monte Carlo simulations (MC, top right), a method proposed by Schlegel *et al.* (2012) (bottom left) and Equation (9.49) (bottom right). Only Equation (9.49) proposes uncertainties for interpolated values, which present a correct trend: (i) being higher than the measurement ones, and (ii) a continuity at the edge of the interpolation area.

9.8 DATA-DRIVEN METHODS

The discussion of methods related to machine learning (ML) – hereinafter referred to as *data-driven methods* – has been intentionally discarded here for two main reasons. Firstly, those methods have only very recently been applied, in various fields with differing success. Their application in the UDSM field has not yet reached the level of maturity that would allow an objective judgement of their usefulness. Secondly, data-driven methods are often based on a black-box approach: the method parameters often lack physical interpretation, data may be rejected (e.g. as false positive) without explicitly providing a

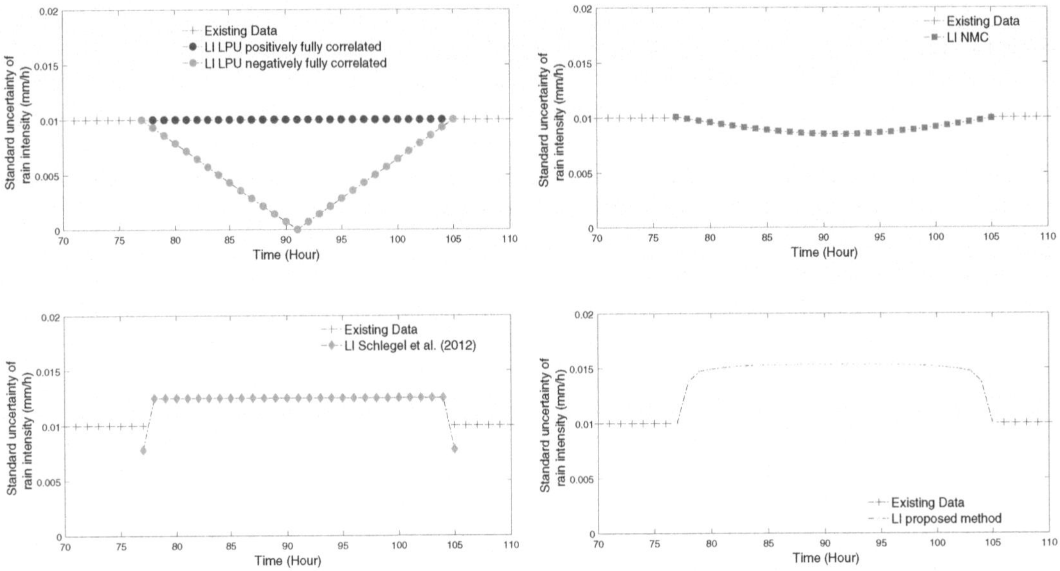

Figure 9.36 Uncertainties of interpolated values *vs.* uncertainties of existing data (black +) for a linear interpolation. *Source*: adapted from Lepot *et al.* (2017).

reason, i.e. the input to understand this rejection and better understand the monitored system. This approach is contrary to the one we want to propose in this book. However, data-driven approaches are rapidly developing.

It is acknowledged that the suite of data-driven concepts may gain further popularity once domain knowledge has sufficiently been integrated into purely machine/data-driven approaches. In the following section, a brief, but clearly limited review of why such concepts can be useful and what challenges are associated with their application is presented.

9.8.1 Motivation

Existing guidelines (Bertrand-Krajewski & Muste, 2007; DWA, 2011) and automated data validation pipelines (e.g. Alferes & Vanrolleghem, 2016; Branisavljević *et al.*, 2010) specifically developed for UDSM applications provide very useful solutions. Generally, these approaches suggest a consecutive application of standard rule-based methods for basic data validation. Rule-based methods – some of which are discussed in Sections 9.3 and 9.4 – imply the use of parameters defined based on expert knowledge about sensors and system behaviour (min/max ranges, acceptable changes, pre-defined correlations). Integrating this expert knowledge in the form of manual intervention and subjective judgement is rather expensive, and it does not necessarily lead to reproducible results. Other limitations become obvious in the case of real-time applications requiring minimum latency, e.g. if it is necessary to simultaneously check several signals of different types in real time to trigger control, and/or if computationally expensive analysis methods are used.

Conducting *automated* data validation for identifying abnormal behaviour of the deployed sensors is therefore becoming increasingly critical. Data-driven methods promise easy help here. They suggest a timely, coherent, complete and, with increasing data volume, more efficient assessment of data quality

(Aggarwal, 2017). The algorithms ability to 'learn' suggests higher efficiency with minimal human intervention while increasing flexibility.

While data-driven methods are experiencing a boom in the field of image processing, natural language processing, speech recognition, stock portfolio management, and other fields, so far only very few applications are known in the field of urban hydrology. Troutman *et al.* (2017) combine Gaussian processes (dry-weather flows) and dynamical System Identification (wet weather discharge) aiming to simulate rainfall-run-off dynamics in a combined sewer network purely based on sensor data. Although the detection of novelty in monitored data had not been the primary objective, this approach could be applied to do so. Russo *et al.* (2019) present an anomaly detection method based on a convolutional neural network (CNN), i.e. a deep autoencoder, for validating urban drainage monitoring data. However, the suggested methods come along with some deficiencies: immense data pre-processing is required, and a high false positive rate is still present, although the latter aspect is partially justified with a somewhat high complexity of the data, i.e. in-sewer flows, used in the study. Rodriguez-Perez *et al.* (2020) applied a range of artificial neural networks on water-quality data (turbidity and conductivity) with high temporal resolution to evaluate the 'best' performing model depending on the variable-, environment-, and anomaly type. Common anomaly types present in online water quality data (with characteristics very similar to variables monitored in urban drainage) were previously categorized by Leigh *et al.* (2019). Rodriguez-Perez *et al.* (2020) in turn found that semi-supervised classification was better able to detect instantaneous faults (e.g. spikes), whereas supervised classification had higher accuracy for predicting long-term anomalies, such as drifts.

Despite the fact that results of these studies look rather promising, further systematic evaluation on *different* real-life data sets applying *different* data-driven approaches is required to show the usefulness and likewise the limitations of such approaches.

9.8.2 Challenges and constraints

Direct application of purely data-driven methods for the validation of urban hydrological data is challenging for several reasons: (i) system-determining rainfall events with random occurrence do not result in easily recognizable patterns; (ii) the range of values of measured state variables is sometimes limited on one side (e.g. in the case of flow restrictions caused by a throttle) resulting in unilaterally constrained data; (iii) processes are inherently non-linear – the boundary conditions are difficult to define; and (iv) the complexity of some urban drainage signals, i.e. the decomposition of overlapping fluxes of different dynamics, can be challenging.

Current research in the field of ML mostly focuses on new and signal-specific methods, or method comparisons with limited representativeness. Comparisons and proofs of application are often carried out with synthetic, i.e. modelled data sets or data from other fields. Description or necessary steps for the preceding data preparation is often omitted, or their effect on the final result is often unclear. Against this background, it is obvious that more studies are required, applying various methods to different data sets (benchmarking of algorithms applied on open-access data). The repetition of experiments, i.e. reproduction of results using new methods, would increase the trustworthiness of data-driven methods.

Existing and future studies in the UDSM field should be carefully evaluated. The specific urban drainage context (constrained/unconstrained, dry/wet weather) must be taken into account. If the community manages to integrate sufficient domain knowledge into machine learning concepts (basically shifting from black box to grey box models), data validation for urban drainage applications may experience a significant push. However, it has to be kept in mind that ML may be a convincing hammer, but not every problem is a nail!

9.9 SUMMARY AND TRANSITION

Validation of field measurements is crucial to ensure data consistency and to allow for optimal interpretation of the data. High-quality data increases the trustworthiness of the derived information enabling informed decisions, but obtaining high-quality data is not an easy task.

This chapter attempts to remedy this situation. It discusses individual aspects of checking the plausibility of data points, assessing their quality, and it strives for options to curate data as an inherent part of the validation process, and/or as a subsequent step. A brief excurse on the use of machine learning techniques is given due to its increasing popularity, also in the field of data validation.

The process of data validation should be understood as a stepwise approach, which can be split into a basic check of consistency and plausibility, and a more subjective assessment. If applied, the latter should clearly be dependent on the purpose the data is used for. Meta-data is considered decisive for correct data interpretation. This additional, often non-numeric information should be collected systematically and archived with a distinct relation to the corresponding data point(s).

Various tests for data quality assessment are introduced: from the simpler to the more complex ones. Once the outputs of individual tests are calculated, results may be concatenated to obtain a single metric per data point. Despite the fact that methods are mathematically founded, there can be substantial subjectivity associated with their application. Some of the statistical techniques described require – in a strict mathematical sense – data properties that are fulfilled. Still, these techniques are successfully applied and widely accepted in practice. In such cases, this has been indicated, but scholastic correctness has been considered as subordinate in favour of practicality. End users must be aware of such subjectivities. In any case, pedantic documentation and transparent communication on parameters, weights, and methods applied is highly recommended.

With the rise of new sensors and data communication technologies, and as we are adopting the Internet of Things (IoT), collecting data has become less cumbersome, even in such challenging environments as UDSM systems. This will inevitably lead to a substantially increased amount of data. But it is anticipated that the quality of that data will not necessarily increase in the course of this trend. This, in turn, raises the importance of a *quantitative* data quality assessment, as it is for the automation of this process.

In a positive sense, this trend will stimulate the development of new methods, and their integration into automatized data validation pipelines. But it will also come with new challenges: higher complexity, higher diversity, an increased risk of confusion, and a lack of transparency in the process. The following strategies will help to efficiently tackle present and future challenges:

- Introducing a basic level of harmonization, effectively resulting in somewhat standardized approaches, including commonly agreed interfaces in the data assessment pipeline. What does it bring? It allows the comparison of the quality of different data sets, for instance in the form of benchmarking between different data providers and different systems. It enables performance assessment of different evaluation approaches. Defined interfaces enable straightforward integration of new methods.
- Development of data literacy across different qualification levels, that is from a technician that does sewer maintenance to the CEO that manages a wastewater utility.

- Establishing a culture of open-data, i.e. data sharing internally and between organizations. The evaluation of anonymized data sets in the form of a benchmarking process leads to objective performance assessment and continuous improvement.
- Increasing the degree of automation in the data validation process.

Along these lines, it can be expected that, in the (near) future, subjectivities in the data validation process can be minimized, and the efficiency increased by using advanced data analysis tools. From the perspective of scientific importance as well as for practical applications, all this would be beneficial to further facilitate the use of validated data for evidence-based decision making in the field of urban drainage and stormwater management.

REFERENCES

Aggarwal C. C. (2017). *Outlier Analysis, 2nd edn*. Springer, New York (USA), 488 p. ISBN 978-3319475776.

Al Janabi M. A. M. (2005). *Financial Risk Management: Application to the Moroccan Stock Market*. Al Akhawayn University Press, Ifrane (Morocco), 137 p. ISBN 9789954413470.

Alferes J. & Vanrolleghem P. A. (2016). Efficient automated quality assessment: dealing with faulty on-line water quality sensors. *AI Communications*, **29**(6), 701–709. doi: 10.3233/AIC-160713.

Barnett V. & Lewis T. (1996). *Outliers in Statistical Data, 3rd edn*. John Wiley & Sons, Chichester (UK), 604 p. ISBN 978-0471930945.

Bayley G. V. & Hammersley J. M. (1946). The 'effective' number of independent observations in an autocorrelated time series. *Supplement to the Journal of the Royal Statistical Society*, **8**(2), 184–197. doi: 10.2307/2983560.

Bertrand-Krajewski J.-L. & Muste M. (2007). Data validation: principles and implementation. In *Data Requirements for Integrated Urban Water Management*, T. Fletcher & A. Deletic (eds.), Taylor & Francis, London (UK), Urban Water series – UNESCO IHP, 103–126. ISBN 9780415453455.

Branisavljević N., Prodanović D. & Pavlović D. (2010). Automatic, semi-automatic and manual validation of urban drainage data. *Water Science and Technology*, **62**(5), 1013–1021. doi: 10.2166/wst.2010.350.

Choi B. (1992). *ARMA Model Identification*. Springer, New York (USA), 212 p. ISBN 978-1461397472.

Conover W. J. (1999). *Practical Non-Parametric Statistics, 3rd edn*. John Wiley & Sons, New York (USA), 584 p. ISBN 978-0-471-16068-7.

DWA (2011). *Merkblatt DWA-M 181 Messung von Wasserstand und Durchfluss in Entwässerungssystemen [Measurement of water level and flow in urban drainage systems]*. DWA – Deutsche Vereinigung für Wasserwirtschaft, Abwasser und Abfall, Hennef (Germany), 10 p. ISBN 978-3-941897-94-6. Available at https://webshop.dwa.de/de//dwa/download/?link=TV8xODFfMDlfMjAxMS5wZGY= (accessed 21 Dec. 2020).

Emmert-Streib F. & Dehmer M. (2019). Large-scale simultaneous inference with hypothesis testing: multiple testing procedures in practice. *Machine Learning & Knowledge Extraction*, **1**(2), 653–683. doi: 10.3390/make1020039.

ESS (2018). *ESS Handbook – Methodology for data validation 2.0 – Revised edition 2018*. European Commission, Eurostat, European Statistical System, Brussels (Belgium), 85 p. Available at https://ec.europa.eu/eurostat/cros/system/files/ess_handbook_-_methodology_for_data_validation_v2.0_-_rev2018_0.pdf (accessed 21 Dec. 2020).

Gray K. L. (2007). *Comparison of Trend Detection Methods*. PhD thesis, University of Montana, Missoula (MT), USA, 97 p. Available at https://scholarworks.umt.edu/cgi/viewcontent.cgi?article=1247&context=etd (accessed 21 Dec. 2020).

Grubbs F. E. (1969). Procedures for detecting outlying observations in samples. *Technometrics*, **11**(1), 1–21. doi: 10.1080/00401706.1969.10490657.

Hamilton J. D. (1994). *Time Series Analysis*. Princeton University Press, Princeton, NJ (USA), 820 p. ISBN 978-0691042893.

Iglewicz B. & Hoaglin D. C. (1993). *How to Detect and Handle Outliers*. American Society for Quality Control, Milwaukee (USA), 87 p. ISBN 9780873892476.

Leigh C., Alsibai O., Hyndman R. J., Kandanaarachchi S., King O. C., McGree J. M., Neelamraju C., Strauss J., Talagala P. D., Turner R. D. R., Mengersen K. & Peterson E. E. (2019). A framework for automated anomaly detection in high frequency water-quality data from in situ sensors. *Science of the Total Environment*, **664**, 885–898. doi: 10.1016/j.scitotenv.2019.02.085.

Lepot M., Aubin J.-B. & Clemens F. H. L. R. (2017). Interpolation in timeseries: an introductive overview of existing methods, their performance criteria and uncertainty assessment. *Water*, **9**(10), 796, 20 p. doi: 10.3390/w9100796.

Lettenmaier D. P. (1976). Detection of trends in water quality data from records with depended observations. *Water Resources Research*, **12**(5), 1037–1046. doi: 10.1029/WR012i005p01037.

Leutnant D., Hofer T., Henrichs M., Muschalla D. & Uhl M. (2015). Model-based time drift correction of asynchronous measurement data in urban drainage. *Proceedings of the 10th International Conference on Urban Drainage Modelling*, 20–23 September, Mont Sainte Anne, Quebec, Canada, 131–134.

Lindenmayer D. F. & Likens G. E. (2018). Maintaining the culture of ecology. *Frontiers in Ecology and the Environment*, **16**(4), 195. doi: 10.1002/fee.1801.

Mourad M. & Bertrand-Krajewski J.-L. (2002). A method for automatic validation of long time series of data in urban hydrology. *Water Science and Technology*, **45**(4–5), 263–270. doi: 10.2166/wst.2002.0601.

Pratama I., Erna Permansari A., Ardiyanto I. & Indrayani R. (2016). A review of missing values handling methods on time-series data. *Proceedings of the 2016 International Conference on Information Technology Systems and Innovation (ICITSI)*, 24–27 Oct., Bandung, Bali (Indonesia), 1–6. doi: 10.1109/ICITSI.2016.7858189.

Rodriguez-Perez J., Leigh C., Liquet B., Kermorvant C., Peterson E., Sous D. & Mengersen K. (2020). Detecting technical anomalies in high-frequency water-quality data using artificial neural networks. *Environmental Science and Technology*, **54**(21), 13719–13730. doi: 10.1021/acs.est.0c04069.

Russo S., Disch A., Blumensaat F. & Villez K. (2019). Anomaly detection using deep autoencoders for in-situ wastewater systems monitoring data. *Proceedings of the 10th IWA Symposium on Systems Analysis and Integrated Assessment (Watermatex 2019)*, September 1–4, Copenhagen, Denmark, 7 p. Available at https://arxiv.org/abs/2002.03843 (accessed 21 Dec. 2020).

Schilperoort R. P. S. (2011). *Monitoring as a Tool for the Assessment of Wastewater Quality Dynamics*. PhD thesis, Delft University of Technology, Delft, The Netherlands, 330 p. ISBN 978-90-8957-021-5.

Schlegel S., Korn N. & Scheuermann G. (2012). On the interpolation of data with normally distributed uncertainty for visualization. *IEEE Transactions on Visualization and Computer Graphics*, **18**(12), 2305–2314. doi: 10.1109/TVCG.2012.249.

Troutman S. C., Schambach N., Love N. G. & Kerkez B. (2017). An automated toolchain for the data-driven and dynamical modeling of combined sewer systems. *Water Research*, **126**, 88–100. doi: 10.1016/j.watres.2017.08.065.

Upton G. J. G. & Rahimi A. R. (2003). On-line detection of errors in tipping-bucket raingauges. *Journal of Hydrology*, **278** (1–4), 197–212. doi: 10.1016/S0022-1694(03)00142-2.

van Bijnen J. A. C. (2018). *The Impact of Sewer Condition on the Performance of Sewer Systems*. PhD thesis, Delft University of Technology, Delft, The Netherlands, 202 p. ISBN 978-94-6233-987-3.

Villez K. & Habermacher J. (2016). Shape anomaly detection for process monitoring of a sequencing batch reactor. *Computer and Chemical Engineering*, **91**(4), 365–379. doi: 10.1016/j.compchemeng.2016.04.012.

Wilcox R. R. (2016). *Introduction to Robust Estimation and Hypothesis Testing, 4th edn*. Elsevier Academic Press, London (UK), 810 p. ISBN 978-0128047330.

Wongoutong C. (2020). Imputation for consecutive missing values in non-stationary time series data. *Advances and Applications in Statistics*, **64**(1), 87–102. doi: 10.17654/AS064010087.

Chapter 10

Data archiving and meta-data – saving the data for future use

Dušan Prodanović and Nemanja Branisavljević

Faculty of Civil engineering, Department of Hydraulic and Environmental Engineering, University of Belgrade, Belgrade, Serbia

ABSTRACT

This chapter covers the main aspects of data archiving, as the last phase of data handling in the process of urban drainage and stormwater management metrology. Data archiving is the process of preparing and storing the data for future use, usually not executed by the personnel who acquired the data. A data archive (also known as a data repository) can be defined as storage of a selected subset of raw, processed, validated and resampled data, with descriptions and other meta-data, linked to simulation results, if there are any. A data archive should be equipped with tools for search and data extraction along with procedures for data management, in order to maintain the database quality for an extended period of time. It is recommended, mostly for security reasons, to separate (both in a physical and in a digital sense) the archive database from the working database. This chapter provides the reader with relevant information about the most important issues related to data archive design, the archiving process and data characteristics regarding archiving. Also, the importance of good and comprehensive meta-data is underlined throughout the chapter. The management of a data archive is evaluated with a special focus on predicting future resources needed to keep the archive updated, secure, available, and in compliance with legal demands and limitations. At the end, a set of recommendations for creating and maintaining a data archive in the scope of urban drainage is given.

Keywords: Archive, backup, database, data reuse, data security, meta-data.

doi: 10.2166/9781789060119_0391

10.1 INTRODUCTION

Measured data value(s) is the aim of each measurement process. Therefore, it may be assumed that there exists a purpose of the measurement, the reason why data are needed and therefore acquired (see Chapter 6). On the other hand, it can be assumed that, apart from the main reason that the data were collected, there are potentially some other purposes for the obtained data to be used in the future. That *potential purpose may not be known at the moment of data acquisition*, but it is sometimes possible to predict that purpose in advance. For example, the hydro-meteorological data are being collected and stored for a current project or for on-line forecasts, but the historical value of the acquired data could become apparent only after many years. For the archived data to show their full usefulness potential, the state of the system and detailed information about the conditions at the time when the data are acquired needs to be reconstructed as precisely as possible, only on the basis of the data and meta-data stored in the archive. As another example, consequent research projects that are planned to be conducted over considerably long periods of time, with a considerable number of researchers involved, may be considered. The data collected by researchers and practitioners during the projects conducted previously may be of great value for the following ones, and therefore with proper data archiving, data management in the latter project stages may be considerably simplified.

Since the data archiving implies a lengthy process, the information technology used to acquire data, process and store them, will evolve over time. Therefore, for the sake of information persistence, a well-designed procedure for proper and documented data archiving is needed.

It is important to understand that large SCADA (supervisory control and data acquisition) systems, with powerful databases (like Historian database, GE DIGITAL, 2019) which can hold a vast volume of raw data and have a specialized SQL with data processing ability, cannot be considered as an archive in the context of this chapter. They lack the validation, resampling, extensive meta-data management and possibility to be used and shared among users from various organizations. On the other hand, the SCADA is a valuable and powerful data source to create the archive. By extracting data for selected periods of time (e.g. one year), adding meta-data and validation tests, an optimal archive can be created.

The data lifecycle shown in Figure 10.1 includes:

- Data creation through acquisition (measurement) or data generation (logical assumption, calculation, simulation, prediction, etc.).
- Data storing for present use.
- Using of data to create information and knowledge (Prodanović, 2007a).
- Exchange of data and information with stakeholders and interested parties.
- Archiving of selected and prepared data and information.
- Destroying (erasing) unnecessary data.

The data are usually associated with some natural or technical variable (see Chapter 1). Also, the data may become available using either measurement or some logical or experimental assumptions. The type of data origin may provide essential information about the data validity and the level of confidence one should have about the data value. Therefore, in addition to the data value itself, it is necessary to keep available the information on how the data originated, and the environment at the time of its origination. This information is usually recorded *in the form of meta-data – data about the data.*

Further, during their lifetime, the data and accompanying meta-data are usually stored for present/immediate use (see Chapters 5, 7, 8 and 9). Their use refers to the reason why the data were generated in the first place. For example, SCADA systems are used to collect data, with limited (if any) sharing options. The time span over which the data are stored for immediate use is limited and can extend from a few seconds, for real-time observations, up to years, in cases when dealing with data

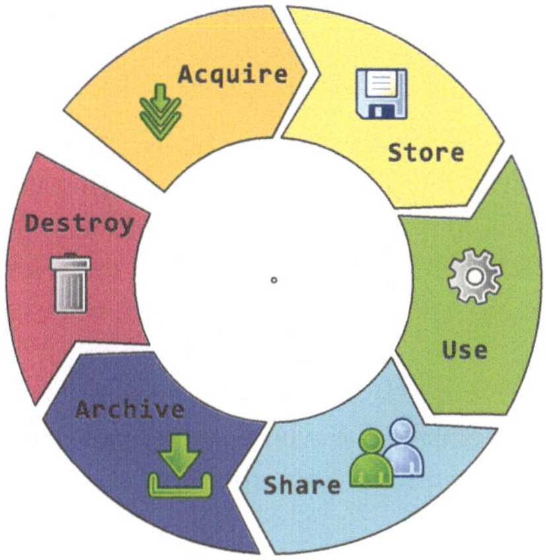

Figure 10.1 Data lifecycle. *Source*: Dušan Prodanović and Nemanja Branisavljević (University of Belgrade).

about the slowly changing environmental processes or for legal purposes. During that period of data lifetime, data is often associated with calculated information that can sometimes exceed the volume of directly collected raw data. Since this kind of data storage has its own rules and constraints, it is necessary to *clearly distinguish the data storage for immediate use and data storage for archiving of data for future use.*

Data sharing and data-exchange are important parts of the urban drainage data life span (see Chapter 5). During the process of data exchange, the data often become much more expressive and sometimes get changed, thus losing their integrity and relation with the original data. Therefore, it is suggested that, if possible, genuine (or raw) data should be kept and shared with processed data that are calculated by the various procedures (Prodanović, 2007b). All changes and updates, along with assessed quality of data (see Chapters 8 and 9) should be stored and explained in shared meta-data.

Archiving of data comes as the next phase in the lifetime of the data, and *it implies the process of preparing and storing the data* for use in the future. The data archive can be defined as storage of a *selected subset of raw data, processed, validated, resampled data, with suitable descriptions and meta-data, linked with simulation results, especially for later use by users not involved in the data acquisition and preparation*. Often there is no well-defined period for which the data has to be kept. In the general case, it may be assumed that the data will be stored infinitely, which is why it is necessary to carefully evaluate all the issues associated with the scope of the data, the storage medium and the data format that will be used. Also, it is important to consider that partial sets of the data may be lost at some point of time in the future, mostly due to technical reasons, so a certain level of redundancy in data should be considered, as also mentioned in Section 6.2 on the design of monitoring networks. The progress in data storage technology can make some data records obsolete and inaccessible (e.g. data recorded on magnetic tapes is no longer readily available for access), therefore an important part of data archiving is to plan the regular backup, and to provide the proper and timely format update and storage medium conversion mechanism.

In the previous paragraph, the archive is defined as a *selected subset of … data …* This means that an *archive is not meant to be used as a working database*, where monitored processes are continuously loading the new measured values. To create an archive, we have to extract a set of measured quantities from the working database, for a selected period of time (mostly for one year), with maximal resolution, and prepare (process, validate, resample, etc.) the data, the required meta-data and other documents, and store them in a distinct archive. The data archive which is produced in the process of archiving, as seen in this chapter, is a *synonym for data repository*: both will keep a subset of data for the purpose of future search and extraction, will have to manage the databases and preserve them for a long period of time.

This chapter covers the main aspects of data archiving, providing the reader with relevant information about the most important issues. We tried to consider data archiving and principles that need to be taken into account, whether it is a small amount of data collected for short-term projects, or large, data collection campaigns. Although the scope of this book is the metrology related to urban drainage and stormwater management (UDSM), most parts of this chapter are relevant and applicable for metrology in all water-related fields.

At the end of the data lifecycle (Figure 10.1), but not the least important, it is necessary to predict the data end of life and create a plan for erasing the whole set of data or only selected parts. In this chapter, a few scenarios will be considered regarding when the data are no longer needed and are allowed to be deleted.

Key messages on data archiving and meta-data

- KM 10.1: *An archive is not just a huge, rarely used database* – An archive is not just a database used by SCADA holding all raw data. It is a carefully selected subset of data for a certain past time, with raw and processed data and all the required meta-data to allow archive usage in the future.
- KM 10.2: *Timely design of the archive!* – Design the outline of your archive simultaneously with SCADA, before the start of a data collection campaign. Think about data security and check the possible issues with data usage rights.
- KM 10.3: *The present-day system is not the same as the one for which data are archived* – The archive will be used by someone in the (far) future, who may not be familiar with some specific detail which could be crucial, so please think about this when preparing meta-data. Archive the system description, network layout, and system geometry. Remember that the urban system will change in the future.
- KM 10.4: *An archive needs regular maintenance* – An archive has to be maintained and updated yearly or biyearly due to: storage medium change, used archiving software change, used data format change, and periodic data validation.
- KM 10.5: *Do not archive all video and high-resolution data* – Try to prevent digital data pollution by archiving too many potentially unnecessary data (video footages, oversampled data, unnecessary events, etc.). Use data reduction techniques and store statistical data.
- KM 10.6: *Running the archive is a costly business!* – Running and maintenance of the archive is costly. Try to assess the costs, foresee the potential data users and other sources of incomes, and make the archive sustainable.
- KM 10.7: *Check the ownership of data and the archive itself* – Data ownership can be an important issue: who owns the data within the archive? Who owns the information (processed data)? Who owns the archive itself? Link the ownership with permission issues and charging policy (if any).
- KM 10.8: *Select and prepare data for archive* – It is not the optimal strategy to archive all collected data: select and resample data, process and validate data, create some summary data.
- KM 10.9: *Optimal design of an archive implies no obsolete data* – An archive should be kept forever. If some data in the archive could become obsolete, then the archive is not optimally designed.

10.2 DATA ARCHIVING PROCESS

Although UDSM is a rather small, focused branch of technology and science, it relies on large amounts of data from various sources. The hydro-meteorological data, urbanization data, census data, data on built and planned infrastructure, results from simulations, radar images, citizen's complaints or video survey recordings are only a subset of the required data. Originating from different sources (sometimes distributed over a number of organizations, private as well as public), the data are usually stored in a range formats (text files, databases, geospatial formats, etc.) and mostly kept on different kinds of media. Due to the large volume of data and various data sources and formats, it is necessary to carefully *plan the data archiving in order to preserve all the potentially needed information* for the future.

In addition to collected data, it is necessary, in the best possible way, to preserve information on the conditions during the data acquisition. This information helps the future data user to understand the environment in which the data were obtained, the potential risks of errors, and the link to other types of data. Information about the environmental conditions is usually recorded in the form of so-called meta-data 'information that is given to describe or help you use other information' (Cambridge English Dictionary, 2019), or shorter, *data about the data*. Meta-data are usually an integral part of an archive database.

The processes of data archiving can be seen as an interaction of *three parties*. The first party (individuals or organizations) produces data (*data producers*), the second party keeps the stored data and distributes data on demand (*data providers*), while the third party is the party that uses the data (*data users*):

- *Data producers* are experts that are involved in data production (e.g. experts that are using monitoring equipment or experts that are involved in UDSM processes simulation and prediction) and the first nine chapters of this book are about them. Beside data production, data producers usually create the information out of collected data and use it to design, operate, analyse, etc. UDSM systems. In most modern systems, specialized SCADAs are utilized to monitor and control certain process (like in a wastewater treatment plant), where the optimal and uninterrupted operation of the purification process is of primary concern. Such SCADA systems often have a vast amount of interesting acquired data which is internally stored for a certain period of time and cannot be easily shared or extracted to become a part of an archive. In most cases, an additional effort is needed to persuade the SCADA developers to 'open' the system and extract selected data.
- Data producers should have all necessary information about the data they are producing, such as the used equipment, the environmental and working conditions and all the circumstances when the data is acquired. Therefore, it is of great importance that the *data producer is the one who prepares the plan for data archiving*, designs the database and algorithms for filtering and resampling, data processing and data validation, and chooses formats of meta-data and needed documentation, always having in mind the possibilities of how the data may be used in the future.
- *Data providers* only provide (keep and share) data that data producers archived. They do not produce data and it is not required for them to be experts in UDSM or data production. On the other hand, it is expected that they should be experts in data management, since they have to keep the archived data, manage the used database(s) and manage the data requests from various users.
- *Data users* are experts that are going to use the data at some point in the near or far future. At the time of data usage, the information about the environmental conditions during data acquisition, equipment used, etc. can be reconstructed from the saved meta-data only. When designing the archive and accompanying meta-data records, it has to be taken into account that the data user is not expected to be able to communicate with the data producer to obtain explanations. Therefore, the data producer is the one that should provide a data archiving plan, and consider all the information that

the data user would need in the future. The archiving plan should follow the existing standards or good practice if any exist (for example, in the field of hydrometry, the CEN Technical Committee 318 has developed the guidance document CEN TS 17171 (CEN, 2018) for observed hydrometric data, taking into account the foreseen future needs and technology developments, with all the necessary meta-data that would provide the data user with sufficient information about the system and data acquired.

A *data archive represents the ultimate destination of data* and often represents an important source of information about the state of the system in the past. Given that, the data archives are invaluable for numerous base-line analyses of the natural or built system states, allowing us to evaluate the effect of any intervention on the system or system change. Examples of analyses that include previous states are numerous. For example, the analysis of the impact of climate change on the urban environment (Smith *et al.*, 2009), or the prediction of the outflows increase due to urbanization (Han *et al.*, 2016), etc.

Data archive design greatly depends on the technological development at the time it is created. Figure 10.2 shows the timeline of technological development and technical possibilities for data archiving.

Collection and archiving of data essential for UDSM started long before the design and implementation of the first UDSM modern systems. According to Schlanger (2017), the oldest historical meteorological data are from Galileo's time, but generally it is accepted that start of the modern global record-keeping is 1880, since the '*earlier available climate data doesn't cover enough of the planet to get an accurate reading*' (Schlanger, 2017). It may be assumed that people started to collect the data related to UDSM systems at the time of the construction of the first urban drainage systems. Although the first organized drainage of rainwater and wastewater may be recognized in the earliest civilizations of the Middle East, the Mediterranean and the Orient (Delleur, 2003), important paradigm change took place in the field of urban drainage in the 1960s (Harremoës, 2002) with the development of computer models able to simulate, analyse, and predict the system performance. That change, along with development of measurement sensors and data acquisition techniques, led to the organized monitoring of system performance (mainly rain intensity and water levels/flows) and acquired data storage and distribution.

Used technology for data recording at that time was mostly analogue – usually pen writing on stripe charts. That kind of analogue data recording enabled easy and reliable data comparison, minimum and maximum detection and data quality estimation during a single measurement campaign. On the other hand, comparison of the data from various measurement campaigns, from various locations or data exchange with other users required comparing two or more paper strips or rewriting data in numerical form. A that time paper-based tables were usually used for data archiving, with meta-data that were recorded in written form directly on the paper strip (Figure 10.3).

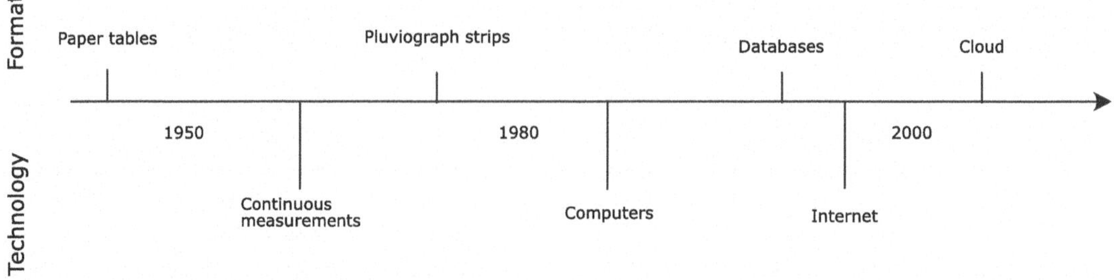

Figure 10.2 Timeline of used technology and formats in data archives. *Source*: Nemanja Branisavljević (University of Belgrade).

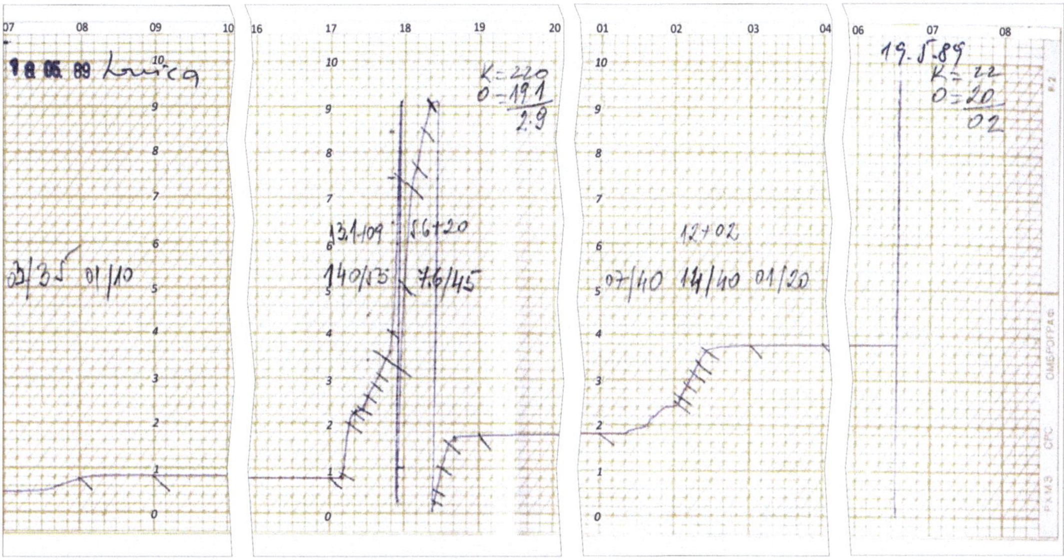

Figure 10.3 Analogue pluviograph strip with handwritten intensities, Loznica, Serbia, 18.05.1989. *Source*: Hydromet (Serbia).

After the invention of computers, the digital form of data format emerged, which allowed for easier use, manipulation, copying and sharing than previously possible. The first digital plain text format was followed by spreadsheets, and the spreadsheets were followed by relational databases, that are still in wide use. By the invention of the internet, a new standard for data sharing was introduced. Internet protocols enabled development of so-called cloud systems for data storing and sharing all over the network. So, nowadays, standards of data archiving are developed to be more decentralized, easily accessible and more comprehensive than ever before.

Development of internet, databases and cloud storage has made the archiving process easier. However, the total life cost of such an approach is arguable, since a lot of resources (energy, manpower) are needed to keep the archive up-to-date and accessible. On the other hand, keeping the archive in the old paper format has its advantages, too. For example, a book about rainfall-runoff data from worldwide experimental catchments, collected and printed in Urban Drainage Catchments by Maksimović & Radojković (1986), had high initial costs to prepare and print it, while it will be accessible and can be used indefinitely almost free-of-charge. In contrast, if the same data were kept in a digital format, which was at that time the IBM magnetic tapes, and processed on Digital PDP (programmed data processor) computers, it would need constant investments in updates of data formats and storage medium along with an organizational effort do effectuate these updates in time.

One of the drawbacks of keeping data in a digital format is the frequent change of technology for data collection, storing and processing. During the lifetime of an archive, any change of storage technology, the medium on which data are stored or change in used data format, necessitates updating of the archive by adopting the latest data formats or by copying data to a new type of medium. By using standardized native data formats (like text, Extended Mark-up Language XML or TIFF, Figure 10.4) the potential loss of data decreases. Also, it may be assumed that the available storage volume will surely increase in the future, either by updating the existing devices like IBM (2017) is doing with magnetic tapes or by

(a) (b)

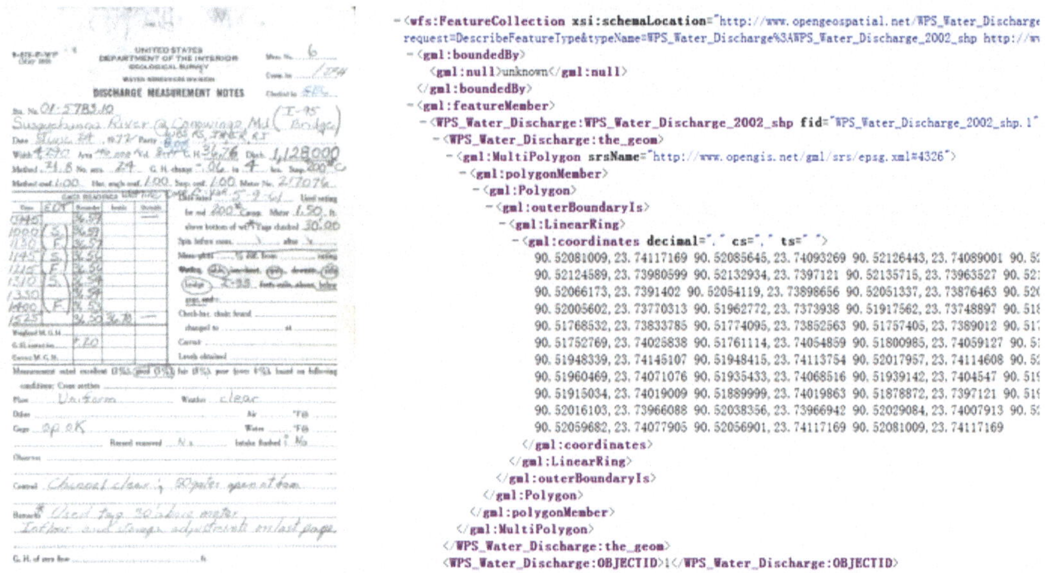

Figure 10.4 Manually filled form (a) and XML based file (b) for river discharge data. *Sources*: (a) from https://md.water.usgs.gov/floods/Agnes/Conowingo/9-275_1.jpg; (b) from https://www.researchgate.net/figure/GML-format-for-water-discharge-in-case-study-area-Demra-of-Bangladesh-in-2002_fig5_260291805.

creating new methods such as optical storage in a single nanocrystal (Riesen *et al.*, 2018). The process of adjusting and updating the archive data format and used storage medium must be planned in advance, especially on the cost bases. This will ensure the usability and sustainability of the data archive in the future.

The latest generation of data storage systems, so-called cloud systems (Dhief *et al.*, 2018), provide significant flexibility in accessing data from anywhere, at any time via the internet. However, this feeds a serious concern about data security and the problem of data corruption, since the data is (physically) kept outside the organization that manages the data. Therefore, it is recommended to achieve a special agreement with cloud data storage providers that guarantees data security, data integrity, availability, redundancy which reduces the risk of permanent loss of data, and authorization which controls the rights to access selected parts of the data by different data users. Since the cost of cloud storage depends mostly on used storage volume, it is necessary to carry out a cost-effective analysis on the level of data redundancy that is needed. To further increase the redundancy, it is possible to mix cloud (for convenient usage) and a proprietary in-house database (for backup). If archiving is carried out in such a decentralized way, apart from synchronization issues, it is important to ensure the security and persistence of all data sources.

10.3 DATA CHARACTERISTICS REGARDING DATA ARCHIVING

UDSM covers various scientific and technical fields, and therefore it comprises a broad spectrum of data types. The data that are usually considered to be used in urban drainage designs, projects, studies and

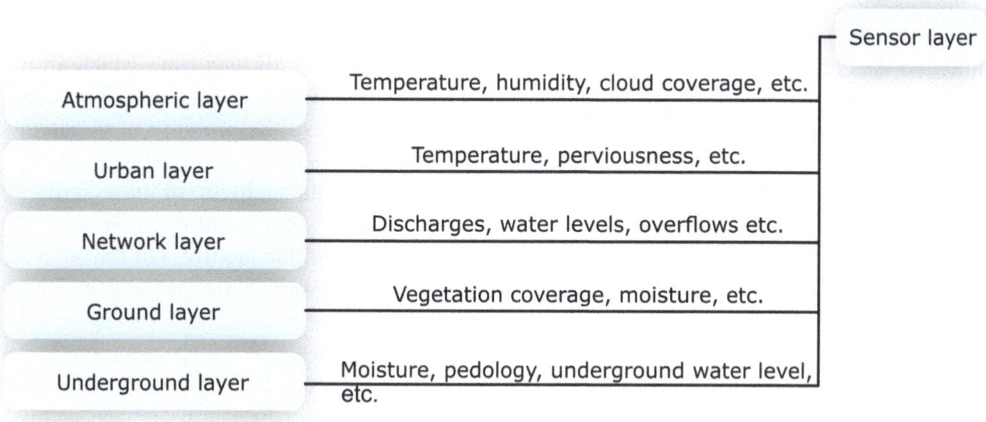

Figure 10.5 Sensor layer comparing to other layers. *Source*: Nemanja Branisavljević (University of Belgrade).

analyses can be classified into several categories, that may be presented as layers. The classification follows the vertical distribution of layers of natural and built environments (Figure 10.5):

- Atmospheric layer – rainfall, wind, etc.
- Urban layer – roofs, buildings, etc.
- Network layer – pipes, manholes, source control measures, objects, etc.
- Ground layer – land use, census data, etc.
- Underground layers – ground water, pedology, etc.

Each layer implies a large number of parameters that are of interest in the scope of UDSM, and therefore might be considered for monitoring and eventually archiving. An additional layer may be introduced in this constellation – the sensor layer (Chapters 2 to 4), with all the sensors and equipment used for measuring the values of parameters of interest over time. The sensor layer may be understood as the layer that comprises all the presented layers, as shown in Figure 10.5. For example, sensors are used for measuring moisture in the ground layer, ground water characteristics in the underground layer, level of urbanization by cameras on satellites or drones in the urban layer, etc.

Acquired data are usually *grouped in datasets*. Datasets may be formed according to the period when the data were created (time series), according to the location of data origination (often, but not always, the location of the sensor) or by the type of data (numerical data, textual data, positional data, images etc.). Data grouped in time series can have a constant time interval between the data values, variable time interval or measurement times can be determined by some event (e.g. rain event – wet or dry period, by time of bucket tip in rain gauge or by constant increment in change of measured quantity). However, during the data archiving, this form of grouping data should be preserved as much as possible as it keeps the information about the time and space relation between the data values in one dataset.

Apart from grouping, the acquired raw data are often pre- and/or post-processed using various tools and methods. Averaging, resampling, and computation of statistical parameters from data are just some of the examples of how data is calculated and adapted for use. In the process of *designing the data archive it is important to adopt a strategy for how calculated data are stored in addition to the original measured data*. In most cases it is not necessary to store the data that can be subsequently calculated. By the rule of thumb, the data archive designer may estimate the minimum amount of data from which all required information, necessary in the future, can be extracted. But it is always *a good practice to store some redundant data too*, like commonly used statistics, which will improve the usability of the archive and reduce data processing and traffic. For example, in complex pumping stations on sewer systems, with several inlets, pumps, pump sumps and outlets (Figure 10.6), data archives of time series with a time interval of one second can become huge. To analyse some simple information, like monthly overflow in recent decades, a huge set of data has to be retrieved from the archive. On the other hand, if we archive also the hourly and daily volumes of pumped/overflowed water, we can easily get the required information.

In urban drainage, data is mostly acquired in harsh, hostile environments. Working conditions are usually far from ideal, and therefore collected data may be compromised (Figure 10.7). Data validation and error assessment (see Chapter 9) of acquired data have to be performed prior to data usage and archiving, as real-time (or on-line) or near-real-time (off-line) processes. The validation process has to be documented and all the steps of validation have to be archived together with the data.

Sometimes, the *additional validation of long-term data stored in archives* is needed as well. By the subsequent analysis of the long-term changes of the environment, it is sometimes possible to spot errors in data which were not detected during real-time validation, *usually in the form of some subtle systematic*

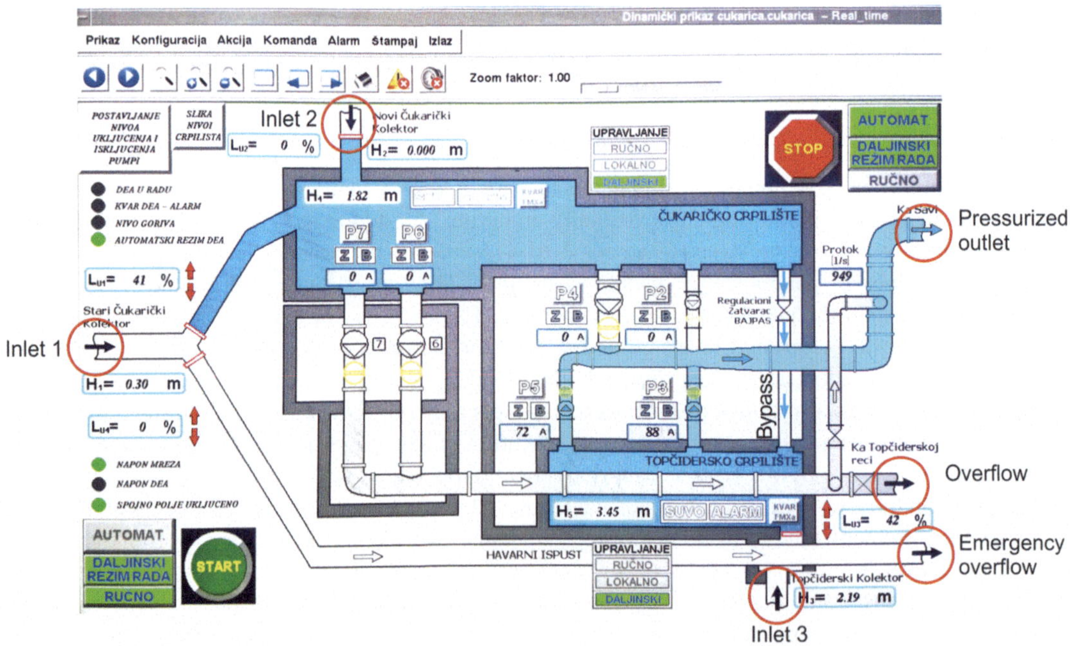

Figure 10.6 Typical SCADA screen of complex sewer pumping station, with three inlets, two pump sumps with bypass, one outlet and two overflow outlets (controlled and emergency). *Source*: screen copy of Belgrade Sewer system's SCADA, made by Dušan Prodanović (University of Belgrade).

error. Therefore, it is necessary to provide the methodology for subsequent data validation of archived datasets, as well as the *possibility to document those subsequent data validation* results in the data archive. This important job of validation of archived data can hardly be done by data providers, since in most situations they will not have sufficient expert knowledge. So, it is up to the data users to perform the subsequent data validation of selected parts of the data archive and store sufficient information about the changes.

10.4 META-DATA CHARACTERISTICS

Generally speaking, sensors could be considered as the interface to the environment where the measurement takes place. In an ever-changing world, the environment where the sensor is located and/or where data is created (it is important to separate those two locations since, for example, the camera as a remote sensor can be installed in a completely different environment to the measured quantity) is constantly changing, so for proper data interpretation, it is essential to be able to reconstruct the overall environmental conditions.

Two examples, where the information about the local environment conditions is crucial to understand the results of a water level measurement, are illustrated in Figure 10.7. In the first example (Figure 10.7(a)), the water level of a small river is measured using a submerged pressure sensor in the upstream reach and a bubbling technique sensor in the downstream reach. The level (pressure) sensor of the downstream bubbler is located in a metal house-container on the riverbank with a bubbling pipe of 15 m length. Five level peaks (indicated in Figure 10.7(a)) were observed in the downstream reach during the first 5 days (red line, downstream). After analysing the rainfall data, only the 4th peak was recognized as rainfall runoff (it can be seen on the upstream level, too). To understand the other four peaks, the data about the internal temperature of the level logger have to be considered: since the metal house-container was closed and exposed to direct sun light, during daylight the temperature rose to over 65 °C, which in turn

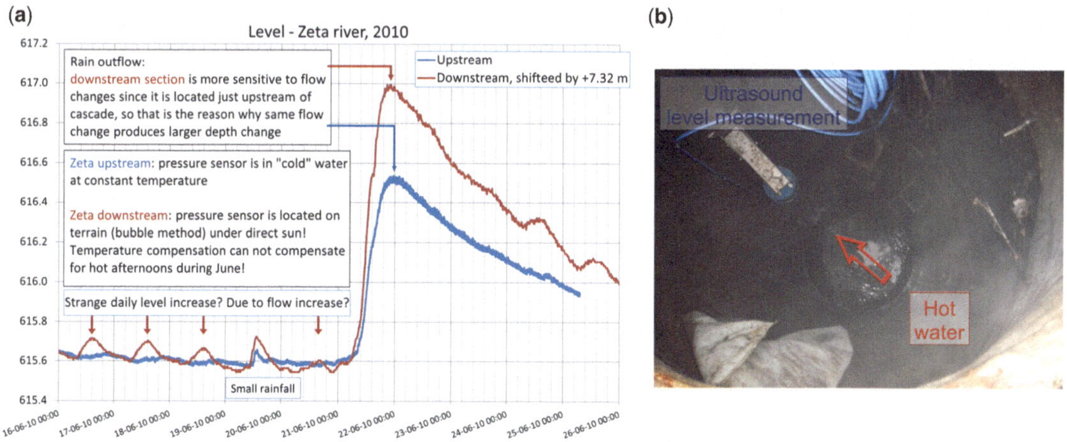

Figure 10.7 Water level measurement using bubbling technique (a) and ultrasound (b). *Sources*: (a) Measurements on HE Perućica, Montenegro; Institute Jaroslav Černi, 2010; (b) Dušan Prodanović's lecture notes: http://hikom.grf.bg.ac.rs/stari-sajt/web_stranice/KatZaHidr/Predmeti/Merenja/Literatura/Skripta/SlikePng/2006-02-03-06-MM8-Visnjica-S.png.

changed the pressure sensor offset, resulting in small peaks (they can be seen also for the last two days) which one can misinterpret as runoffs! By comparing the measured level downstream with measured level upstream (dark blue line), where environmental conditions for the pressure sensor were more stable, the interpretation of the measurement results may be confirmed.

The second example (Figure 10.7(b)) is related to a level measurement in a sewer system using a standard ultrasonic level sensor, on a location where the upstream user may regularly dispose of hot water (car washing service in the given example). The ultrasound sensor has an in-built temperature measurement for air sound velocity compensation. The temperature sensor measures the air temperature near the ultrasound 'head', which is mostly beneath the manhole cover. Flushing hot water through the sewer produces a non-homogenous air temperature and humidity field, which in turn produces a systematic offset in water level measurements. In such situations, one should use the ultrasound device capable of measuring two reflected levels at the same time: one from the water level and the second from the fixed reference which can be used to compensate for variable ultrasound speed.

The reconstruction of the environmental conditions, once the data is archived, becomes even more challenging since the data user is not the same as the data producer. Using meta-data (according to Cambridge English Dictionary – 2019 meta-data is '*information that is given to describe or help you use other information*'), one can reconstruct those environmental conditions. For example, some of the meta-data regarding sensors are:

- Type of sensor.
- Working principle.
- Producer, model, serial number.
- Calibration data and date of calibration, link to calibration history.
- Power level, status of sensor operation, temperature, humidity (if important).
- Sensor position at certain time (for portable equipment), sensor micro-position can be important for certain systems (upstream or downstream of the pipe connection), etc.

In some situations, the data from the same data source can be considered both as data and as meta-data. For example, in the case of satellite imagery of the area that is being analysed, the image of the urbanized area taken during clear sky can be used as an input for estimating the area of the impermeable surfaces, while the image recorded during a storm event can be used to validate the spatial distribution of ground rainfall measurement using recorded cloud coverage.

Meta-data should be an integral part of the archived data collection, and *archived data should be considered only in the context of meta-data*. Meta-data can be stored in various forms and formats, just like the data, and therefore there is an unclear boundary between data and meta-data. We can split meta-data into two groups:

- Meta-data that is actively used in the process of interpretation, validation and analysis of collected data (battery voltage of measuring equipment, measured signal strength, assessed quality of data, parameters used for validation, parameters used for running certain hydraulic model, etc.), and
- Meta-data that is stored to save information about the environment where the data was generated (location of equipment, calibration history, operational history, communication line problems, important events, etc.).

The first set of meta-data is usually recorded in a structured form following current standards (DCC, 2019), recommendations (WMO, 2019) or custom formats, such as basic numeric or textual format, or more complex formats such as GIS formats (ISO, 2014) for storing geospatial data. The other group of data is often kept in an unstructured form, such as images, links or text (e.g. freely written comments).

10.5 DATA ARCHIVES MANAGEMENT

10.5.1 Digital formats

One of the first steps when planning data archive is the choice of storage formats that is going to be used. *The data and meta-data format selection has to be considered during the early stages of data archive design* since it has a direct implication on the system implementation, its robustness and later usage, including data sharing, searching, filtering, sorting, revalidation and data processing.

There are many formats suitable for storing UDSM data. Some formats can be easily read without any transformation, while some need to be transformed before they become understandable by humans. The first group includes alphanumeric ASCII files in different formats:

- Formats for storing various data (text, numbers, attribute-number pairs, arrays, etc.) – CSV, XML, JSON, or free unstructured TXT formats.
- Formats for storing graphic data – uncompressed SVG.
- Formats for storing geospatial data – GML, TAB, KML, etc.
- Formats for storing database files – DBF.

A main drawback of text files is the volume of stored data: to maintain human readability a large number of unneeded characters are used to satisfy the form. A much more compact form of data storage is the binary form, but it cannot be read without data conversion. The reading method has to be known, well documented and accepted like ZIP or RAR compression routines, PDF for text and drawings, or shape SHP and Google KMZ for GIS data (which is simply zipped KML format) where it is assumed that in the far future, we can at least reproduce the routines to convert the data.

Another option is to use specialized, custom and better optimized formats, but these are often poorly documented and can be subject to changes by the vendor from one version to another. In that case it is advisable to also archive the reading software along with the data, otherwise the data may become inaccessible in the future. The specialized formats are usually used by specific software packages, so when the software changes or stops supporting a previous format version (like CorelDraw CDR formats 4 or older) and there is no adequate converter from one format to another, data can be lost indefinitely.

More complex data types, such as geospatial data, can also be stored in either text or binary form or may use a combination of the two (e.g. a shape file format). The binary form is usually applied to record a larger amount of data and, as with text, there are standard formats such as MIF for storing vector data, which is used in the MapInfo GIS software package or the GeoTIF format for raster data or USGS DEM specialized digital elevation model format. The text format is mostly used to store table structures and meta-data.

At present, spreadsheets are commonly used in everyday work. More advanced spreadsheet formats, such as Microsoft Excel XLSX can be used for data storage, along with data authorization and transformation procedures. The XLSX format is in fact a ZIP-ed Office Open XML (Extended Markup Language) format, or Open XML (Microsoft-OpenXML, 2020), a file format for documents intended to be used with all Microsoft Office programs after the 2007 version. The file format should be freely available, and it could be implemented by anyone with recent versions of Windows, which guarantees the long life of the format and the possibility for it to be accessible in the future. Using the software package Microsoft Excel, spreadsheet files can be linked, can use data from other sources, or just invoke data when calculating over real time, allowing significant flexibility in data processing and management.

10.5.2 Databases

Digital data are commonly stored in databases. The most convenient database type for data archive storage is the relational database. A relational database provides a compact and optimized way of storing data, with minimal data transfers, easy data administration and access to data, as well as data manipulation and connection to data from other databases, both local and remote. Some of the main characteristics of relational databases are (Prodanović, 2007b, and Chapter 5):

- Database software allows working with data on both local and remote levels (in clouds, too) using different protocols (SSH, HTTP, etc.).
- Data can be easily stored and retrieved using SQL language.
- By indexing, the data is accessed using fast binary search algorithms instead of sequential search algorithms.
- Database administration is carried out using a specific language or user interface.
- Authorization and authentication are embedded in the database, enabling enhanced security of administration (roles, groups, procedures, etc.).

There are numerous software database solutions on the market, such as Microsoft SQLServer, ORACLE, MySQL, PostgreSQL, MariaDB, etc. All these solutions rely on a subset of Standard Query Language (SQL) and various types of connectors and libraries. Being standard, there is a high probability that in the future the SQL data format will still exist and that there will still be a considerable community using it.

Apart from standard databases, there are a number of specialized database solutions, well suited for different users, with advanced SQL incorporating data processing capabilities. One of the examples is the Historian database management software from GE (GE DIGITAL, 2019), which can be used to collect massive raw data, aggregate data and make data archives. In the field of hydrology, the SODA and WISKI software solutions are used in some informational systems: SODA is a data acquisition program that can poll data from many different types of loggers and is used to insert the raw data into WISKI, the software which compiles, stores, validates and allows the user to analyse the data in various ways, prepare the archives and manage them (WISKI, 2019). Also, open-source database solutions can be found on the market, specially designed for time series (for example, the InfluxDB, 2019 or Prometheus, 2019) which allow full customization for specific needs. However, although the open-source databases are usually free of charge, substantial knowledge and effort for customization, implementation and maintenance are needed.

The ability to access data over the network, usually through internet protocols, offers the potential for data to be stored outside of the data provider's institution (who is in charge for data archive management) and the service for the users to easily access the data from any part of the world. This allows greater flexibility and expands the access to data, but also increases the cost of data security. With the introduction of cloud-based data exchange systems, *the role of a data provider is changed from being simply a manager of data storage and hardware to an advanced manager of distributed parts of an archive, taking care of availability and security issues.*

In both cases, the usage of local dataset storage or of a specialized cloud distributed database system for archive, it is necessary to *perform a thorough analysis of the total data management costs*. This in general implies the cost of database administration, database software pricing, provision of support to users, management of backup copies, periodic update of databases to a more modern format, and periodic validation and update of database. Also, if a database is stored locally, the expected lifetime of the used media and associated (computer) systems has to be considered.

Often, the data that have to be archived are collected during a certain project, which has a limited scope and time span, 3–4 years mostly (like most EU projects these days or the plan cycles with the departments in e.g. a waterboard or a municipality) or even shorter (for most national research projects or incidental monitoring projects for specific purposes). To maintain the research continuity, it is a smart decision to create the project data archive at the end of the project, plan the archive, add all necessary meta-data, have a web interface for data search and preview, and even plan some funds for short-term future archive management. But the true question is how to plan the long-lasting archive, how to transfer the data to the far-future users. Or to make it simpler, how to plan the future budget for such archives. This question has to be foreseen by the research community. And always *consider a paper print-out of the most valuable data and results!*

10.5.3 Archive security, availability and legal use considerations

The security of an archive can be more easily guaranteed if the storage is on local servers housed by the data provider, where only a limited group of users can access it. If a decentralized system is used, where parts of the archive are stored in remote locations, and if these locations are connected to the internet, *security issues are to be thoroughly considered.*

The internet is a potentially hostile environment. Despite the fact that data can be shared and administered in a more convenient way, the use of the internet introduces a significant risk to data security. The fact that the data are located on servers outside the institution that owns the data archive may lead to a situation where the data-keeping company may cease to exist and therefore (parts of) the data archive gets lost, or where, due to the political pressure in the country in which the data are stored, a part of the sensitive data could be compromised.

On the other hand, the use of specialized distributed services reduces the costs of managing databases. By careful planning of the data archive redundancy, it is even possible to reduce the risk of data loss and enhance protection from hacker attacks and viruses. This can be achieved by distributing either mirrors of databases (like the RAID 10 standard used in computer storage, see Wikipedia RAID, 2019) or required recovery information (like the RAID 5 standard). Since data archive management and usage are a multi-stakeholder service, it is necessary to have an agreement in advance with all parties on the availability of data, incorporating the estimated time necessary for the database maintenance (backup creation, etc.) and the available internet resources. The *availability* can be defined as 'the probability that the system is operating properly when it is requested for use' (Weibull, 2020) and is related to a total system operation. On the other hand, the *reliability* is the probability which does not account for any repair actions needed to return the archive into working condition.

Regarding legal issues, archive data management also has to follow current regulations. The recently published General Data Protection Regulation (GDPR) regulation of the European Union (OJEU, 2016) implies a much more attentive attitude towards data than previously required (Wikipedia GDPR, 2019). The GDPR regulation focuses on the protection of personal data from the company that collects it. In the case of data related to UDSM, there is a set of personal data that has to be considered:

- Data on sewage users and quantities of water used.
- Data on user habits and features.
- Data on user commuting during the day or year.
- Data on the location of the data collecting operators, etc.

These data must be closely monitored by the data protection officer and all the instructions from the GDPR regulations must be followed. It is interesting to note that *most UDSM data are under the GDPR for fine resolution analysis* (at the level of a house, for example) and not for coarser resolutions (level of a catchment or larger).

10.5.4 Archive lifetime considerations

In the data archive design process, in order to efficiently manage it during the whole lifetime, it is important to consider the following issues:

- Overall archive lifetime – some parts of the archive may be considered as 'more important' and should be saved 'forever' and some may be considered as 'lesser important' and could have a certain lifetime.
- The expected lifetime of used (mostly custom) archived data and meta-data formats and the need for update/refresh of the data due to technology changes.
- The expected need for data re-validation in some future time, after gaining new knowledge about monitored processes.
- Regular update of storage media and used database software.

The process of archiving data, collected in a UDSM project or research, is certainly not just a permanent storage of all files, records, photos and spreadsheets. It is always *recommended to prepare a subset of data, keeping some averaged and/or cumulative data for certain periods of time or per certain events and archive it separately*. For example, for continuous monitoring of sewer overflow, this could be the statistical data of one overflow event, like start/end of overflow, total volume of water, total mass of monitored water quality parameters, maximal flow, maximal level, maximal concentrations and time of maximal values. Or for continuously operating systems (like a sewer pumping station, Figure 10.6) the statistical data can be calculated per one hour (or per one day, if hourly variations are not significant) with hourly average (or cumulative), minimal and maximal value for flow rate and similar continuous quantities (level, pressure), and total operating time, number of pump starts, maximal flow rate, maximal power current, average power current during the on-period, and average and maximal temperature of pump's motor. Archiving this subset of data will dramatically improve its efficient usage.

One of the main issues to be resolved in archive design is *whether to permanently store all raw measured data or just a processed time series specially prepared for archiving* (filtered and resampled, normalized and verified). There are pros and cons in both approaches. If only raw data are stored, we will have a larger volume of archived data with high potential to re-evaluate them in the future when new techniques and knowledge emerge, but also with the possibility to wrongly extract and interpret the information if we are not experts (for example, if only the river stage as raw data is stored, but we cannot evaluate the changes in the zero level of a used gauge, or to select the right rating curve for flow calculation when the cross section is unstable). And vice versa, if we store only processed and resampled data, the overall volume of data can be much smaller, but we cannot easily re-validate or re-assess data if needed. Of course, there is a third option, to store both raw data with lots of meta-data and processed data, with the higher expense of archiving.

When considering what data are to be archived and how much redundancy to allow, it is important to think about possible future users, to allow them to easily search within the archive and extract data they need. In most cases this will lead to preparation and storage of separate (pre)processed datasets, each dataset having a detailed header – explanation, link to documentation, accompanying figures and photographs, handwritten notes and similar. For example, the archive of a runoff monitoring campaign should have, apart from the raw outflow, stage, quality parameters, meteorological data, etc., also the smaller dataset with total rain, outflow volumes and masses measured per rain event. One of the criteria when preparing this kind of dataset is to allow the future user to inspect it and then to decide whether the whole (large, full) dataset is holding data/information of value or not, prior to purchasing/extracting.

The file format of archived data is usually related to the hardware that is used for the measurement process. For example, raw flow velocity and water level data from a Doppler ultrasound sensor, together with meta-data from built-in sensor temperature and Doppler signal strength are often saved into the database with a predefined structure and possibly with different sampling rates. The applied database structure is in most cases predefined by the sensor manufacturer or data providers while reading of acquired data is only possible using custom made software. For day-to-day use, this approach is acceptable as it proves to offer a reliable sensor operation. But for long-lasting data archiving purposes, it is necessary either to enable a change to the database structure when the reading software is changed, or to export/convert the data into some standardized data format, either textual or binary.

Archived data may contain undetected errors. After validating the data, it is assumed that most of the data errors have been eliminated, but it must be foreseen that certain, *most common systematic errors may remain in the data* (for example, that the rating curve was wrong due to a backwater effect which was not considered at the time of data preparation for archiving). Therefore, it is necessary to provide the tools and procedures for subsequent data validation of archived data. It is recommended to establish procedures for when the user of the archive detects an error in data. When the data user reports the error to the data provider, who should analyse the problem but in most cases has not enough knowledge to do so, it should be decided whether the data needs to be corrected, only commented with findings or simply deleted.

During the archive lifetime, the technology used for data storage and management will evolve. Therefore, when archiving data, it is necessary to *consider the potential advances in technology and anticipate the costs of adapting to the new technology* (the format and media for data storage). For example, over the past 30 years, several digital storage media have been standard during certain periods of time (Figure 10.2). Magnetic tapes were superseded by floppy disks, CDs, DVDs, flash memory. Followed by the reintroduction of magnetic tapes with much higher storage volumes. The technology advance will also drive the changes in database software and, potentially, in database structure. Both changes (or regular updates) of storage media and used database software require considerable effort and financial resources for proprietary, in-house archives, which has to be taken into consideration. It might seem that moving the archive to cloud storage can resolve those two issues since cloud providers will always keep the system up-to-date, but care has to be taken regarding the agreement with the provider to cover those options, and potential loss of data during updates.

10.5.5 Archive backup considerations

Regular backup and backup procedures are part of each database system. The backup strategy considers full backups after selected time intervals, regular incremental backups, management of replicated copies of the database and management of distributed databases, if used. The database backup strategy minimizes the risk of data loss due to failure or database corruption.

In the course of this chapter, the data archive is not considered as a huge working database which is continuously loaded with new raw data. The data archive is considered as a selected subset of raw data, together with processed, validated, resampled data, added descriptions and other meta-data, linked with simulation results if they exist, and stored for later use. According to this definition, the archive is mostly used in 'read-only' mode, so instead of regular backup, exact database copies should be created. And for safety reasons, a list of all copies has to be maintained.

As mentioned in Section 10.5.4, there is a chance that in the future, some systematic error would be detected in an existing archive. In such a situation, the archiving strategy has to define if data in the archive will be replaced by the correct data, new data will be added or only a description of the detected error as new meta-data will be added. In any case, the backup strategy needs to take care of *synchronization between the main database and all copies*.

In some cases, a part of the data archive may be used as a source of valuable information of high importance. For example, in the case of cadastral data, where the technical data, legal data and data about ownership is stored, the data is frequently used for legal issues. In those cases, for safety reasons, only parts of archive can also be saved as backup copies. So, the backup strategy for distributed archives can be rather complex.

10.5.6 Archive destroy considerations

In general, to archive something means to keep it for some future use, for a yet unknown user and basically forever. It is however realistic to expect that someone will argue that certain parts of archived data could become, at some point in time, obsolete or at least have too high a resolution and that substantial expenses could be saved if, for example, full time series are reduced by a resampled one. This may even be followed by simple analysis which will try to compare the benefits that data bring and the cost of further data retention/reduction.

The main drawback of such 'economic' analysis of archive relevance is that *future benefits of archived data are impossible to predict*. We have to consider such questions and arguments during archive design and try to keep it as compact as possible, to avoid keeping 'all possible data' and to extract 'enough' data and meta-data for all foreseen users. This approach will lead to a sustainable archive, which will not need such a 'test of time'.

Of course, given the importance of historical data in some disciplines such as hydro meteorology, the value of stored data will never fade and it is unlikely this value will even be subjected to debate. These types of archives are unlikely to become obsolete. Therefore, special attention should be paid to archiving both in terms of media, format and the amount of data being stored.

10.6 DATA ARCHIVING RECOMMENDATIONS

In this chapter, we covered a broad range of issues related to data archiving, which should eventually help someone to design, plan and implement a successful, long lasting and optimal data archive. In the following paragraphs, we provide several questions, mostly for data producers and data providers as a guide for developing and implementing a successful data archiving strategy for UDSM data.

10.6.1 General questions

It is suggested to answer the following questions prior to planning the data archiving.

10.6.1.1 Are the collected data in digital or analogue format?

Nowadays, it is not common that the data is recorded using analogue measuring devices, but in some cases, especially when the data is collected in the field, it is more convenient to write the data, make a sketch or create a note on a piece of paper than on a digital device. On the other hand, sometimes it is necessary to use some historical data that are still in the form of written data or diagrams. In those cases, digitization should be considered before the data archive is planned.

10.6.1.2 Who is the data for?

Data may be intended to be used by professionals or by the public that is not so well-informed about UDSM issues. Generally, professionals would be more skilled in data management and analysis, and therefore some less advanced data formats, like text CSV files, would be appreciated as much as some advanced formats that may automatically be used by sophisticated data analysis software, like GIS, database or spreadsheet files. On the other hand, data for public use have to be properly presented and prepared for those individuals that are not so familiar with advanced data formats or software solutions. For public use, diagrams, maps or any other visual formats would probably be the best manner to inform the audience.

10.6.1.3 What is the volume of archived data?

The volume of data that have to be archived may vary from a few numbers to gigabytes of time series. Therefore, a data management system may vary from a collection of text files to large distributed relational databases with suitable database administration systems. This issue is usually related to financial resources. To reduce the volume (and hence the cost) of data, it is crucial to properly anticipate future data requirements and to archive only source data without any derived, calculated or simulated data. Also, data redundancy should be carefully planned and managed.

It is good practice to keep an archive per a certain, selected period of time (per year, for example, or per event, or per project) creating a *set of archives* during longer time periods.

10.6.1.4 What kind of changes and adjustments to the data archive are expected in the future?

A data archive may undergo significant changes in the future, especially considering vertical and horizontal scaling. If the archiving system is going to scale vertically (number of data archives in one set is continuously increasing) usually new data storage units have to be installed. If the data archives are on the cloud system, the vertical scaling may be implemented by adopting a new payment plan, but if the data is kept in-house, new storage hardware usually have to be acquired, installed and maintained.

Horizontal scalability usually refers to introduction of new tables/files or new fields with additional data, mostly due to subsequent detection of errors in data. If the data archive has to be horizontally scalable, the data storing technology should be properly selected. NoSQL databases like MongoDB or CouchDB provide easy horizontal scalability at the expense of reduced querying capabilities in the case of relationally linked data tables. This could also relate to incorporation of the data into an asset management system of e.g. a municipality in which data on roads, water supply, etc. are stored. Such linages can be useful when evaluating the effectiveness of e.g. a maintenance policy over sustained periods of time (see Sections 7.4, 7.6 and 9.3.4).

10.6.1.5 What is requested data availability?

Data providers must maintain the selected (or contracted) level of data availability. The availability and reliability are not the same features: the reliability is the probability that an archive will operate without failure, but it will not account for the 'downtime' period when the archive is not operational. On the other hand, the availability is the probability that the archive is operational. The availability can be classified according to Weibull (2020) as point, mean, steady state or operational which includes all experienced sources of downtime (administrative downtime, logistic downtime, etc.).

10.6.1.6 How often will the data archive be used?

Archived data may be intended for personal use only, for a small number of users (company) or for public use. It is assumed that the volume of data requests for personal use will be relatively small compared with the volume of data requests for public use. Therefore, for personal use, data may be stored in some less accessible format, like text files with relatively unstructured data. For a small number of users, data may be stored in some shared folders in the company server, while for public use some professional data management system with web interfaces should be considered.

10.6.1.7 Are data in the archive valid?

Data validation is one of the most important steps in the data collection procedure (see Chapter 9). The aim of data validation is to detect errors in data and to eventually correct them or discard the data. Therefore, it is necessary to document all the data validation tools and procedures that were used and to provide, if necessary, raw original data along with the data with improved quality. Nevertheless, some errors may still remain in the data even when the data is archived. Therefore, it is recommended to enable postponed validation in the data archive itself.

10.6.1.8 How to anticipate the possible future users and their needs?

One of the most important questions regarding making a data archive for future use is how to anticipate the future users and their needs. One can always start with present needs and possibilities for data use, considering that future user will not have prior knowledge about the technical system and environmental conditions, so adding a detailed description, reports, site-specific information (e.g. on land use or the presence of vegetation, industrial or construction activities) will help. Keep in mind that more meta-data is better than less meta-data.

10.6.1.9 Archive vs. repository

In this chapter the data archive is seen in a broader way and is considered to be the same as a data repository. Both archives and repositories save the data and information in an organized way, for future data users, having data providers to take care of it. However, a stricter definition is given by USGS (2019) where archiving is '*a process that supports long-term storage of scientific data and methods used to read or interpret it*', and a data repository is a '*centralized place to store and maintain data;* a repository can consist of one or more databases or files which can be distributed over a network; data repositories are often managed by data curation personnel who ensure that files are managed and preserved for the long-term.'

10.6.2 Meta-data choice

10.6.2.1 Are there any recommendations for sensor location meta-data?

The location where the data are acquired is one of the most important meta-data. Therefore, one should record details about the measurement location, especially if there are some anomalies from the regular state. The level of deposit in the measurement location, strange objects that may influence the measurement outputs, poor location access that prevents regular equipment maintenance, or weak GPRS communication signal are some of the examples that should prevail in meta-data.

10.6.2.2 Are there any special hydro-meteorological conditions?

Hydro-meteorological parameters are among the most important ones in practising UDSM. Also, hydro-meteorological data carry important information about the local environmental conditions where the measuring equipment is installed. Sub-zero or very high temperatures, high humidity or strong winds are just some of the examples of meta-data that should be recorded in order to preserve the impression of the circumstances when the measurement took place.

10.6.2.3 Are there any special issues regarding the measurement devices?

Measurement devices play a crucial role in data collection and therefore some information about them should be preserved in the data archive for later data interpretation. Battery level, calibration curves, sensor positioning issues, signal strength, etc. are some of the meta-data that should be recorded along with the data values.

10.6.2.4 Did any special event trigger the data capture?

Sometimes the data is collected due to some event, like flood, overflow or the start of a rainfall. In those cases, it is necessary to provide a broader insight into circumstances that are used to trigger the data collection. As also mentioned in Chapter 6, this policy (for events like e.g. exceeding a certain water level triggering registration) was utilized in the past to save storage capacity. As this is no longer an issue, it is not recommended as this manner of data acquisition is very prone to failures.

10.6.2.5 Are there any deviations from the standard measurement procedures and good measurement practice, what are they and why did the deviation happen?

Sometimes, the standard data collection procedure could not be applied. For example, a velocity-based flowmeter sensor needs long straight upstream/downstream sections, a condition which is hard to fulfil in most cases. Therefore, it is necessary to deviate from standard procedures and provide the data by the best measurement practice. Procedures applied should be extensively described in the data archive for future data interpretation. For example, if a flowmeter has to be installed in an irregular position (see Chapter 3) we have to add a sketch with explanation, add a comment about streamlines, add some photographs and assume the possible overall error.

10.6.3 Data security and legal issues

10.6.3.1 Are there any captured personal data?

If the data possess some personal characteristics (names, phone numbers, water consumption, etc.) and we intend to use them at the level of a house/user, they should be treated according to GDPR regulations. On the other hand, averaged consumption of 100 houses for example, with inhabitants grouped in several social categories and depersonalized is, in most countries, free to use. Due to the ongoing development of a broad spectrum of water quality sensors that can be employed in UDSM (for forensic purposes, see e.g. Bannwarth *et al.*, 2019), the question of privacy can potentially become an issue.

10.6.3.2 Are there any security issues regarding the data?

Data about crucial urban infrastructure, financial data or personal data of the utility's clients represent sensitive data that should be stored in-house (utility, company, public institutions, state institutions) with a controlled sharing policy. In the case of dealing with sensitive information for data archiving it is recommended to use some professional data archiving software installed on the in-house servers and avoid using any remote or cloud solutions. Special attention should be paid to security issues.

REFERENCES

Bannwarth A., Morelato M., Benaglia L., Been F., Esseiva P., Delemont O. & Roux C. (2019). The use of wastewater analysis in forensic intelligence: drug consumption comparison between Sydney and different European cities. *Forensic Sciences Research*, **4**(2), 141–151. doi: 10.1080/20961790.2018.1500082.

Cambridge English Dictionary (2019). https://dictionary.cambridge.org/dictionary/english/meta-data (accessed 23 January 2019).

CEN (2018). *CEN TS 17171:2018 Management of Observed Hydrometric Data – Guidance*. CEN (Comité Européen de Normalisation – European Committee for Standardization), Brussels (Belgium), 38 p.

DCC (2019). *Digital Curation Centre: List of Meta-data Standards*. http://www.dcc.ac.uk/resources/meta-data-standards/list (accessed 9 April 2019).

Delleur J. W. (2003). The evolution of urban hydrology: past, present, and future. *Journal of Hydraulic Engineering*, **129**(8), 563–573. doi: 10.1061/(ASCE)0733-9429(2003)129:8(563).

Dhief Y. B., Djemaiel Y., Rekhis S. & Boudriga N. (2018). Cloud-Based Global Monitoring System for Smart Cities. *Proceedings of the 32nd International Conference on Advanced Information Networking and Applications Workshops (WAINA)*, 16–18 May, Krakow, Poland, pp. 307–312. doi: 10.1109/WAINA.2018.00103.

GE DIGITAL (2019). https://www.ge.com/digital/applications/hmi-scada/historian (accessed 23 January 2019).

Han L., Xu Y., Lei C., Yang L., Deng X., Hu C. & Xu G. (2016). Degrading river network due to urbanization in Yangtze River Delta. *Journal of Geographical Sciences*, **26**, 694–706. doi: 10.1007/s11442-016-1293-0.

Harremoës P. (2002). Integrated urban drainage, status and perspectives. *Water Science and Technology*, **45**(3), 1–10. doi: 10.2166/wst.2002.0041.

IBM (2017). *IBM Systems Technical White Paper: Defining the future of tape*. Armonk (USA): IBM, TSW03548-USEN-00, 8 p. Available at https://www.ibm.com/downloads/cas/Z3RYV0BR (accessed 21 December 2020).

InfluxDB (2019). https://www.influxdata.com/ (accessed 9 April 2019).

ISO (2014). *ISO 19115-1:2014 Geographic Information – Metadata*. Geneva (Switzerland): ISO, 4 parts.

Maksimović Č. & Radojković M. (1986). *Urban Drainage Catchments: Selected Worldwide Rainfall-Runoff Data from Experimental Catchments*. Pergamon Press, Oxford (UK), 374 p. ISBN 0-08-034086-5.

Microsoft-OpenXML (2020). https://docs.microsoft.com/en-us/office/open-xml/open-xml-sdk (accessed 2 July 2020).

OJEU (2016). Regulation (EU) 2016/679 of the European Parliament and of the Council of 27 April 2016 on the protection of natural persons with regard to the processing of personal data and on the free movement of such data, and repealing Directive 95/46/EC (General Data Protection Regulation). *Official Journal of the European Union*, 04 May, L119/1-L119/88. Available at https://eur-lex.europa.eu/eli/reg/2016/679/oj (accessed 25 January 2019).

Prodanović D. (2007a). Use of data to create information and knowledge. In *Data Requirements for Integrated Urban Water Management*, T. Fletcher & A. Deletić (eds), Taylor & Francis and Unesco, Leiden (The Netherlands), pp. 139–158. ISBN 978-0-415-45344-8.

Prodanović D. (2007b). Data handling and storage. In *Data Requirements for Integrated Urban Water Management*, T. Fletcher & A. Deletić (eds), Taylor & Francis and Unesco, Leiden (The Netherlands), pp. 127–138. ISBN 978-0-415-45344-8.

Prometheus (2019). https://prometheus.io/ (accessed 9 April 2019).

Riesen N., Pan X., Badek K., Ruan Y., Monro T. M., Zhao J., Ebendorff-Heidepriem H. & Riesen H. (2018). Towards rewritable multilevel optical data storage in single nanocrystals. *Optics Express*, **26**(9), 12266–12276. doi: 10.1364/OE.26.012266.

Schlanger Z. (2017). *Forensic Meteorology: Why Do All Our Climate Data Start in 1880?* https://qz.com/1055629/why-does-all-our-climate-data-start-in-1880 (accessed 25 January 2019).

Smith J., Howe C. & Henderson J. (eds) (2009). *Climate Change and Water – International Perspectives on Mitigation and Adaptation*. IWA Publishing and American Water Works Association, London (UK), 309 p. ISBN 9781583217306.

USGS (2019). United States Geological Survey. https://www.usgs.gov/products/data-and-tools/data-management/archive-vs-repository-there-a-difference?qt-science_support_page_related_con=0#qt-science_support_page_related_con (accessed 9 April 2019).

Weibull (2020). *Relationship Between Availability and Reliability*. https://www.weibull.com/hotwire/issue26/relbasics26.htm (accessed 2 July 2020).

Wikipedia GDPR (2019). *General Data Protection Regulation*. https://en.wikipedia.org/wiki/General_Data_Protection_Regulation (accessed 25 January 2019).

Wikipedia RAID (2019). *Standard RAID Levels*. https://en.wikipedia.org/wiki/Standard_RAID_levels (accessed 25 January 2019).

WISKI (2019). *Kisters North America*. https://www.kisters.net/NA/products/wiski/ (accessed 24 January 2019).

WMO (2019). *Discovery Meta-data*. http://www.wmo.int/pages/prog/www/WIS/meta-data_en.html (accessed 25 January 2019).

Chapter 11

Data collection in urban drainage and stormwater management systems – case studies

Alma Schellart[1], Frank Blumensaat[2,3], Francois Clemens-Meyer[4,5,6], Job van der Werf[4], Wan Hanna Melina Wan Mohtar[7], Salwa Ramly[8], Nur Muhammad[7], Jérémie Bonneau[9,10], Tim D. Fletcher[11], Justin F. Costelloe[11], Robert James[11], Matthew Burns[11], Peter Poelsma[11], Susana Ochoa-Rodriguez[12], Daniel Bourne[12], Zoe Hancock[12], Gilles Wallwork[12], James Hale[12], Nadia Nikolova-Peters[13], Stefan Kroll[14], Johan van Assel[14], David McCarthy[15], Baiqian Shi[15], Simon Bloem[3] and Christian Ebi[3]

[1]University of Sheffield, Civil and Structural Engineering, Sheffield, UK
[2]ETH Zurich, Institute of Environmental Engineering, Zurich, Switzerland
[3]Eawag, Swiss Federal Institute of Aquatic Science and Technology, Dübendorf, Switzerland
[4]Faculty of Civil Engineering and Geosciences, Water Management Department, Delft University of Technology, Delft, The Netherlands
[5]Faculty of Engineering, Dept. Civil & Environmental Engineering, Norwegian University of Science & Technology, Trondheim, Norway
[6]Unit Hydraulic Engineering, Dept. Experimental Facility Support, Deltares, Delft, The Netherlands
[7]Department of Civil Engineering, Faculty of Engineering and Built Environment, Universiti Kebangsaan Malaysia, Selangor, Malaysia
[8]SMART Control Center, Kuala Lumpur, Malaysia
[9]University of Lyon, INSA Lyon, Laboratory DEEP, Villeurbanne, France
[10]INRAE, UR HHLy, Villeurbanne, France
[11]University of Melbourne, School of Ecosystem and Forest Sciences, Burnley, Australia
[12]RPS Group, Abingdon, UK
[13]Anglian Water Services, Huntingdon, UK
[14]Aquafin NV, Aartselaar, Belgium
[15]EPHM Lab, Department of Civil Engineering, Monash University, Melbourne, Australia

doi: 10.2166/9781789060119_0415

ABSTRACT

Data collection in urban drainage systems comes with many challenges. However, many examples already exist, containing numerous useful lessons learned. This chapter therefore contains several urban drainage and stormwater management metrology case studies, selected to cover a wide range of scopes, scales, objectives, climates, data validation methods, and data storage approaches. The case studies are initiated by academics as well as by institutions from the water industry.

Keywords: Costs, data collection, organization, lessons learned, project planning.

11.1 INTRODUCTION

Challenges in urban drainage monitoring are numerous. Many types of monitoring devices are deployed in urban drainage and stormwater management (UDSM) systems, and there are different ways of designing, maintaining and operating monitoring networks. People in different types of organizations may pursue various approaches to validate, store and manage their data collected in the field. This chapter describes a handful of diverse case studies to serve as illustrations of common challenges as well as lessons learned when implementing urban drainage metrology projects. These case studies were selected to cover a wider range of different types of metrology projects: from different countries, climates, monitoring objectives – from research-driven to full-scale applications in practice – organizational arrangements and local guidelines. Each monitoring initiative seems unique – all come with their own challenges and opportunities.

The following case studies are described and their links to the most relevant chapters are detailed in Table 11.1.

- **Real-time control for improvement of receiving water quality, Eindhoven, The Netherlands.** This is an example of a long-term urban drainage monitoring initiative, started in 2006 and still ongoing in 2020, including rain gauges, flow gauges and quality monitoring in sewer networks and an urban river. An example of a collaboration between three parties for the installation and management of sensors (a waterboard, a municipality and a consultant), where the technical questions of the consultant (can I do RTC – real time control?) also led to successful research within universities.

Table 11.1 Case studies and main related chapters.

Case study/chapter	1	2	3	4	5	6	7	8	9	10
Eindhoven		X	X		X	X			X	X
SMART		X				X	X	X	X	
The Basin			X				X		X	
Anglian Water		X	X		X	X	X		X	
Impakt			X							
Nextgen		X	X	X	X	X	X			
UWO		X	X		X	X	X		X	X

- **Let's SMARTly combat flood in Kuala Lumpur, Malaysia** describes the operation of the SMART Tunnel, which has a dual function of accommodating road traffic as well as flood diversion. This illustrates real-time decision-making based on rainfall, urban river depth and CCTV (closed circuit television) monitoring. The tunnel has been operational since 2007.
- **Wicks Reserve bioretention basin, The Basin, Victoria, Australia** describes monitoring of shallow groundwater and flows into a bioretention basin. This illustrates tipping bucket rain gauge monitoring, challenges that come with monitoring shallow flows in stormwater facilities and fully capturing hydrograph tails.
- **Flow monitoring campaign for company-wide integrated urban drainage model upgrade, Anglian Water Services, United Kingdom.** This illustrates the systemic application of industry standard flow monitoring guidelines and model verification guidelines in the United Kingdom. It covers the approach of a UK private water company for the upgrade and maintenance of sewer network models, with 3500 flow monitors and 800 rain gauges installed by subcontractors for shorter periods over two years in different catchments. It includes rainfall monitoring using rain gauges and radar, as well as sewer system flow monitoring. It describes its own machine learning tool for quality checking of flow data and tips for placing flow monitors and health and safety considerations.
- **'Impakt' – Optimization of the urban drainage systems in the Dommel and Warmbeek river subbasins, from a river quality point of view, Flanders, Belgium**. This case study describes rainfall and flow monitoring in sewer systems to obtain a sound understanding of system hydraulics and eventually inform integrated water quality models for scenario testing. A mixture of short- and long-term flow surveys, managed by sub-contractors and an in-house team, with long-term campaigns to capture seasonal patterns and identify dry weather flow variations. A clear meta-data and data legacy strategy is described, with helpful tips for sensor locations and lessons learned.
- **'NextGen' Urban Water Monitoring – A highly distributed field monitoring of an urban drainage network with affordable sensors and real-time data communication, Australia.** This is an example of low-cost telemetered water depth sensors, EC (electrical conductivity) sensors, and rainfall monitors using Arduino technology. The sensors are aimed at being easy to install, and low cost, and thereby achieving very high-resolution data. The raw data are publicly accessible via a website.
- **The UWO – A field laboratory for distributed real-time monitoring with low-power sensor and data communication technology, Fehraltorf, Switzerland.** This case study describes a long-term initiative to monitor the urban water balance using low-power sensors and low-power wide area networks (LPWAN) technology. It combines traditional monitoring techniques with Internet of Things (IoT)-driven approaches. It provides lessons learnt from a practical point of view regarding data communication issues, large-scale sensor management, design of networks and data validation. Validated data is publicly accessible through a data dashboard.

In the remainder of this chapter occasionally the costs of materials and/or installations are mentioned in a variety of currencies, these are converted to amounts in Euro. However, as it is not exactly known which price level was used in the original amounts, the conversion to Euro is based on the average exchange rate over 2019, therefore the figures mentioned are to be regarded as an indication only and are *certainly not meant* to be used as a basis for budget estimations.

Key messages from urban drainage metrology case studies

- KM 11.1: *Challenge* – Keeping urban drainage and stormwater management metrology projects going long term is very challenging.
- KM 11.2: *Harmonization* – Existing urban drainage metrology projects tend to have bespoke data quality management and data/meta-data storage systems. The quest to achieve a more harmonized approach to data quality management and data storage remains open, with questions as to what level of harmonization would be desirable, and what such an approach should look like.
- KM 11.3: *Maintenance efforts* – The person-costs for management and maintenance of both sensors and databases should not be underestimated. These usually exceed the costs of the actual sensors by far and are ongoing throughout the project. Successful metrology projects foresee sufficient budget for operation and an efficient maintenance strategy.
- KM 11.4: *Guidelines* – A range of practical urban flow monitoring guidelines exists. Some countries have nationwide recognized practitioner guidelines, whereas in other countries this information is held within individual drainage authorities or companies. Practitioners will have to adhere to such guidelines, whereas academic researchers would need to know *what* these guidelines are *when* setting out on a collaborative metrology project. Researchers would need to first work within these guidelines before they are able to compare existing and new methods, and gather evidence, that other methods may work better. Without going through this step, new methods are very unlikely to be widely adopted.
- KM 11.5: *Water utility planning cycles* – Researchers should bear in mind water utilities planning and management cycles. This can help to find the optimal time to set up new metrology projects, as well as identify and prepare for continuity issues in longer-term projects.
- KM 11.6: *Low-power and low-cost* – There are emerging opportunities for distributed extensive urban drainage and stormwater management metrology deploying lower cost sensors. This is attractive for researchers, utilities, and authorities to gain a bigger picture by wider monitoring of infrastructure and capturing a more complete range of events. Data quality checking and data management become even more crucial when working with such a technology at larger scales, whereas the design of monitoring networks includes redundancy of sensors and assumes relatively short lifespans of sensors.

11.2 REAL-TIME CONTROL FOR IMPROVEMENT OF RECEIVING WATER QUALITY, EINDHOVEN, THE NETHERLANDS

11.2.1 Scope and objectives

11.2.1.1 Scope

The Dutch Waterboard 'Waterschap de Dommel' (WDD) operates a large-scale monitoring network with 83 sensors in the sewage transport network of Eindhoven, The Netherlands and in 181 adjacent, cooperating municipalities (as of June 2019). The monitoring initiative was commissioned in 2006 and is still ongoing. The initiating moment has been the European Water Framework Directive (WFD) (OJEU, 2000) targets set for the River Dommel, and the need to identify the most cost-effective measures to reach these targets (Langeveld *et al.*, 2013).

11.2.1.2 Objectives

The main objective of this campaign was to collect data to calibrate a set of detailed sub-models of a drainage system, WWTP (wastewater treatment plant), and river. Subsequently, sub-models were in parts simplified and combined into an integrated model, in order to study potential impacts of sewer real time control (RTC) strategies on receiving water quality. The data are also utilized to steer the rule-based RTC system, whereby the control stations and pumping stations manage the flow to a predefined amount, given the height measured before the station. More information about this case study and its RTC system is given by Schilperoort (2011), Langeveld *et al.* (2013) and van Daal-Rombouts *et al.* (2017a).

Initial findings concluded that the receiving water quality was expected to improve significantly through receiving water impact-based RTC. However, the application of this RTC scheme did not lead to full compliance with WFD requirements. van Daal-Rombouts *et al.* (2017b) described the problems that arise when trying to demonstrate how well an RTC system performs, based only on relatively short duration datasets. Hence, this project is still ongoing, with further improvements to the RTC strategy, as well as its robustness, currently being tested. Moreno-Rodenas *et al.* (2017) and Camacho-Suarez *et al.* (2019) describe research on checking uncertainty in the models used and the impact of spatial and temporal variability of input data and models used to describe the system.

11.2.2 Measured variables and location of monitors

The monitoring locations coincide with existing pumping stations and flow control stations, the latter also known as vital sewerage infrastructure (VSI). VSIs are regulated based on upstream head, with level sensors installed upstream, and the flow downstream is monitored. Rain gauges were initially installed very close to VSIs. Additional level sensors have been placed at prominent CSOs (combined sewer overflows) in the system. An overview of the number of sensors and their installation dates is given in Figure 11.1 and Table 11.2.

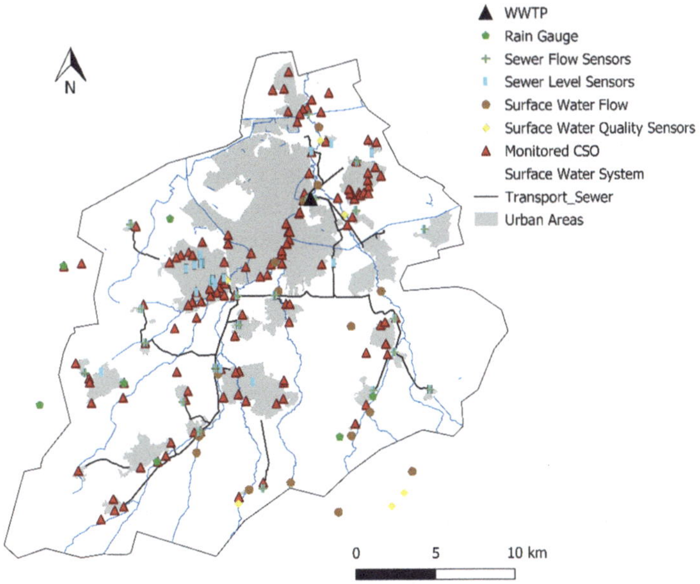

Figure 11.1 Monitoring network for Dommel River and Eindhoven urban drainage system. *Source*: Job van der Werf (TU Delft).

Table 11.2 Monitoring network in Dommel River and Eindhoven urban drainage system.

Type of measurement	Availability (from–to)	Monitoring frequency	Remarks
Precipitation	1951–now	1/h	Rainfall measurement of the Royal Netherlands Meteorological Institute
	2006–2009	1/5 min	25 rain gauges of Waterboard de Dommel
	2010–now	1/5 min	8 rain gauges of Waterboard de Dommel and Municipality Eindhoven combined with rain radar
Water level	2006–now	1/min	Water level sensors in all pumping stations and control structures (Figure 11.1) Water level sensors at 26 CSOs of Municipality Eindhoven
	2010–now	1/min	Water level sensors at all 200 CSOs
Flow	2006–now	1/min	Flow monitoring at all pumping stations, control structures and Dommel River
	2006–2009	1/min	Flow sensors at connections of municipal sewers to transport/interceptor sewer

Source: adapted from Langeveld *et al.* (2013).

11.2.2.1 Rainfall

The Royal Dutch Meteorological Institute (KNMI) operates several rain gauges (Hellman 200 cm^2, 0.1 mm resolution) within the catchment. WDD installed 25 tipping-bucket rain gauges (TBR, 0.1–0.15 mm resolution by Observator Instruments OMC-201-2 Rain Gauge) to increase the spatial distribution of rainfall information. However, the data quality of *some* WDD TBRs was too poor to be useful. Rain gauge installations had been realized very close to tree cover or large infrastructures (Schilperoort, 2011), leading to largely deviating rain depths that could not attributed to spatial rainfall variability. Hence, some of these locations were abandoned, and since 2009 eight rain gauges have been operated by WDD in the area. Merging nationwide C-band radar rainfall estimation (5 min aggregated at 1 km^2 resolution) with rain gauge data, and the effect of rainfall resolution on integrated water quality model simulations have recently been investigated and described in Moreno-Rodenas *et al.* (2017).

11.2.2.2 Flow

Flow control stations in this context refer to stations with a bypass. When a certain threshold is reached, a moveable weir shuts the main pipe and redirects the flow through a smaller pipe, causing filling of the pipes upstream. An electromagnetic flowmeter is connected to the bypass, only showing readings while the main pipe is shutoff.

'Gemaal Aalst' is the main pumping infrastructure in the RZ transport system (RZ is an acronym for 'Riool Zuid'; the southern main branch of the transport system), with a maximum flow of 12,000 m^3/h through two pressure mains. Pressure mains and pumping stations are monitored using electromagnetic flow monitors.

In pumping stations different types of submersible pressure transmitter are used for measuring flow depth. Rather than submersible pressure sensors, an ultra-sonic option was adopted for the CSOs due to them not being constantly submerged. Some newer measuring points, where municipalities connect into the main transport lines, are now equipped with radar sensors.

11.2.3 Sensor operation and maintenance

The sensors are validated (see Chapters 7 and 9) within the Delft-FEWS (Flood Early Warning System) WIS (Water Information System) platform (Deltares, https://oss.deltares.nl/web/delft-fews, accessed 17 June 2021). This platform is mainly used as a central database, in order to assess if a particular sensor needs recalibration or other calamities are occurring with the sensor. If this is the case, the maintenance team can assess the situation on site. Cleaning maintenance is done regularly; the frequency of cleaning depends on the type of sensor, location and importance. Rain gauges are cleaned and recalibrated yearly, whereas oxygen sensors in the surface water are checked twice a month.

Service levels are given to sensors that are maintained by third-party service providers. They detail the maximum response time to calamities with the sensor and the need for cleaning maintenance. Continuity of the monitoring at sites allows for the analysis of trends. This was not necessarily recognized in the earlier monitoring plans, where often a rotary scheme was adopted to cover more ground with less investment cost. These campaigns have been useful for short term analysis and understanding of the system, but for research purposes often did not yield enough information.

Sensor power management does not pose significant problems for the main measuring points near VSI. However, other measuring points need to be either battery powered or have a separate power set-up constructed near to them. This is also an imperative when choosing the location for the sensors. Battery replacement is a key task of the maintenance team, batteries are changed approximately yearly for all the data loggers, or sooner in the case where investigation of a communication black-out has shown a need for replacement.

11.2.4 Data management and data accessibility

All the sewer-related data from WDD are currently logged within a *GE Historian* database (https://www.ge.com/digital/applications/proficy-historian, accessed 17 June 2021). This is connected to the central SCADA (supervisory control and data acquisition) at the WWTP, where all real-time information is used for the operation of the system. On a daily basis (at 6 a.m.), the data are read into the *FEWS Water Information System (WIS)* system. WIS is similar to GIS (Geographic Information System) for water related data. It follows the same principles for more instinctive representation of data. *GE Historian* does not have an easy GIS interface and is used as a background storage place for the data, where the FEWS allows interaction.

Front-ends: There are various platforms, through which the data can be retrieved. The main forms of data collection (when larger datasets are required) are through Microsoft Excel® (through the Proficy Historian Add-on) or Matlab. These forms require knowledge of the 'tagnames', which are scattered throughout the WDD systems. The data can also be approached through Z-Info and a FEWS WIS system. The FEWS WIS system is a GIS-based system which calls the data from *GE Historian*, allowing the user to find the relevant tags (and data, if desired) for known locations. It also instantly visualizes the requested data.

Two others external 'data-tools' are in use by WDD: TMX and HydroNET. TMX (http://www.tmx.nl, accessed 28 April 2021) is a larger, system-wide framework for both soft- and hardware. The TMX system is used for surface water quality readings, but the sewer related data points are also read into the TMX environment. HydroNET (https://www.hydronet.com/, accessed 28 April 2021) uses the databases for operational use and is not linked to research.

11.2.5 Data validation

There are set requirements to the quality and availability of data set by WDD. These depend on the location, information need/use of information and type. WDD requires that (i) raingauges have a 0.1 mm resolution and (ii) 95% confidence intervals for measured water levels are 24 mm wide. Other accuracies are reported as % rather than absolute values. The availability of data for the *Kallisto* project (Langeveld *et al.*, 2013), and later adopted throughout WDD, was 90%, meaning that the sensors should operate successfully 90% of the time. In the past, these requirements have not been met due to shortage of dedicated staff time (reports suggesting 50% for particular sites) but are likely to be more obtainable as monitoring of the system becomes a more integral part of the WDD duties.

All new sensors will be placed using internal protocols, as specified by the manufacturer. To ensure the correct operation, two tests are performed: a Factory Acceptance Test (FAT) to check if the sensor functions according to the factory specification and a Site Acceptance Test (SAT) which involves checks for time-synchronization with the atomic clock and sensor calibration according to internal protocols and depending on sensor types. The SAT is repeated after a sensor has been installed for some time and is expected to have drifted.

There are two streams of data available in all platforms: raw data (RD) and processed data (PD). *GE Historian* only stores raw data, with validation starting at the *FEWS* level. The validated data are then available in parallel to the raw data, which enables researchers to re-validate raw data based on the specific requirements for their project.

Primary validation is completely automated, looking if the values are within certain limits: range check (min/max) and noisiness (see Section 9.3). This validation is based on the local parameters and therefore both site specific (local minima and maxima) and factory specific (noisiness). Secondary validation is non-automated or semi-automated (see Section 9.5). This process can override whatever decisions are made by the automation in the first step. It is highly site specific and based on the knowledge of the system. This also includes cross-comparison of different related parameters within the system, such as the rainfall with hydrodynamic responses of the system (van Bijnen & Korving, 2008). Further details of the data-yield of the monitoring network, which was poor at the start of the project for various reasons, are described in Schilperoort (2011).

There are several performance indicators (PIs) in use to improve the monitoring system. These are based on interviews with different departments in the WDD, to ensure PIs are aligned with what is required. The PIs used are data availability, fraction usable data, and accessibility.

Data availability refers to the fraction of data points that arrive in the database (dependent on quality of data transmission). The fraction usable data is the part of the arrived data that passes the validation stages (quality of the sensors), and accessibility is the ability for the data to be used effectively and rapidly (therefore dependent on the database and data communication structure).

There is a high variability in the quality of the data stream, for both data availability and fraction usable data. Over the 2007–2008 period, the average data availability per sensor cluster ranged from 70–99%, with the fraction usable data ranging from 0–99%. In general, sensors installed in the WWTP influent pump performed well, whereas UV/Vis sensors performed most poorly.

11.2.6 Data transfer and communication system

Depending on the purpose of the measurement, the dataflow goes through different communication platforms. If there is a need for a bidirectional communication because of a control measure, the communication needs are different. The communication from the SCADA to the actuators in the system is through TCP/IP (transmission control protocol/internet protocol). Given that these are large buildings,

TCP/IP or Ethernet is viable, as opposed to more remote monitoring locations. For monitoring purposes only, GPRS is sufficient for data transmission. Vodafone and/or T-mobile SIM cards are in place to allow for the GPRS connection. When two are in place, the logging can continue during calamities with the parallel SIM card.

11.2.7 Reporting, management and availability of data files for research

Within WDD, five-year cycles are defined, where monitoring targets are defined, including parameters, monitoring frequency, confidence intervals (% or absolute value) and availability (operational % of the time).

This cycle 'starts' by creating a monitoring cycle plan (MCP), which sets out how the monitoring should be carried out both financially and operationally, and determines the requirements for information. Based on this plan (expansion of) the monitoring network is set up and maintenance plans are drawn up. Based on the new (maintained) monitoring networks, data are harvested, validated, and ultimately used for analysis of the system. These analyses are reported along with whatever actions deemed necessary by the operator. Based on this, new information might be required, which is then set out in the new monitoring cycle plan (Sections 6.1 and 6.2).

The MCP is an integrated report and requires input from several departments within WDD (as it pertains to surface water (hydrobiology, hydrodynamics, etc.), sewerage (quality and quantity of flow) and groundwater (hydrogeology). This integrated approach to the MCP allows for 'optimized' data flow.

11.2.8 Challenges, lessons learnt

Monitoring is not always seen as a priority by WDD management, which is elected for 4 years, and monitoring is likely to become more fruitful over a longer time horizon. This is a potential hurdle, which needs to be overcome for other Waterboards. For the entire WDD governed area (e.g. not just Eindhoven urban drainage area), the cost for maintaining and operating the monitoring network is roughly € 2.5 million annually (on an overall total annual budget of € 110 million).

A key suggestion for improvement, based on the opinions of researchers working on ongoing collaborative research projects with WDD, would be to include meta-data (type of sensor, calibration results and maintenance issues) within the FEWS system, rather than have it scattered in disjointed Microsoft Excel® files (Section 5.3). This way, the FEWS database takes in all the necessary data, with reporting on data and ease of analysis expected to improve significantly.

The use of several platforms has led to a lacuna of knowledge, causing a lack of ability to use the entire range of options within each different platform. With the external tools (HydroNET and TMX mainly), the knowledge tends to reside outside WDD, which can lead to trouble for the maintenance or changes in the system. Although the current logging of the data is adequate for all purposes (both retrieval of past datasets and operation), data from before the change to the current database architecture seems to have been lost. Before the use of the *GE Historian* database, a database from another company (GBS) was used. Theoretically, the data within this database should still be there. However, despite numerous efforts from various operators and other relevant WDD employees, these data points have not yet been recovered. *Transitioning to new databases should therefore at all times consider future needs for data and ensure the availability of past data is preserved.*

A collaboration between several organizations (a waterboard, several municipalities, at least four consultants, several manufacturers and construction firms, and a university) whereby especially the waterboard, the municipalities and the university have different requirements regarding data quality and different aims in mind, needs careful project management that is considerate of these different viewpoints. Achieving high data quality is often not seen as financially viable or desirable by practitioners, hence collaboration and input of PhD students proved essential to achieve good data validation in this case study. The amount of work needed to maintain a monitoring network of this scale was initially underestimated, with 0.5 FTE (full time equivalent) available for all tasks during the first year. This resulted in, euphemistically formulated, a less than optimal installation of rain gauges and a lack of sensor maintenance and numerous changes in the set-up during the first years of the project.

11.3 LET'S SMARTLY COMBAT FLOOD IN KUALA LUMPUR, MALAYSIA

11.3.1 Scope and objectives

The Stormwater Management and Road Tunnel, famously known as SMART Tunnel, is an innovative approach to mitigate flooding in the heart of Kuala Lumpur City Centre. The capital of Malaysia experiences frequent flood events, the first recorded event was in 1881 (Williamson, 2015), due to the high intensity of precipitation within a short duration. The booming development of Kuala Lumpur saw changes in the land types, decreasing infiltration capacity, increasing high surface runoff and subsequently more frequent flood events. The intensity of a rainfall episode in Kuala Lumpur can be as high as 120 mm/h (Abdullah *et al.*, 2019), which results in high volume discharge into the river as well as immediately exceeding the urban drainage capacity. SMART Tunnel serves not only as the solution to flooding but also provides an alternative route to reduce traffic congestion both inbound and outbound of Kuala Lumpur.

The Department of Irrigation and Drainage (DID) operates the facility, which came into operation on 30th June 2007 after almost four years of construction worth close to RM 1.93×10^9 (i.e. 1.93 billion) (being equivalent to ~€ 0.39 billion). The main objective of this mega structure is to divert high flow discharge at the critical confluence of the Klang and Ampang Rivers. The design aims to reduce or minimize the probability of inundation at the downstream confluence of the Klang and Gombak Rivers,

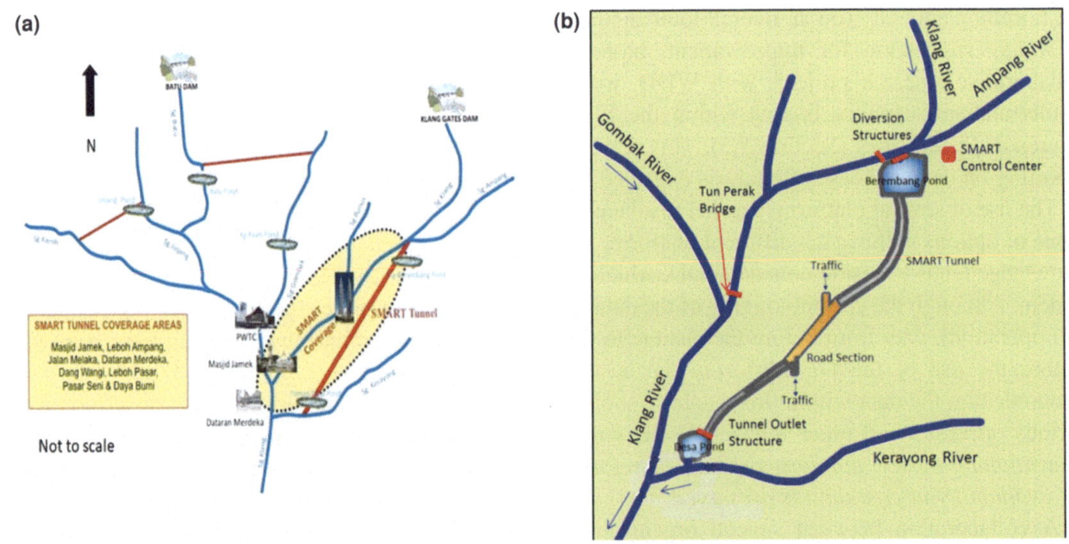

Figure 11.2 The location of SMART Tunnel alignment (a), diverting the discharge from Klang/Ampang Rivers to Kerayong River (b), minimizing flood occurrence within the marked SMART coverage. *Source*: Department of Irrigation and Drainage, Malaysia.

where the Masjid Jamek is (as shown in Figure 11.2(a)). The design consideration is that the peak flow here should not exceed 180 m^3/s (Lai, 2016). At this point, the discharge is from an approximately 160 km^2 watershed of the upper Klang River catchment. The SMART was designed to cater to a 100-year average recurrence interval (ARI) flood event with a maximum diversion volume of 300 m^3/s.

Shown in Figure 11.2(b), the flow from the upper catchment of the Klang and Ampang Rivers is diverted to the Berembang holding pond before flowing through the 9.7 km bypass, 13 m diameter tunnel into the Desa attenuation reservoir located at the downstream of the Kerayong River.

11.3.2 Operation of SMART

The facility operates in four modes, as illustrated in Figure 11.3 depending on the flow discharge from the Klang River. Figure 11.4 indicates the frequency with which the different operational modes for SMART have occurred from 2007 to July 2020.

11.3.2.1 Mode 1

Activated when the weather is fair with little or no rain and the traffic is allowed in the tunnel.

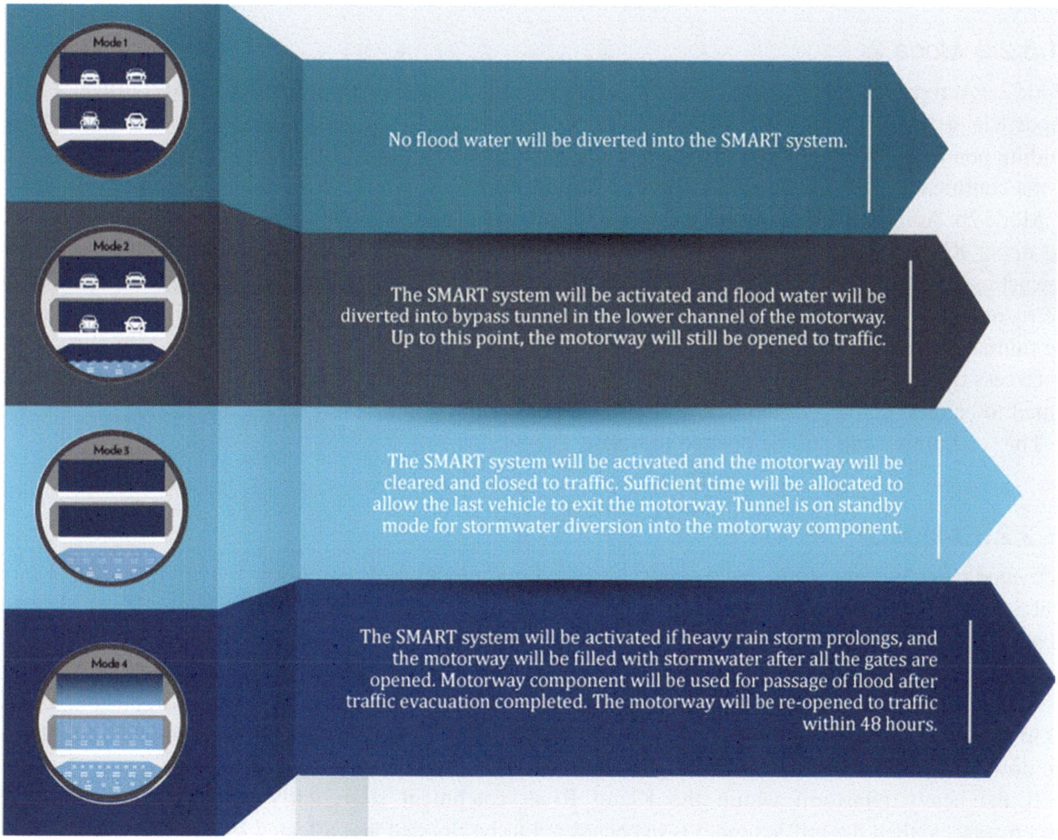

Figure 11.3 Four modes of SMART operation. *Source*: Department of Irrigation and Drainage, Malaysia.

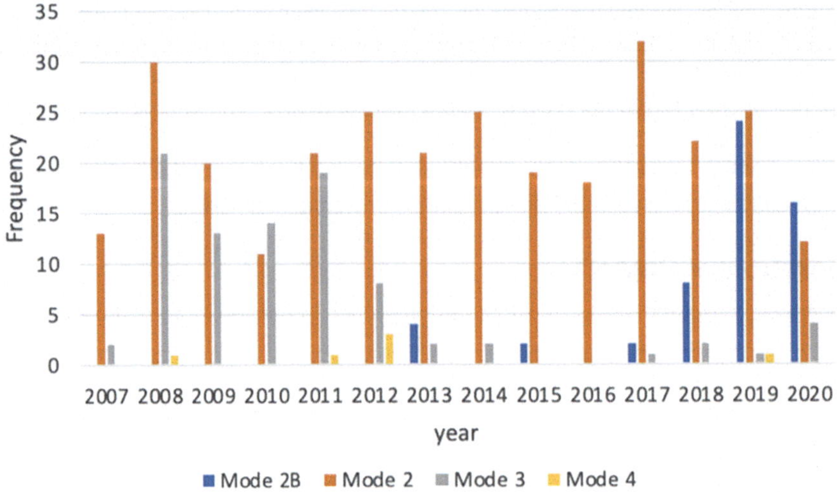

Figure 11.4 Frequency of operational modes for SMART from 2007 to July 2020. *Source*: Department of Irrigation and Drainage, Malaysia.

11.3.2.2 Mode 2

Mode 2a: Activated when a moderate rainfall event occurs and the flow rate recorded at the confluence of the upper Klang River/Ampang River is 30 m^3/s. In this mode, all the discharge will be diverted into the holding pond. This mode will be activated when there is heavy rainfall in the city centre and downstream of the confluence of the upper Klang River/Ampang River.

Mode 2b: Activated when a moderate rainfall event occurs and the flow rate recorded at the confluence of the upper Klang River/Ampang River is 70–150 m^3/s. In this mode, only 50 m^3/s is allowed to flow downstream of the Klang River, and the rest of the discharge will be diverted into the tunnel. Based on the records, normally Mode 2 is activated for rainfall events with less than 10 year ARI and depends on the rainfall duration.

Excess flood water will be diverted to the Berembang Pond. Mode 2 allows only the lower drains of the tunnel to be used conveying the flow to the Desa attenuation pond.

The road tunnel will still be opened to traffic.

11.3.2.3 Mode 3

Activated when a major storm event occurs and the flood model forecasts a flow rate of 150 m^3/s or more (but not exceeding the designed Q_{100} of 300 m^3/s) at the confluence of the upper Klang River/Ampang River during this particular storm event. The flood detection system (FDS) predicts the flow (and water level) at the city centre (i.e. Tun Perak Bridge) based on the rainfall data within the catchment.

Traffic will be evacuated from the road tunnel, which normally takes about one hour. In this mode, only 10 m^3/s is allowed to flow downstream of the Klang River to provide storage for the high incoming flow at the downstream. The rest of the discharge will be diverted to the tunnel.

If the heavy rainstorm within the Klang River catchment stops early or due to some specific circumstances, then the traffic tunnel is expected not to be flooded and affected by the inundation.

The road tunnel will be re-opened to traffic within one to three hours after closure.

11.3.2.4 Mode 4

Activated if a heavy rainstorm is prolonged, usually will be confirmed 1–2 hours after Mode 3 is declared. Results from the FDS determine the extent of the flood.

The road tunnel will be used for passage of flood after traffic evacuation has been completed. Due to the high discharge, only 10 m^3/s is allowed to flow downstream of the Klang River and the rest of the flow will be diverted to the tunnel.

The road tunnel will be re-opened within four days of closure.

Following deactivation of SMART operations, the excess water will be pumped to the Taman Desa attenuation pond as a dewatering process.

The frequency of SMART operational modes from the year 2007 to 2020 is shown in Figure 11.4. A total of 445 SMART activated events were recorded with Mode 2 being the most common. The worst-case scenario of Mode 4 only occurred in the years 2008, 2011, 2012 and 2019. The visible trend of high-frequency Mode 3 was regularly observed for the years 2008–2011 and started to diminish from the year 2012, maintaining a lower frequency until 2020.

11.3.3 Location of monitors and measured variables

As detecting flood (through rainfall prediction) and monitoring flow (to avoid flood) are the main focuses of this facility, SMART employs a rigorous monitoring system of rainfall gauges, and water level and river velocity measurement. There are 22 rainfall stations dedicated to the SMART facility distributed within the Klang River catchment, the distance between each station is less than 3 km. The rainfall data are measured using tipping bucket rain gauges at an interval time of 5 minutes. The control centre monitors real-time data of rainfall and water level, assisting in making the decision of when to activate the SMART facility. All measured information is transferred every 5 minutes to the centralized supervisory control and data acquisition (SCADA) by transmitting the telemetry data. To ensure safety in the data transfer and storage, the communication of data to the SCADA is backed up by the GSM (global system for mobile communication) system when the radio system fails. Due to excessive data size, the data are not stored in the SIM card.

A total of 16 compact bubbler sensors are installed at strategic locations within the catchment to measure water level. A bubbler system is employed as it is more robust, especially for rivers carrying large amounts of debris such as the Klang and Ampang Rivers, in particular during high flows. The velocity of 16 strategic locations within the Klang River stretch and its tributaries is measured using Doppler current meters (DCMs). The river discharge is calculated based on the developed rating curve and validated with the record (of autocorrected discharge) from the DCM. Strategic and crucial measuring locations include the confluence of the upper Klang River/Ampang River, the SMART Control Centre (at Berembang Pond), and downstream of SMART at Kerayong River. The measurements for both river flow velocity and the water level are taken at 5-minute intervals, and data are transmitted via the same system as rainfall, through SCADA and GSM.

All instruments are powered using a solar panel with a battery pack as a supporting energy plan. In addition to advocating green technology, the solar panel also minimizes dependence on the battery simultaneously reducing operation and maintenance costs. The calculation of power consumption for a solar system for rainfall and water level stations is described in the Hydrological Procedure 32 and 33, respectively (DID, 2018a, 2018b).

Twenty CCTVs are installed at strategic locations (Figure 11.5) to provide real-time observations of water level. The critical confluence of the Klang River and Gombak River (where the popular tourist hotspot of Jamek Mosque is located) has three monitoring CCTVs. The other CCTVs are installed at the

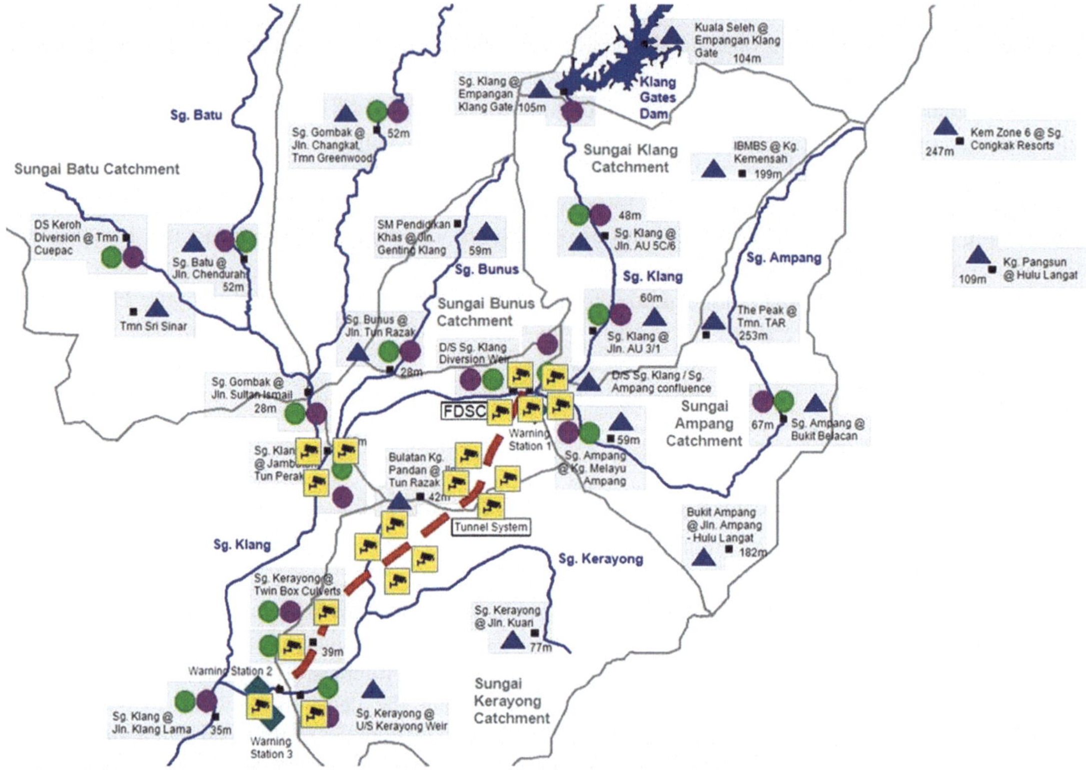

Figure 11.5 Location of the CCTVs installed within the vicinity of SMART. FDSCC – Flood Detection System Control Centre. *Source*: Department of Irrigation and Drainage, Malaysia.

SMART Control Centre, Berembang Pond, along the stretch of the Klang and Ampang Rivers and the downstream Desa attenuation pond at Kerayong River. The installation of these CCTVs is used as observed verification of the SCADA data for operational decisions.

Utilizing the fine temporal scale of rainfall data, a sophisticated and modern flood detection system (FDS) provides real-time flood forecasting information. This enables efficient and safe management of the operation of the tunnel.

11.3.4 Data management and data accessibility

All transferred raw data are stored in the SCADA database according to each measuring station, listed based on parameters and the recorded date. All recorded data are stored at the SMART data centre, internally managed by the DID. Retrieval of data is obtained using the ClearSCADA SCX MySQL relational database management system (RDBMS), allowing quick response and specific data requirements. Where the data need to be processed (usually for operational purposes), specific data will be retrieved and analysed elsewhere. In-house data quality analysis is done at the SMART Control Centre. Data availability is checked by a network connectivity test between the SCADA database and remote telemetry unit (RTU) at sites. The quality of the data is tested using hydrological modelling, by comparing the observed record with modelling results. Although at the moment (i.e. 2020), the data are only accessible to DID, upon receiving permission from DID, the data are allowed to be distributed to the interested parties. DID acknowledges that with updated technology, highly accurate algorithms and mathematical modelling, data sharing is the way forward to ensure sustainable operation of SMART. Climate-change induced rainfall patterns obviously may alter the local rainfall intensity, and subsequently the runoff volume within the Klang River watershed. By sharing the data with others, more rigorous data analysis is feasible, which provides benefits to SMART operations and the DID. Realizing the importance of data availability, all recorded data are kept in the storage server (with no data discarded after a number of years), allowing a comprehensive data review when needed.

Data obtained from the measuring gauges are of high quality, with more than 95% accuracy and minimum missing data. Even so, if there is any requirement for obvious missing data (during the transmission), data may be retrieved from the data logger on-site which has a one-month storage capacity. SMART puts heavy emphasis on flood detection and control, whereby particular attention will be paid to missing data during storm events. Accuracy checking through regular calibration procedures for the measuring equipment is done in-house, in accordance with the manual provided by the manufacturer. Well-calibrated equipment is the top priority is for operational purposes and preventive maintenance. Given the huge amount of available data, SMART engages in cleaning up the data for important and specific tasks such as analysis, prediction of storm events, and operations during high flows.

As high accuracy data are imperative, a quick response for repairing a faulty rain gauge is a must. The response team is deployed when there is a glitch in data transferring and obvious unusual SCADA data. When one rainfall gauge has malfunctioned, calculation of average area rainfall for the catchment using the inverse distance method based on the neighbouring rainfall gauges is conducted. As the rainfall gauges are within a radius of 3 km from each other, the availability of fine spatial data allows for an accurate approximation of rainfall.

11.3.5 Design of monitoring networks

An overview of the locations of rainfall gauges and measuring stations dedicated to SMART operation is shown in Figure 11.6. Data at the upper catchment of the Klang River serve as an indication of a downstream flooding scenario, which provides ample time for appropriate SMART operational actions. The time between peak flow from upper catchment to the critical confluence of the Klang/Ampang Rivers takes about 30–60 minutes, depending on the rainfall intensity of a storm event.

The annual cost of operation, including flood detection, flood flow components, and housekeeping can become close to RM 4–6 million (€ 0.80–1.21 million). A significant fraction of the operational cost is attributed to the pumping and control gates. There are 25 gates installed within the SMART system to control the systematic conveyance of excess stormwater within the tunnel, including two gates (i.e. NJB and SJB gates) specified for the traffic tunnel (Figure 11.7). During Mode 3, the NJB (north junction box) and SJB (south junction box) gates are closed. The maintenance of the instruments can reach up to RM 700–800 thousand (€ 140–160 thousand) in a year.

11.3.6 Operation and maintenance

The SMART facility runs 24/7, all year long. In ensuring minimum problems in data monitoring and storm management, monthly preventive maintenance is mandatory for all instruments discussed in Figure 11.6. Maintenance for each measuring station follows the checklist prepared by DID to guarantee a thorough inspection and correct protocol. The systematic calibration procedure for each instrument is conducted quarterly by an accredited calibration agency. Calibration work for water level sensors must comply with the procedures set and is controlled by the quality management system ISO 9001:2015 (ISO, 2015; Cert

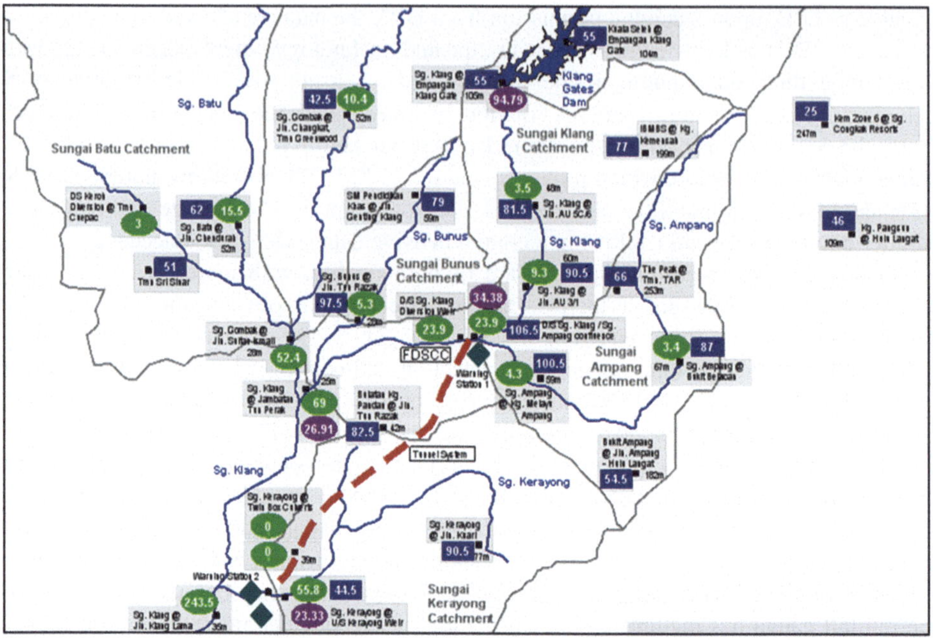

Figure 11.6 A snapshot of the monitoring network within the Klang River catchment, specifically for the operation of SMART. Data for rainfall, water level, and flow velocity are displayed in blue, green, and purple boxes, respectively. The dashed thick red line indicates the SMART Tunnel alignment. *Source*: Department of Irrigation and Drainage, Malaysia.

no: QMS 02621). Immediate corrective action is administered for repairing malfunctioned equipment and pressing issues such as missing data during storm events.

Guidelines for safety and health relating in operating and maintaining the hydrometeorological equipment are available in the HP32 (for rainfall – DID, 2018a) and HP33 (for water level – DID, 2018b) documents. Maintenance work of the telemetric unit, testing, and validation of water level sensor, repair, and services of hydrological equipment is properly recorded using standardized forms.

The safety of the instruments at the station is safeguarded by constructing non-climbing, anti-cut and wired gates. The locking system is installed at both the entrance gate and station door to minimize illegal public trespassing.

11.3.7 Uncertainty assessment and data validation

There are 22 rainfall gauges in the network, which are dedicated to cover an area of 596 km^2. These rainfall and water level gauges are point measurements. Therefore, the extensive rainfall and water level gauges are designed to capture the spatial and temporal variations of any storm events, given that Malaysia is a tropical country that is likely to experience frequent storms resulting in a significant amount of rainfall, especially in April and October, i.e. inter-monsoon seasons (see Muhammad *et al.*, 2016).

SMART's hydrological monitoring system is vulnerable to error due to wind and the type of instruments. The tipping bucket may underestimate the rainfall amount due to the bucket tipping action, especially during extreme storm events. Other than that, systematic wind field deformation above the gauge orifice may cause a 2% to 10% systematic error (WMO, 2008). Although these errors cannot be eliminated completely,

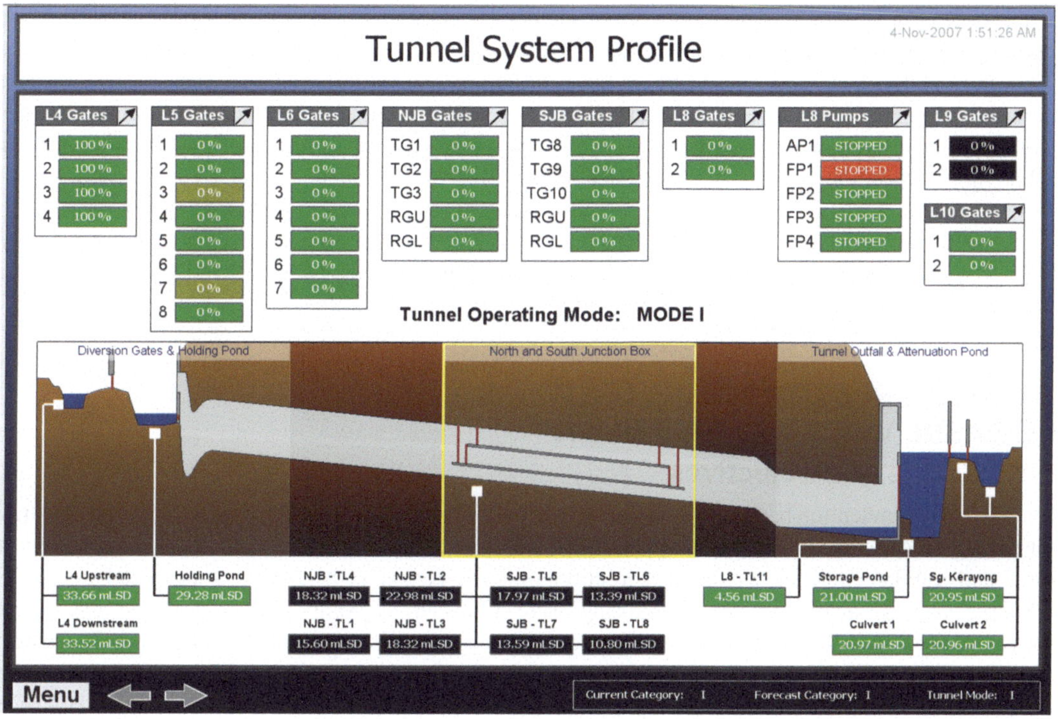

Figure 11.7 A snapshot of operational display Mode 1, showing the operational status of the gates installed along the tunnel. Note: NJB and SJB gates are for the traffic tunnel. *Source*: Department of Irrigation and Drainage, Malaysia.

periodic instrument maintenance and calibration and continuous assessment of observed hydrological data are highly crucial to ensure the error and data uncertainty are minimized.

11.3.8 Challenges, lessons learnt

The booming development of Kuala Lumpur and rapid expansion of the urbanized area is increasing the runoff (and with a shorter time) into the main stem of the Klang River. The climate change phenomenon has also contributed to more frequent storm events with higher rainfall intensities. Due to the scale, magnitude and sensitivity of SMART, the number of rainfall and water level gauges available in the catchment should follow the recommended minimum station densities for hydrology given by WMO (2008), i.e. one station per 10–20 km^2 of the catchment area. The use of advanced hydrological monitoring equipment and technology, such as radar, remote sensing, and satellite observations should also be explored to increase the accuracy and quality of observed data and minimize data uncertainty, particularly for the GIS data. An updated map of Kuala Lumpur and Selangor, and details of changes of land use within the Klang River catchment is essential to ensure a reliable and high accuracy runoff volume (for modelling).

An increase in the frequency of flood events (both fluvial and pluvial) within Kuala Lumpur poses challenges to the local authorities in the adaptation of managing stormwater and minimize flooding scenarios at the critical strategic areas of Kuala Lumpur, particularly the commercial zones. Data indicate that the operation of SMART is facing more challenges during the El-Nino season due to the high occurrence of extreme rainfall events. Higher frequency of flood operations was observed compared to the lower rainfall events and subsequently fewer flooding incidents during the drier La-Nina season. Extreme high precipitation events in Malaysia correspond with the occurrence of El-Nino-Southern Oscillation (ENSO) (commonly known as El-Nino) whilst the significant decrease in wet events is due to the La-Nina, during the months of December-January-February (Tangang *et al.*, 2017). ENSO is the irregular, periodic variation of sea temperature and winds over the tropical eastern Pacific Ocean, which has been extensively studied to be well correlated with extreme precipitation events (Casanueva *et al.*, 2014). Learning from experience, particularly during the El-Nino season, rigorous hydrological modelling is conducted to assist in smoother SMART operations.

11.4 WICKS RESERVE BIORETENTION BASIN, THE BASIN, VICTORIA, AUSTRALIA

11.4.1 Scope and objectives

Wicks Reserve bioretention basin was built in 2011 in the eastern suburbs of Melbourne, Victoria, Australia to protect the local stream against the impact of urbanization. This basin (Figure 11.8) was monitored to inform and validate policies as the construction of such basins is becoming widespread across the Melbourne area. Construction and monitoring were financed by Melbourne Water (the local water manager) and Knox City Council. Monitoring has been conducted by the Waterways Ecosystems Research Group (WERG) of The University of Melbourne. The aims of the monitoring programme were (i) to assess the hydrologic performance of the basin (reduction of stormwater volumes and peaks), (ii) to assess its water quality treatment performance and (iii) to monitor the fate of infiltrated stormwater from the basin to the nearby stream. Research was the primary driver for setting up the monitoring system, which will be dismantled when the research project ends (monitoring started in March 2013, with probes

Figure 11.8 Wicks Reserve bioretention basin. *Source*: Jérémie Bonneau (INRAE).

being installed in 2013 and 2014. Flow monitoring was stopped in 2018 and groundwater monitoring was stopped in December 2019). Monitoring has not led to direct policy changes.

11.4.2 Recorded data

11.4.2.1 Rainfall data

Rainfall was monitored onsite using an *Odyssey* Logger with a tipping bucket rain gauge (NZ$ 369 ~€ 215 – all prices are given as of June 2019), clear from canopy interception. The rain gauge was inspected weekly to fortnightly to verify that the hole of the rain gauge bucket was not blocked (i.e. by leaves). If the rain gauge was found to be partially blocked, the cumulative rainfall was compared to another nearby rain gauge. Blockages are obvious on such graphs. If the hole was found to be completely blocked with no water reaching the tipping pivot, data since the previous download were considered not usable. Rainfall was recorded at a 1-min timestep in the built-in logger. Total blockages were rare. The rain gauge did need calibration: the volume of rainfall was collected every month, weighed to 0.1 g accuracy and compared to the number of tips given. While the manufacturer-claimed measured depth was 0.2 mm of rainfall per pivot tip, calibration showed that a factor of 1.1 needed be applied to the data (i.e. the actual tip was 0.22 mm). This calibration factor was checked monthly to check for any drift. Most data (>95%) were usable.

11.4.2.2 Flow data

Flows going into and out of the basin were monitored using area-velocity sensors connected to *Hach Sigma 950* flowmeters (hereafter referred to as HACH). The probes recorded velocity with a Doppler ultrasonic sensor and water level with a pressure diaphragm. Monitoring of flows proved very challenging. Small weirs had to be installed in two pipes to make sure the probes remained submerged to better monitor low flows, which was very important to close the water balance. Both inlet pipes were large and likely oversized (750 and 525 mm in diameter) to accommodate high storm flows, but this proved challenging for flow monitoring as large pipes resulted in low water levels. Catchment hydrographs had long tails, with substantial and long-lasting low flows (<1 L/s). Flows (level and velocity) measured were within manufacturer-claimed accuracy ranges (−1.52 to 6.1 m/s for velocity, 0–3 m for level). Flowmeters were manually calibrated regularly (twice a year), using an electro-magnetic flowmeter for high flows and a bucket and a stopwatch for low flows. Rating curves were constructed to correct the probe readings, which demanded a substantial amount of work. A fire hydrant was connected to an electro-magnetic flowmeter and water poured into the stormwater pipes where the flowmeters were located. As such it was possible to compare the flow value given by the electro-magnetic flowmeter and readings (flow, level, velocity) given by the flowmeters in the pipes. This allowed the construction of rating curves (e.g. measured flow or level or velocity vs. observed flow) to correct the values given by the flowmeters. For low flows (<5 L/s), the rating curves were completed with manual measurement of the actual flow, measured by a bucket and a stopwatch. Flow data were stored in a built-in logger, at a 6-min timestep. Sediments caused issues with flowmeters at the inlet of the basin, resulting in very large chunks (>50%) of data deemed unusable, after discussion between researchers, for one inlet probe. The sediments were cleaned regularly, when they had accumulated on top of the probe.

11.4.2.3 Water quality (autosampling)

Two autosamplers (Sigma 900 MAX portable, with 24 × 1 L samples) were installed at the inlet and at the outlet of the basin to monitor the water quality treatment performance of the basin. The autosamplers were connected to the flowmeters and samples were taken every given volume to cover a large range of hydrographs. Such samplers were required to be turned on manually before a rainfall event and samples were collected and sent to the laboratory for analysis within 24 hours of sampling.

11.4.2.4 Water and groundwater level data

Water level at the surface of the basin was monitored with four *Odyssey* capacitance probes, chosen primarily for their price and availability (NZ$ 248 ~ € 140). The water level within the filter media of the basin was monitored using three of the same *Odyssey* capacitance probes. The fate of infiltrated stormwater was monitored with 15 water level probes (the same *Odyssey* capacitance probes), installed around the basin to monitor shallow groundwater (ranging 2–5 m deep). All these probes were calibrated in the laboratory before deployment. Calibration was checked and adjusted when needed. Because of their relatively short length, it was easy enough to remove every probe from their bore during download to clean them and check the calibration with a home-built 2-m long PVC pipe filled with water. The operation took about 2 hours for all probes, so missing data during downloads were minimal. Water levels were read with an accuracy ranging from 2–5 mm, and most data were usable, except for times with flat batteries or unexplained probe issues (around 10% of the dataset).

11.4.3 Maintenance, operational cost

The site was visited weekly or fortnightly for routine checks and maintenance operations (replacement of flat batteries, faulty probes, collection of water quality samples, calibration of probes, cleaning of probes, checking flowmeters, etc.). Overall monitoring was costly in operational time. The site and data were mostly managed by a PhD student (50% of his/her time) assisted by two technicians (around 0.5 days a week each).

11.4.4 Database, accessibility and data management

Data were downloaded from the loggers, on site, every month, the operation taking about half a day to a day of work. All data were downloaded on a computer and then stored into a *Dropbox* folder shared by all members of the WERG. The folder was only accessible by researchers from the group. The *Dropbox* folder contains data of all projects of the WERG and is managed by a data manager, funded by the partnership between the WERG and Melbourne Water. The primary responsibility of the data manager is to look after data from all WERG projects. Data validation was done by or with researchers who spend time on sites. Data format and data storage were systematic and the same for all probes (flow, water level, …). Data were stored in three folders:

- Raw data, containing the files obtained from the sensors. These files were never touched, never modified.
- Compiled data, containing data being processed (validation, correction…).
- Final data, containing validated data (in either.csv files or .r data format), ready to be read and used in R for data analysis.

Additional but crucial data validation was done manually using field notes and observations to adjust data when needed. A *quality code system* was set up as: 1 (good data), 2 (data might have a problem) or 3 (unusable), based on field observations (for example, if a wooden stick was observed on the probe but data still made sense and visually looked OK, the quality code would be 2). Researchers shared a spreadsheet where timings, all field observations and actions taken for the probes were recorded so that anomalies in the data could be explained later on (for example: 'on this day, probe nr x has been removed from its casing to sample groundwater', can explain a couple of timesteps containing surprising data). Data are available upon request pending agreement of the researchers who collected them and pending specific conditions on the use of data. Some data collected at this site have been shared with research teams across the world (USA, Chile, Australia, France) for potential collaborations.

11.4.5 Power management

Water level probes were powered by two built-in AA 3.6 V lithium-ion batteries that were replaced when needed (every 1–2 years on average). Flow meters and autosamplers were powered by 12 V batteries that were checked and replaced when needed with batteries charged in the WERG workshop. Due to the workload involved, a battery charger was later connected to the main electric grid, and 12 V cables were run from this point to the batteries. Such a solution is ideal because running 12 V cables does not involve licensed expensive and dangerous electrical work.

11.4.6 Health and safety

Site visits were performed under the health and safety guidelines of both *Melbourne Water* and the *University of Melbourne*. *Melbourne Water* mandates its collaborators to possess a permit obtained after taking an online class and a subsequent test (even though this basin is not a *Melbourne Water* asset – it is a council asset). The University of Melbourne employees and students are mandated to complete a Risk Assessment specific to dangers of the site visited and the tools used or the tasks performed on site. Each new staff or student must review and sign this Risk Assessment before coming on site. This document covers safety issues for general fieldwork and work specific to this particular site and project (never be alone on site, wear personal protection equipment, snake bites, aggressive sun, heat). For more detailed, rare or high-risk tasks (e.g. going into a pit), researchers are required to complete a Task Risk Assessment (TRA) specific to this particular task.

11.4.7 Reporting

Contact between the University's researchers, council engineers and *Melbourne Water* officers (the overarching authority) was very frequent with regular update meetings and very frequent emailing. Scientific papers were published (Bonneau *et al.*, 2018, 2020) or are in the making. Data were presented at international conferences, and at several presentations with industry partners. An intern was hired in 2017 to put together a monitoring report for *Melbourne Water*.

11.4.8 Lessons and suggestions

The monitoring set-up at this site required constant human inputs and labour with at least weekly site visits, indicating that a remote way to communicate with probes would save operational time.

Meta-data (information about sensors, calibration, field notes) are crucial to data management and analysis, as they allow understanding of why chunks of data look suspect. The way meta-data will be recorded and used should be discussed during planning a monitoring strategy. Similarly, processing and analysing data as early as possible after collection (instead of storing long periods of data without looking at them) can avoid further trouble as many issues with probes were identified during data analysis. In this case study, weeks of data were lost before realizing that low flows were important and that therefore weirs had to be installed, and that the flowmeters needed substantial calibration effort to provide reliable data. This was all found by trial and error.

Original aims of the monitoring were achieved and stakeholders (research, council, water authority) were pleased with the monitoring results. Reliable experimental assessment of the hydrological and water quality performance of the basin was obtained. To date, monitoring of this site has not yielded actual policy changes, but has enhanced confidence amongst stakeholders to implement such systems across the Melbourne area.

11.5 FLOW MONITORING CAMPAIGN FOR COMPANY-WIDE INTEGRATED URBAN DRAINAGE MODEL UPGRADE, ANGLIAN WATER SERVICES, UNITED KINGDOM

11.5.1 Overview

Anglian Water Services (AWS) have adopted an integrated supply chain and holistic approach for their AMP6 (AMP6 stands for Asset Management Plan 6. This is the UK water industry financial cycle running from 2015 to 2020) Integrated Urban Drainage (IUD) Modelling Programme. The key objectives of the IUD Programme comprise working collaboratively with all partners and within the business to deliver 100% coverage of urban drainage models to a level of confidence appropriate for intended use. Focusing on the key metrics of flooding and pollution which are common performance commitments across all water companies, these assessments have included amongst others, catchment performance and risk assessment as part of the preparation of Drainage and Wastewater Management Plans, asset health assessments at critical ancillaries such as pumping station failure analysis, the impacts of future catchment changes such as growth, climate change and urban creep and ultimately, to define a catchment strategy to mitigate all of these risks.

The IUD Programme entailed systematic assessment of existing models and risks across all AWS's catchments, risk-based planning and execution of asset and short-term flow surveys, and subsequent model upgrade and re-calibration based upon the base data and the data collected through recent surveys.

This case study focuses on the flow survey component of the IUD Programme. Information about other components of the programme can be found in Brayshaw & Wilkes (2016). Note that, in addition to the short-term flow surveys undertaken as part of this programme, see WRc (1987) for information on this type of survey, permanent monitoring at strategic sewer locations, overflows and other ancillaries is also undertaken by AWS. The permanent monitoring falls outside of the scope of this case study.

11.5.2 Risk-based flow survey planning

Survey planning entailed initial screening of all Anglian Water catchments, based on assessed risk, followed by actual flow monitoring scoping following cost-benefit principles. Figure 11.9 shows a map of the region where Anglian Water Services operates.

The initial catchment screening (referred to as Model Delivery Milestone 1 – MDM1) considered factors such as growth, pollution and flooding occurrence. This resulted in 447 urban drainage catchments to be considered in the second stage (MDM2).

MDM2 consisted of a high-level scoping of flow monitoring needs for the catchments of interest. Monitor numbers were derived by looking at the total length of all the sewers in each catchment and the number of ancillaries with the potential to cause pollution, mainly combined sewer overflows (CSOs) and pumping stations. The rough criteria used was one monitor for every 1 km of pipe, plus one additional monitor per overflow. For example, a catchment with circa 10 km of sewer pipes and five CSOs would be assigned a minimum of 15 monitors.

Monitoring needs were then compared against available budget, in turn based on a preliminary cost estimate, assuming a 12-week duration per survey. The time required to complete all the surveys (including planning, installation, monitoring and decommissioning) was also considered and when programmed it was obvious that it was not time efficient to undertake many small surveys. At this point

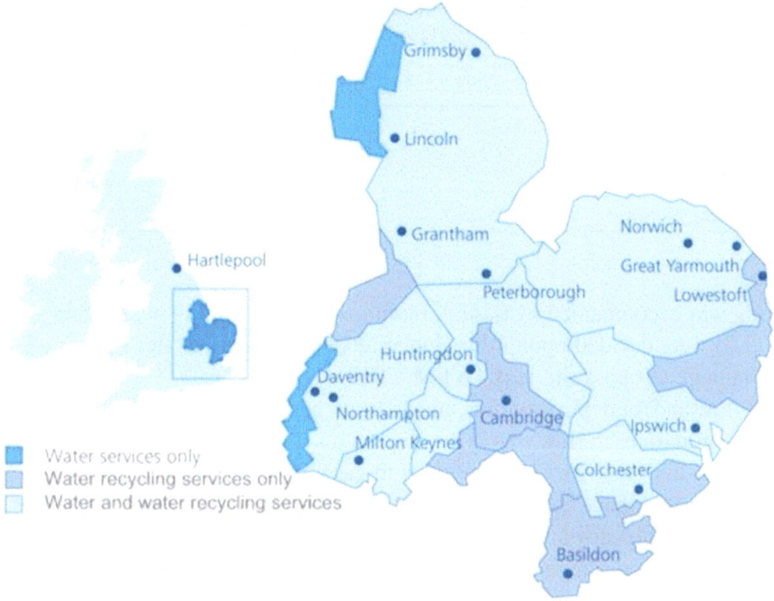

Figure 11.9 Map showing the operational area of Anglian Water Services, which comprises 28,000 km^2. *Source*: Anglian Water Services (2019) Anglian Water's Cross Sector Infrastructure Access Statement, March 2019. https://www.anglianwater.co.uk/siteassets/developers/development-services/cross-sector-infrastructure-access-statement---march-2019.pdf (accessed 28 April 2021).

a decision was made to remove small, low priority catchments from the monitoring campaign. Instead, for these low priority catchments, existing telemetry (SCADA) data at ancillaries (AWS's SCADA system comprises telemetry sensors at key ancillaries, which provide real-time performance metrics e.g. water levels at wet wells, pump status, treated effluent flow rate) alongside radar rainfall data were adopted for model calibration. These catchments were mainly in locations where drivers such as growth were not as highly ranked at MDM1 stage, but which were still relevant to AWS's business needs for the next AMP cycle (AMP7). The final number of catchments that had a new flow survey carried out was 143.

For the selected 143 catchments, site selection for each flow monitor was made with potential hydraulic conditions, access requirements and health and safety (H&S) in mind. For example, locations on bends in the network and chambers that may have turbulent conditions were avoided. The initial selection was automated through SQLs in Infonet (asset database). Accounting for the above-mentioned considerations improved the likelihood that, when the site engineers came to install the monitors, the location would be suitable for accurate data capture. It was also crucial during the selection of sites that traffic management and site access were accounted for, avoiding locations where costly traffic management would need to be put into place to install and remove the monitor, or where access to manholes would be difficult due to access restrictions on private land. In catchments where we had access to information on historic flow surveys, we installed monitors in the same locations where we could confirm that the data captured previously were of an acceptable standard. This increased the likelihood that it would be possible to install the monitor, and the hydraulic conditions would lend themselves to accurate data capture. Details on H&S factors considered during survey planning can be found in Section 11.5.4 below.

Table 11.3 WaPUG code of practice recommended rain gauge densities for short term flow surveys (updated version: CIWEM, 2017).

Type of terrain	Typical number of rain gauges
Flat	$1+1$ per 4 km^2
Average	$1+1$ per 2 km^2
Mountainous	$1+1$ per 1 km^2

Flow monitoring was accompanied by rainfall monitoring using tipping bucket rain gauges. Rain gauge densities (Table 11.3) were decided upon following WaPUG (UK Wastewater Planning Users Group, now CIWEM Urban Drainage Group) criteria (CIWEM, 2017).

A total of 3,568 flow and/or depth monitors and 801 rain gauges were installed between March 2017 and November 2019. The aim for each flow survey (with each one of the selected 143 urban drainage catchment areas having one individual flow survey done) was to capture three WaPUG compliant storms (WaPUG compliant storm events must have total rainfall depth $\geq$5 mm, rainfall intensity $\geq$6 mm/h for more than 4 minutes, and should ideally display limited spatial variability, to ensure that the rainfall field can be well represented based on rain gauge measurements) and two dry weather days, one on a weekend and one on a week day. This enabled urban drainage model calibration and verification in line with WaPUG guidelines. Achieving this target across the vast number of monitors installed was not easy and required adhering to a detailed programme of works. In some cases, when the flow survey had been in the ground for a period of more than 12 weeks, storm events that had only a partial number of monitors pass the WaPUG criteria were then accepted as viable events. This helped to ensure that prolonged dry spells did not significantly impact the programme of works.

11.5.3 Monitoring system – technical specifications

As indicated above, a total of 3,568 monitors and 801 rain gauges were installed between March 2017 and November 2019 as part of AWS's monitoring campaign. Technical specifications of the sensors, data loggers and data transmission, storage and management system are given next.

11.5.3.1 Flow monitoring

Generally, two in-sewer flow attributes were monitored: depth and velocity. Based on depth and velocity measurements, alongside conduit geometry, it was possible to estimate flow rates. It is worth noting that, depending on the monitoring purpose and site characteristics, at some locations only depth monitoring was undertaken.

Three types of in-sewer sensors were used in the monitoring campaign:

- *Detectronic multi-sensor flow meter* (MSFM), which includes ultrasonic velocity, pressure depth and (optional) ultrasonic level.
- *Technolog Cello 4S with depth pressure sensor.*
- *Technolog Cello CSO ultrasonic depth recorder.*

The three sensors are encapsulated to withstand harsh environments and can be used to monitor raw sewage, industrial effluent and stormwater. Likewise, the recording frequency is programmable between 1 s and 1 h.

In the case of the AWS's surveys, a logging frequency of 2 minutes, which is the UK industry standard, was adopted.

The MSFM sensor was used at approximately 90% of the locations, as it is the only sensor out of the three sensors used in the survey which can monitor velocities and depth, thus allowing flow calculation which was required in the vast majority of sites. The Cello depth sensors were used at locations where only depth monitoring was needed (e.g. at some CSOs and tanks) or where the MSFM monitor was unsuitable, with the limiting factor of the MSFM monitor being the range in which it can accurately monitor depths (i.e. a maximum of 3.5 m). Generally, the Cello pressure sensor was used more often than the ultrasonic one. In fact, the ultrasonic sensor was only used at a handful of locations where the site conditions did not allow for the installation of a pressure sensor (e.g. due to access and/or flow conditions).

11.5.3.2 Rainfall monitoring

Rainfall across the catchments subjected to flow monitoring was measured by means of tipping bucket rain gauges equipped with GPRS data loggers. Key rain gauge and data logger specifications are summarized in Table 11.4. As mentioned earlier, rain gauge densities were decided upon following WaPUG guidance (Table 11.3).

The reader must be reminded that for catchments not subjected to short-term flow monitoring, available telemetry data (i.e. SCADA depth and flow records at ancillaries) and radar rainfall data were used for model re-calibration. The radar data used for this purpose were the UK Met Office data, available at 1×1 km^2 and 5 min resolution from the CEDA archive (http://ceda.ac/uk/, accessed 17 Dec. 2020).

11.5.3.3 Data transmission, storage and retrieval

In-sewer monitors and tipping bucket rain gauges were equipped with GPRS data logging and transmission units.

Data from each logger were transmitted on a daily basis via a 2G network to a central physical server. All loggers were set to transmit data from 6 a.m. every day. The logger would attempt to send data for up to 3 hours after the first attempt but would then cease in order to conserve battery life. Once in the server, data could be visualized via the RPS Flow Survey Online Viewer. The reason data were transmitted only once per day, rather than at a higher frequency, was to preserve battery life.

It is worth noting that data transmission issues – related to poor signal – were encountered in approximately 35% of all monitoring sites. Data from the problem sites were retrieved manually on a weekly basis; this entailed renewing traffic permits at traffic-sensitive sites. The weekly retrieval frequency allowed timely data review, and therefore timely maintenance, and avoided filling up the

Table 11.4 Technical specifications of tipping bucket rain gauges and data loggers.

Description	Units
Bucket size	0.2 mm
Catchment area	400 cm^2
Accuracy	$\pm$1% at 26 mm/h
Logging frequency	Variable – time of tip
Storage capacity	In a practical sense unlimited

loggers' internal memory. The memory capacity at 2 minutes logging rate, as is the case for flow monitors, is approximately three weeks.

Final datasets were available in FDV format. Data in this format can be readily imported into RPS' data processing and hydraulic modelling software packages, i.e. FlowBot data processing toolbox (described below) and InfoWorks.

11.5.4 Health and safety management

Health and safety (H&S) considerations were kept in mind throughout all survey stages, including survey planning, installation, operation, maintenance and decommissioning.

During survey planning a desktop-based hazard screening was undertaken for every monitor location, including preferred and alternative locations. In the case of in-sewer monitors, to be installed via manholes, consideration was given to the specific risk of the location. For example, no locations on roads with speed limits of over 30 mph (48 km/h) (unless locations were rural) were planned and locations such as major junctions, pedestrian crossings and roundabouts were avoided. Likewise, consideration was given to access, proximity to emergency services and schools, manhole depth (with too large depths avoided and/or marked for inspection) and pipe size, amongst other factors. In the case of ancillaries (e.g. CSOs, pumping stations, treatment works), consideration was given to factors such as site configuration and presence of electrically-powered elements.

Once potential monitoring locations were identified, pre-installation surveys were carried out. These consisted of visiting the potential monitoring sites to determine if they were suitable and safe to install in. For in-sewer monitors, only manholes classed as low or medium risk and having direct line of sight to the person entering were selected for installation (i.e. NC1 and NC2 sites as per Water UK National Classifications (NC) for Confined Space Entries (Water UK, 2019)). In the case of complex ancillaries within AWS compounds, survey contractors were required to complete site induction upon arrival and to consult with operatives to decide on the safest plan of action. During these visits, gas monitoring was performed at all times, even if a confined space entry was not going to be made. The gas monitor would be lowered into the manhole and/or around the ancillary to ensure that it was safe to stand over it to look inside to assess the site's suitability for flow monitoring. In addition, one operative would have PPE (personal protective equipment) on for a confined space entry as it is not always possible to assess the site from above. During those visits' access/egress was assessed, including review of dimensions to ensure safety of the configuration, and depth of flow was estimated with sites with water levels deeper than 500 mm being discarded on safety grounds.

At the end of the pre-installation stage, an installation plan was formulated for each catchment. Based on assessed risk, monitoring sites were classed as 'generic' or 'site specific'. The 'generic' category generally covers NC1 and NC2 manholes (for installation of in-sewer monitors) and rain gauge locations. The 'site specific' category generally covers complex ancillaries. For 'generic' sites, standard risk assessments and method statements (RAMS) were prepared for installation, while for 'site-specific' monitoring locations customized RAMS were prepared for each site. Installation was generally carried by 3-man crews, following the corresponding RAMS.

Once the surveys went live, survey contractors and hydraulic modellers assessed monitor performance on a weekly basis (more details in following section). Any sites identified as performing poorly, either due to sensor malfunctioning or site characteristics, were flagged up and maintenance was undertaken following the corresponding RAMS.

Survey progress in terms of storm event and dry day recording was also reviewed on a weekly basis by hydraulic modellers. Once three WaPUG-compliant storm events and two dry days had been successfully recorded at a given catchment, a decision was made to terminate the given survey. Monitor decommissioning was carried out following the corresponding RAMS.

11.5.5 Data quality assurance during and after monitoring period

Flow and rainfall monitoring data were assessed on a weekly basis as well as at the end of the survey period using RPS' in-house FlowBot toolbox.

FlowBot is a bespoke software solution that can be utilized by both flow survey data analysts and hydraulic modellers to visualize, manage and understand flow survey data. FlowBot allows undertaking a range of checks on the depth and flow measurements, including flow monitoring data comparison against theoretical depth vs. flow rating curves, flow monitor comparison including volumetric checks between upstream and downstream monitors, and automatic identification of dry weather and storm events based upon user-defined criteria. Many of these checks are in line with those recommended in the Guide to Short-Term Flow Surveys (WRc, 1987). In addition, FlowBot allows for the deployment of a machine learning algorithm (decision tree-based) to classify the performance of flow and depth monitors and ultimately identify faulty measurements. The machine learning algorithm has been trained using 30,000 days of previous human classifications, mined from previous RPS flow surveys. The outputs that are generated have a high degree of accuracy, with assessments that would usually take hours being completed in minutes. Automating this user-intensive operation reduced the time spent by engineers classifying data and allowed for more time to be spent proactively managing a flow survey, maximizing the quality of the final data. In addition to depth and flow checks, FlowBot includes features for rain gauge data quality assurance via cumulative rainfall plots.

On the whole, the use of the FlowBot toolbox led to 80%-time savings in the processing of flow survey data, in relation to conventional (more manual) methods.

By undertaking weekly assessments while the survey was still on the ground, it was possible to detect and rectify any problems with given monitoring locations throughout the survey and ultimately ensure that good quality data were collected at the locations of interest. Problems encountered with monitors included power failure, transmission problems due to poor network coverage and/or location of the monitor inside sewer network, ragging, flow/depth data below monitor resolution, and rain gauge blockages, amongst others. Some of the problems could be rectified through sensor maintenance. Others required sensor re-location. Ultimately, quality assurance of flow survey data ensured that the best data possible were adopted for model calibration and verification.

11.5.6 Conclusions and outlook

As part of Anglian Water Services' AMP6 Integrated Urban Drainage Modelling Programme, short-term (average of 12-weeks duration) flow and rainfall surveys were undertaken across the majority of the company's catchments, resulting in a total of 3,568 flow and/or depth monitors and 801 rain gauges being installed and operated between March 2017 and November 2019. The data collected through these surveys enabled standardized re-calibration of hydraulic sewer models to a level of confidence appropriate for intended use. The resulting models, alongside other company datasets (e.g. permanent monitoring data at strategic locations), will enable a range of catchment performance assessments and implementation of management strategies, which will ultimately lead to improved catchment management.

The magnitude of this project required a systematic approach and enabled development of survey planning and execution, and data quality-assurance tools, which delivered significant efficiencies and ensured collection of high-quality flow survey data. In particular, the magnitude of the project enabled extensive validation with Machine Learning (ML) tools for data quality assurance, which were shown to provide an efficient alternative to time-consuming manual data review.

Likewise, essential to the delivery of this project was the collaborative work across different business units and with external partners, including survey and hydraulic modelling contractors.

Problems encountered throughout these surveys included, amongst others, poor signal at approximately 35% of the sites (in which cases data had to be manually retrieved) and difficulty in capturing the three WaPUG-compliant storm events within the envisaged survey duration (i.e. 12 weeks) in some of the catchments. Likewise, maintenance of and data removal from traffic sensitive sites required constant renewal of traffic permits, which was time-consuming. These problems were however tackled with proactive data review and effective work-planning and collaboration between the modelling and survey teams.

The data collected as part of these short-term surveys will be stored centrally by AWS, both to provide an audit trail as well as to enable future studies such as detailed flooding and pollution investigations, further development of ML-based data assurance tools, preliminary testing of real-time analytics, amongst others. UK Water Utilities normally keep their short-term flow data in-house, and not publicly available. Depending on the type of data and the sensitivity of the catchment, access may be arranged for research on a case-by-case basis, through individual legal agreements with the researchers. Some higher-level aggregate longer-term monitoring data are available publicly through OFWAT's annual performance reports (https://www.ofwat.gov.uk/regulated-companies/company-obligations/annual-performance-report/, accessed 28 April 2021).

11.6 IMPAKT! – OPTIMIZATION OF THE URBAN DRAINAGE SYSTEMS IN THE DOMMEL AND WARMBEEK RIVER SUBBASINS, FROM A RIVER QUALITY POINT OF VIEW, FLANDERS, BELGIUM

11.6.1 Scope and objectives

The IMPAKT! Project (https://www.grensregio.eu/projecten/impakt, in Dutch, accessed 28 April 2021) aims at defining measures to reduce the impact of wet weather urban discharges on the receiving rivers' water quality and ecology. The two river subbasins Dommel and Warmbeek comprise five drainage areas and five municipalities. The definition of the optimal measures is to result from scenario analysis on a set of integrated and linked models (sewer system, WWTP, receiving water). An extensive sewer flow and sewage quality monitoring campaign was set up with a view to both validating the different models and keeping them up-to-date in the years to come. Furthermore, it is expected that thorough analysis of the many measurement data will allow the optimization of the many pumping stations and storage basins.

11.6.2 Measured variables and location of monitors

A total of 75 monitoring sites have been defined throughout the project. As budget constraints did not allow for parallel monitoring at all locations, the project was designed to work with a rotation system for (part of) the sensors. This is reflected in the design of the used database. The average period for which monitors had been operated at one individual site for rotated devices was between 9 and 12 months. This period was chosen to capture seasonal variations and as a result of practicalities. The absolute minimum for monitoring at one location is deemed to be around 3 months. First sensors were installed in March 2017. As of July 2020, about 35 locations were still active without any plans for de-installation.

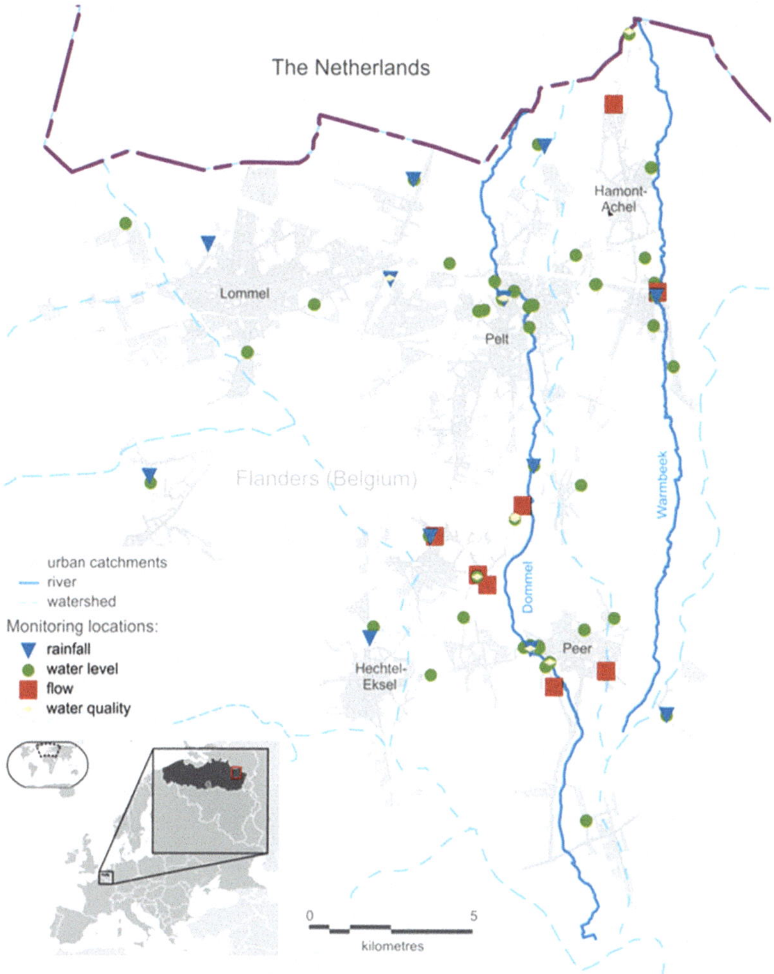

Figure 11.10 Monitoring locations in the project area of the municipalities Lommel, Hechtel-Eksel, Pelt, Peer and Hamont-Achel (Flanders, Belgium); lower left: case study location. *Source*: adapted from https://nl.wikipedia.org/wiki/Vlaanderen#/media/Bestand:Flemish_Community_in_Belgium_and_Europe.svg (accessed 28 April 2021).

Figure 11.10 shows an overview on the finally selected locations for flow, water level and precipitation, but also continuous water quality monitoring (temperature, conductivity, turbidity, TSS (total suspended solids), BOD (biological oxygen demand), COD (chemical oxygen demand), NO_3, NH_4, all backed up by lab samples) not detailed here any further.

11.6.2.1 Flow and water level

Flow is monitored at 13 locations using 10 devices based on proven technology: hydrostatic depth transducer and ultrasound Doppler velocity sensor. Such devices have been successfully used throughout many short-term campaigns over the last 20 years (using an external contractor). Usually the focus of short -term campaigns, which typically run for seven weeks with preferably a minimum of three reasonable rain events being monitored (WRc, 1987), is on the characterization of the system's most relevant locations and ancillaries (e.g. large CSO structures, throttle locations, pumping stations and joining points of collectors) to develop a sound understanding of systems hydraulics. In the IMPAKT! Project this classic approach was maintained for much longer (approx. 1.5 years) to allow for the identification of seasonal variations in dry weather flows, and to allow rotation between a number of locations.

The data gathered by the flow monitoring campaign are extended through the permanent installation of 20 additional water level monitoring devices (pressure gauges) at CSOs (in collectors and/or storage tanks), and 15 data loggers capturing existing level measurements in pumping stations. At these locations full flow monitoring is of less or no relevance given the poor velocity conditions, but levels monitored at pumping stations can be converted using the wet well's geometry to estimate flows as done by Fencl *et al.* (2019).

Desirable monitoring locations are selected based on detailed analysis of the existing sewer models or asset databases. Ideally, more than one potential monitoring location is identified for the same purpose. All potential monitoring locations to be maintained in-house are visited by a team of modeller(s), desk personnel in charge of sensor operation and the technician(s) responsible for installation and maintenance in order to evaluate their suitability for the monitoring task at hand and possibilities for sensor installation. Discussed criteria were, depending on locations: required safety measures to be taken for installation/maintenance, (lack of) options for sensor positioning, relevance and representativeness of the location for the process to be monitored, expected disturbances, requirements for sensor configuration (offset set-up).

Monitoring location choice is paramount. During dry weather, monitoring locations situated several kilometres downstream of a point of high impact will still reflect upstream influence as shown in Figure 11.11.

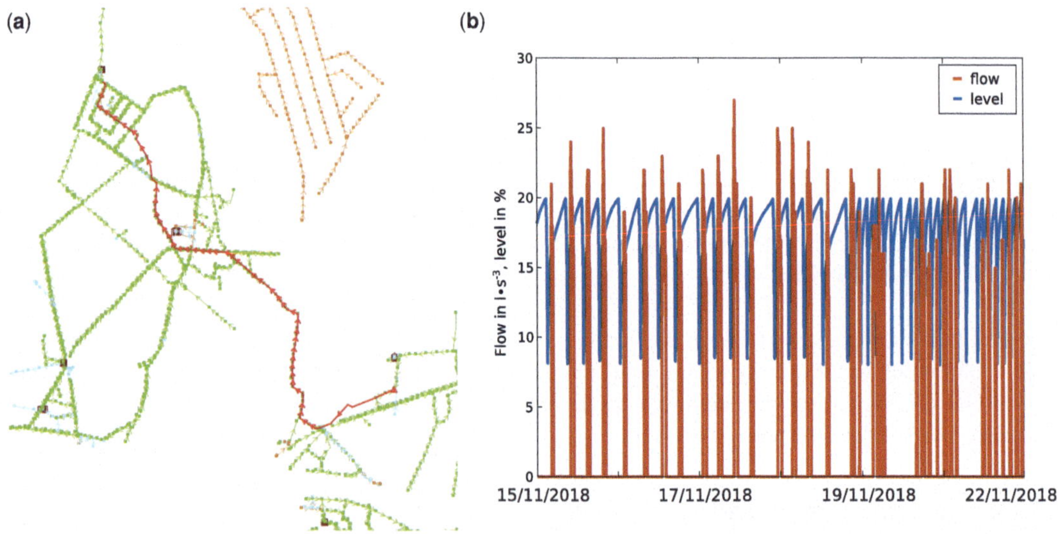

Figure 11.11 The change in the dry weather pattern in a pumping station is traceable 3.5 km downstream in a downstream flow monitoring location; (a) flow path between discharge point and monitoring location; (b) change in patterns recorded on 19/11/2018. *Source*: Aquafin NV.

The upstream pump logging clearly shows a sudden change in the duration of the pump cycle during dry weather. This is confirmed in the flow pattern at the downstream monitoring location. This example proves that the choice of locations should ideally take into account the potential interactions between different parts of the system.

The nature of the system (flat with fast surcharging or steep with mostly free-flowing conditions) defines whether specific monitoring locations and measured hydraulic parameters should be focusing on both dry and wet weather verification or only one of them. In Flanders, most models should be verified from upstream to downstream using low flows and from downstream to upstream using levels in surcharged conditions. Very often velocity measurements appear to be missing in surcharged conditions whereas the model can prove that in reality they drop below detection limit. It is therefore important in such systems not to focus on just level and flows, but equally on velocity.

Another typical point for attention in dry weather is the potential unbalance between level and velocity where flows show a good correspondence. A backup monitor at a nearby location might be useful to determine whether the unbalance is a result of wrong roughness estimates or of local gradient deviations.

11.6.2.2 Precipitation

For standard flow campaigns at Aquafin, at least two or three precipitation monitoring locations are selected, depending on the density of urbanization (usually one per residential area). As these short-term campaigns are usually carried out by a contractor, roofs of public buildings are preferred installation sites. For the project presented here the size of the studied catchment resulted in 12 tipping bucket rain gauges. These are preferred over more expensive options as the loss in accuracy is deemed acceptable and allows for a higher number of monitoring locations. The devices are installed on the ground on fenced company facilities (e.g. wastewater treatment facilities, pumping stations or storage tanks) as they are maintained by in-house personnel. To select appropriate locations, all available fenced sites in the project area have been assessed for the distance of the mounting point to the site fence (vandalism!) and any obstructions and GPRS signal strength. All sites deemed suitable after this test were geographically mapped and final locations were selected based on their coverage of the study area through Thiessen polygonization. If multiple sites remained as candidates in close proximity to an optimal monitoring location, preference was given to sites that would also host other monitoring equipment to optimize maintenance schedules.

11.6.3 Data communication

Flow data are manually transferred to an external data server on a weekly basis (service provided by the contractor) and downloaded from there. For all other data sources, data acquisition is done wherever possible in real-time or at different frequency intervals (10 min … 1 day) depending on the type of device used and preferably via wireless machine-to-machine (M2M) communication via public cell phone networks (GPRS, 3G). This is also true for devices recording data of pumping stations that are equipped with an Ethernet connection to the local SCADA system: company-internal security guidelines prohibit the installation of uncertified devices in such production environments.

As the project area is situated close to country borders, the use of SIM cards with enabled roaming was planned, tests for actual improvement of signal strength are still ongoing. For all locations not requiring data roaming, virtual-private-network SIM cards have been used to allow for high data security paired with ease of use in a fixed IP address (Internet Protocol address) range (no firewall issues). Data loss due to failed communication was limited to $<1\%$ as a result of sufficiently large data storage in the loggers, timely manual intervention and the use of high-performance antennae where required.

The different types of monitoring devices are reflected in different ways for data communication and had to be accounted for in the data import routines for the individual sensor types. Communication is carried out either as a push or pull service, directly to the sensor or via a sensor specific data server in the form of files or API (application programming interface) calls by a number of device-specific Matlab™ scripts. Sensors offering neither of these data acquisition methods or requiring manual intervention for data read-out have not been considered for use in the project presented here (except these handled by the flow survey contractor).

11.6.4 Data management, validation and accessibility

The data management system is designed around a three-level hierarchy of monitoring site (location or mounting place), monitoring device, and data channel (a time series of any variable, e.g. flow or pump switch-on level). To accommodate this structure, the system has been designed in-house as part of the project presented here. As the structure and performance of the system have proved satisfactory, the system will be used for future projects. All data are stored in one central relational *PostgreSQL*™ database (http://www.postgresql.org, accessed 28 April 2021) running on a company-wide accessible dedicated virtual server. The decision for *PostgreSQL*™ was based on the ease of use of vectorized time series data (data with multiple measurements per timestamp) as they are frequently used for spectrometric water quality measurements.

A history is kept for all channels connected to a device and all devices located at a site. This way, the user can transparently select, e.g. a full time series of the water level monitored at a certain monitoring site even though the initially installed device has been replaced after some time. As the data are still recorded per device, rather than per site, the user can also analyse monitoring data for individual devices (or channels) allowing to e.g. trace back device-dependent calibration errors.

Currently, only a basic automatized data validation (handling of broken data files, erroneous data due to interrupted communication) and unit and daylight saving time (DST) conversion is implemented, subject to extension over time. After data import, additional calculated channels such as corrected rainfall or Poleni based spill flow are determined. Formulae required for these calculations are defined in the channel's meta-data, to allow for easy modification.

All time series and meta-data can be consulted and manipulated by a feature-rich and flexible querying tool and free programming based on Matlab™. This tool also allows the import, visualization and analysis of frequently used data from external sources such as historians (e.g. GE Historian™), web-based data servers (e.g. waterinfo.be) and simulation results (here e.g. from Infoworks ICM™, SWMM (Storm Water Management Model). Alternatively, a read-only web interface exists for quickly creating views of time series data and a real-time overview on monitoring performance and device connectivity in a GIS-enabled, browser-based environment implemented as overlays on Openstreetmap™ (Figure 11.12).

Time series data are dynamically queried using PHP (Hyper text Pre-processor, also known as Personal Home Pages) and visualized with an existing JavaScript charting library (http://dygraphs.com/, accessed 28 April 2021).

The fairly detailed and expansible data structure for meta-data is inspired by DWA-M 151 (DWA, 2014) and is aimed at rendering the database self-explanatory to users not accustomed to the project itself. For example, all monitoring locations can be visualized in Google Streetview™ through a link, allowing easy recognition of the location when on site.

Each monitoring channel, device and site is assigned a state calculated in real-time which reflects the fitness of the data it represents: the data logger shown in Figure 11.12(b) occurs in red (intervention required) as the database did not receive any data for one of its channels (Pump 5).

Daily backups of the entire database ensure it can be restored in case of irreparable data loss or damage.

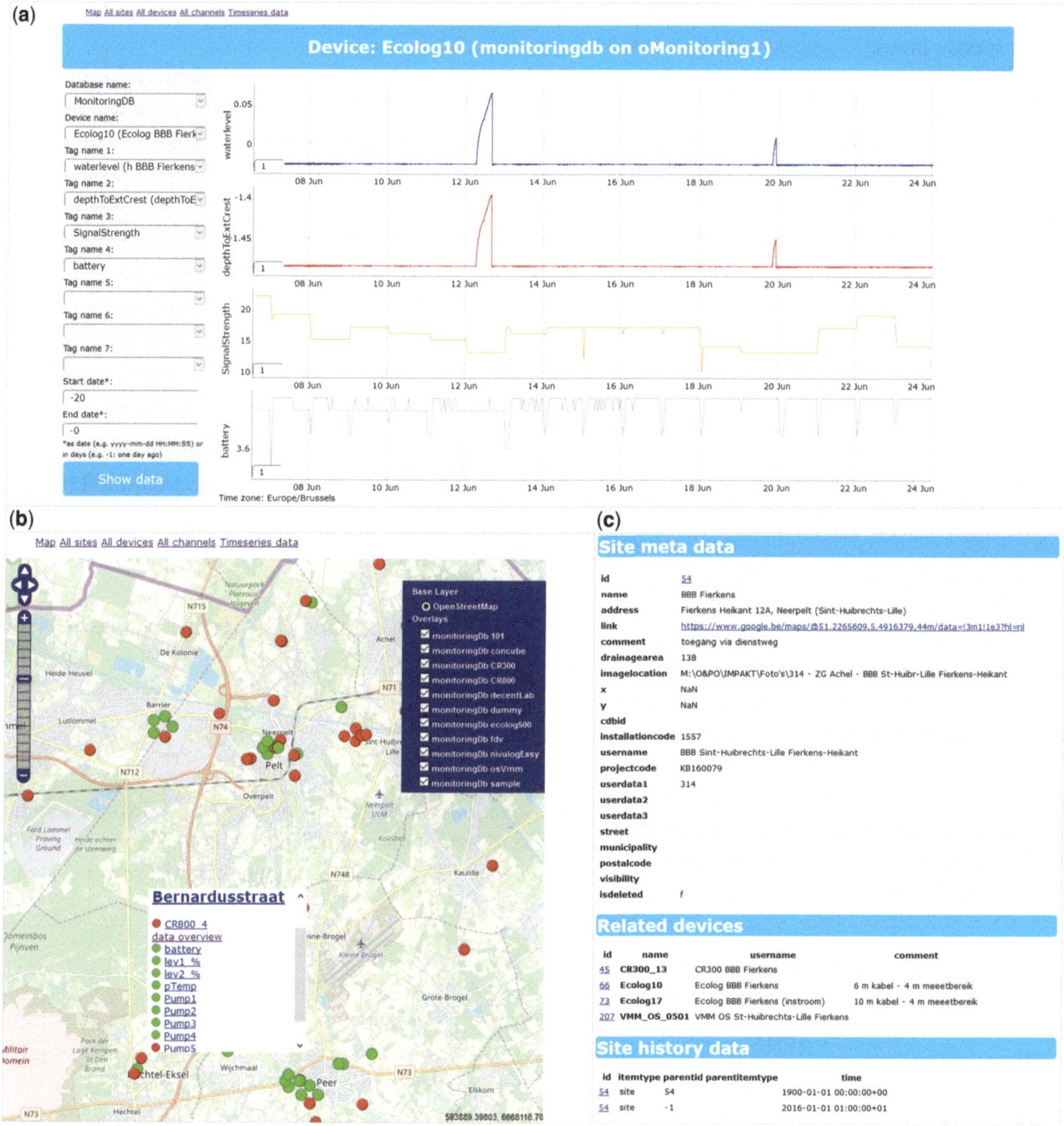

Figure 11.12 In-house browser-based real-time overviews on monitoring data; (a) time series data of three raw data channels and one calculated channel 'depthToCrest'; (b) map of sensor locations for monitoring performance and connectivity; (c) meta-data for one monitoring site. *Source*: Aquafin NV.

11.6.5 Sensor operation, maintenance and budget

11.6.5.1 Operation and maintenance

All monitoring read-outs are visually inspected at least weekly by the personnel who will later be working with the data, e.g. for model calibration to ensure sufficient data quality and plausibility. Implausible data, high noise ratio or comparable phenomena potentially caused by poor monitoring quality will lead to device inspection. Sensor maintenance is therefore accounted for in the weekly planning of technicians to ensure short response times.

All flow sensors are visually inspected on a weekly basis and calibrated under lab conditions once every year. Batteries are changed, if required. Water velocity and level are manually measured on site using a hand-held screw current meter and folding rule, respectively. Small corrections are possible on-site, large deviations will lead to sensor replacement. If necessary, the wetted section will be re-surveyed. In the case of doubtful results for flow, water level and especially velocity measurements can still lead to important insights into system characteristics and are used for data verification. For intervals of ambiguous flow data, at the very least a cross-comparison of level and velocity is carried out.

Pressure gauges for water level monitoring typically receive no regular maintenance. If a drift is noted in the data, the sensor is cleaned and carefully re-installed to its exact position in the location. In-house rain gauges are cleaned when personnel are on-site and if necessary. Rain gauges maintained by the contractor are inspected and cleaned on a weekly basis.

Sensors that work on non-rechargeable batteries require regular and in-time replacement of the batteries (lifetime varies depending on many parameters). Therefore, preference is given to equipment and locations that have power supply (e.g. pumping stations). The level sensors chosen for this project required battery replacement roughly every 3–4 months.

11.6.5.2 Budget and cost considerations

Only rough estimates of CAPEX (capital expenditure) and OPEX (operating expenditure) can be given in Table 11.5, all listed costs are excl. VAT.

11.6.6 Challenges and lessons learnt

11.6.6.1 Planning

Desirable monitoring locations are not always suitable locations: during this project, a sensor installed in the influent channel of one of the WWTPs led to an increased risk of clogging. This provoked flooding of the (unmanned) installation (Figure 11.13). This could have had the same effect anywhere in the sewer system: flooding caused by monitoring!

Table 11.5 CAPEX and OPEX estimates for monitoring equipment and maintenance.

	CAPEX	OPEX
Flow survey contract		80,000 €/year for 10 sites, weekly maintenance
Rain gauge	2500 € pp (incl. logger, modem)	
Water level sensor	1500 € pp (incl. logger, modem) 75 € pp (replacement battery)	25 person days per year for 20 devices, mostly for battery change, incl. some re-configuration
Data logger for pumping stations	1000 € pp (including modem)	20 person days for 25 locations, mostly modem re-sets, incl. cleaning of rain gauges
M2M SIM cards		5 €/month/device in a bundle within a company-wide contract with the network provider

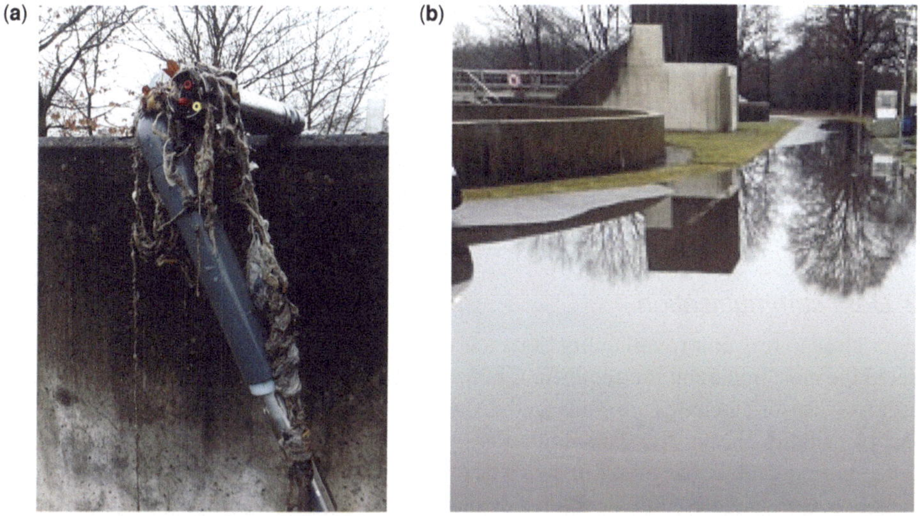

Figure 11.13 Flooding caused by monitoring; (a) monitoring device and accumulated debris after removal; (b) resulting flooding throughout the WWTP. *Source*: Aquafin NV.

Even though desk studies are valuable and frequently deemed the most significant source for the identification of optimal monitoring locations, the value of experience of operational staff and maintenance technicians cannot be rated highly enough, for systems characteristics as well as sensor installation. These personnel should be involved in the planning of the monitoring campaign as early as possible.

While the traffic situation at monitoring locations can play a major role, another important aspect is the accessibility of monitoring locations. This is especially true for public parking lots, where parking prohibition (depending on regional law) cannot be granted for the full period of the monitoring campaign. Temporal parking prohibition for regular intervention can then lead to significant administrative overheads making the choice of a different monitoring location preferable.

The installation of monitoring equipment under sealed manhole covers should be avoided. Most devices – even when tested for IP code 68 – are only water resistant for a limited amount of time.

Installations at locations with high turbulences in the conduit should be avoided for flow monitoring. Ideal locations are the incoming conduits of manholes with one single ingoing and one single outgoing pipe in line with one another and no nearby invert steps. Installation of flow meters directly in manholes is to be avoided to prevent the interference of debris accumulating in the manhole.

Preferably, flow sensors should be installed at locations without sedimentation. If this is impossible, locations with firm sediment layers should be preferred over locations with soft sediment, to prevent sediment accumulation on the sensor. While the installation of flow sensors outside the centre of the sewer invert is possible using an offset configuration, it can lead to less accurate results, possibly to asymmetric velocity patterns in the wetted section.

For flow measurements in conduits with large and complex cross-sections it can be beneficial to work with two sensors simultaneously to increase the accuracy of average velocity measurements. Most devices support this without requiring additional post-processing of the monitoring data for the user.

The installation of monitoring equipment in close vicinity to high-voltage cables should be avoided as the electromagnetic field can – in rare cases – influence sensor performance and wireless communication.

Especially when using legacy devices, flow meters should be installed in conduits with standard shapes, if possible, as the configuration of user defined profiles can be tedious and error prone.

One frequently employed way of installing flow monitors in sewer pipes is through the use of mounting rings built from metal plates that can be installed in the pipe without the need for mounting holes in the pipe walls. While this is the only way of installing a flow measurement in re-lined conduits (where drilling holes into the pipe walls is not an option), at all other locations, the flow meter should be directly mounted to the conduit wall as mounting rings might untighten, especially in harsh hydraulic conditions and at locations prone to sedimentation.

11.6.6.2 Data communication

Real-time (or close to real-time) availability of data is commonplace for almost all types of sensors and can be considered the most important pre-requisite for efficient sensor maintenance. It should be insisted on for all new devices if sensor purchases are envisaged. Download intervals should be chosen as a compromise of urgency for plausibility checks and data validation (and resulting counter measures) and battery lifetime. Data security considerations can make it necessary to rely on GPRS connections for data acquisition even at locations were wired communication is available.

While the adoption of standardized protocols for data communication should be encouraged for new devices, the use of legacy devices will make it necessary to design an extensible, flexible system, which can handle arbitrary (future) data formats.

11.6.6.3 Operational aspects including data management

Flexible plastic tubing, as used for the protection of network or electricity cables in building constructions, can be applied for additional protection of cables subject to in-sewer installations. They then serve as an effective prevention measure of sharp bends in the pressure compensation lines of pressure gauges.

Responsibility – sensor stewardship: for each device there should be a contact person, a staff member who knows where and why the sensor was installed, and what its typical mode of operation is. Ideally, this person will later have to work with the sensor data to ensure a natural interest in the quality of these data resulting in frequent (≤weekly) plausibility checks and – if required – maintenance.

Data that cannot be placed in the correct context are of little value. The use of meta-data should thus be enforced wherever possible. This can be easily realized by the sensor network administrator by only allowing sensors to be added into the monitoring system if a minimum amount of meta-data (e.g. proper naming, address or X-Y-coordinates of installation site) are provided so that the monitored signals are self-explanatory to colleagues and project partners.

Aside from flexible data analysis tools, also low-threshold, ubiquitous access to the data should be provided to ensure project-wide adoption. The usability of the web interface (initially purely conceived for maintenance personnel) of the project presented here has led to other projects migrating their data into this database purely for its ease of data access.

For any monitoring campaign that accommodates multiple sensors for more than a couple of weeks, the use of an adequate database architecture should be preferred over data storage in individual files: the amount of data currently maintained in the database (80 GB as of July 2020, after 3.5 years of operation) would be challenging when using files but is inconsequential when using database systems. Once a suitable database system is established, it can easily be extended to allow for the connection of new sensors and device types. This, in turn, will lead to increased acceptance of the system over time and ultimately allow for more professional and sustainable data handling and create a platform for the accumulation of knowledge and experience with respect to monitoring campaigns.

Data backup (for time series and meta-data) will become relevant at some point in every monitoring project. Proper data backup and restoration procedures should be in place and regularly tested (a copy of time series data into a spreadsheet/text file is no backup!).

Company-/project-wide communication and operator involvement are of the utmost importance for good operation: sensor data that are visible to the operating staff in real-time, and help explaining system characteristics, will be cared for. The inverse is true also.

The currently used sensor network and data management and storage system(s) is/are never complete. Easy integration of external data from other platforms or files and future monitoring campaigns should be anticipated.

11.7 'NEXTGEN' URBAN WATER MONITORING – A HIGHLY DISTRIBUTED FIELD MONITORING OF URBAN DRAINAGE NETWORK WITH AFFORDABLE SENSORS AND REAL-TIME DATA COMMUNICATION, AUSTRALIA

11.7.1 Scope and objectives

In conventional urban water monitoring, a sampling station is usually established at the catchment outlet to measure the flow and to analyse pollution levels. Researchers relied on data collected at this single point to develop models and design mitigation strategies, however, most of them end up with a poor model performance (Bonhomme & Petrucci, 2017; Dotto *et al.*, 2010). One of the main reasons for this is that by lumping the study area into a simple system, researchers ignored the inconsistency (e.g. land uses and randomly-occurring dry weather discharges) happening at the site-specific level within the catchment. To overcome these, the NextGen Urban Water Monitoring is a long-term data collection approach aiming at providing real-time information about flow and water quality in urban catchments at high spatiotemporal resolution (i.e. depends on the monitoring purpose – usually at the street or lot scale).

The high cost of installing multiple conventional sampling stations (>AU$ 12,000 ~ € 7,300) for a flowmeter and autosampler) in an urban catchment hinders data collection at a higher spatial resolution. Making the sensor and data transmission affordable but also providing reliable readings have become the top priority of NextGen monitoring initiatives. From the pollution source tracking perspective, the proposed sensor and data logger must be smaller than conventional sampling techniques which do not require any surface assets. In addition, the sensor must be easy to install and relocate within a catchment, i.e. avoiding any confined space entry within the catchment to minimize the budget and time to do one installation.

Based on the fundamental requirements of NextGen Urban Water Monitoring, this project was inspired by the recent development and innovation of Arduino and Arduino-compatible hardware components. The proposed Arduino data logger with 3G shield, 7.2 V battery, depth and EC sensors (Figure 11.14) were tested under lab and field conditions and eventually installed in a study catchment to track the highly polluted discharge area. The installed Arduino loggers with telemetry functionality continuously upload live data to a development website for data adjustment (e.g. calculating water depth and EC from raw data) and storage (Figure 11.14). Based on pre-set triggers, alarms will be sent to corresponding persons via SMS, which allows an immediate inspection when unusual discharges appear during a dry weather period. The collected spatiotemporal dataset can also generate an FDI (fault detection and isolation) database which characterizes the appearance frequency, flow duration and intensity of various land uses and urban activities. Such a database allows researchers to consider the impact of non-rainfall driven pollution sources in urban water models.

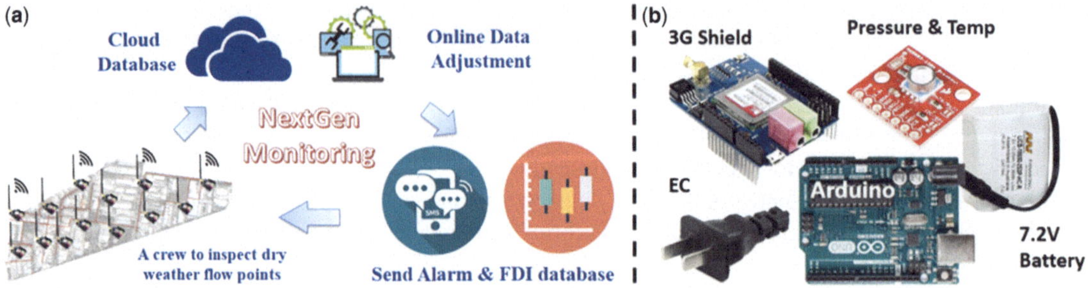

Figure 11.14 (a) data collection process of NextGen Urban Water Monitoring network; (b) Arduino logger package including cheap US power cable, 3G shield, MS5803 pressure sensor and 7.2 V battery. *Source*: EPHM Lab and BoSL.

11.7.2 Measured variables

11.7.2.1 Water depth measurement

The water depth of stormwater flow is measured by a high resolution and low power altimeter sensor – MS5803-01BA from TE Connectivity. As MS5803 measures the absolute environment pressure, another sensor needs to be installed in the catchment surface to correct for air pressure changes and temperature for density changes. A 3D printed sensor case was designed for easy installation in the urban drainage network. A potting compound gel was utilized to fill the sensor case, which ensures all the electrical connections and the sensor itself will be waterproofed under 3 m of water.

To verify the accuracy of the Arduino depth sensor MS5803, we installed one Arduino sensor adjacent to a HACH submerged probe inside a stormwater drain (Figure 11.15(a)). At this specific sampling location,

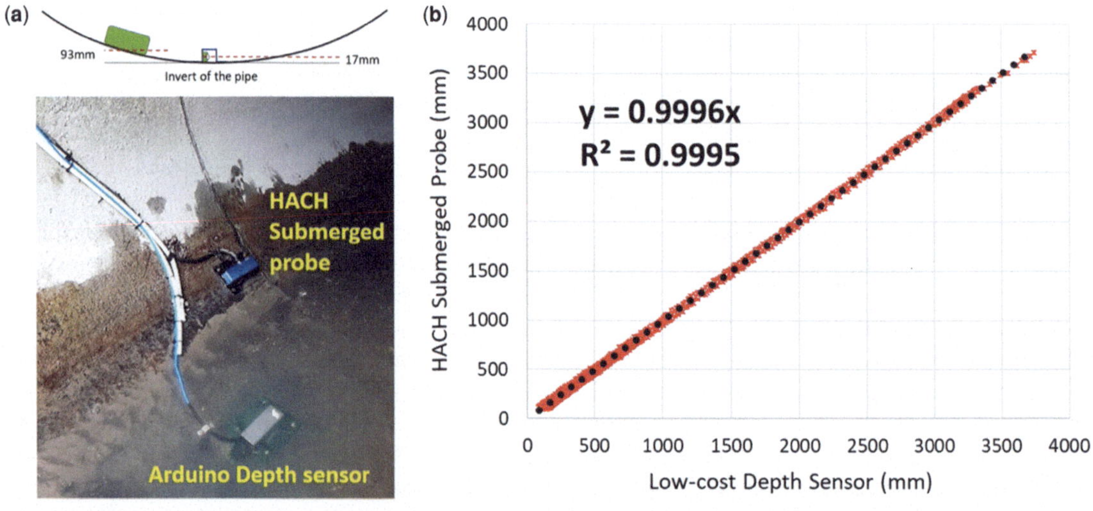

Figure 11.15 Arduino and HACH sensors installation location inside an urban stormwater drain; (a) The water level readings of a rain event; (b) Correlation between Arduino and HACH sensors. *Source*: EPHM Lab and BoSL.

although the pipe diameter is only 1.8 m, the water level could go up to 3.8 m and become the overland flow of a creek. The HACH sensor was installed towards the side of the pipe due to the high sediment accumulation, which always buries the sensor, thereby affecting the level and velocity readings. The comparison between the low-cost Arduino depth sensor and the HACH submerged probe over a one-month deployment (Figure 11.15(b)) shows a linear correlation, which indicates that the low-cost sensor is capable of measuring the water level as accurately as high-end products.

11.7.3 Study catchment, location of monitors and installation methods

In order to verify the performance of NextGen Urban Water Monitoring system for detecting urban dry weather discharges, more than 20 Arduino-based loggers have been installed in stormwater drains of Old Joes Creek catchment (Figure 11.16). Old Joes Creek is a suburban catchment with mixed industrial and residential land uses towards the east of Melbourne with a total catchment area of 854 ha. The red-shaded area is a 200 ha industrialized region close to the catchment outlet, which has been recognized as the major pollution contribution of the downstream waterways. Hence, 17 Arduino loggers are installed inside the industrial area mainly to capture the discharge events. A few sensor modules are also installed outside the area to capture different commercial activities like the shopping precinct, local community centre and even residential input.

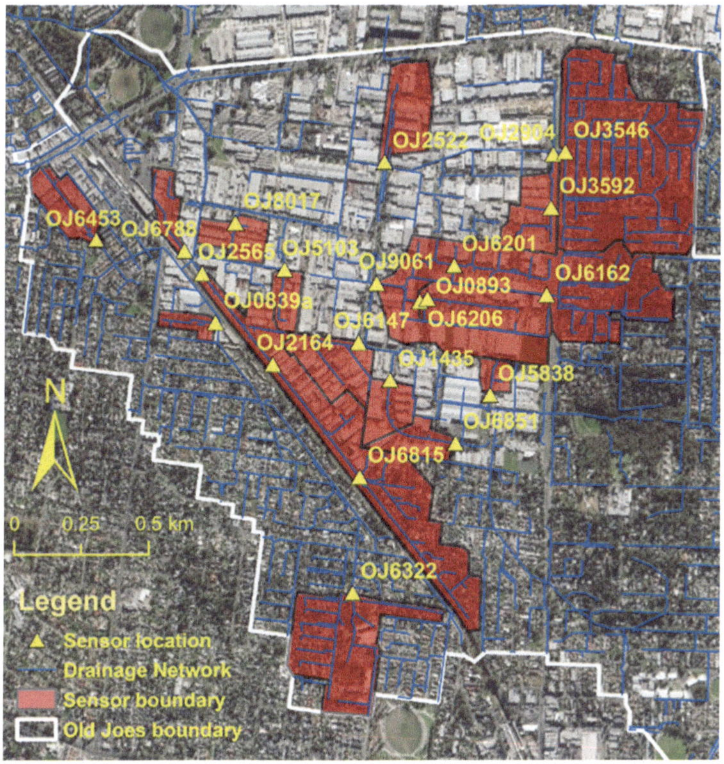

Figure 11.16 NextGen Urban Water Monitoring system in Old Joes Creek – locations of installed monitors, catchment boundary and drainage networks. *Source*: EPHM Lab and BoSL.

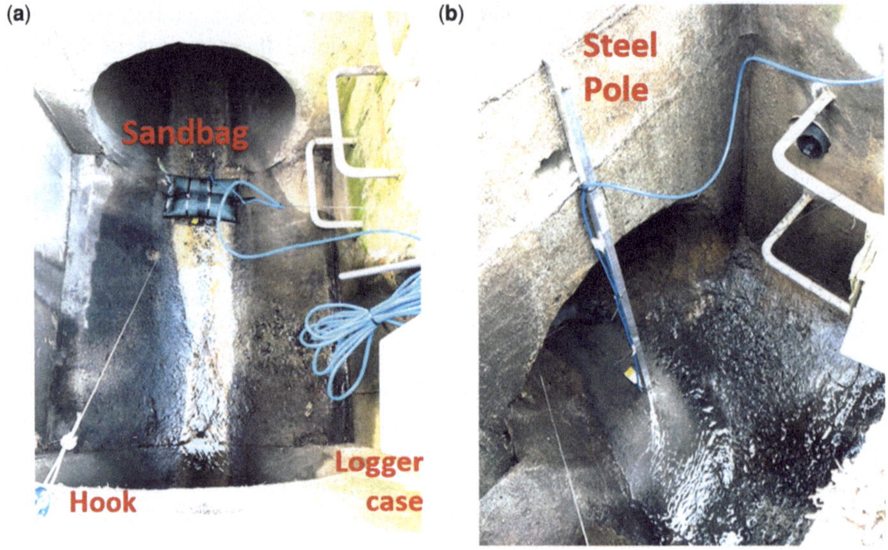

Figure 11.17 Typical installation methods of NextGen Urban Water Monitoring sensors in drainage network – attaching to a sandbag (a), fixing a pole with anchors and connecting the sensor to the bottom of the pole (b). *Source*: EPHM Lab and BoSL.

To minimize the installation time of one sensor module and make it easy to maintain and relocate, confined space entry into stormwater drains is not permitted in this project. Two installation methods were trialled in this study: attaching the sensor to a sandbag or fixing the sensor to a steel pole (Figure 11.17).

The sandbag is used to keep the sensor in the right position and prevent it being washed away during the wet weather event. Two stainless steel ropes are attached to the front corners of the sandbag, and the other ends are attached to installed hooks closed to the pit lid. The significant advantage of the sandbag method is the quick installation – usually, it takes 30 minutes to install one sensor. However, the research team has also experienced two major issues: (i) sandbag can easily flip over after a large rain event, leaving the sensor on top of the sandbag (in the air) and detecting nothing; (ii) the sandbag will create a pool of water, which is acceptable when we only want to know whether there is flow coming down from the upstream, but means it is not possible to record the true water level.

As an enhanced option, a steel or aluminium pole has been used to hold the sensor at a fixed location, without requiring a worker to enter the drain. This installation method has been trialled at 10 sampling sites, which all showed reliable sensor readings over a sampling period longer than 3 months. The sandbag is not required in this method anymore, so the measured water level will not be affected by pooling water. During the field trial of this installation method, the only issue is that the installed pole tends to capture leaves and other plastic rubbish, and so requires regular manual cleaning. The research team was worried the stick might be bent during a rain event, but it was strong enough to even hold the sensor in the right place under two major flood events where the stormwater overflowed the pit. In conclusion, the steel/aluminium pole method will be the standard method of installation for future urban drainage network monitoring projects. The low-cost sensor is also easy to install for other water assets such as the wetland and biofilter.

Table 11.6 Detailed costs of Arduino-based sensor of NextGen urban water monitoring.

	Components	Cost (AU$)
1	Tosduino Uno R3	16.00 (€ 9.5)
2	SIM5320 Shield	70.00 (€ 42)
3	MS5803–01BA Module	28.00 (€ 16.8)
4	Waterproofed box	2.60 (€ 1.70)
5	Cables and connection glands	6.00 (€ 3.60)
6	Marine epoxy and araldite	10.00 (€ 6)
7	7.2 V lithium battery 9500mAh	105.00 (€ 63)
8	SIM card	5.00 (€ 3)
Total		242.60 (€ 145.5)

11.7.4 Sensor cost and maintenance

The net present value of one Arduino-based monitoring module is approximately AU$ 242.6 (€ 145.5). The detailed cost of individual parts is listed in Table 11.6. The lithium battery and SIM5320 shield for 3G connection are the two most expensive parts, which account for 72% of the total cost of one module. The cost of the steel or aluminium pole is slightly variable per site depending on the depth of the manhole.

Currently, the sensor network maintenance including sensor checking, re-calibration, re-installation, battery changing and site cleaning runs on a bi-weekly basis which takes a full day of 8 hours of two employees. With increased numbers of sensors being installed in the catchment, the cost and time taken for network maintenance may further increase. In order to minimize the maintenance cost, apart from achieving a stable performance of the sensor network (i.e. minimize sensor malfunction), the power consumption of the current Arduino-logger should be further reduced to make the sensor run for longer than three weeks. Our newly developed Arduino-compatible module (in testing) is aiming to have a lifetime of more than half a year by minimizing unnecessary functionalities of the existing Arduino and 4G internet connection technology.

11.7.5 Data storage and website management

The data collected in this case study are stored in the web-based cloud of the Enhancing our Dandenong Creek project (EoDC, http://www.eodc.com.au, accessed 28 April 2021). The website is designed and maintained by BoSL of Monash University, Australia (http://www.bosl.com.au, accessed 28 April 2021, for further information about our recent low-cost sensor technology development). The EoDC project website has the following key functions and features:

- Stores raw data that uploaded from each deployed sensor.
- Automatically runs scripts to clean the data (data quality check and removal of poor quality data).
- Automatically backs up all data, twice daily by sending to the correspond manager's email address.
- Collects rainfall data based on the sensor location and the radar image from Bureau of Meteorology, Australia.
- An alarm system which sends SMS to the corresponding person when improper stormwater discharge is detected.

Figure 11.18 Web-based platform of NextGen Urban Water Monitoring – EoDC project. *Source*: Google Maps, EPHM Lab and BoSL.

In the current website of the EoDC project (see Figure 11.18), catchment boundaries, drainage network and sensor locations are marked on top of the Google Map. By clicking each site, live water depth, temperature and EC data are directly plotted for users to check the trend over last 48 hours. A raw-data (in csv format) download link is provided at the bottom of the pop-up window. The cleaned data are only accessible through the website's backstage database.

11.7.6 Data quality check, cleaning, and validation

The raw data collected from the low-cost sensor need to be cleaned and validated to remove all poor-quality readings (e.g. negative readings due to sensor malfunction). A data checking algorithm is used to automatically create a cleaned data file. By checking each scan in the time series data, the algorithm can assign different error codes based on the following criteria and possible data issues:

- Abnormal water pressure readings (water level <-0.5 m or >5 m, based on individual case study's condition).
- Unusual temperature readings (<-5°C or >40°C).
- No reference air pressure readings within one-hour time.
- When the sensor is under maintenance.

A meta-data file will be automatically created for each sensor site at the same time the sensor is being cleaned. The meta-data includes the description of each error code, how many poor-quality data points are detected under each error code, and when these errors occur in the time series.

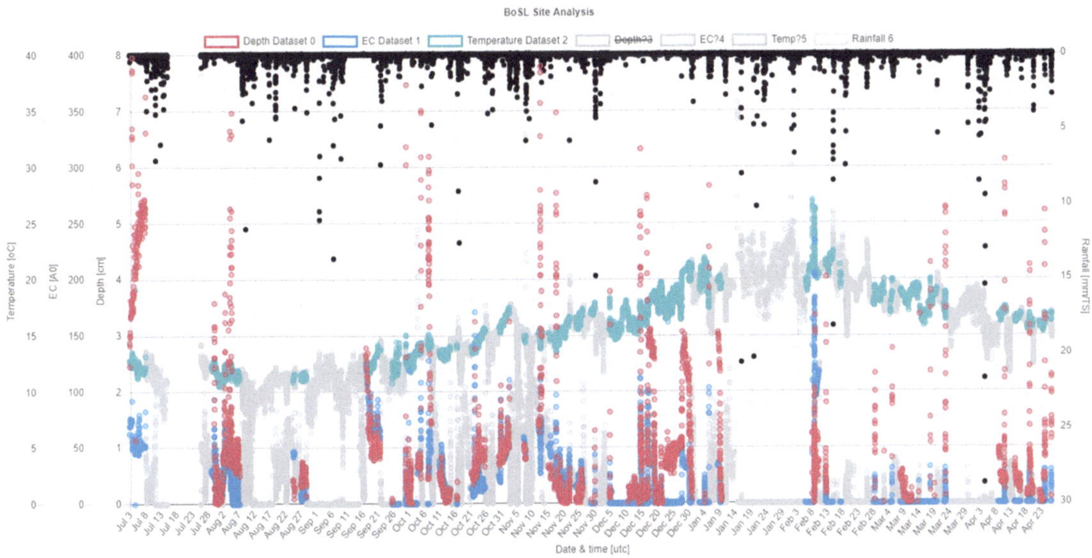

Figure 11.19 Time series plot of the clean, validated and dry weather data for one of the sensors in this case study; poor quality data or wet weather periods are greyed out in this figure; red – depth readings, blue – conductivity readings and green – temperature readings. *Source:* EPHM Lab and BoSL.

Apart from cleaning the poor-quality data, the data checking algorithm is also designed to correct the potential depth sensor drift by using the sensor recalibration results collected during each manual site maintenance. In addition, the actual weather condition of each data point (either wet or dry) is determined based on the rainfall intensity data collected from radar imageries every five minutes. When the cumulative rainfall during the previous 6 hours is more than 1 mm, the data are considered in a wet period and will then remain to be considered in a wet period for the next 48 hours.

The final cleaned data are automatically plotted on EoDC website's data analysis portal (Figure 11.19 – under testing and not available to the public). This plot includes water depth, EC, temperature, and rainfall data collected during the sampling period. Users can zoom in and out to check the data of a specific period of their interest. The poor-quality data are automatically greyed out, and the users can click onto the legend bar to turn off the data they are not interested in (e.g. data collected during wet weather period).

11.8 THE UWO – A FIELD LABORATORY FOR DISTRIBUTED REAL-TIME MONITORING WITH LOW-POWER SENSOR AND DATA COMMUNICATION TECHNOLOGY, FEHRALTORF, SWITZERLAND

11.8.1 Scope and objectives

The Urban Water Observatory (UWO, http://www.eawag.ch/uwo, accessed 28 April 2021) is a dedicated long-term monitoring initiative aiming at (i) collecting a consistent dataset on water and matter fluxes in an urban area at very high spatiotemporal resolution (1 sensor / ha; 5-minute recording interval), and (ii) testing and developing a low-power sensor and data communication technologies for efficient environmental and infrastructure monitoring. The UWO field laboratory was commissioned in early 2016, and it is expected to run at least until the end 2021. At the time of writing (i.e. 2020), the UWO sensor network consisted of more than 120 different sensors, deployed across various compartments of the urban water cycle. Besides typical urban hydrology parameters such as precipitation, water level and flow, data on (waste-)water temperature and conductivity (i.e. capacitive sensors, inductive conductivity) were collected in sewers, in rivers, and in the groundwater compartment.

The project was inspired at a time at which miniaturization of hardware components, increasing computational capacities and the integration of various types of digital technology in our everyday life became more and more prevalent, and this digital transformation started to expand into the urban water sector. A key objective was to illustrate benefits and limitations of a so far unseen sensor density in the context of distributed, long-term, real-time urban drainage monitoring. Furthermore, pressing challenges like an increasing regional water scarcity, increasing flood risk due to a changing climate, and a growing cost-pressure for operation and maintenance (for instance due to high sewer infiltration rates), motivated the establishment of the field laboratory in the municipality of Fehraltorf, Switzerland (Figure 11.20).

Essentially, the project attempts to collect consistent data for water research, and to coherently study, field-test and advance four interrelated aspects in the process of data collection (Figure 11.21): 1–sensor application, 2–data communication, 3–data management, 4–semi-automated data validation.

Common to most urban drainage monitoring initiatives are: (i) often a considerable effort for installation in underground locations that are difficult to access and explosion endangered increasing maintenance intervals (associated with costs), and (ii) a limited scalability and flexibility in terms of deployment and operation. Hence, the following operational requirements can be formulated:

- A cheap, reliable and scalable wireless communication enabling data collection in real-time.
- A solid energy supply for sensor and transmission technology with long battery lifetime up to several years.

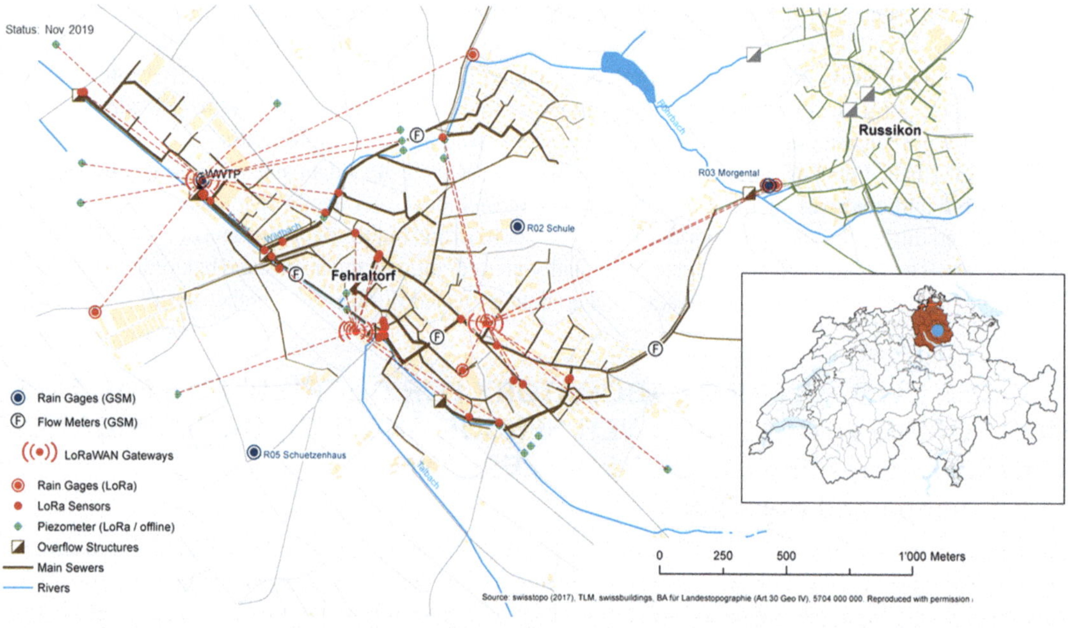

Figure 11.20 Overview on the UWO sensor network in Fehraltorf including sensors and low-power wide-area network (LPWAN) infrastructure (Status: November 2019). Small figure: location of the Canton Zurich and the case study location, 12 km north-east of the city of Zurich within Switzerland. *Source:* Eawag.

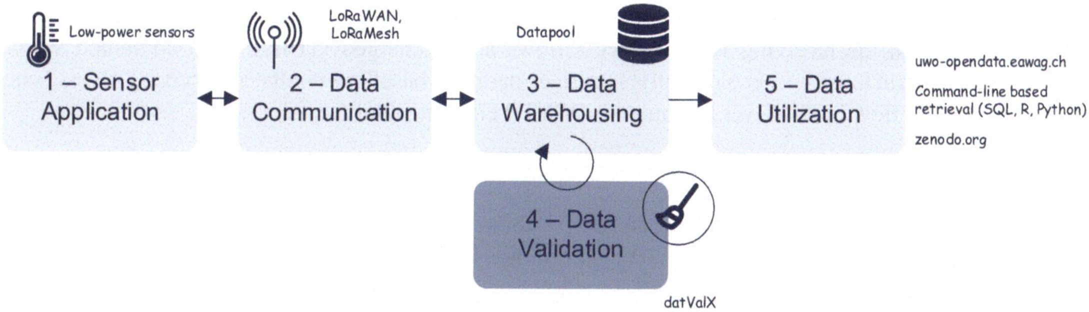

Figure 11.21 Principal understanding of data collection as a coherent process, including individual contributing components, i.e. projects embedded within the UWO initiative. *Source*: Blumensaat *et al.* (2018).

- A simple and fast installation without structural modifications of sewer infrastructure, such as the removal of antenna cables or the installation of special radio-enabled manhole covers.
- Fully time-synchronous sensor readings.
- A high standard regarding data encryption and security.
- Sparse, flexible and non-proprietary data management solution.
- Devices that are suitable for use in explosion-proof environments, solely due to their hardware components.

These particular aspects motivated the UWO research, which eventually lead to the development of the following components for a 'From-the-Sensor-Signal-to-the-Data-Point' pipeline (Figure 11.21):

- 1 – Low-cost and ultra-low-power sensors for CSO detection and wastewater characterization.
- 2 – A new wireless data communication protocol based on the LoRa technology (Ebi *et al.*, 2019), fully compatible with the renowned LoRaWAN (long range wide-area network) standard (LoRaWAN, 2015) to overcome range critical situations, as they occur when transmitting data from underground.
- 3 – A mature data warehouse application named Datapool (Blumensaat *et al.*, 2018) allowing for flexible integration of various data sources (own sensors, SCADA systems, meteorology services and foreign servers) from multiple data providers.
- 4 – An automated data validation pipeline (Disch & Blumensaat, 2019) enabling real-time data curation and timely sensor maintenance (datValX).

11.8.2 Catchment area, measured variables and location of monitors

11.8.2.1 Catchment area and operation challenges

The investigated urban catchment Fehraltorf (Figure 11.20) is a Swiss-typical settlement with a modified combined sewer network (13 km combined sewers, 4.6 km foul sewers, 10.9 km storm drains) where sanitary sewage and some stormwater flows are carried to a central wastewater treatment plant (WWTP; design capacity: 12,000 PE). A small proportion of the stormwater is directly discharged into receiving creeks without any treatment. The total settlement area adds up to 127.3 ha, however, only 40 ha can be accounted for as areas connected to the combined system. A significant share of the sewer pipes lies below the groundwater table. Thus, sewer infiltration is – depending on the season – considerable and contributes to the WWTP inflow with an estimated varying rate of 35% up to 55%. Sewer rehabilitation planning is ongoing but constrained by limited municipal budgets. Four sewer retention basins, adding up to an average specific storage volume of 36.1 m^3 per hectare runoff-efficient area, are implemented to mitigate impacts on the receiving waters. Excess flows are discharged via five main combined sewer overflow (CSO) structures into ecologically sensitive, partially baseflow-regulated receiving rivers; one quarter of the base flow in the River Luppmen is WWTP effluent (dilution ratio 3:1).

11.8.2.2 Sensor network evolution

Network architecture: The sensor network can be partitioned into two parts: (i) a 'monitoring backbone' consisting of four conventional rain gauges and seven high-precision flow monitors operated on batteries and equipped with cellular data loggers, and (ii) a low-power wireless sensor network (LPWSN) which – in 2020 – collected data from more than 90 low-power sensor nodes using wireless low-power communication, i.e. LoRa-based techniques. In 2020 we collected 123 monitoring signals covering the system dynamics at 1 to 5 min temporal resolution. This corresponds to more than 50,000 data points, i.e. sensor observations, per day, excluding operational parameters such as battery voltage. Data from rain gauges are collected in 1-minute intervals. Sensors are positioned according to the monitoring objectives, i.e. primarily across the central part of the drainage network, along the main collectors, and at overflow structures. Groundwater piezometers are established around the municipal area and next to main collectors. Many locations are equipped with two or more sensors of the same or different types to intentionally pursue the concept of signal redundancy and signal diversity. Sensor positioning is clearly motivated by aspects related to urban drainage and flow topology, but not necessarily related to an optimum wireless communication network coverage (see discussion below).

Rain gauge network: In total 14 rain gauges, 4 conventional (weighing principle; OTT Pluvio II L) and 10 low-power rain gauges (R.M. Young, Model No. 52203; LUFFT, WS700) are established in an urban area of 127.3 ha. This rain gauge density in particular allows the capture of spatial rainfall variability, e.g. during convective summer storm events. Rainfall data are used to feed hydraulic sewer models and to align with radar information provided by the Swiss National Weather Service. The seven low-power tipping bucket rain gauges have been found to be generally within a similar range of accuracy as the high-precision weighing gauges (<5%). Still, for extreme rainfall events, these rain gauges tend to underestimate rainfall depths (~15%). The gauge deployment revealed that the adequate positioning of rain gauges in urban areas is not trivial. Compliance with WMO standards, which do not specifically address requirements in urban areas, on the one hand, and finding operational solutions on the other hand, lead to compromise solutions.

Flow monitoring: In-sewer flow rates are monitored at four strategic locations, i.e. at main collectors within the network, at connecting sewers carrying transfer flows from adjacent settlements and at the inlet of the central WWTP. Three of the four locations are equipped with redundant flow sensors but of different monitoring principle (radar and US Doppler techniques; Nivus, Flodar, Sommer).

11.8.2.3 Low-power wireless sensor network

A key component of the UWO is a fully scalable, low-power wide-area network (LPWAN) communication system, consisting of sensor nodes and gateways. This concept is based upon the new, yet renowned standard LoRaWAN™ allowing for a bidirectional radio communication link between battery-driven sensor nodes and central gateways over a long range (~10 km, for above ground applications).

Whereas other LPWAN techniques exist (Raza *et al.*, 2017), LoRaWAN was chosen for three main reasons: (i) low-power data transmission on a license-free bandwidth, in Europe 868 MHz, (ii) a standard with an open-source-like character with a growing community of developers and users, and (iii) availability in 2015, when the UWO was initiated. The LoRaWAN architecture is laid out in a star-type

Figure 11.22 Hardware components in LoRaWAN sensor node prototype, embedded in an ATEX compatible chassis. *Source*: Eawag.

network topology, i.e. each individual sensor node directly communicates with the best available gateway through a contention-based approach, meaning that data are sent without receiving acknowledgement messages. In the case of the UWO, three 'privately' operated gateways (Kerlink, Wirnet Station 868), one solar-powered and two AC-powered, currently collect data from more than 90 low-power sensors (Figure 11.22 – status: November 2019).

Battery-powered sensor nodes, i.e. LoRa-enabled loggers which various sensor types can be connected to, are (i) self-designed and -manufactured prototypes (Figure 11.23), and (ii) manufactured by an external supplier (http://www.eawag.ch/uwo, accessed 28 April 2021) according to our design specifications.

Figure 11.23 LoRaWAN prototype including water level sensor (MaxBotix) installed in a sewer manhole. *Source*: Eawag.

Currently, the following sensors are integrated into our LPWSN: 34 ultrasonic level sensors (MaxBotix MB7389, MB7369, MB7386), 10 dielectric conductivity sensors (Decagon, 5TM/5TE), 2 pressure gauges (Keller, 36 XKY), 15 multi-parameter probes for groundwater quantity and quality monitoring (*STS*, DL/N 70; Keller, 36XiW-CTD,DCX-22-CTD), 5 humidity sensors, 26 dual in-sewer temperature sensors (DS18B20), 1 very low-cost pluviometer (Davis), and 7 low-power tipping bucket rain gauges (R.M. Young, Model 52203).

11.8.3 Data management, validation, availability

11.8.3.1 Data management

Existing data management solutions are generally designed with a very specific work- and data-flow in mind – which is most unlikely to correspond exactly with your needs. So even while the underlying database platforms are commonly well-established, the full sensor data warehouse needs careful technical specification, a design concept and individual configuration. For the UWO we aimed at a solution that (i) can handle very diverse data (types) of potentially yet unknown sources and formats, (ii) stores and links data and meta-data simultaneously, (iii) ensures consistency of the data whenever possible, (iv) makes data usage at the front-end easy, and (v) does not require in-depth skills to add novel data sources. Note, the flexibility of point (i) limits the potential for automation and makes point (v) inevitably harder.

The result of our efforts is a data warehouse application named Datapool. This solution is based on separation of concerns: a field scientist (cf. data provider – see Figure 11.24) adding a new sensor is responsible for providing (i) a file with meta-data and (ii) a conversion script to transform raw files from the sensor into a standardized text format. As soon as a new raw file arrives from a sensor the Datapool applies the corresponding (conversion) script automatically and imports the data into a relational database (PostgreSQL). Finally, data users can either directly query data and meta-data, use packages (available in R, Python, Julia, Matlab) with a basic set of query functions, or use a web interface to

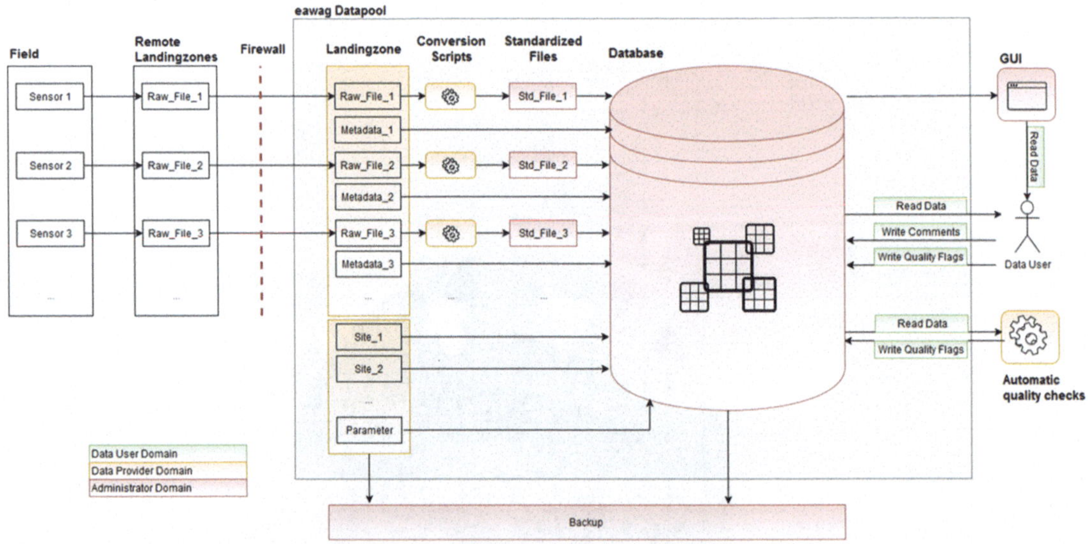

Figure 11.24 Concept of and data flow in the data warehouse application Datapool. *Source*: adapted from Blumensaat *et al.* (2018).

retrieve data (http://www.uwo-opendata.eawag.ch, accessed 28 April 2021), the Datapool application carries out consistency checks, provides error logs, it is designed to guarantee scalability, and it manages read/write-permission of the different user types. Incremental backups of the underlying SQL database are regularly written; warehouse administration is reduced to a minimum.

While the Datapool (Figure 11.24) development was initially driven by UWO requirements only, over time it was adapted to serve a number of other data collection initiatives. Today, it can be used as a stand-alone tool, the code is made freely available, and it comes with comprehensive documentation (https://datapool.readthedocs.io/en/latest/, accessed 28 April 2021) (credit: ETH Zurich, SIS and Eawag).

11.8.3.2 Data availability and access

All data and meta-data are made publicly available through the web-based platform (TMX http://www.tmx.nl, accessed 28 April 2021) accessible from anywhere. While researchers from external institutions will retrieve the full data as a batch from an open-data repository (in preparation in 2020), the broader public can view collected data series and associated information about monitoring sites in quasi-real-time through this platform.

11.8.3.3 Semi-automated data validation

Due to the shear amount of data currently collected, an automated data validation that goes beyond the standard protocols is required. Firstly, a system-inherent consistency check, i.e. verification of time and data format and a check for real duplicates, is performed when signals arrive at the landing zone, thus avoiding disparate values to be integrated. Secondly, a semi-automatic flagging concept feeds a traffic-light maintenance system allowing for efficient sensor maintenance and quasi real-time data quality control. The principal validation routine includes range, gradient and outlier checks that are individually parameterized depending on signal/sensor type and monitoring site. Thirdly, a data source specific anomaly detection algorithm (Disch & Blumensaat, 2019) allows further identification of doubtful sensor values, indicating sensor failures or abnormal operation situations.

11.8.4 Sensor network operation

11.8.4.1 Maintenance and operational costs

Regular and on-demand maintenance includes sensor cleaning, calibration, troubleshooting, de-/re-installation in case of sewer cleansing, battery change, firmware updates, and LPWAN infrastructure maintenance. Unexpected events requiring on-demand maintenance are condensation of ultrasonic level sensors, corrosion of electrical parts, sensor pollution due to sewer surcharge, poor radio connectivity, and sensor manipulation through third-parties. All efforts add up to 0.25 FTE, i.e. one skilled technician is in duty for more than one day per week. While aiming at minimizing the number of field trips for sensor upkeeping seems a rational idea, it became clear that an excessive reduction compromises a stable sensor network operation and spoils data quality. Day-to-day sensor network maintenance can be considered to make up the highest share of total costs, but clearly guarantees adequate data quality. Hence, planning should focus on strategies to minimize maintenance effort, such as *ex situ* and contact-free measurement techniques, long battery lifetimes, and predictive maintenance through continuous data collection and real-time data validation.

Initial efforts to manage data required an additional 0.25 FTE, in particular during the phase of warehouse development and testing. While sensor network maintenance increases with an increasing number of sensors, costs for data management may level out (0.1 FTE) with a robust and fully established data management framework. Our experience is that initially invested time in conceptual thinking that goes beyond current needs but follows parsimonious thoughts about how to manage data very much pays off in the long-term.

Generally, costs for external services including fees for cellular communication, server storage, LPWAN fees but also rentals (e.g. antenna installation at rooftops) should not be underestimated. They can easily add up with an increasing number of devices. For larger deployments or permanently operated systems in larger utilities, it may be efficient to establish in-house ('private') infrastructure and services.

11.8.4.2 Data communication

Flow sensors and weighting rain gauges transmit their data through conventional cellular modems, accepting all common bottlenecks of this technique: energy-hungry operation requiring battery changes every four weeks, service fees to be paid to the GSM network provider, reception problems for difficult-to-reach locations.

For the LPWAN system, we observed a non-negligible data loss at an early stage of the monitoring programme. Quantifying the radio performance of 25 LoRaWAN sensor nodes – of which 18 were positioned underground – for a 5-month test period revealed the following (Ebi *et al.*, 2019):

- A 12% packet error rate (PER) averaged over the 25 monitored sensor nodes.
- An increasing PER with increasing distance to the central gateway.
- A significantly different PER increase with growing distance depending on the radio node position (above/below ground).
- A threshold of approximately 500 m distance between sensor and gateway for which reception of packets from below ground nodes is good, i.e. with very few packets lost.

Based on a quantitative evaluation in a real-life environment, we concluded that the LoRaWAN technique provides either long-range coverage above ground or medium-range underground connectivity, but not both at the same time. Deployment of additional gateways – one possible solution to overcome range limitations for underground applications – was not an option mainly for three reasons: (i) gateways usually require AC mains power, (ii) costs for gateway installation, management, and internet access increase with the number of installed gateways, and (iii) options to place gateways at adequate locations are *per se* limited (location suitability; legal permission requirements at private properties).

Instead, we developed a new LPWAN concept based on LoRa technology, named synchronous LoRa mesh (i.e. LoRaMesh), a multi-hop data communication (Figure 11.25). With the synchronous LoRa mesh technique we (i) enhance transmission reliability, efficiency, and flexibility in range-critical situations through meshed multi-hop routing, and (ii) ensure a precise time-synchronization through optional GPS or DCF77 long-wave time signalling. Battery consumption at individual nodes is even further reduced since data rates can be higher due to better radio link quality. In two independently conducted long-term field tests under real-world conditions, we analysed the synchronous LoRaMesh sub-network performance with standard LoRaWAN through redundantly installed radio nodes. For the UWO, we compared the performance of 11 synchronous LoRa mesh nodes with 5 standard LoRaWAN reference nodes over an evaluation period of 45 days. We observed a very low packet loss of 2.2%, averaged over all synchronous LoRa mesh systems deployed in this test, whereas some of the standard LoRaWAN nodes did not even connect to the existing gateways. Details including hardware specification, firmware code and test results are provided in Ebi *et al.* (2019).

11.8.4.3 Power management and time synchronicity

Key advantages of the LoRa-based, low-power monitoring are (i) perfect time synchronicity since the network is centrally operated and is synchronized with the internet time, and (ii) extremely low energy

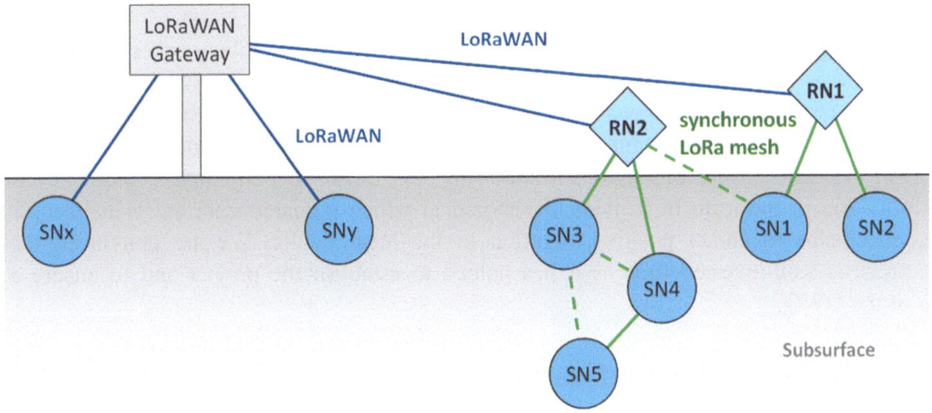

Figure 11.25 Synchronous LoRa mesh topology (RN: repeater node; SN: sensor node). Dashed lines between synchronous LoRa mesh sensor nodes indicate alternative link paths. *Source*: adapted from Ebi *et al.* (2019).

demands, still depending on sensor type, measuring and transmission frequency, data packet size and the signal strength at the given location. More specifically: for a low-power ultrasonic water level sensor measuring and transmitting every 5 minutes, the radio nodes powered by a standard Lithium-Polymer battery – 3.7 V, 6700 mAh – have an estimated lifetime of approximately two years. The same nodes powered by two standard LR20 Alkali-Mangan mono cells – 1.5 V, 18000 mAh – last approximately six years. Practice in the UWO shows that after 3 years of operation (May 2016–May 2019), the very first water level sensor nodes operated at 5-minute intervals still ran on the same set of batteries from installation. Event-triggered adaptive sampling further reduces energy demands, but also decreases the robustness of communication. On the other hand, current research and existing prototypes (https://www.eawag.ch/swp, accessed 28 April 2021) indicate that in the near future we will have energy-autonomous sensor nodes. Initial results show that energy harvesting can be successfully applied to operate underground sensor nodes without any battery.

Given the fact that real-time clocks in sensors and data loggers tend to drift, sometimes considerably, time synchronicity becomes a very relevant issue (see also Section 9.3.5), in the case of large sensor networks and a highly dynamic rainfall-runoff response. For instance, for a few sensors that were operated offline, we observed timestamp deviations of up to 20 min in a period of 6 months. Loggers connected the LoRaWAN inherently provide data with correct time stamps as time syncing with the internet time is ensured through gateways. The maximum latency from the time of the sensor reading to the time the reading is registered at the server is estimated at 6 seconds.

Most of the conventional, off-the-shelf equipment for in-sewer monitoring was certified to meet explosion-proof standards (e.g. ATEX). On the other hand, LPWAN sensor nodes (own prototypes and later externally manufactured) did not have an ATEX certification but were designed and manufactured to be ATEX compliant. A significant advantage of low-power techniques is that the voltages and currents present are inherently so low that, in the event of a fault, the ignition capability cannot be achieved and the equipment can be designed for intrinsic safety according to the ignition protection type. Notwithstanding that certification of commercially available LoRaWAN sensor nodes is under way, it can be debated if equally strict ATEX requirements need to be fulfilled, despite the inherent low-power characteristics of LPWAN devices.

11.8.4.4 Involving the public

Despite there being quite a few sensors positioned above ground and in publicly accessible areas, there are only very few cases (2) of vandalism. Information campaigns, for instance through announcements and articles in local newspapers, give-away flyers, a well-developed project website, informative display panels installed next to sensors/antennas and – at a later stage – public information events very much help to make residents aware of ongoing monitoring and create a 'participative moment'. Returning useful information in the form of a living dashboard at which prepared data are visualized, at best in real-time, also helps to attract positive attention in the local public. We are convinced that Public Relations work is well-invested time as it has helped to establish the project and to ensure a smooth operation of the UWO.

11.8.5 Hardware cost considerations

The net present value of the currently installed sensors and wireless communication infrastructure adds up to an estimated CHF 120,000 (€ 106,000) for conventional sensors (rain, flow) and CHF 60,000 (€ 53,000) for the LPWSN including LPWAN infrastructure. An additional CHF 20,000 (€ 17,500) are accounted for groundwater multi-parameter monitors, and a further CHF 5,000 (€ 4,400) for installation material is required. Investments for the LPWSN are comparably low, in total and specific for the sensor node. Based on a conservative estimation for an in-house prototype node with a standard low-power water level sensor built in 2017 we estimated specific costs of CHF 125 (€ 110) per year. In other words, a water level pattern for one day assuming a 5-minute measurement interval costs € 0.30. This estimate includes (i) the costs for hardware components for the LoRaWAN sensor node (Eawag prototype), the water level sensor (Maxbotix), batteries (standard D cells), 5% proportionate costs for the LoRaWAN gateway device (assuming a depreciation over five years for all), and (ii) the costs for external services, such as LPWAN backend management and data dashboard services. While the costs for hardware components are expected to decrease significantly in the future, service fees may remain in the same order of magnitude. Overall, the total costs for LPWAN based measurements is expected to further decrease.

11.8.6 Main findings and lessons learnt

The main findings from the UWO monitoring project can be summarized as follows:

(1) Low-power sensor technology combined with a very efficient LPWAN technique (Ebi *et al.*, 2019) considerably facilitates data collection in underground infrastructure. Low energy demands, nearly unlimited scalability, time synchronicity and very low specific costs increase the efficiency when collecting data, and enables harvesting more reliable information through signal redundancy and diversity. Generating truly spatially distributed information in quasi real-time becomes a reality. Still, the availability of such new data sources opens up new challenges regarding data management, and institutional and regulatory requirements that need to be adapted likewise. In the preparation of a sensor node deployment, exploratory analyses of available link budgets on site are essential to avoid upsets regarding high packet error rates due to poor radio connection.

(2) Dynamic and flexible data management solutions come at a price. It is not the pure amount of data that is most challenging, but the diversity of different signals. Within this context, full meta-data integration in the data management process is laborious but clearly consolidates data interpretation in the long-term. Changing data formats and server settings, and the lack of

standardized protocols hamper efficient data management. Standards and protocols that link operational and infrastructure data need to be established/harmonized based on research and practical experience.

(3) The use of simple and robust capacitive sensors for overflow activity monitoring is an efficient method to reliably estimate urban drainage emissions. Such information can also be utilized to calibrate hydrodynamic models (Wani *et al.*, 2017). Information content increases when censored signals, i.e. signals that are constrained e.g. to provide 0/1 information, are combined with uncensored information from parallel installed water level sensors.

(4) The performance of (emerging) data validation techniques, and hence future use of the data, is dependent on the quality of the collected data and the corresponding pre-processing. No matter which data validation technique is applied, careful sensor set-up and configuration is priceless in order to obtain a meaningful dataset (Disch & Blumensaat, 2019).

(5) Direct access to sites, short travel times between headquarters and the study catchments, real-time data communication and automated validation is key to keep the maintenance effort low. Early and transparent information to the public, and sound and trustful collaboration with the local operator and municipality is key to success.

REFERENCES

Abdullah J., Muhammad N. S., Muhammad S. A. & Julien P. Y. (2019). Envelope curves for the specific discharge of extreme flood events in Malaysia. *Journal of Hydro-Environment Research*, **25**, 1–11. doi: 10.1016/j.jher.2019.05.002.

Blumensaat F., Scheidegger A., Ebi C., Dicht S., Schaltegger F., Rüst A., Schmitt U. & Maurer M. (2018). Digitalization Meets Reality – Concept and Experiences from Long-Term Wireless Data Collection with 50+ Sewer Monitors. *Proceedings of the 11th International Conference on Urban Drainage Modelling*, 23–26 September, Palermo, Italy, pp. 360–363.

Bonhomme C. & Petrucci G. (2017). Should we trust build-up/wash-off water quality models at the scale of urban catchments? *Water Research*, **108**, 422–431. doi: 10.1016/j.watres.2016.11.027.

Bonneau J., Fletcher T. D., Costelloe J. F., Poelsma P. J., James R. B. & Burns M. J. (2018). Where does infiltrated stormwater go? Interactions with vegetation and subsurface anthropogenic features. *Journal of Hydrology*, **567**, 121–132. doi: 10.1016/j.jhydrol.2018.10.006.

Bonneau J., Fletcher T. D., Costelloe J. F., Poelsma P. J., James R. B. & Burns M. J. (2020). The hydrologic, water quality and flow regime performance of a bioretention basin in Melbourne, Australia. *Urban Water Journal*, **17** (4), 303–314. doi: 10.1080/1573062X.2020.1769688.

Brayshaw P. & Wilkes R. (2016). Integrated Delivery to Provide Stakeholder Benefit – Anglian Water Integrated Urban Drainage Programme. *Proceedings of the Urban Drainage Group Autumn Conference*, 9–11 November, Blackpool, UK.

Camacho Suarez V. V., Brederveld R. J., Fennema M., Moreno-Rodenas A., Langeveld J., Korving H., Schellart A. & Shucksmith J. D. (2019). Evaluation of a coupled hydrodynamic-closed ecological cycle approach for modelling dissolved oxygen in surface waters. *Environmental Modelling and Software*, **119**, 242–257. doi: 10.1016/j.envsoft.2019.06.003.

Casanueva A., Rodriguez-Puebla C., Frias M. D. & Gonzalez-Reviriego N. (2014). Variability of precipitation extremes over Europe and its relationships with teleconnection patterns. *Hydrology and Earth System Sciences*, **18**(2), 709–725. doi: 10.5194/hess-18-709-2014.

CIWEM (2017). *Urban Drainage Group – Code of Practice for the Hydraulic Modelling of Urban Drainage*. CIWEM – Chartered Institution of Water and Environmental Management, London (UK), 169 p. Available at https://www.ciwem.org/assets/pdf/Special%20Interest%20Groups/Urban%20Drainage%20Group/Code%20of% 20Practice%20for%20the%20Hydraulic%20Modelling%20of%20Ur.pdf (accessed 05 December 2020).

DID (2018a). *Hydrological Procedure No 32. Hydrological standard for rainfall station instrumentation.* Kuala Lumpur (Malaysia): DID – Department of Irrigation and Drainage, 224 p. ISBN 978-983-9304-38-1. Available at http://h2o.water.gov.my/man_hp1/HP32.pdf (accessed 22 December 2020).

DID (2018b). *Hydrological Procedure No 33. Hydrological standard for water level station instrumentation.* Kuala Lumpur (Malaysia): DID – Department of Irrigation and Drainage, 64 p. ISBN 978-983-9304-39-8. Available at http://h2o.water.gov.my/man_hp1/HP33.pdf (accessed 23 December 2020).

Disch A. & Blumensaat F. (2019). Messfehler oder Prozessanomalie? – Echtzeit-Datenvalidierung für eine zuverlässige Prozessüberwachung in Kanalnetzen [Measurement error or process anomaly? – Real-Time Data Validation for Reliable Process Monitoring in Sewer Networks]. In *Proceedings of Aqua Urbanica 2019*, 9–10 September, Rigi-Kaltbad, Switzerland, M. Burkhardt & C. Graf (eds), pp. 73–78. doi: 10.5281/zenodo.3384207.

Dotto C. B. S., Kleidorfer M., Deletic A., Fletcher T. D., McCarthy D. T. & Rauch W. (2010). Stormwater quality models: performance and sensitivity analysis. *Water Science and Technology*, **62**(4), 837–843. doi: 10.2166/wst.2010.325.

DWA (2014). *Merkblatt DWA-M 151 – Messdatenmanagementsysteme (MDMS) in Entwässerungssystemen [Measurement Data Management Systems (MDMS) in sewer systems].* Hennef (Germany): DWA – Deutsche Vereinigung für Wasserwirtschaft, Abwasser und Abfall, 51 p. ISBN 978-3-944328-67-6. (in German).

Ebi C., Schaltegger F., Rüst A. & Blumensaat F. (2019). Synchronous LoRA mesh network to monitor processes in underground infrastructure. *IEEE Access*, **7**, 577663–557677. doi: 10.1109/ACCESS.2019.2913985.

Fencl M., Grum M., Borup M. & Mikkelsen P. S. (2019). Robust model for estimating pumping station characteristics and sewer flows from standard pumping station data. *Water Science and Technology*, **79**(9), 1739–1745. doi: 10.2166/wst.2019.176.

ISO (2015). *ISO 9001:2015 quality management systems — Requirements.* ISO – International Organization for Standardization, Geneva (Switzerland), 29 p.

Lai K. X. O. (2016). *Impact of the SMART Tunnel Outflow on the Hydraulics of the Kerayong River, Malaysia.* Fort Collins (USA): Colorado State University, Technical Report.

Langeveld J. G., Benedetti L., de Klein J. J. M., Nopens I., Amerlinck Y., van Nieuwenhuijzen A., Flameling T., van Zanten O. & Weijers S. (2013). Impact-based integrated real-time control for improvement of the Dommel River water quality. *Urban Water Journal*, **10**(5), 312–329. doi: 10.1080/1573062X.2013.820332.

LoRaWAN (2015). *LoRaWAN: what is it? A technical overview of LoRA and LoRaWAN.* San Ramon, CA (USA): LoRa Alliance, Technical Marketing Workgroup 1.0, 20 p. Available at https://lora-alliance.org/sites/default/files/2018-04/what-is-lorawan.pdf (accessed 07 Dec 2020).

Moreno-Rodenas A., Cecinati F., Langeveld J. G. & Clemens F. H. L. R. (2017). Impact of spatiotemporal characteristics of rainfall inputs on integrated catchment dissolved oxygen simulations. *Water*, **9**(12), 926. doi: 10.3390/w9120926.

Muhammad N. S., Julien P. Y. & Salas J. D. (2016). Probability structure and return period of multiday monsoon rainfall. *Journal of Hydrologic Engineering*, **21**(1), paper 04015048, 11 p. doi: 10.1061/(ASCE)HE.1943-5584.0001253.

OJEU (2000). Directive 2000/60/EC of the European Parliament and of the Council establishing a framework for Community action in the field of water policy. *Official Journal of the European Union*, 22 Dec., L327/1-L327/73. Available at https://eur-lex.europa.eu/legal-content/EN/TXT/HTML/?uri=CELEX:32000L0060&from=EN (accessed 28 April 2021).

Raza U., Kulkarni P. & Sooriyabandara M. (2017). Low power wide area networks: an overview. *IEEE Communications Surveys & Tutorials*, **19**(2), 855–873. doi: 10.1109/COMST.2017.2652320.

Schilperoort R. P. S. (2011). *Monitoring as a Tool for the Assessment of Wastewater Quality Dynamics.* PhD thesis, Delft University of Technology, Delft, The Netherlands, 330 p. ISBN 978-90-8957-021-5.

Tangang F., Farzanmanesh R., Mirzaei A., Supari Salimun E., Jamaluddin A. F. & Juneng L. (2017). Characteristics of precipitation extremes in Malaysia associated with El Niño and La Niña events. *International Journal of Climatology*, **37**(Suppl. 1), 696–716. doi: 10.1002/joc.5032.

van Bijnen M. & Korving H. (2008). Application and Results of Automatic Validation of Sewer Monitoring Data. *Proceedings of the 11th International Conference on Urban Drainage*, 31 August–5 September, Edinburgh, Scotland, UK, 9 p.

van Daal-Rombouts P., Benedetti L., de Jonge J., Weijers S. & Langeveld J. (2017a). Performance evaluation of a smart buffer control at a wastewater treatment plant. *Water Research*, **125**, 180–190. doi: 10.1016/j.watres.2017.08.042.

van Daal-Rombouts P., Gruber G., Langeveld J., Muschalla D. & Clemens F. (2017b). Performance evaluation of real time control in urban wastewater systems in practice: review and perspective. *Environmental Modelling & Software*, **95**, 90–101. doi: 10.1016/j.envsoft.2017.06.015.

Wani O., Scheidegger A., Carbajal J. P., Rieckermann J. & Blumensaat F. (2017). Parameter estimation of hydrologic models using a likelihood function for censored and binary observations. *Water Research*, **121**, 290–301. doi: 10. 1016/j.watres.2017.05.038.

Water UK (2019). *Occasional guidance note – the classification & management of confined space entries – edition 3.0.* Water UK, London (UK), 27 p. Available at https://www.water.org.uk/guidance/the-classification-and-management- of-confined-space-entries-2019/ (accessed 28 April 2021).

Williamson F. (2015). *Historic Flood Records in Colonial Malaysia: Sources and Context.* National University of Malaysia, Kuala Lumpur (Malaysia). Slide presentation, 19 p. Available at https://blog.nus.edu. sg/environmentcluster/files/2015/03/FUTURE-FLOODS-1tp9vk9.pptx (accessed 28 April 2021).

WMO (2008). *Guide to Hydrological Practices – Volume 1: Hydrology – From Measurement to Hydrological Information.* Geneva (Switzerland): World Meteorological Organization, report WMO No 168, 6th edition, 296 p. ISBN 978-92-63-10168-6. Available a http://www.wmo.int/pages/prog/hwrp/publications/guide/ english/168_Vol_I_en.pdf (accessed 28 April 2021).

WRc (1987). *A Guide to Short Term Flow Surveys of Sewer Systems.* WRc (Water Research Centre) Engineering, Swindon (UK), 131 p.

Chapter 12

Appendices

Jean-Luc Bertrand-Krajewski[1], Francois H. L. R. Clemens-Meyer[2,3,4] and Mathieu Lepot[4,5]

[1]*University of Lyon, INSA Lyon, Laboratory DEEP, Villeurbanne, France*
[2]*Unit of Hydraulic Engineering, Deltares, Delft, The Netherlands*
[3]*Faculty of Engineering, Dept. Civil & Environmental Engineering, Norwegian University of Science & Technology, Trondheim, Norway*
[4]*TU Delft, CITG – Civil Engineering and Geosciences, Delft, The Netherlands*
[5]*Un Poids Une Mesure, Lyon, France*

12.1 BASIC DEFINITIONS EXEMPLIFIED IN THE FIELD OF UDSM

In order to ensure that communication on metrological terms is free from misunderstanding, this book adopts the 'BIPM (Bureau International des Poids et Mesures) International Vocabulary of Metrology', abbreviated as VIM (Joint Committee for Guides in Metrology [JCGM], 2012) and available here.

As a series of definitions is not the most attractive way to introduce them, we propose a guided tour of the VIM to present, step by step, the most important definitions along with short examples related to UDSM (urban drainage and stormwater management) systems. Each definition from the VIM is given in a frame in italic characters.

Let us consider an operator who needs to know the discharge in a sewer pipe at a given time. The discharge is the *quantity* the operator is interested in. The operator will use *measurement instruments* and *sensors*, parts of the whole *measuring system*, and apply a *measurement procedure* to estimate the discharge, which is named the *measurand.*

doi: 10.2166/9781789060119_0471

quantity

Property of a phenomenon, body, or substance, where the property has a magnitude that can be expressed as a number and a reference.

measuring instrument

Device used for making measurements, alone or in conjunction with one or more supplementary devices.

sensor

Element of a measuring system that is directly affected by a phenomenon, body, or substance carrying a quantity to be measured.

measuring system

Set of one or more measuring instruments and often other devices, including any reagent and supply, assembled and adapted to give information used to generate measured quantity values within specified intervals for quantities of specified kinds.

measurement procedure

Detailed description of a measurement according to one or more measurement principles and to a given measurement method, based on a measurement model and including any calculation to obtain a measurement result.

measurand

Quantity intended to be measured.

This implies that, when reporting on monitoring results, apart from the experimental set-up applied, the applied methods on data pre- and/or post-processing should be mentioned and described.

Most sensors are designed to deliver an output as a response to an input which may differ from the measurand. For example, a piezo-resistive sensor delivers a water depth *d* (output) corresponding to a given pressure *P* exerted on its membrane (input). If the sensor membrane is positioned horizontally at the same level as the sewer pipe invert, then the water depth *d* is equal to the water level *h* in the sewer pipe (Figure 12.1(a)) If the sensor membrane is positioned at a different level z_0 above or below the sewer pipe invert, the corresponding offset shall be accounted for in the estimation of *h* (Figure 12.1(b)).

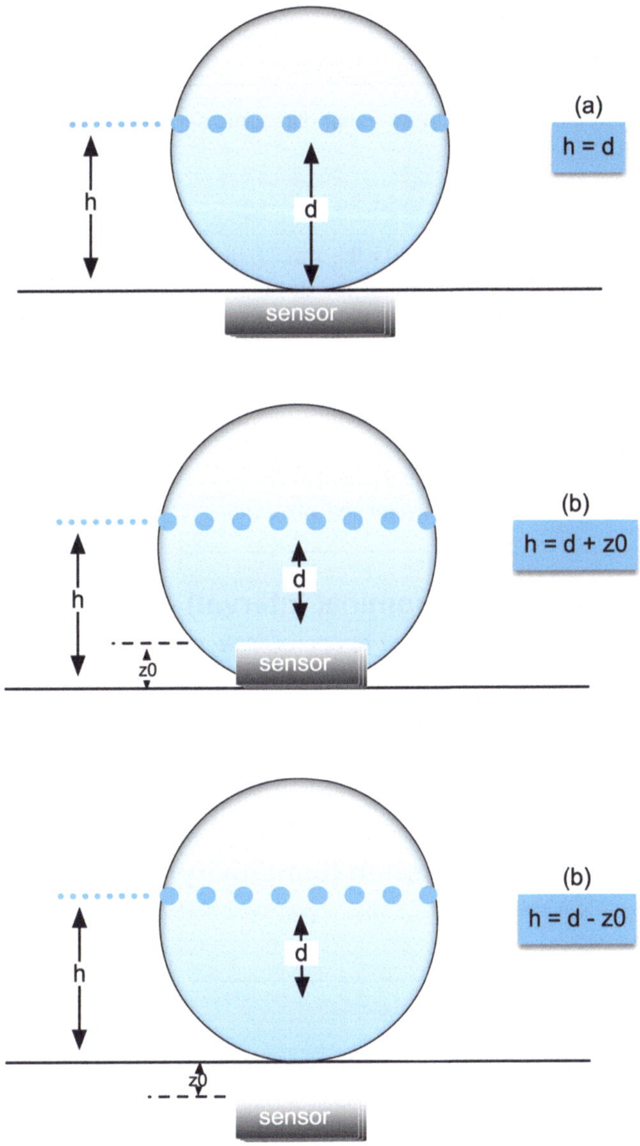

Figure 12.1 Water level measurement with a piezo-resistive sensor, (a) without offset, (b) with positive or negative offset. *Source*: Francois Clemens-Meyer (Deltares/TU Delft/NTNU).

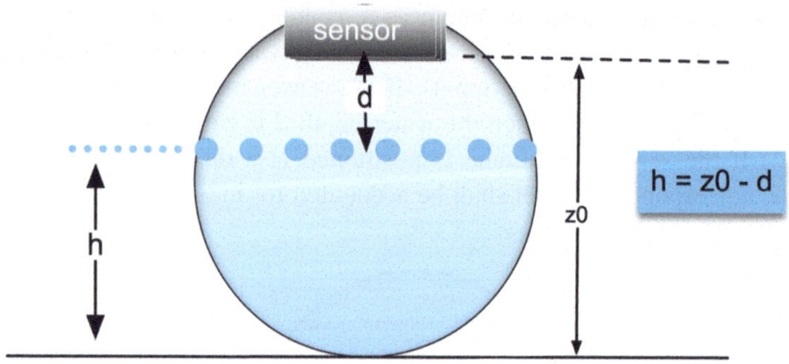

Figure 12.2 Water level measurement with an aerial ultrasonic sensor. *Source*: Francois Clemens-Meyer (Deltares/TU Delft/NTNU).

Similarly, an aerial acoustic sensor converts the time t an acoustic wave in the air travels from the sensor to the free surface and return (input) into the distance d between the sensor and the free surface and then into the water level h (output) determined from the position of the sensor within the cross section (Figure 12.2).

The component of the sensor converting the input into the output is the *transducer*. Transducers are very diverse and based on various laws of physics: mechanics, electricity, magnetism, wave propagation, etc. (see Chapters 2, 3 and 4).

A sensor and by extension a measuring instrument or system is characterized by various specifications including *nominal indication interval*, *sensitivity* and *resolution*.

nominal indication interval (nominal interval)

Set of quantity values, bounded by rounded or approximate extreme indications, obtainable with a particular setting of the controls of a measuring instrument or measuring system and used to designate that setting.

sensitivity of a measuring system (sensitivity)

Quotient of the change in an indication of a measuring system and the corresponding change in a value of a quantity being measured.

resolution

Smallest change in a quantity being measured that causes a perceptible change in the corresponding indication.

More basic measuring instruments exist, e.g. *detectors* of overflows.

detector

Device or substance that indicates the presence of a phenomenon, body, or substance when a threshold value of an associated quantity is exceeded.

Before the measurement, the value of the discharge is unknown (if not, there is no need to do a measurement…). As no measurement can be absolutely perfect under ideal conditions with a truly flawless sensor, it implies that (i) the true value of the discharge will remain unknown, and (ii) each measurement is considered as an estimation of the true value. The aim of the operator is to obtain an estimation with a required or acceptable level of *accuracy*.

measurement accuracy (accuracy of measurement, accuracy)

Closeness of agreement between a measured quantity value and a true quantity value of a measurand.

All measurements are affected, to various degrees, by *errors*, which can be either *systematic* or *random*. In the case where a *reference quantity value* is available, the systematic error can be estimated: this is the *bias*. In other cases, errors cannot be quantified.

measurement error (error of measurement, error)

Measured quantity value minus a reference quantity value.

systematic measurement error (systematic error of measurement, systematic error)

Component of measurement error that in replicate measurements remains constant or varies in a predictable manner.

random measurement error (random error of measurement, random error)

Component of measurement error that in replicate measurements varies in an unpredictable manner.

reference quantity value (random error of measurement, random error)
Quantity value used as a basis for comparison with values of quantities of the same kind.

measurement bias (bias)
Estimate of a systematic measurement error.

If the operator uses a water level sensor and the sensor is not positioned correctly in the sewer (e.g. too low or too high compared to the pipe invert or pipe ceiling), all water level measurements will be affected by an offset: this is a systematic error. If a reference quantity value is available (i.e. if it is possible to use e.g. some reference shim), the bias can be estimated and the *correction* can compensate the systematic error. Using reference quantity value thus allows correction of systematic errors and estimating biases. But there is no means to correct random errors which, by definition, differ for each measurement: they are due to sensors flaws and defects, their use under non-nominal conditions, noise and artefacts in components, the operator behaviour and actions, and variable environmental conditions affecting the measurement without being controlled (temperature, pressure, humidity, electromagnetic disturbances, radiations, non-stationarity of power supply, etc.).

correction (bias)
Compensation for an estimated systematic effect.

Let us assume that the pipe where the discharge shall be measured is a circular pipe. Its diameter is a quantity that needs to be known to estimate the wet cross section and then the discharge. The operator can obtain the value of the diameter from various sources: reports, maps, data bases, GIS software tools, etc. However, the most recommended option is to measure the diameter *in situ* with a meter, a laser or any other appropriate instrument, as it is the operator's responsibility (quality assurance) to check and verify all quantity values to be used in the measurement procedure.

The operator may do a single measurement of the diameter. But as measurements in sewers are difficult due to harsh conditions (obscurity or reduced light, humidity, odours, lack of space, use of thick gloves, presence of water, sediments, pipe structure defects, etc.), he/she may decide to proceed to multiple consecutive measurements, carried out by the same person under unchanged conditions: in this case, the operator is working with *repeatability conditions*.

repeatability condition of measurement (repeatability condition)
Condition of measurement, out of a set of conditions that includes the same measurement procedure, same operators, same measuring system, same operating conditions and same location, and replicate measurements on the same or similar objects over a short period of time.

If the operator decides to do multiple measurements but with various measuring systems (various instruments or sensors), carried out by different persons under various conditions, he/she is working with *reproducibility conditions*.

reproducibility condition of measurement (reproducibility condition)

Condition of measurement, out of a set of conditions that includes different locations, operators, measuring systems, and replicate measurements on the same or similar objects.

Intermediate precision conditions may also be encountered.

intermediate precision condition of measurement (intermediate precision condition)

Condition of measurement, out of a set of conditions that includes the same measurement procedure, same location, and replicate measurements on the same or similar objects over an extended period of time, but may include other conditions involving changes.

In the case of repeated measurements, the values of the pipe diameter will show some dispersion around the mean value due to random errors, as it is unexpected that all of them will be exactly the same. The dispersion may be quantified by the standard deviation of the repeated measurements results (see Chapter 8). Moreover, the dispersion usually tends to increase with the number of repeated measurements. As additional sources of errors and variability are present in reproduced measurements, the dispersion of the diameter values will be larger than with repeated measurements.

It may happen that all successive repeated measurements of the pipe diameter are slightly different and rather close to each other, with a small dispersion. In that case, the measurements show a high *precision*. If the successive values are more dispersed, the measurement precision is low.

measurement precision (precision)

Closeness of agreement between indications or measured quantity values obtained by replicate measurements on the same or similar objects under specified conditions.

However, a high measurement precision does not necessarily mean that the pipe diameter value is close to the true value if significant systematic errors are present. The closeness between the repeated diameter values and the true diameter value is described by the *measurement trueness*.

measurement trueness (trueness of measurement, trueness)

Closeness of agreement between the average of an infinite number of replicate measured quantity values and a reference quantity value.

Unfortunately, the definition of the trueness involves an infinite number of measurements, which is not convenient in practice. In the case of a single measurement or a finite number of repeated measurements, one thus considers the trend toward trueness and also the measurement *accuracy* if the true value is known or assumed to be known (reference value).

Precision, trueness and accuracy are different concepts that may be related to systematic and random errors. Four typical cases are possible, conceptually represented in Figure 12.3:

(a) If the successive repeated measurements of the pipe diameter are both close to each other and to the true or reference value, the measurements are precise, true and accurate. There are no significant systematic (trueness and accuracy) or random (precision) errors. A limited number of measurements is sufficient to obtain an accurate estimation of the true value of the diameter. In the ideal case, one single measurement could be enough (but remember that in practice, measurements in UDSM systems are far from being ideal and perfect).

(b) If the successive repeated measurements of the pipe diameter are close to each other but not close to the true or reference value, the measurements are precise, not true and inaccurate. There are significant systematic errors and no significant random errors. The significant bias should be detected and corrected in order to estimate the true value of the diameter. An increasing number of repeated measurements will not be sufficient to improve the estimation if systematic errors are simultaneously neither detected nor corrected.

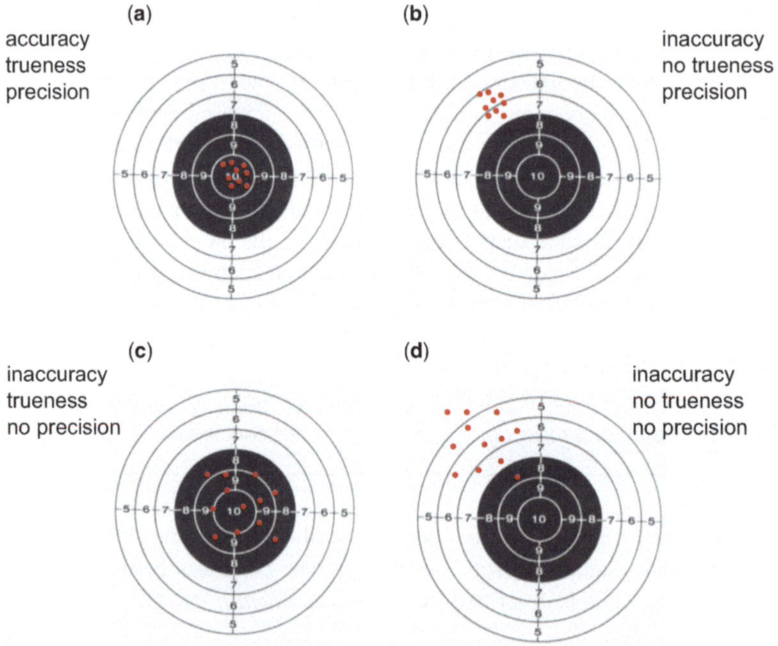

Figure 12.3 Conceptual representation of measurement precision, trueness and accuracy. The true value is equal to 10 and is represented by the centre of the target. Repeated measurements are represented by arrow impacts. Systematic errors in measurements are negligible if the impacts are in the central dark area with scores ranging from 8 to 10. Random errors are represented by the dispersion of the impacts. *Source*: Jean-Luc Bertrand-Krajewski (INSA Lyon).

(c) If the successive repeated measurements of the pipe diameter are not very close to each other but remain close to the true or reference value, the measurements are imprecise but rather true and accurate. Systematic errors are not significant and random errors are predominant. In this case, increasing the number of repeated measurements will lead to a mean value of the diameter very close to the true value. Repeated measurements provide an accurate mean value of the pipe diameter.

(d) If the successive repeated measurements of the pipe diameter are not close to each other and not close to the true or reference value, the measurements are imprecise, not true and inaccurate. This is the worst case in practice, with high systematic and random errors. Improving the quality of the estimation of the pipe diameter would require (i) estimating and correcting the bias and (ii) increasing the number of repeated measurements and applying the bias correction to the mean value of the diameter.

In cases b and c, detecting systematic errors is the most challenging task and requires reference values. The above elements indicate two main topics in metrology:

- Estimation of random errors.
- Detection, estimation and correction of systematic errors.

The latter is the most difficult.

If the quality of the *measurement result* is lower than the quality either required (by law, regulation, standards, contract, etc.) or set as acceptable (by the user of the measurement result for further calculation, modelling, decision making, etc.), the operator has to explore and find, if possible, means to improve it.

measurement result (result of measurement)

Set of quantity values being attributed to a measurand together with any other available relevant information.

Such means may include changes and/or adaptations of the measuring system, the *measuring chain*, the measuring instruments, the measurement procedure, etc.

measuring chain

Series of elements of a measuring system constituting a single path of the signal from a sensor to an output element.

For example, if the discharge measurement in a sewer pipe does not comply with the required or acceptable level of precision, the operator may repair or replace the sensor (Chapter 7), change the sensor technology (Chapter 3) or any other element of the measuring chain (transmitter, logger, etc.), the sensor location in the sewer (Chapters 3, 4 and 6), the operation and maintenance conditions (Chapter 7), the measurement procedure, protocols and conditions, etc.

In order to evaluate the quality of any measurement result, it is of crucial importance to quantify the *measurement uncertainty* (Chapter 8).

measurement uncertainty (uncertainty of measurement, uncertainty)

Non-negative parameter characterizing the dispersion of the quantity values being attributed to a measurand, based on the information used.

The uncertainty is expressed in practice as an interval or range of values, named the *coverage interval*, containing the true value of the measurand with a given level of probability, named *coverage probability*.

coverage interval

Interval containing the set of true quantity values of a measurand with a stated probability, based on the information available.

coverage probability

Probability that the set of true quantity values of a measurand is contained within a specified coverage interval.

Chapter 8 presents in detail the methods the operator may apply to evaluate measurement uncertainties. Three methods are available, depending on the context of each measurement:

- The Type A evaluation is based on replicate measurements of the quantity of interest and applies statistical inference to evaluate its uncertainty.
- The Type B evaluation, also named the law of propagation of uncertainties (LPU), is based on a mathematical relation, named the measurement model, allowing evaluation of both the quantity of interest and its uncertainty from other known quantities and their respective uncertainties.
- The Monte Carlo evaluation is based on stochastic simulations of the measurement model followed by statistical analysis of their results.

Type A evaluation of measurement uncertainty (Type A evaluation)

Evaluation of a component of measurement uncertainty by a statistical analysis of measured quantity values obtained under defined measurement conditions.

Type B evaluation of measurement uncertainty (Type B evaluation)

Evaluation of a component of measurement uncertainty determined by means other than a Type A evaluation of measurement uncertainty.

measurement model (model of measurement, model)

Mathematical relation among all quantities known to be involved in a measurement.

A global description of the uncertainty and its components is given in the *uncertainty budget*.

uncertainty budget

Statement of a measurement uncertainty, of the components of that measurement uncertainty, and of their calculation and combination.

With the Type A evaluation, the operator first calculates the *standard uncertainty* of the quantity of interest, which, in first approximation, is equivalent to the standard deviation.

standard uncertainty (standard uncertainty of measurement, standard measurement uncertainty)

Measurement uncertainty expressed as a standard deviation.

With the Type B evaluation, the standard uncertainty is obtained by applying the measurement model and the LPU, which provides a *combined standard measurement uncertainty*.

combined standard measurement uncertainty (combined standard uncertainty)

Standard measurement uncertainty that is obtained using the individual standard measurement uncertainties associated with the input quantities in a measurement model.

The operator then sets the coverage probability. Most frequently, a 95% probability is used. The last step consists of calculating the coverage interval by multiplying the standard uncertainty u by a coefficient k, named the *coverage factor*, corresponding to the chosen coverage probability. The result is the *expanded measurement uncertainty* $U = ku$. The final result is then expressed as $y \pm U$, or $[y - U, y + U]$.

coverage factor (standard uncertainty of measurement, standard measurement uncertainty)

Measurement uncertainty expressed as a standard deviation.

expanded measurement uncertainty (expanded uncertainty)

Product of a combined standard measurement uncertainty and a factor larger than the number one.

For example, if the discharge in a circular pipe is $Q = 0.45\ \mathrm{m^3/s}$, the combined standard uncertainty is $u(Q) = 0.02\ \mathrm{m^3/s}$ and the coverage factor is $k = 2$ for a 95% probability, then the coverage interval is expressed as $Q = 0.45 \pm 0.04\ \mathrm{m^3/s}$. For the sake of simplicity, this is interpreted as 'the true value has 95% probability to be between 0.41 and 0.49 $\mathrm{m^3/s}$'. It may also be convenient to express the result as a fraction or a percentage of the measured value. In this case, the *relative standard uncertainty* is 0.044 or 4.4% and the *expanded relative uncertainty* is 0.089 or 8.9%.

A note on reporting digits: when presenting the outcome of some measurement, care has to be taken that this is communicated in the right manner. Suppose in the previous example the average value for the discharge is $Q = 0.4536541\ \mathrm{m^3/s}$ (being a typical result as presented in the double precision computer outcome) with a combined standard uncertainty of $u(Q) = 0.02\ \mathrm{m^3/s}$, the last five digits in the average result should not be reported as they would suggest an accuracy that is not there. In this case only the last two digits have meaning and are to be reported accordingly. Another example: 45700 ± 300 should be reported as $(4.57 \pm 0.03) \times 10^4$.

relative standard measurement uncertainty

Standard measurement uncertainty divided by the absolute value of the measured quantity value.

With the Monte Carlo evaluation, the stochastic simulations of the measurement model result in (i) the mean value of the quantity of interest, and (ii) the boundaries of the shortest interval containing a given percentage of the simulated values. The percentage corresponds to the coverage probability.

In the previous discharge example, the Monte Carlo evaluation gives the mean discharge $\bar{Q} = 0.45\ \mathrm{m^3/s}$ and the 95% coverage interval [0.41, 0.19].

It is worth noting that both Type A and Type B methods give symmetric coverage intervals, with the measurement result expressed as $y \pm U$. This is correct if some hypotheses are verified, especially with the Type B evaluation. The Monte Carlo evaluation is more general and may give non-symmetric coverage intervals defined by their boundaries.

Two aspects of uncertainty assessment shall be distinguished. The first one consists of estimating uncertainties of measured values, as just briefly presented above. The second aspect consists of comparing the obtained uncertainties with the *target measurement uncertainty* either required by law, regulation, standards, contracts, etc., or set as acceptable by the operator according to the use of the measurement results for design, control, modelling, decision making, etc.

target measurement uncertainty (target uncertainty)

Measurement uncertainty specified as an upper limit and decided on the basis of the intended use of measurement results.

If the target uncertainty is not reached, the measurement system, chain and process shall be improved or revised.

It is of crucial importance to ensure that measurements are free of systematic errors or that systematic errors are detected and corrected. This includes two steps: (i) ensuring under controlled conditions that the sensor itself has no systematic error by means of a comparison with a reference value, and (ii) ensuring that the *in situ* use of the sensor in field conditions is free of systematic error by means of a comparison with reference values.

In the first step, the operator compares the values given by the sensor to *measurement standards*. This is the sensor *calibration* (see Chapter 7). No sensor shall be used without initial calibration, followed by periodic *verifications* and a new calibration if a verification fails.

measurement standard

Realization of the definition of a given quantity, with stated quantity value and associated measurement uncertainty, used as a reference.

calibration

Operation that, under specified conditions, in a first step, establishes a relation between the quantity values with measurement uncertainties provided by measurement standards and corresponding indications with associated measurement uncertainties and, in a second step, uses this information to establish a relation for obtaining a measurement result from an indication.

verification

Provision of objective evidence that a given item fulfils specified requirements.

For the example of discharge measurement in a circular pipe, the water level sensor shall be calibrated. The measurement standards depend on the sensor technology (piezo-resistive sensor, ultrasonic sensor, radar sensor, etc.).

In the case where the calibration shows that sensor outputs are not as close to standards values as required or set as acceptable, the operator may proceed to the *adjustment* of the sensor or of the measuring system. It is also possible to establish a mathematical relation (calibration function) to correct the sensor outputs without adjusting the sensor. In the simplest cases, the adjustment includes two steps: *zero adjustment* and span (or gain) adjustment.

adjustment of a measuring system (adjustment)

Set of operations carried out on a measuring system so that it provides prescribed indications corresponding to given values of a quantity to be measured.

> **zero adjustment of a measuring system (zero adjustment)**
> *Adjustment of a measuring system so that it provides a null indication corresponding to a zero value of a quantity to be measured.*

In the second step, the operator compares the values given by the calibrated sensor to *in situ* reference values. For example, checking that the position of the sensor in the sewer pipe corresponds to its theoretical position, that there is no offset, etc. This may require a sequence of calibrations, defined as *calibration hierarchy*. This is necessary to ensure a full *metrological traceability*.

> **calibration hierarchy**
> *Sequence of calibrations from a reference to the final measuring system, where the outcome of each calibration depends on the outcome of the previous calibration.*

> **metrological traceability**
> *Property of a measurement result whereby the result can be related to a reference through a documented unbroken chain of calibrations, each contributing to the measurement uncertainty.*

Measurement results shall be systematically submitted to verification and *validation* before any further use (see Chapter 9).

> **validation**
> *Verification, where the specified requirements are adequate for an intended use.*

When a measured value is zero, what does it mean? Here we touch upon definitions regarding the Detection Limit (also referred to as LoD, Limit of Detection) and Quantification Limit (also referred to as LoQ, Limit of Quantification). These terms have a close relation with *nominal indication interval*, *sensitivity* and *resolution* as discussed earlier and are mostly used and applied in e.g. analytical chemistry, but they can have meaning as well in the field of UDSM.

For example, in some countries, the number of CSO events has to be reported for which a range of definitions is applied, all have one thing in common: namely the number of times a spill has occurred in a certain time window. Now, suppose the operator applies a water level measuring device to check whether or not the water level has risen above the crest level of the weir. When the measured water level h is smaller or equal to the crest level, no spill occurs; when h is larger than the crest level, a spill has occurred. Here the operator has to deal with the Detection Limit. In Figure 12.4, the blue curve

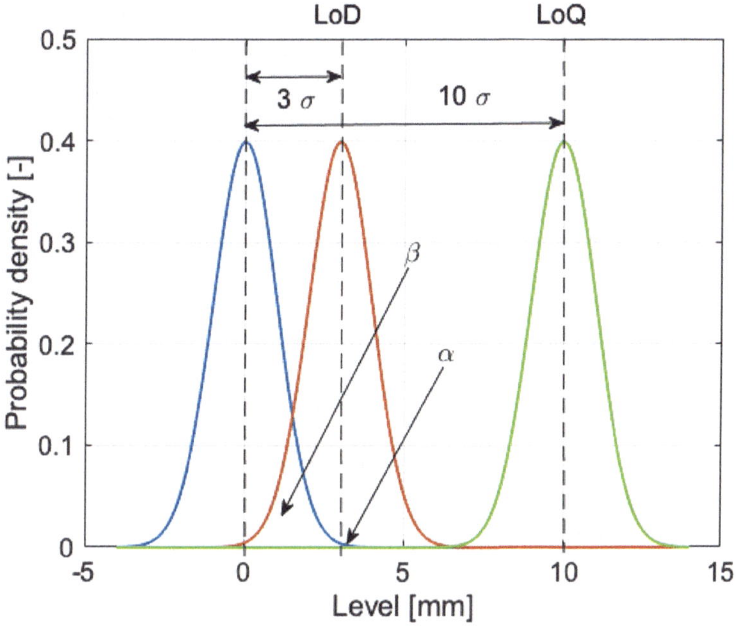

Figure 12.4 Schematic figure illustrating the concepts of LoD and LoQ. *Source*: Francois Clemens-Meyer (Deltares/TU Delft/NTNU).

symbolizes the probability density function of the crest level of the weir (the mean value is used as the estimate of 'the crest level'), the red curve symbolizes the probability density of the water level measurement. When the standard deviation of both the crest level and the water level measurement are equal, then a simple definition of the Detection Limit is that the mean values of crest level and water level should be three standard deviations apart to be sure that the water level is either under or above the crest level. A more advanced approach is to set a certain maximum to the probability β (in the example 0.1336), or set a value for the minimum difference of the measured value to zero, commonly 3 times the standard deviation is used for defining the LoD.

When the volume or the discharge of spilled CSO water is of importance, the operator has to deal with the Quantification Limit. In many cases a stage-discharge relation is used to determine the spilled discharge as a function of measured water level (see Section 3.4).

In Figure 12.4, the green curve depicts the uncertainty of the water level measurement. In this case, the Quantification Limit is defined as follows: the distance between the crest level and the measured water level is such that the probability that the values overlap is much less than 1% when the distance between the values is 10 standard deviations (overlap = ~6×10^{-7}). In this case, when the green graph is valid, the measured value can be used to quantify the difference between the crest of the weir and the measured water level, and therefore the discharge can be quantified.

So, in the case where the red graph represents the measurement, one can only state that 'a CSO event has occurred' but the volume was less than the LoQ. Of course one is free to choose different (possibly more relaxed) values for α and β provided this is clearly documented along with reported results, see e.g. Armbruster & Pry (2008) for more details.

12.2 LIST OF DATA AND MATLAB FILES

The following paragraphs list the Matlab® code files (.m files) and the data files (.csv or.mat files) which are used in Chapters 6 to 9 as examples of application of methods and algorithms explained in the main text. These files are available for download on the book companion webpage at https://doi.org/10.2166/9781789060102.

In Chapter 3:

- ap_1.txt
- av_1.txt
- Calculation steps for tracing experiments.xlsx
- D_10-100.txt
- D_10-1000.txt
- dilution.txt
- P.dat
- pic_1.txt
- pic_1_Ascending_part.fig
- pic_1_Descending_part.fig
- pic_1_result-tracer.dat
- pic_1_Start&End.fig
- regw123etalo.dat
- sans_dilution.txt
- tracer.m

In Chapter 6:

- IWA_hydro.mat
- IWA_Jacobian.m
- loc1loc2.mat
- meas_freq_exp.m
- v2_k15_C2.csv

In Chapter 7:

- ciaponi1.csv
- OLS123.m
- piezo1.csv
- turbi251.csv
- WLS123cal.m

In Chapter 8:

- eggshape1.csv
- hV1.csv
- manning1.csv
- raingauge1.csv
- uMCM.m
- uTypeA.m
- uTypeB.m
- uTypeBsum.m
- vol1.csv

In Chapter 9:

- ARMA_SPACE.m
- lin_step_power.m
- linear_trend.m
- min_max_sample.m
- outlier_tests.m
- outliers_example.mat
- piece_lin_fit.m
- step_trend.m

REFERENCES

Armbruster D. A. & Pry T. (2008). Limit of blank, limit of detection and limit of quantitation. *The Clinical Biochemist Reviews*, **29**(1), S49–S52. Available at https://www.ncbi.nlm.nih.gov/pmc/articles/PMC2556583/pdf/cbr29_s_pgs49.pdf (accessed 12 January 2021).

JCGM (2012). *JCGM 200:2012 – International vocabulary of metrology – Basic and general concepts and associated terms (VIM) (Vocabulaire international de métrologie – Concepts fondamentaux et généraux et termes associés (VIM)*. Joint Committee for Guides in Metrology, Geneva (Switzerland), 108 p. Available at https://www.bipm.org/documents/20126/2071204/JCGM_200_2012.pdf (accessed 26 April 2021).

IWA Publishing's authorised EU representative for General Product Safety Regulations is Diane D'Arras, 15 rue Duret, 75116 Paris, France, e-mail: safety@iwap.co.uk.

Printed and bound by CPI Group (UK) Ltd, Croydon, CR0 4YY
12/05/2026
02108869-0017